# Lecture Notes in Computer S

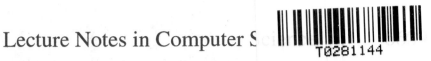

*Commenced Publication in 1973*
Founding and Former Series Editors:
Gerhard Goos, Juris Hartmanis, and Jan van Leeuwen

Alexander Gelbukh (Ed.)

# Computational Linguistics and Intelligent Text Processing

12th International Conference, CICLing 2011
Tokyo, Japan, February 20-26, 2011
Proceedings, Part II

 Springer

Volume Editor

Alexander Gelbukh
Instituto Politécnico Nacional (IPN)
Centro de Investigación en Computación (CIC)
Col. Nueva Industrial Vallejo, CP 07738, Mexico D.F., Mexico
E-mail: gelbukh@gelbukh.com

ISSN 0302-9743                          e-ISSN 1611-3349
ISBN 978-3-642-19436-8                  e-ISBN 978-3-642-19437-5
DOI 10.1007/978-3-642-19437-5
Springer Heidelberg Dordrecht London New York

Library of Congress Control Number: 2011921814

CR Subject Classification (1998): H.3, H.4, F.1, I.2, H.5, H.2.8, I.5

LNCS Sublibrary: SL 1 – Theoretical Computer Science and General Issues

Typesetting: Camera-ready by author, data conversion by Scientific Publishing Services, Chennai, India

Printed on acid-free paper

Springer is part of Springer Science+Business Media (www.springer.com)

# Preface

CICLing 2011 was the 12th Annual Conference on Intelligent Text Processing and Computational Linguistics. The CICLing conferences provide a wide-scope forum for the discussion of the art and craft of natural language processing research as well as the best practices in its applications.

This set of two books contains four invited papers and a selection of regular papers accepted for presentation at the conference. Since 2001, the proceedings of the CICLing conferences have been published in Springer's *Lecture Notes in Computer Science* series as volume numbers 2004, 2276, 2588, 2945, 3406, 3878, 4394, 4919, 5449, and 6008.

The set has been structured into 13 sections:

- Lexical resources
- Syntax and parsing
- Part-of-speech tagging and morphology
- Word sense disambiguation
- Semantics and discourse
  Opinion mining and sentiment analysis
- Text generation
- Machine translation and multilingualism
- Information extraction and information retrieval
- Text categorization and classification
- Summarization and recognizing textual entailment
- Authoring aid, error correction, and style analysis
- Speech recognition and generation

The 2011 event received a record high number of submissions. A total of 298 papers by 658 authors from 48 countries were submitted for evaluation by the International Program Committee, see Tables 1 and 2. This two-volume set contains revised versions of 74 papers, by 227 authors, selected for presentation; thus the acceptance rate for this set was 25%.

The books feature invited papers by

- Christopher Manning, Stanford University, USA
- Diana McCarthy, Lexical Computing Ltd., UK
- Jun'ichi Tsujii, U. of Tokyo, Japan, and U. of Manchester and NacTeM, UK
- Hans Uszkoreit, Saarland University and DFKI, Germany

who presented excellent keynote lectures at the conference. Publication of extended full-text invited papers in the proceedings is a distinctive feature of the CICLing conferences. Furthermore, in addition to the presentation of their invited papers, the keynote speakers organized separate vivid informal events; this is also a distinctive feature of this conference series.

**Table 1.** Statistics of submissions and accepted papers by country or region

| Country or region | Authors Subm. | Papers[1] Subm. | Accp. | Country or region | Authors Subm. | Papers[1] Subm. | Accp. |
|---|---|---|---|---|---|---|---|
| Australia | 17 | 7 | 3 | Korea (South) | 10 | 4.29 | 1 |
| Austria | 2 | 1.33 | 0.33 | Macao | 4 | 1 | – |
| Belgium | 4 | 2 | 1 | Malaysia | 5 | 2 | – |
| Brazil | 5 | 2 | 1 | Mexico | 13 | 6.92 | 2 |
| Canada | 11 | 6.33 | 2 | Myanmar | 7 | 2 | – |
| China | 47 | 17.67 | 5.67 | Nigeria | 3 | 1 | – |
| Colombia | 3 | 2 | – | Norway | 7 | 2.17 | – |
| Croatia | 3 | 2 | – | Pakistan | 6 | 3.57 | – |
| Cuba | 2 | 0.67 | – | Peru | 2 | 0.50 | – |
| Czech Rep. | 14 | 8.50 | 3 | Poland | 2 | 2 | – |
| Egypt | 9 | 2.67 | 1.67 | Portugal | 25 | 9.67 | 2 |
| Finland | 3 | 2 | – | Romania | 7 | 3.33 | – |
| France | 36 | 16.68 | 4.83 | Russia | 8 | 2.33 | – |
| Georgia | 1 | 1 | – | Saudi Arabia | 1 | 1 | – |
| Germany | 29 | 12.58 | 3.50 | Singapore | 7 | 2.50 | 1 |
| Greece | 6 | 2 | 1 | Spain | 39 | 14.30 | 4.30 |
| Hong Kong | 5 | 2 | 1 | Sweden | 5 | 1.39 | 1 |
| India | 85 | 41.75 | 6.42 | Taiwan | 13 | 5 | – |
| Iran | 23 | 18 | 3 | Thailand | 6 | 3 | 1 |
| Ireland | 14 | 7 | 1 | Tunisia | 9 | 3.15 | – |
| Israel | 3 | 1.75 | – | Turkey | 8 | 4.17 | 1 |
| Italy | 17 | 6.25 | 2.25 | UK | 13 | 6.67 | 0.50 |
| Japan | 71 | 29.67 | 14 | USA | 39 | 17.87 | 4.53 |
| Jordan | 1 | 0.50 | – | Viet Nam | 8 | 4.33 | 1 |
| | | | | *Total:* | 658 | 298 | 74 |

[1] By the number of authors: e.g., a paper by two authors from the USA and one from the UK is counted as 0.67 for the USA and 0.33 for the UK.

With this event we introduced a new policy of giving preference to papers with verifiable and reproducible results: we encouraged the authors to provide, in electronic form, a proof of their claims or a working description of the suggested algorithm, in addition to the verbal description given in the paper. If the paper claimed experimental results, we encouraged the authors to make available to the community all the input data necessary to verify and reproduce these results; if it claimed to advance human knowledge by introducing an algorithm, we encouraged the authors to make the algorithm itself, in some programming language, available to the public. This additional electronic material will be permanently stored on CICLing's server, www.CICLing.org, and will be available to the readers of the corresponding paper for download under a license that permits its free use for research purposes.

In the long run we expect that computational linguistics will have verifiability and clarity standards similar to those of mathematics: in mathematics, each claim is accompanied by a complete and verifiable proof (usually much

**Table 2.** Statistics of submissions and accepted papers by topic[2]

| Accepted | Submitted | % accepted | Topic |
|---|---|---|---|
| 13 | 40 | 33 | Lexical resources |
| 13 | 47 | 28 | Practical applications |
| 11 | 39 | 28 | Clustering and categorization |
| 11 | 44 | 25 | Other |
| 10 | 28 | 36 | Acquisition of lexical resources |
| 10 | 29 | 34 | Statistical methods (mathematics) |
| 10 | 52 | 19 | Machine translation & multilingualism |
| 9 | 25 | 36 | Syntax and chunking (linguistics) |
| 9 | 31 | 29 | Semantics and discourse |
| 9 | 58 | 16 | Information extraction |
| 7 | 46 | 15 | Text mining |
| 6 | 12 | 50 | Symbolic and linguistic methods |
| 6 | 50 | 12 | Information retrieval |
| 5 | 13 | 38 | Parsing algorithms (mathematics) |
| 5 | 16 | 31 | Noisy text processing and cleaning |
| 5 | 18 | 28 | Summarization |
| 4 | 11 | 36 | Text generation |
| 4 | 16 | 25 | Opinion mining |
| 4 | 17 | 24 | POS tagging |
| 3 | 7 | 43 | Speech processing |
| 3 | 8 | 38 | Cross-language information retrieval |
| 3 | 15 | 20 | Word sense disambiguation |
| 3 | 20 | 15 | Formalisms and knowledge representation |
| 2 | 6 | 33 | Emotions and humor |
| 2 | 13 | 15 | Named entity recognition |
| 1 | 5 | 20 | Spelling and grammar checking |
| 1 | 7 | 14 | Anaphora resolution |
| 1 | 7 | 14 | Textual entailment |
| 1 | 8 | 12 | Question answering |
| 1 | 11 | 9 | Natural language interfaces |
| 1 | 12 | 8 | Morphology |
| – | 6 | 0 | Computational terminology |

[2] As indicated by the authors. A paper may belong to several topics.

greater in size than the claim itself); each theorem —and not just its description or general idea—is completely and precisely presented to the reader. Electronic media allow computational linguists to provide material analogous to the proofs and formulas in mathematics in full length—which can amount to megabytes or gigabytes of data—separately from a 12-page description published in a book. A more detailed argumentation for this new policy can be found on www.CICLing.org/why_verify.htm.

To encourage the authors to provide algorithms and data along with the published papers, we introduced a new Verifiability, Reproducibility, and Working Description Award. The main factors in choosing the awarded submission were technical correctness and completeness, readability of the code and documentation, simplicity of installation and use, and exact correspondence to the claims of the paper. Unnecessary sophistication of the user interface was discouraged; novelty and usefulness of the results were not evaluated—those parameters were evaluated for the paper itself and not for the data.

The following papers received the Best Paper Awards, the Best Student Paper Award, as well as the Verifiability, Reproducibility, and Working Description Award, correspondingly (the best student paper was selected from the papers of which the first author was a full-time student, excluding the papers that received a Best Paper Award):

1st Place:      *Co-related Verb Argument Selectional Preferences*, by Hiram Calvo, Kentaro Inui, and Yuji Matsumoto;

2nd Place:      *Self-Adjusting Bootstrapping*, by Shoji Fujiwara and Satoshi Sekine;

3rd Place:      *Effective Use of Dependency Structure for Bilingual Lexicon Creation*, by Daniel Andrade, Takuya Matsuzaki, and Jun'ichi Tsujii;

Student:       *Incorporating Coreference Resolution into Word Sense Disambiguation*, by Shangfeng Hu and Chengfei Liu;

Verifiability:  *Improving Text Segmentation with Non-systematic Semantic Relation*, by Viet Cuong Nguyen, Le Minh Nguyen, and Akira Shimazu.

The authors of the awarded papers (except for the Verifiability Award) were given extra time for their presentations. In addition, the Best Presentation Award and the Best Poster Award winners were selected by a ballot among the attendees of the conference.

Besides its high scientific level, one of the success factors of CICLing conferences is their excellent cultural program. The attendees of the conference had a chance to visit Kamakura—known for the Kamakura period of ancient history of Japan—where they experienced historical Japanese cultural heritage explained by highly-educated local volunteers and saw Shinto (traditional religion of Japan) shrines and old Buddhist temples characteristic of Japan. They recalled recent history at the Daigo Fukuryu Maru Exhibition Hall, which tells the story of a Japanese boat exposed to and contaminated by nuclear fallout from a thermonuclear device test in 1954. Finally, the participants familiarized themselves with modern Japanese technology during guided tours to Toshiba Science Museum and Sony Square; the latter can only be accessed by special invitation from Sony. And of course they enjoyed Tokyo, the largest city in the world, futuristic and traditional at the same time, during an excursion to the Japanese-style East Gardens of the Imperial Palace and a guided tour of the city, by bus and boat (see photos on www.CICLing.org).

I would like to thank all those involved in the organization of this conference. In the first place these are the authors of the papers that constitute this book: it is the excellence of their research work that gives value to the book and sense to the work of all other people. I thank all those who served on the Program Committee, Software Reviewing Committee, Award Selection Committee, as well as the additional reviewers, for their hard and very professional work. Special thanks go to Ted Pedersen, Grigori Sidorov, Yasunari Harada, Manuel Vilares Ferro, and Adam Kilgarriff, for their invaluable support in the reviewing process.

I thank the School of Law and Media Network Center of Waseda University, Japan, for hosting the conference; the Institute for Digital Enhancement of Cognitive Development (DECODE) of Waseda University for valuable collaboration in its organization; and Waseda University for providing us with the best conference facilities. With deep gratitude I acknowledge the support of Professor Waichiro Iwashi, the dean of the School of Law of Waseda University, and Professor Toshiyasu Matsushima, Dean and Director of Media Network Center of Waseda University. I express my most cordial thanks to the members of the local Organizing Committee for their enthusiastic and hard work. The conference would not have been a success without the kind help of Professor Mieko Ebara, Ms. Mayumi Kawamura, Dr. Kyoko Kanzaki, and all the other people involved in the organization of the conference and cultural program activities.

My most special thanks go to Professor Yasunari Harada, Director of DECODE, for his great enthusiasm and infinite kindness and patience; countless nights without sleep, after a whole day of teaching and meetings, spent on the organization of the conference, from the strategic planning to the finest details. I feel very lucky to have had the opportunity to collaborate with this prominent scientist, talented organizer, and caring friend. From him I learnt a lot about human relationships as well as about planning and organization.

With great gratitude I acknowledge the financial support of the Kayamori Foundation of Information Science Advancement, which greatly helped us to keep the fees low. I would like to express my gratitude to the Kamakura Welcome Guide Association for making our visit to this historical city of Japan a memorable and enjoyable one. Thanks are also due to Sony and Totsusangyo Corporation, Toshiba Science Museum, and Daigo Fukuryu Maru Exhibition Hall, for arranging special visits and guided tours for CICLing 2011 participants. I would like to specifically recognize the help of Mr. Masahiko Fukakushi, Executive Officer and Corporate Senior Vice President of Toshiba Corporation, in arranging our visit to Toshiba Science Museum and the help of Dr. Atsushi Ito, Distinguished Research Engineer at KDDI R&D Laboratories, in providing wireless Internet access to the attendees of the conference.

The entire submission and reviewing process was supported for free by the EasyChair system (www.EasyChair.org). Last but not least, I deeply appreciate the Springer staff's patience and help in editing this volume and getting it printed in record short time—it is always a great pleasure to work with Springer.

February 2011                                             Alexander Gelbukh
                                                           General Chair

# Organization

CICLing 2011 was co-hosted by the School of Law and Media Network Center of Waseda University, and was organized by the CICLing 2011 Organizing Committee, in conjunction with the Natural Language and Text Processing Laboratory of the CIC (Center for Computing Research) of the IPN (National Polytechnic Institute), Mexico, and the Institute for Digital Enhancement of Cognitive Development of Waseda University, with partial financial support from the Kayamori Foundation of Information Science Advancement.

## Organizing Chair

Yasunari Harada

## Organizing Committee

Glen Stockwell
Ryo Otoguro
Joji Maeno
Noriaki Kusumoto
Akira Morita

Kyoko Kanzaki
Kanako Maebo
Mieko Ebara
Mayumi Kawamura
Alexander Gelbukh

## Program Chair

Alexander Gelbukh

## Program Committee

Sophia Ananiadou
Bogdan Babych
Sivaji Bandyopadhyay
Roberto Basili
Pushpak Bhattacharyya
Nicoletta Calzolari
Sandra Carberry
Kenneth Church
Dan Cristea
Silviu Cucerzan
Mona T. Diab
Alex Chengyu Fang

Anna Feldman
Daniel Flickinger
Alexander Gelbukh
Roxana Girju
Gregory Grefenstette
Iryna Gurevych
Nizar Habash
Sanda Harabagiu
Yasunari Harada
Ales Horak
Eduard Hovy
Nancy Ide

## Software Reviewing Committee

## Award Committee

# Additional Referees

Naveed Afzal
Rodrigo Agerri
Alexandre Agustini
Laura Alonso Alemany
Rania Al-Sabbagh
Maik Anderka
Paolo Annesi
Eiji Aramaki
Jordi Atserias
Wilker Aziz
João B. Rocha-Junior
Nguyen Bach
Vít Baisa
Jared Bernstein
Pinaki Bhaskar
Arianna Bisazza
Eduardo Blanco
Bernd Bohnet
Nadjet Bouayad-Agha
Elena Cabrio
Xavier Carreras
Miranda Chong
Danilo Croce
Amitava Das
Dipankar Das
Jan De Belder
Diego Decao
Iustin Dornescu
Kevin Duh
Oana Frunza
Caroline Gasperin
Jesús Giménez
Marco Gonzalez
Amir H. Razavi
Masato Hagiwara
Laritza Hernández
Ryuichiro Higashinaka
Dennis Hoppe
Adrian Iftene
Iustina Ilisei
Chris Irwin Davis
Yasutomo Kimura
Levi King
Olga Kolesnikova

Natalia Konstantinova
Sudip Kumar Naskar
Alberto Lavelli
Yulia Ledeneva
Chen Li
Nedim Lipka
Lucelene Lopes
Christian M. Meyer
Kanako Maebo
Sameer Maskey
Yashar Mehdad
Simon Mille
Michael Mohler
Manuel Montes y Gómez
Rutu Mulkar-Mehta
Koji Murakami
Vasek Nemcik
Robert Neumayer
Zuzana Neverilova
Ryo Otoguro
Partha Pakray
Santanu Pal
Michael Piotrowski
Natalia Ponomareva
Martin Potthast
Heri Ramampiaro
Luz Rello
Stefan Rigo
Alvaro Rodrigo
Alex Rudnick
Ruhi Sarikaya
Gerold Schneider
Leanne Seaward
Ravi Sinha
Amber Smith
Katsuhito Sudoh
Xiao Sun
Irina Temnikova
Wren Thornton
Masato Tokuhisa
Sara Tonelli
Diana Trandabat
Stephen Tratz
Yasushi Tsubota

Jordi Turmo                                Clarissa Xavier
Kateryna Tymoshenko                        Jian-ming Xu
Sriram Venkatapathy                        Caixia Yuan
Renata Vieira                              Rong Zhang
Nina Wacholder                             Bing Zhao
Dina Wonsever

## Website and Contact

The website of the CICLing conference series is www.CICLing.org. It contains
information about the past CICLing conferences and their satellite events, including published papers or their abstracts, photos, video recordings of keynote
talks, as well as the information about the forthcoming CICLing conferences and
the contact options.

# Table of Contents – Part II

## Machine Translation and Multilingualism

## Information Extraction and Information Retrieval

## Text Categorization and Classification

# Summarization and Recognizing Textual Entailment

# Authoring Aid, Error Correction, and Style Analysis

## Speech Recognition and Generation

# Table of Contents – Part I

## Lexical Resources

## Syntax and Parsing

## Part of Speech Tagging and Morphology

## Word Sense Disambiguation

# Semantics and Discourse

# Opinion Mining and Sentiment Detection

# Text Generation

# Ontology Based Interlingua Translation

Leonardo Lesmo, Alessandro Mazzei, and Daniele P. Radicioni

University of Turin, Computer Science Department
Corso Svizzera 185, 10149 Turin
{lesmo,mazzei,radicion}@di.unito.it

**Abstract.** In this paper we describe an interlingua translation system from Italian to Italian Sign Language. The main components of this systems are a broad coverage dependency parser, an ontology based semantic interpreter and a grammar-based generator: we provide the description of the main features of these components.

## 1 Introduction

In this paper we describe some features of a system designed to translate from Italian to Italian Sign Language (henceforth LIS). Many approaches have been proposed for automatic translation, which require different kinds of linguistic analysis. For instance, the *direct* translation paradigm requires just morphological analysis of the source sentence, while the *transfer* translation paradigm requires syntactic (and sometimes semantic) analysis too [1]. In contrast, our architecture adheres to the *interlingua* translation paradigm, i.e. it performs a deep linguistic processing in each phase of the translation, i.e. (1) deep syntactic analysis of the Italian source sentence, (2) semantic interpretation, and (3) generation in LIS of the target LIS sentence. These three phases form a pipeline of processing: the syntactic tree produced in the first phase is the input for the second phase, i.e semantic interpretation; similarly, the semantic structure produced in the second phase is the input of the third phase, i.e. generation. In order to work properly, Interlingua pipeline requires good performances in each phase of the translation. Moreover, since the semantic interpretation in crucially related to the world knowledge, the state-of-the-art computational linguistic techniques allow the interlingua approach to work only on limited domain [1]. In our work, we concentrate on the classical domain of weather forecasts.

A challenging requirement of our project is related to the target language, the LIS, that does not have a *natural* written form (which is typical of the signed languages). In our project we developed an *artificial* written form for LIS: this written form encodes the main morphological features of the signs as well as a number of non-manual features, as the gaze or the tilt of the head. Anyway, for sake of clarity in this paper we report a LIS sentence just as a sequence of *GLOSSAS*, that is the sequence of the names[1] of the signs, without any extra-lexical feature.

---

[1] A name for a sign is just a *code* necessary to represent the sign. As it is customary in the sign languages literature, we use names for the signs that are related to their rough translation into another language, Italian in our work.

A. Gelbukh (Ed.): CICLing 2011, Part II, LNCS 6609, pp. 1–12, 2011.

The building blocks of our architecture are the dependency parser for syntactic analysis, the ontology-based semantic interpreter, the CCG-based generator. In the paper, we first introduce the dependency parser (Section 2), then we focus on the description of the main issues of the semantic interpretation and provide a case study on ordinal numbers (Section 3). A key point in the semantic interpretation is that the syntax-semantics interface used in the analysis is based on an ontology, similar to [2]. The knowledge in ontology concerns the domain of application, i.e. weather forecasts, as well as more general information about the world. The latter information is used to compute the sentence meaning. The result of the semantic interpretation is a complex fragment of the ontology: predicate-argument structures and semantic roles describing the sentence are contained in this fragment. In Section 4 we describe the generation phase, and illustrate the a combinatory categorial grammar that we devised for LIS. Finally, in Section 5 we conclude the paper and point out some future developments to the system.

## 2 Syntactic Analysis

In limited domains (as the one of weather forecasts) it is possible to obtain a "deep understanding" of the meaning of texts. To get this result, we need the detailed syntactic structure of the input sentences and specific information about the meaning of the words appearing in the sentences. The syntactic structure is produced by the TULE parser [3]. It uses a morphological dictionary of Italian (about 25, 000 lemmata) and a rule-based grammar. The final result is a "dependency tree", that makes clear the structural syntactic relationships occurring between the words of the sentence. After two preliminary steps (the *morphological analysis* and *part of speech tagging*, necessary to recover the lemma and the part of speech (PoS) tag of the words), the sequence of words goes through three phases: *chunking*, *coordination analysis*, and *verbal subcategorization*.

Let us consider the following sentence: "Locali addensamenti potranno interessare il settore nordorientale" (*Local cloudiness could concern the northeastern sector*). By looking for chunks (i.e. sequences of words usually concerning noun substructures), we get "Locali addensamenti" and "il settore nord-orientale". Then verbal subcategorization is used to attach these chunks to the verbs "potere" and "interessare" and for inserting the trace. Each word in the sentence is associated with a node of the tree. Actually, the nodes include further data (e.g., the gender and number for nouns and adjectives and verb tenses) which do not appear in the figure for space reasons. The nodes are linked via labeled arcs that specify the role of the dependents with respect to their governor (the parent). For instance, "addensamento" (*cloudiness*) is the subject of the verb "potere" (to *can*: verb-subj), while "il" (*the*) is the direct object of "interessare" (*to interest*: verb-obj). In the Figure, there is also a special node (framed by a dashed line and labeled *t*), which is a "trace". It specifies that the subject of "interessare" is "addensamento", although the node associated with it syntactically depends on "potere". In other words this node, which does not correspond to any word in the sentence, enables us to specify that "addensamento" is a subject shared by "potere" and "interessare".

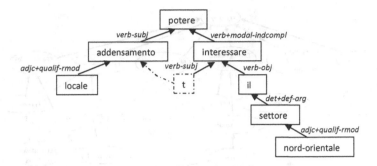

**Fig. 1.** Syntactic structure of the sentence "Locali addensamenti potranno interessare il settore nord-orientale" (*Local cloudiness could concern the north-eastern sector*)

## 3 Knowledge Representation and Semantic Interpretation

In the overall architecture of the system the ontology is accessed to build a semantic representation of the input sentence, which is then used by the generative process. Basically the system searches for a match between the (annotated) syntactic trees and the concepts in a domain ontology. We now introduce two preliminary issues, that are the notion of *semantic proximity* and the problem of *linguistic mediation*; then in Section 3.1 we describe the taxonomy of the entity *description*, and in 3.3 we provide an example to illustrate how the system copes with the linguistic phenomenon of ordinals; we then illustrate the use of the *ordinal-description* entity which is central to the interpretation process.

In the present setting we build on the notion of semantic *proximity* between concepts that are present in a given region. Intuitively, the proximity between two concepts can be defined as the number of intervening steps between them in a concept hierarchy [4]. The process of semantic interpretation can be cast to the problem of finding a path between pairs of words. A shortest path is searched that represents the strongest semantic connection between the words; and in turn, the strongest semantic connection is that minimizing the (semantic) distance between the considered words.

In general, an ontology can collect two kinds of rather different entities. On the one side entities that are concerned with the application domain, such as temporal entities and geographic regions, weather status. On the other side we deal with the *description* of such entities, which is rooted in the linguistic mediation that has to do with 'talking of' things, rather than with 'things themselves'. Accordingly, in devising ontologies (and in particular ontologies of intrinsically linguistic and communicative acts, such as weather forecasts) one has to deal with two problems: how to represent the knowledge about the world, and how is that knowledge connected to language that is needed to talk about the knowledge about the world. This problem is sometimes referred to as *ontological stratification*, and it has received different answers in the scientific community [5]. A possible solution to the ontological stratification problem consists in considering multiple ontological levels, such as a *material* level of constitution and the *'objects themselves'* level [6]. Under a different perspective, the dichotomy between

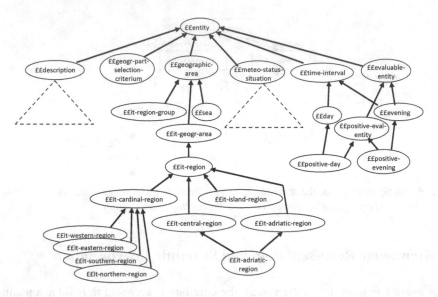

**Fig. 2.** The top level of the weather forecasts ontology. Dashed triangles represent regions of the hierarchy that are illustrated separately in Figure 3 and 4.

the world and its description inspired the so called *D&S* (so named after Descriptions and Situations) and its constructive counterpart *c.DnS* [7,8]. In particular, *c.DnS* can be used to extend the DOLCE foundational ontology [9] by providing it with an epistemo-logical perspective "from which the entities of the domain are considered". In particular, the mentioned approach proposes to describe *conceptualizations* from some given do-main through *descriptions* and the settings (also called states of affairs) relevant to the considered domain through *situations*. In our ontology descriptions are entities sepa-rated from entities representing concepts themselves. For example, if today is October 29, 2010 in the ontology we distinguish the (deictic) description 'today', from the re-ferred instance of day. Similarly 'October 29, 2010', would be represented like another (absolute) description of the same instance of day.

### 3.1   The Ontology

The top level of the ontology is illustrated in Figure 2.[2] In the following we denote *concepts* (classes) with the *££* prefix; *instances* have a *£* prefix, and *relations* and their instances are prefixed with *&*. We start by considering the classes most relevant to weather forecasts, that is *££meteo-status-situation*, *££geographic-area* and *££description*.

- *££meteo-status-situation*. This is the most relevant subclass in the present setting, since it refers about the possible weather situations, thus providing the starting point

---

[2] We defer to a future work the investigation of how the present ontology could be connected to a foundational ontology, such as DOLCE.

–in principle– to every weather forecast. It may concern the sea status (and the ££sea-current), a generic weather status (in particular if it is stable or not) or possible atmospheric events such as snow, rain or clouds. Three subclasses are rooted in ££meteo-status-situations: ££sea-status-situation, ££weather-event and ££weather-status-situation.

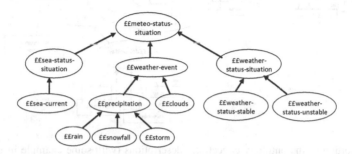

**Fig. 3.** The portion of the ontology describing weather situations

- ££time-interval. Any weather situation holds in a specific temporal interval, thereby making time a fundamental element in weather forecasts. Such time interval could last one or more days or a part of a day.
- ££geographic-area. Any weather situation holds in a specific place; in particular, the relevant places are geographic areas. A ££geographic-area can be an Italian region, a group of regions, a sea, or may be identified by specifying a cardinal direction (North, South, . . . ).
- ££description. In the hierarchy rooted in the concept ££description, particular relevance have the deictic descriptions (see Figure 4), since most temporal descriptions (*today, tomorrow*, but also the weekday names, as *Monday, Tuesday*, . . . ) are deictic in nature.

Further relevant subclasses of ££entity are ££degree, which is used to specify, for instance, that the weather is more or less stable; ££reified-relation, about which we elaborate in the following.

**Relations.** The last relevant portion of the ontology concerns *relations*. Although the ontology has no axioms, class concepts are connected through relevant relations. In turn, relations constitute the basic steps to form paths. All relations in the ontology are binary, so that the representation of relations of arity greater than 2 requires them to be reified. In Figure 5 we report two example relations that occur in the weather forecast domain. Relations are represented as arrows with small boxes. The *domain* of the relation is the node that the arrow leaves, while the *range* is the node that the arrow enters. The name of the relation is reported near the small box. The functionality information has the usual meaning, and is used to add constraints on the fillers of a relation with respect to some class. Namely, 1:1 means that both the relation and its inverse are *functional*; 1:N means that each individual of the domain can be associated with N individuals of the

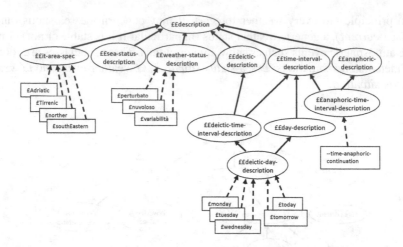

**Fig. 4.** The portion of the ontology concerning descriptions (with some example instances, depicted as boxes)

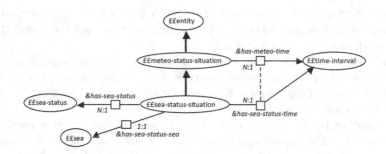

**Fig. 5.** Use of relations to represent the features of *££sea-status-situation*

range, but not viceversa. The converse is expressed by the notation N:1. N:M refers to the absence of functionality constraints. The dashed link connecting *&has-meteo-time* and *&has-sea-status-time* specifies that the latter relation restricts the former one.

## 3.2   Semantic Interpretation

The basic assumption underlying the semantic interpretation is that the meaning of words is expressed in terms of ontology nodes, and that a central component of the overall meaning of the sentence is a complex path on the ontology that we call *ontological restriction*. In this Section we define the *meaning interpretation function* $\mathcal{M}_\mathcal{O}$: we start from the dependency tree of the sentence, and on the basis of the lexical meaning of the words (given in terms of an ontology $\mathcal{O}$) we compute the ontological restriction.

Given a sentence $S$ and the corresponding syntactic analysis expressed as a dependency tree $depTree(S)$, the meaning of $S$ is computed by applying the meaning interpretation function to the root of the tree, that is $\mathcal{M}_\mathcal{O}(root(depTree(S)))$. In procedural terms, the meaning corresponding to a sentence is computed in two steps: (*i*) we annotate each word of the input sentence with the corresponding lexical meaning; (*ii*) we build the actual ontological representation in a quasi-compositional way, by joining paths found in the ontology in a single representation which is a subgraph (with possible redundancies) of the ontology itself. These two steps can be formalized as a meaning interpretation function $\mathcal{M}_\mathcal{O}$ defined[3] as:

$$\mathcal{M}_\mathcal{O}(n) := \begin{cases} \mathcal{L}\mathcal{M}_\mathcal{O}(n), & \text{if } n \text{ is a leaf} \\ \dot{\cup}_{i=1}^{k}(\mathcal{C}\mathcal{P}_\mathcal{O}(\mathcal{L}\mathcal{M}_\mathcal{O}(n), \mathcal{M}_\mathcal{O}(d_i))), & \text{otherwise} \end{cases}$$

where $n$ is a node of the dependency tree and $d_1, d_2, \ldots, d_k$ are its dependents. $\mathcal{L}\mathcal{M}_\mathcal{O}(w)$ is a function that extracts the lexical meaning of a word $w$: that is, a class or an individual in the ontology $\mathcal{O}$. The meaning is determined by accessing the dictionary. $\mathcal{C}\mathcal{P}_\mathcal{O}(y, z)$ is a function that returns the shortest path on $\mathcal{O}$ connecting $y$ to $z$. The search for connections is based on the idea that the shortest path that can be found in the ontology between two nodes represents the stronger semantic connection between them; consequently, such path must be used to build the semantic representation. Finally, the operator $\dot{\cup}$ is used to denote a particular merge operator. As a general strategy, shortest paths are composed with a union operator, but each $\mathcal{C}\mathcal{P}_\mathcal{O}(y, z)$ conveys a set of ontological constraints: the merge operator takes all such constraints into account in order to build the overall complex ontological representation. A particular case of ontological constraints is present in the interpretation of ordinal numbers, which is discussed in next Section.

### 3.3 A Case Study: The Ordinal Numbers

In order to translate from Italian to LIS, we have to account for a number of semantic phenomena appearing in the particular domain chosen as pilot study, i.e. weather forecast. One of the most frequent constructions are ordinal numbers. Let us consider the simple phrase *l'ultimo giorno del mese* (*the last day of the month*). The (simplified) dependency structure corresponding to this phrase is depicted in Figure 6: the head word *giorno* (*day*) has two modifying dependents, *ultimo* (*last*) and *mese* (*month*). Since the

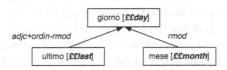

**Fig. 6.** The dependency analysis of *ultimo giorno del mese* (*last day of the month*) enriched with lexical meaning (in bold face)

---

[3] For sake of simplicity in this definition we do not describe the mechanism used for ambiguity resolution.

interpretation relies heavily on the access to the ontology, we first describe the part of the ontology used for the interpretation and then we illustrate the application of the function $\mathcal{M}_\mathcal{O}$ on the given example.

The relevant fragment of the ontology is organized as shown in Figure 7, where it has been split in two parts. The upper part –labeled *TEMPORAL PARTS*– describes the reified *££part-of* relation and its temporally specialized subclasses. The lower part –labeled *ORDINALS*– is constituted by some classes that account just for ordinal numbers. In the *TEMPORAL PARTS* region of the Figure we find the *££temporal-part-of* (reified) sub-relation, which, in turn, subsumes *££day-month-part-of*. This specifies that days are parts of months, so that *day of the month* can be interpreted as *the day which is part of the month*. The *££part-of relation* has two roles: we use the term *role* to refer to the binary relation associated with a participant in a reified relation. These roles are "value-restricted" as *&day-in-daymonth* and *&month-in-daymonth* respectively, for what concerns *££day-month-part-of*. The most relevant class in the ORDINALS part of Figure 7 is the class *££ordinal-description*. It is the *domain* of three roles, *1*) *&ord-described-item*, *2*) *&reference-sequence* and *3*) *&ordinal-desc-selector*. The range of the first relation *&ord-described-item* is the item whose position in the sequence is specified by the ordinal, that is a *££sequenceable-entity*. The range of the second relation *&reference-sequence* is the sequence inside which the position makes sense, that is an *££entity-sequence*. The range of the third relation *&ordinal-desc-selector* is item that specifies the position, that is a *££ordinal-selector*. In the example, *£last* is an instance of *££ordinal-selector*. Of course, any (true) ordinal (first, second, thirtythird) can fill that role. The two portions of the ontology are connected by two arcs. The first arc specifies that a *££time-interval* is a subclass of *££sequenceable-entity* (so that one can say *the fourth minute, the first year*, and so on). The second arc specifies that *££month* is subclass of *££day-sequence*, which in turn is subclass of *££entity-sequence*. As a consequence it can play the role (can be *range*) of the *&reference-sequence*. Applying the meaning interpretation function to the considered example consists of three steps: *1.* we compute the connection path (*CP* function) between words *giorno* and *ultimo* (i.e., the first branch of the dependency tree in Figure 7) we obtain a connection path connecting *££day* to *££last* passing through *££ordinal-description*; *2.* we compute the connection path between the words *giorno* and *mese* (i.e., the second branch of the dependency tree) and obtain a path connecting *££day* to *££last* passing through *££part-of*; *3.* we compute the overall meaning (*$\mathcal{M}_\mathcal{O}$*) by composing the connection paths previously computed. In this step the presence of the *££ordinal-description* concept is detected in the first ontological restriction; moreover *££day* is recognized as item of this *££ordinal-description*. At this point we need establishing how *££day* fits as the smaller part in a *&part-of* relation. We scan the remaining ontological restriction(s) by looking for a bigger part involved in a *&part-of* relation or in any of its sub-relations. The resulting representation is built by assuming that the larger entity is the reference sequence for the ordering. So, the direct *££day-month-part-of* of the second ontological restriction is replaced by a path passing through *££ordinal-description*. In such final ontological restriction (depicted in Figure 8) *££day* is the *&ord-described-item* and *££month* is the *&reference-sequence*.

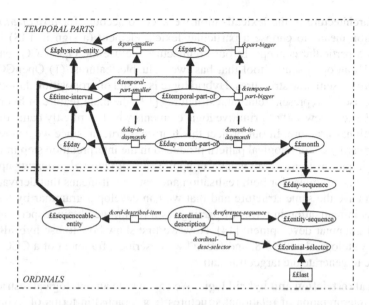

**Fig. 7.** The fragment of the ontology accounting for ordinals

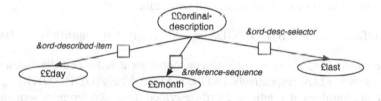

**Fig. 8.** The resulting ontological restriction produced by the semantic interpreter on the dependency tree in Figure 6

## 4   Generation

Natural language generation can be described as a three steps process: text planning, sentence planning and realization [10]. Text planning determines which messages to communicate and how rhetorically to structure these messages; sentence planning converts the text plan into a number of sentence plans; realization converts the sentence plans into the final sentences produced. Anyway, in the context of interlingua translation we think that generation needs only for the realization step. Our working hypothesis is that source and target sentences have exactly the same text and the sentence plans. Indeed, the ontological pattern, that is the output of the semantic interpretation step, contains at the same time the information content as well as the rhetorical and sentence structures of the source messages: our choice is to reproduce exactly the same structures in the generation of the target sentences. As a consequence we use a generation system that performs just realization.

In our architecture in the lexicalization we need to account for *lexicalization* too: lexicalization means to choose a particular lexical element (a sign in LIS) in order to express a particular concept. In our architecture we use the OpenCCG realization system [11], an open source tool that has two valuable features: (1) OpenCCG uses *hybrid logic* for semantic structures. Hybrid logic is a kind of propositional modal logic that can be used to represent relational structures [12]: the main feature of hybrid logic is *nominals*, i.e. a new sort of primitive logic elements which explicitly name the nodes of the relational structure. In our project this feature is crucial, since we can represent straightforwardly the ontological pattern produced in the interpretation step in terms of hybrid logic propositions.[4] (2) OpenCCG applies a *bidirectional grammar* approach, i.e. there is one grammar for both realisation and parsing. It means that derivation and generation have the same structure and that we can develop a grammar by testing its correctness in realization in terms of parsing: as a result, we obtain a speed-up in the process of grammar development [14]. Now we first show how to use hybrid logic to model the ontological path of 8, and second we describe a fragment of a CCG for LIS that is able to generate the target translation.

**Ontological restriction and hybrid logic.** We can rewrite an ontological restriction by using the interpretation of relational structures (e.g. graphs) in terms of hybrid logic given in [12]: each node of the structure will be represented by a distinct *nominal*, and each edge will be represented by using a distinct *modality label*. Applying this procedure to the ontological restriction of 8 we obtain:

$$@_{x_0}(\langle ODI \rangle x_1 \wedge \langle ODRS \rangle x_2 \wedge \langle ODS \rangle x_3) \wedge @_{x_1} \text{day} \wedge @_{x_2} \text{month} \wedge @_{x_3} \text{last} \quad (1)$$

where the nominals $x_0, x_1, x_2, x_3$ represent the ontological nodes ££*ordinal-description*, ££*day*, ££*month*, ££*last* respectively, and the modality labels $\langle ODI \rangle$, $\langle ODRS \rangle$, $\langle ODS \rangle$ represent the ontological relations &*ord-described-item*, &*references-sequence* and &*ordinal-desc-selector* respectively.

**A CCG for LIS.** The target translation in LIS for the Italian source phrase *ultimo giorno del mese* is *MESE GIORNO ULTIMO*: we now describe how realize this LIS phrase by starting from the hybrid logic formula in equation (1). We developed by hand a CCG for LIS that accounts for a number of morphosyntactic phenomena: in particular we account for morphological realization of plural, spatial agreement between verbs and arguments, coordination [15]. In Tab. 1 we present the fragment of the CCG for the lexical elements involved.

Each element in the grammar has four categories: **LEX**, that contains the lexical form of the item; **PoS**, that contains the part of speech category; **SynCAT**, that contains the syntactic category; **SemCAT**, that contains the semantic category. Note that SynCAT e SemCAT are related by using *semantic variables* ($x_i$ and $z_j$ in Tab. 1): these

---

[4] Note that ontological patterns could be written in terms of FOL predicates and, since Hybrid Logic is equivalent to a fragment of FOL, we could rewrite these FOL predicates in terms of hybrid logic, identifying first order variables with nominals of hybrid logic [12]. Moreover our logical interpretation of the ontological pattern does not adhere to the *linguistic meaning* notion that is usually adopted in OpenCCG, i.e. *Hybrid Logic Dependency Semantics* (HLDS) [13].

**Table 1.** Three lexical elements of the CCG for LIS

| LEX | PoS | SynCAT | SemCAT |
|---|---|---|---|
| *GIORNO* | Noun | $n_{x_1}$ | $@_{x_1}$ **day** |
| *MESE* | Noun | $n_{x_0}/n_{z_1}$ | $@_{x_0}(\langle ODI\rangle z_1 \wedge \langle ODRS\rangle x_2) \wedge @_{x_2}$ **month** |
| *ULTIMO* | Adj | $n_{z_2}\backslash n_{z_2}$ | $@_{z_2}(\langle ODS\rangle x_3)@_{x_3} \wedge$ **last** |

$$\frac{\overbrace{n_{x_0}/n_{z_1} : @_{x_0}(\langle ODI\rangle z_1 \wedge \langle ODRS\rangle x_2) \wedge @_{x_2}\,\text{month}}^{MESE}\quad \overbrace{n_{x_1} : @_{x_1}\,\text{day}}^{GIORNO}\quad \overbrace{n_{z_2}\backslash n_{z_2} : @_{z_2}(\langle ODS\rangle x_3) \wedge @_{x_3}\,\text{last}}^{ULTIMO}}{\underset{\displaystyle n_{x_0} : @_{x_0}(\langle ODI\rangle x_1 \wedge \langle ODRS\rangle x_2 \wedge \langle ODS\rangle x_3) \wedge @_{x_1}\,\text{day} \wedge @_{x_2}\,\text{month} \wedge @_{x_3}\,\text{last}}{\overset{\displaystyle n_{x_0} : @_{x_0}(\langle ODI\rangle x_1 \wedge \langle ODRS\rangle x_2) \wedge @_{x_1}\,\text{day} \wedge @_{x_2}\,\text{month}}{>}}}_{<}$$

**Fig. 9.** The realization/derivation of the LIS phrase *MESE GIORNO ULTIMO* by using the lexicon in Table 1

variables appear in the syntactic categories, but are used as pointers to the semantic categories [13,11]. For instance, in the syntactic category $n_{x_0}/n_{z_1}$ there are two semantic variables: $x_0$ and $z_1$. When syntactic categories combine in a derivation, the semantic variables are unified and the corresponding semantic categories are unified too (see below the derivation reported in Tab. 9). Note that the nominal $x_0$ is introduced by the lexical item *MESE*: we are assuming that the semantic ordinal structure is introduced by this lexical element. In Fig. 9 we report the realization of the LIS phrase *MESE GIORNO ULTIMO* based on the lexicon in Tab. 1. For sake of clarity, we are going to describe this realization as a derivation: since we are using a bidirectional grammar, realization applies the same rules of derivation but in the reverse order. The derivation consists of two syntactic steps: in the first step the $n_{x_0}/n_{z_1}$ category (corresponding to *MESE*) combines with the $n_{x_1}$ category (corresponding to *GIORNO*) by a forward application producing a new $n_{x_0}$ category; in the second step, the new $n_{x_0}$ category combines with the $n_{z_2}\backslash n_{z_2}$ category (corresponding to *ULTIMO*) by a forward application producing the final $n_{x_0}$ category. In parallel to these two applications, we have that two semantic variables unify. In the first step, the semantic variable $z_1$ unifies with the semantic variable $x_1$, while in the second step the semantic variable $z_2$ unify with the semantic variable $x_0$. Finally, the last module of our architecture is a virtual actor, i.e. an artificial character, that synthesizes the LIS produced by the OpenCCG.

## 5   Conclusions and Future Work

In this paper we have presented an architecture for the translation from Italian into Italian Sign Language. The implemented system goes all throughout the translation process: we parse the input sentence, we then extract the meaning representation and generate the LIS. Finally, the synthesis of gestures takes place, and it is performed by a virtual character. The architecture tackles presently a restricted domain, that is it is focussed on weather forecasts. We have briefly described the ontology devised, which encodes the knowledge used by the semantic interpreter and we have illustrated the generation phase, a component of the system that relies on the CCG paradigm and adopts a hybrid logic approach for the realization proper. We have provided a working example to illustrate both the semantic interpretation phase and the generation phase. Much work still needs to be done at various levels: *i*) the ontology design needs to be refined to fully account for the richness of weather forecasts and related descriptions; *ii*) the semantic interpreter can be improved, e.g. by focussing on the redundancies resulting from the shortest path procedures, and by adding to the ontology some shortcuts to save computational efforts in computing the meaning representation; *iii*) the generation module still needs refinements, as regards as to consider further syntactic phenomena. All described modules will need substantial improvements in order to extend the coverage of the system.

**Acknowledgments.** This work has been partially supported by the ATLAS project, that is co-funded by Regione Piemonte within the "Converging Technologies - CIPE 2007" framework (Research Sector: Cognitive Science and ICT).

# References

1. Jurafsky, D., Martin, J.H.: Speech and Language Processing: An Introduction to Natural Language Processing. In: Computational Linguistics and Speech Recognition. Prentice Hall, Englewood Cliffs (2008)
2. Nirenburg, S., Raskin, V.: Ontological Semantics. MIT Press, Cambridge (2004)
3. Lesmo, L.: The Rule-Based Parser of the NLP Group of the University of Torino. Intelligenza Artificiale 2, 46–47 (2007)
4. Miller, G.A., Beckwith, R., Fellbaum, C., Gross, D., Miller, K.J.: Introduction to WordNet: An On-line Lexical Database. International Journal of Lexicography 4, 235–244 (1993)
5. Bateman, J.A.: Linguistic interaction and ontological mediation. In: Ontolinguistics. In: How Ontological Status Shapes the Linguistic Coding of Concepts. Mouton De Gruyter (2007)
6. Borgo, S., Guarino, N., Masolo, C.: Stratified ontologies: the case of physical objects. In: Proceedings of the Workshop on Ontology Engineering, ECAI 1996 (1996)
7. Gangemi, A., Mika, P.: Understanding the Semantic Web through Descriptions and Situations. In: Chung, S., Schmidt, D.C. (eds.) CoopIS 2003, DOA 2003, and ODBASE 2003. LNCS, vol. 2888, pp. 689–706. Springer, Heidelberg (2003)
8. Gangemi, A.: Norms and plans as unification criteria for social collectives. Journal of Autonomous Agents and Multi-Agent Systems 16, 70–112 (2008)
9. Masolo, C., Borgo, S., Gangemi, A., Guarino, N., Oltramari, A., Schneider, L.: Wonder Web deliverable D17. Technical Report D17, ISTC-CNR (2002)
10. Reiterand, E., Dale, R.: Building natural language generation systems. Cambridge University Press, Cambridge (2000)
11. White, M.: Efficient realization of coordinate structures in combinatory categorial grammar. Research on Language and Computation 2006, 39–75 (2006)
12. Blackburn, P.: Representation, reasoning, and relational structures: a hybrid logic manifesto. Logic Journal of the IGPL 8, 339–625 (2000)
13. Baldridge, J., Kruijff, G.J.M.: Coupling ccg and hybrid logic dependency semantics. In: Procs. of ACL 2002, pp. 319–326. ACL, Morristown (2002)
14. White, M., Clark, R.A.J., Moore, J.D.: Generating tailored, comparative descriptions with contextually appropriate intonation. Computational Linguistics 36, 159–201 (2010)
15. Volterra, V. (ed.): La lingua dei segni italiana. Il Mulino (2004)

# Phrasal Equivalence Classes for Generalized Corpus-Based Machine Translation

Rashmi Gangadharaiah, Ralf D. Brown, and Jaime Carbonell

Carnegie Mellon University
{rgangadh,ralf,jgc}@cs.cmu.edu

**Abstract.** Generalizations of sentence-pairs in Example-based Machine Translation (EBMT) have been shown to increase coverage and translation quality in the past. These template-based approaches (G-EBMT) find common patterns in the bilingual corpus to generate generalized templates. In the past, patterns in the corpus were found by only few of the following ways: finding similar or dissimilar portions of text in groups of sentence-pairs, finding semantically similar words, or use dictionaries and parsers to find syntactic correspondences. This paper combines all the three aspects for generating templates. In this paper, the boundaries for aligning and extracting members (phrase-pairs) for clustering are found using chunkers (hence, syntactic information) trained independently on the two languages under consideration. Then semantically related phrase-pairs are grouped based on the contexts in which they appear. Templates are then constructed by replacing these clustered phrase-pairs by their class labels. We also perform a filtration step by simulating human labelers to obtain only those phrase-pairs that have high correspondences between the source and the target phrases that make up the phrase-pairs. Templates with English-Chinese and English-French language pairs gave significant improvements over a baseline with no templates.

**Keywords:** Generalized Example-based Machine Translation (G-EBMT), Template Induction, Unsupervised Clustering, data sparsity.

## 1   Introduction

Templates are generalizations of sentence-pairs formed by replacing sequences of words by variables. Like other data-driven MT approaches such as Statistical MT (SMT), EBMT also requires large amounts of data to perform well. Generalization was introduced in EBMT to increase coverage and improve quality in data-sparse conditions [12,4]. If the following sentence-pair (SP: source and its corresponding target sentence) is present in the bilingual training corpus and equivalence classes C1 and C2 are among the clusters available (either obtained automatically or from a bilingual speaker),

```
(SP) source sentence: flood prevention and development plans must
            also be drawn up for the major river basins
target sentence: 各 大 流域 也 要 制定 防洪 、 开发 治理 规划
C1:
flood prevention and development plans ↔ 防洪 、 开发 治理 规划
action plans ↔  动 预案
emergency plans ↔ 应急 预案
```

A. Gelbukh (Ed.): CICLing 2011, Part II, LNCS 6609, pp. 13–28, 2011.

C2:

```
for the major river basins  ↔  各 大 流域
for the major river banks   ↔  各 大 河岸
```

then SP can be converted into a template-pair (TP) by replacing phrase-pairs that belong to any cluster by their class labels (C1 and C2):

```
(TP) source template: <CL1> must also be drawn up <CL2>
target template: <CL2> 也 要 制定 <CL1>
```

TP can be used to translate new sentences such as, emergency plans must also be drawn up for the major river banks, even when these sentences are not present in the bilingual training corpus, especially when the corpus is small. Templates bear resemblance to transfer rules used in Rule-based MT, however, templates use fewer constraints. They are also similar to the rules used in Syntax-based SMT [33] but are not necessarily linguistic-based. These templates can be made into hierarchical rules as in hierarchical phrase-based SMT [8], however, this causes over-generalization in EBMT where a large number of ungrammatical target phrases are generated.

Many of the G-EBMT approaches suggested in the past found patterns in the corpus by few of the following ways: (1) finding similar or dissimilar portions of text in the corpus, (2) finding semantically similar words, (3) used parsers to first linguistically parse the source and target sentences in the corpus and then found syntactic correspondences between source and target phrases (or phrase-pairs). Templates created using (1) ensure that only phrase-pairs that appear in similar contexts are interchangeable. EBMT systems with templates created using (2) and (3) have the advantage that only phrase-pairs with similar parents (syntactically/semantically related) or non-terminals (and context-free) are interchangeable. Also, many of the template-based approaches only perform generalization of words. This paper, takes knowledge from all the three sources to obtain templates where both words as well as phrases are generalized.

[14] and [22] used similar and dissimilar portions, limiting the amount of generalization that can be performed. [30] performed chunking using the marker hypothesis. [16] makes use of syntactic phrases from parsed trees but the templates created are less controllable as the method collapses phrases only by linguistic phrase labels. [1] extracted chunk-pairs from word-alignments to create templates. [5] proposed a recursive transfer-rule induction process to create templates. However, in [4], the optimum value of number of clusters ($N$) needs to be determined.

Earlier EBMT systems lacked statistical decoders that SMT systems used, hence, they had to use the bilingual corpus to generate the translation. If a new sentence to be translated was not found in the corpus, some systems modified closely matching source sentences and their corresponding target sentences to generate a suitable translation [30]. Some systems generalized the new sentences into templates and these templates had to match at least one of the templates completely in the generalized bilingual corpus in order to generate the translation [14]. Others [1,16] adopted simple translation algorithms to join partial target matches to generate the translation. [27] used 'hooks' to indicate words and part-of-speech tags that could occur before and after a fragment.

Recent EBMT systems [3] borrow ideas from Statistical Machine Translation (SMT) systems and use statistical decoders. All possible target phrasal translations for a new sentence to be translated are extracted from the translation model (TM) of the EBMT system using sub-sentential alignment. These partial translations are then decoded using a statistical target *language model* (LM). While EBMT borrowed statistical decoders from SMT, phrase-based SMT (PBSMT) [18] borrowed the TM from EBMT and as a result the boundary between PBSMT and EBMT is not distinct. PBSMT stores all possible phrase-pairs from a corpus as a static phrase-table, while EBMT [4] does the phrase-pair extraction dynamically when a new sentence needs to be translated.

Templates are still useful in present EBMT systems to improve translation quality with computationally restricted decoders. Since finding the best translation among all possible reorderings of the target phrasal fragments for a test sentence is expensive, reordering constraints are placed on decoders. [11] show that it is beneficial to extract longer target phrasal matches from the TM for language-pairs that have very different word orders using templates. In [11], templates were obtained from word-equivalence classes only. This paper shows how to incorporate phrasal equivalence classes as well.

Phrase-pairs for clustering can be extracted using phrase-extraction techniques used in PBSMT, such as, PESA[31], Pharaoh [18] or Moses [19]. Phrase-pairs thus extracted could be clustered to obtain equivalence classes. There are two disadvantages with such a technique. First, these techniques create a large number of phrase pairs (for example, 9 million phrase-pairs were obtained from 200k sentence-pairs using Moses). Clustering and generalizing all these phrase-pairs would result in over-generalized templates (where almost every phrase is replaced by a class marker). This increases decoding time and confusion due to larger number of retrieved target matches when a new sentence has to be translated leading to lower translation quality. Secondly, finding the optimum number of clusters ($N$) [4,10] is expensive where numerous experiments need to be carried out on different $N$ on a development set, and additionally, this expense increases with the number of phrase-pairs to be clustered. Hence, a selection criteria is required to select only the useful phrase-pairs for clustering.

In this paper, we use knowledge about (i) phrase structure (ii) chunk boundaries for the source and target languages under consideration (iii) semantic-relatedness of phrase-pairs and (iv) alignment, to create templates. The purpose of including this knowledge is two fold: first, we use knowledge about the languages to reduce the search space of phrases to be generalized, second, we use semantic similarity to select and cluster phrases, allowing us to generalize only those phrases that will increase coverage when data available is small while not over-generalizing. Although the experiments in this paper are done in EBMT, it can be easily extended to other data-driven MT approaches (like PBSMT). The next section outlines our method with an example.

## 2    Outline of Our Method

In this paper, three models (described formally in Sect. 2.1) are used to extract phrase-pairs. We call these phrase-pairs as segment-pairs as a segment here is formed by combining a group of contiguous chunks and not words. The first model is the word-alignment model which finds correspondences between the source words and the target words (the dark circles in Fig. 1). The second model is the chunk alignment model

| chunk number: | | 1: | 2: | 3: | 4: | 5: | 6: | 7: | 8: | 9: |
|---|---|---|---|---|---|---|---|---|---|---|
| label | | NP | ADVP | VV | VV | NN | PU | NN | NP | PU |
| | | 各 大 流域 | 也 | 要 | 制定 | 防洪 | 、 | 开发 治理 | 规划 | 。 |
| | | $t_1$ $t_2$ $t_3$ | $t_4$ | $t_5$ | $t_6$ | $t_7$ | $t_8$ | $t_9$ $t_{10}$ | $t_{11}$ | $t_{12}$ |
| 1: NP | flood $s_1$ | | | | | | | | | |
| | prevention $s_2$ | | | | | ● | | | | |
| 2: CC | and $s_3$ | | | | | | | | | |
| 3: NP | development $s_4$ | | | | | | | ● | | |
| | plans $s_5$ | | | | | | | ● | ● | |
| 4: MD | must $s_6$ | | | ● | | | | | | |
| | also $s_7$ | | | | | | | | | |
| 5: VB | be $s_8$ | | | | | | | | | |
| 6: VBN | drawn $s_9$ | | | | ● | | | | | |
| | up $s_{10}$ | | | | ● | | | | | |
| 7: PP | for $s_{11}$ | | | | | | | | | |
| | the $s_{12}$ | | | | | | | | | |
| | major $s_{13}$ | ● | | | | | | | | |
| | river $s_{14}$ | | | | | | | | | |
| | basins $s_{15}$ | ● | | | | | | | | |
| 8: PU | . $s_{16}$ | | | | | | | | | ● |

first term, second term, third term, fourth term

**Fig. 1.** Sentence pair with chunks and chunk labels. Dark circles illustrate the primary alignments

which finds correspondences between the source and target chunks using the word-alignment information. Fig. 1 shows both the source ($schk_i$) as well as the target chunks ($tchk_j$) which can be obtained from mono-lingually trained chunkers (a chunk: is a sequence of words that form a meaningful unit in a language). This model limits the alignments to the chunk boundaries, for eg., flood prevention can be aligned to 防洪 whereas, prevention cannot be aligned to 防洪 as prevention does not form a valid chunk. This way alignment errors such as, prevention↔防洪 (should be flood prevention↔防洪) are handled by incorporating the chunk boundaries.

A meaningful unit in one language is not represented in the same way in another language i.e., it is possible that $m$ source chunks correspond to $n$ target chunks, where, $m$ and $n$ are not necessarily equal. Such mismatches are handled by the segment extraction model which extracts all consistent segment-pairs. Thus using knowledge that chunks can be a unit of sentences, we reduce the search space and this allows us to extract much longer consistent phrases. For example, our model extracts only 19 segment-pairs for the sentence-pair in Fig. 1 with no limitations on the length of the segment-pairs extracted, whereas any other standard phrase extraction technique would have extracted about 70 (Moses or Pharaoh) phrase-pairs for the same sentence-pair.

The resulting syntactically coherent segment-pairs are clustered based on their semantic closeness using a clustering algorithm (Sect. 2.4). Features for computing similarity between segment-pairs are modeled as vector space models [29]. The word-context matrix uses the *Distributional Hypothesis* [15] which indicates that words that occur in similar contexts tend to have similar meanings. Similar to word-context matrices, we construct segment*pair*-context matrices. Positive Pointwise Mutual Information (PPMI) is then used to convert the segment*pair*-context matrix into a PPMI matrix. [7] showed that PPMI outperforms many approaches for measuring semantic similarity. Once these clusters are obtained, we proceed to the template induction.

Our EBMT system aligns (source and target word-correspondences in a sentence-pair) and indexes [6] the bilingual training corpus offline. Each sentence-pair is converted into the most general template where all those segment-pairs in the sentence-pair that have been clustered are replaced by their class labels. The replacement is done in the reverse order of length- starting from longer phrases. If two overlapping phrases need to be generalized, the phrase-pair with the most number of alignment correspondences (between the source and target half of the phrase-pair) is generalized. If these overlapping phrases have the same number of correspondences, then the phrase-pair that appears first (from left to right) is generalized. The correspondence table for the sentence-pair is also modified by collapsing the word-alignment scores for every generalized source and its corresponding target phrase by an alignment score of 1.0.

Next we move on to the online phase where a new sentence needs to be translated. The input sentence is converted into a lattice of all possible generalizations and the values (or the translations) of the generalized phrases are stored. Target fragments are obtained for all the source phrases [2] in the input lattice from the indexed corpus. An example of a target fragment for a source phrase, <CL1> must also be drawn up in the input lattice corresponding to the new input sentence. emergency plans must also be drawn up for the major rivers is 也 要 制定 <CL1> with cluster <CL1> containing emergency plans↔应急 预案 as its member.

Class labels in target fragments are replaced by their values that were stored earlier, for e.g., 也 要 制定 <CL1> will be converted to 也 要 制定 应急 预案. All the target fragments for source phrases in the input sentence are then placed on a decoding lattice and the best translation is found by a statistical decoder using a target language model. Our decoder (EBMT system in [3]) performs a multi-level beam search based on the weights on candidate translations and a target language model to select the best translation. The total score for a path is given by,

$$total\ score = \frac{1}{n}\sum_{i=1}^{n}[wt_1 * log(b_i) + wt_2 * log(pen_i) + wt_3 * log(q_i)$$
$$+ wt_4 * log(P(w_i|w_{i-2}, w_{i-1})].$$

where $n$: number of target words in the path, $wt_j$: importance of each score, $b_i$: bonus factor, $pen_i$: penalty factor, $P(w_i|w_{i-2}, w_{i-1})$: LM score. The TM assigns a score to each candidate translation which is computed as a log-linear combination of its alignment score and translation probability. The alignment score indicates the engine's confidence that the right translation has been generated. The translation probability is calculated as the proportion of times each alternative translation was generated while aligning all matches retrieved for that particular source phrase. Each candidate translation is weighted by giving bonuses for longer phrases and penalties for length mismatches between the source phrase and candidate translation. Generalization penalties based on the proportion of words generalized in a path are also used. Hence, if there are two candidate translations: one generated by a lexical source phrase and the other by the same source phrase containing generalizations, then the translation extracted for the lexical source phrase is favored. The weights are tuned using coordinate ascent to maximize the BLEU score [25] on a tune set. We will now give a formal description of our model to extract segment-pairs for clustering and template-induction.

## 2.1 Formal Description of the Model

If the source sentence has $S$ words (as in Fig. 1) : $s_1^S = s_1, s_2, s_3...s_S$ and Target sentence has $\tau$ words: $t_1^\tau = t_1, t_2, t_3, ...t_\tau$, our goal is to define a probability model $P$ and then find the best possible segment boundaries $\hat{B}$ between $s_1^S$ and $t_1^\tau$,

$$\hat{B}(s_1^S, t_1^\tau) = \arg\max_b P(b|s_1^S, t_1^\tau) \ . \tag{1}$$

The source sentence is chunked into $m$ chunks ($schk_1^m$) : $schk_1, schk_2...schk_m$ and the target sentence is chunked into $n$ chunks ($tchk_1^n$) : $tchk_1, tchk_2...tchk_n$, where $m$ and $n$ are random variables. $ca$ represents alignments between the source and target chunks, $wa$ represent alignments between the source and target words. Then, by marginalization,

$$\hat{B}(s_1^S, t_1^\tau) = \arg\max_b \sum_{ca,wa,schk_1^m,tchk_1^n} P(b, schk_1^m, tchk_1^n, ca, wa|s_1^S, t_1^\tau)$$

$$= \arg\max_b \sum_{ca,wa,schk_1^m,tchk_1^n} P(wa|s_1^S, t_1^\tau) P(schk_1^m|s_1^S, t_1^\tau, wa) P(tchk_1^n|s_1^S, t_1^\tau, wa, schk_1^m)$$

$$P(ca|s_1^S, t_1^\tau, wa, schk_1^m, tchk_1^n) P(b|s_1^S, t_1^\tau, wa, schk_1^m, tchk_1^n, ca) \ . \tag{2}$$

In general, (2) is computationally infeasible to compute and so we simplify and make approximations. The source and the target chunks are obtained with chunkers trained on the two languages independently i.e., $P(schk_1^m|s_1^S, t_1^\tau, wa) = P(schk_1^m|s_1^S)$ and $P(tchk_1^n|s_1^S, t_1^\tau, wa, schk_1^m) = P(tchk_1^m|t_1^\tau)$. Source and target chunks are aligned based on word-alignments ($P(ca|wa, schk_1^m, tchk_1^n)$) in the fourth term. The fifth term which is the segment extraction model produces segment-pairs using the chunk alignments ($P(b|schk_1^m, tchk_1^n, ca)$).

$$\hat{B}(s_1^S, t_1^\tau) = \arg\max_b \sum_{ca,wa,schk_1^m,tchk_1^n} P(wa|s_1^S, t_1^\tau) P(schk_1^m|s_1^S) P(tchk_1^n|t_1^\tau)$$

$$P(ca|wa, schk_1^m, tchk_1^n) P(b|schk_1^m, tchk_1^n, ca) \ . \tag{3}$$

We approximate the above equation further by using only the most probable source and target chunk splittings instead of summing over all possible chunk splittings. A beam of different splittings can be considered and not explored in this paper.

$$\hat{B}(s_1^S, t_1^\tau) = \arg\max_b \sum_{ca,wa} P(wa|s_1^S, t_1^\tau) P(ca|wa, \hat{schk}_1^m, \hat{tchk}_1^n) P(b|\hat{schk}_1^m, \hat{tchk}_1^n, ca) .$$

with $P(schk_1^m|s_1^S) = 1$ for $\hat{schk}_1^m = \arg\max_{schk_1^m} P(schk_1^m|s_1^S)$ and 0 otherwise, and with $P(tchk_1^n|t_1^\tau) = 1$ for $\hat{tchk}_1^n = \arg\max_{tchk_1^n} P(tchk_1^n|t_1^\tau)$ and 0 otherwise.

Jointly maximizing the three probabilities in the above equation is computationally expensive. Hence, we model the three probabilities separately recognizing this may lead to sub-optimal $\hat{B}(s_1^S, t_1^\tau)$. We first find the best word-alignments between the source and target sentences. We then align the source and target chunks with the best word-alignments and finally find the best segment boundaries with these chunk alignments. Word alignments can be obtained with GIZA++ [24] or [2]. We proceed to explain the chunk alignment and the segment extraction model.

## 2.2 Chunk Alignment Model

We need to find the best possible alignments, $\hat{ca}$, between the source and target chunks. Say the source chunker [20] generated $m$ chunks and the target language chunker generated $n$ target chunks.

$$\hat{ca} = \arg\max_{ca} \hat{P}(ca|\hat{schk}_1^m, \hat{tchk}_1^n, wa) . \tag{1}$$

We divide the problem into two directions, $P(tchk_l|schk_q)$ and $P(schk_q|tchk_l)$ with $l = 1, 2, ...n$ and $q = 1, 2, ...m$. As a given source (target) chunk could be aligned to more than one target (source) chunk, we select all target (source) chunks with positive alignment probabilities for the given source (target) chunk,

$$SA_q = \left\{ l : P(tchk_l|schk_q) > 0 \right\}; \quad TA_l = \left\{ q : P(schk_q|tchk_l) > 0 \right\} . \tag{5}$$

where, $SA_q$ stores the chunk alignments for the source chunk $(schk_q)$ and $TA_l$ stores the chunk alignments for the target chunk $(tchk_l)$. $P(tchk_l|schk_q)$ is modeled as:

$$Score(tchk_l|schk_q) = \left[ P(t_1^{ls-1}|s_1^{qs-1})^{\frac{\lambda}{ls-1}} \right] \left[ P(t_{le+1}^\tau|s_1^{qs-1})^{\frac{\lambda}{\tau-le}} \right]$$
$$\left[ P(t_1^{ls-1}|s_{qe+1}^S)^{\frac{\lambda}{ls-1}} \right] \left[ P(t_{le+1}^\tau|s_{qe+1}^S)^{\frac{\lambda}{\tau-le}} \right] \left[ P(t_{le}^{ls}|s_{qs}^{qe})^{\frac{1-4\lambda}{ls-le+1}} \right] . \tag{6}$$

where, $ls$ and $le$ are the start and end indices of $tchk_l$, $qs$ and $qe$ are the start and end indices of $schk_q$ (see Fig. 1). $\lambda$ indicates the importance of the five regions in Fig. 1 ($0 \le \lambda \le 0.25$). The first four terms ensure that the boundary is agreed upon not just by the source and target chunks under consideration but also by the neighboring regions. We assume that each target word depends only on the source word that generated it and each source word generated target words independently. Equation (7) looks similar to the equation for extracting phrase-pairs in [31], however, we weigh each of the five probability terms separately to normalize each term by the number of factors that contribute to it. We emphasize that $Score(tchk_l|schk_q)$ is set to zero if none of the source words in $schk_q$ have correspondences in $tchk_l$.

$$Score(tchk_l|schk_q) = \left[\prod_{i=1}^{ls-1} \frac{1}{qs-1} \sum_{j=1}^{qs-1} P(t_i|s_j)\right]^{\frac{\lambda}{ls-1}}$$

$$\left[\prod_{i=le+1}^{\tau} \frac{1}{qs-1} \sum_{j=1}^{qs-1} P(t_i|s_j)\right]^{\frac{\lambda}{\tau-le}} \left[\prod_{i=1}^{ls-1} \frac{1}{S-qe} \sum_{j=qe+1}^{S} P(t_i|s_j)\right]^{\frac{\lambda}{ls-1}} \quad (7)$$

$$\left[\prod_{i=le+1}^{\tau} \frac{1}{S-qe} \sum_{j=qe+1}^{S} P(t_i|s_j)\right]^{\frac{\lambda}{\tau-le}} \left[\prod_{i=le}^{ls} \frac{1}{qe-qs+1} \sum_{j=qs}^{qe} P(t_i|s_j)\right]^{\frac{1-4\lambda}{ls-le+1}} .$$

## 2.3 Segment Extraction Model

Errors caused by automatic text chunkers and mismatches in the number of source and target chunks are handled partly by this model. Union of possible chunk alignments (Fig. 3) is taken in $\chi_{m \times n}$ from (5),

$$\chi_{i,j} = \begin{cases} \frac{1}{2}[Score(tchk_j|schk_i) + Score(schk_i|tchk_j)], & \text{if } j \in SA_i \text{ or } i \in TA_j \\ 0, & \text{otherwise} . \end{cases}$$

$$(8)$$

All consistent segment-pairs of length less than $(S-1)$ words on the source side and $(\tau-1)$ words on the target side are extracted. The procedure is similar to that of [34] and [18] where the boundary $(BP)$ of consistent pairs is defined over words but here we define them over chunks. To form a consistent segment-pair, source chunks within a segment-pair need to be aligned to target chunks within the segment-pair boundary only and not to any target chunks outside the boundary and vice versa. For example, in Fig. 3, the region in the solid-blue box is a consistent segment-pair, whereas, the region in the dotted-red box is not as the target chunk, 防洪 , within the boundary is aligned to a source chunk flood prevention outside the boundary. Segment-pairs extracted for Fig. 3 are given in Fig. 2.

(flood prevention↔ 防洪),(must also ↔也), (be↔ 要), (drawn up↔ 制定),
(development plans ↔开发 治理 规划), (. ↔。 ), (must also be ↔也 要),
(flood prevention and ↔防洪),(flood prevention and↔防洪 、),
(be drawn up↔要 制定),(and development plans ↔、 开发 治理 规划),
(must also be drawn up↔也 要 制定),(flood prevention↔防洪 、),
(for the major river basins ↔各 大 流域), (development plans ↔、 开发 治理 规划),
(must also be drawn up for the major river basins↔各 大 流域 也 要 制定),
(flood prevention and development plans↔防洪 、 开发 治理 规划),
(and development plans↔ 开发 治理 规划),
(flood prevention and development plans must also be drawn up ↔ 也 要 制定
                              防洪 、 开发 治理 规划).

**Fig. 2.** Segment-pairs extracted from Fig. 3

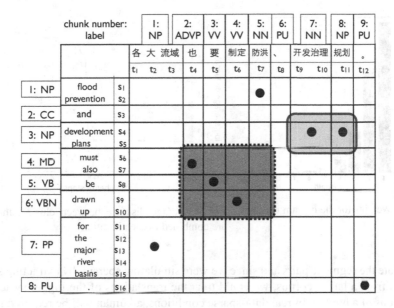

**Fig. 3.** Union of chunk alignments

**Filtering.** Many of the segment-pairs extracted using the segment extraction model were not of good quality. A blame assignment analysis indicated that this was due to poor word-alignments and chunking errors. To counter this we used a filtration step to detect and remove such segment-pairs. We would like to use a classifier that indicates which segment-pairs should be included in our system to maximize the output BLEU score. However, this is a highly non-linear problem and would in general require clustering the data, creating the templates and indexing the corpus many times during the learning phase - a computationally infeasible approach.

Instead, we learn a simple to compute measure of 'goodness' of a segment-pair that serves as a computational surrogate for the output BLEU score of a translation system. We will then train a classifier that given a segment-pair will output a 1 if it is 'good' and 0 otherwise. In order to learn this measure we need an initial source of labeled data. For this a small set of segment-pairs can be chosen randomly and given to a bilingual expert who understands the language-pair. The expert then gives a label of 0 if at least one word needs to be changed in the segment-pair and 1 if there are no changes required. This data can then be used to train a classifier to classify the rest of the segment-pairs.

This method can be extended to situations where an expert is not available by using another MT system trained on a large corpus as an expert black box. Since it would be expensive to translate all the segment-pairs, a small set can be randomly drawn and their source and target-halves can be translated by the MT system. If the translations match the segment-pairs perfectly then a label of 1 can be assigned.

We pause now for a moment to explain why we used the above procedure. The very existence of a good machine translation system would seem to indicate that the language does not suffer from data sparsity. However, in our experiments we did not have a human

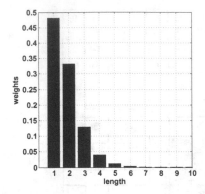

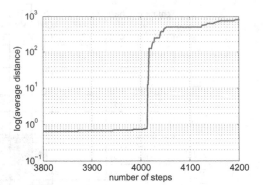

**Fig. 4.** Weights for the n-gram matches

**Fig. 5.** Average distance between clusters that are combined in each iteration

to translate the segment-pairs and since we were simulating sparsity by extracting small data sets from a larger corpus, we could treat the translations of the bigger system as translations of a human. In real data-sparse conditions, a human will be required in the loop to obtain the data for training a classifier. So, our method of using a black box MT system is intended to *simulate* a human labeler of segment-pairs. Our experiments show that this is a more efficient use of the expert resources. In addition, we feel that this is a very interesting method of extracting labels from an expert that may be useful in other cases as well. Consider a case where the phrase-table of an MT system needs to be mounted on a small memory device like a PDA. The above procedure can be used with either the original MT system trained on the larger data set or with a human labeler to decide which translations to store on the device.

Since none of the Machine Translation systems today are 100% accurate, some leniency is required while matching the segment-pairs to the MT translations. We define leniency by the amount the black box MT system diverges from the true translations. For this we used a development set of 200 sentence-pairs and translated the source side and the target side of the language-pair under consideration using the black box MT system. In order to define leniency above, one obvious choice is to measure the difference in translations with the BLEU score. However, the BLEU score is intended for sentence-sized text and is too harsh for segments that have length less than four words resulting in low scores for almost all extracted segments.

We design an alternate quality score by linearly combining all the n-gram matches between the translations and the references. We computed the distribution of the length of the segment-pairs that were extracted previously and used the probabilities as weights for the quality score. For example, if most of the segment-pairs have only two words on their target side, then we want the black box MT system to be more accurate at obtaining $bi$-gram translations with respect to the reference translations. Hence, the weight sets the importance of a particular $n$. The quality score can then be used as a threshold (separate thresholds for translating source to target $th_{s \to t}$ and target to source $th_{s \leftarrow t}$) to decide if a segment-pair should receive a label of 0 or 1. As an example, in our experiments with the 30k Eng-Chi data set, $th_{s \to t}$ ($s \to t$: while comparing references and translations in Chinese) was found to be 0.714. This implies that for a segment-pair to obtain a label of 1, it is enough if 71.4% of the target words of the segment-pair match with that of the black box MT system. Similarly, $th_{s \leftarrow t}$ ($t \rightarrow s$: while comparing references and translations in English) was found to be 0.769. The features used for classification are based on alignment scores, length of segment-pairs and source-target labels that are good indicators of the 'goodness' of a segment-pair. Say a consistent segment-pair contains source chunks: $schk_j....schk_{j+h}$ and target chunks: $tchk_i....tchk_{i+w}$.

**Feature 1:** average of chunk alignment scores ($\chi$ defined in (8)) of the segment-pair

$$\text{Feature1} = \frac{\sum_{x=i}^{i+w} \sum_{y=j}^{j+h} \chi_{x,y}}{(h+1)*(w+1)}.$$

**Feature 2** and **Feature 3:** fraction of chunk alignments within the segment-pair

$$\text{Feature 2} = \frac{\sum_{g=i}^{i+w} sgn[\mathbb{1}_{h+1}^{T} sgn(\chi_{j:j+h,g})]}{w+1}.$$

$$\text{Feature 3} = \frac{\sum_{g=j}^{j+h} sgn[sgn(\chi_{g,i:i+w})\mathbb{1}_{w+1}]}{h+1}.$$

where $\mathbb{1}_{h+1}^{T}$ is a row vector of ones of size $h+1$ and $\mathbb{1}_{w+1}$ is a column vector of ones of size $w+1$, $\chi_{j:j+h,g}$ is a column vector corresponding to rows $j$ to $j+h$ and column $g$ of $\chi$, $\chi_{g,i:i+w}$ is a row vector corresponding to columns $i$ to $i+w$ and row $g$ of $\chi$.

**Feature 4, Feature 5:** Number of words in the source and target-half of the segment-pair.

**Feature 6, Feature 7:** Number of chunks in the target ($= w+1$) and source-half of the segment-pair ($= h+1$).

**Feature 8** and **Feature 9:** Since syntactic labels for the source and target chunks are available, we could compute the probability of observing the source-chunk label sequence and the target-chunk label sequence. Maximum likelihood estimates for these probabilities are obtained from a labeled corpus.

$$\text{Feature8} = \frac{0.5 * P(label_{schk_j}....label_{schk_{j+h}})}{P(label_{schk_j})*P(label_{schk_{j+1}})*.....*P(label_{schk_{j+h}})} +$$
$$\frac{0.5 * P(label_{tchk_i}....label_{tchk_{i+w}})}{P(label_{tchk_i})*P(label_{tchk_{i+1}})*.....*P(label_{tchk_{i+w}})}.$$

$$\text{Feature 9} = 0.5 * [P(label_{schk_j},....label_{schk_{j+h}} \mid label_{tchk_i},....label_{tchk_{i+w}}) +$$
$$P(label_{tchk_i},....label_{tchk_{i+w}} \mid label_{schk_j},....label_{schk_{j+h}})].$$

Once these features have been extracted for all segment-pairs, they are normalized (mean=0,variance=1). The length bias in **Feature 8** and **Feature 9** is removed by normalizing the scores separately based on the length of the segment-pairs. We used Support Vector Machines to train and classify the segment-pairs. For training the classifier, 2000 segment-pairs were picked randomly and were labeled 1 if the fraction of matches of the target side of the segment-pair and the translation of the black box MT was greater than $th_{s \to t}$ or if the fraction of matches of the source side of the segment-pair and the translation of the black box MT (when translating the target to its source) was greater than $th_{s \leftarrow t}$. The classifier gave an accuracy of 83% with leave-one-out cross-validation.

## 2.4    Clustering Based on Semantic-Relatedness of Segment-Pairs

In order to cluster the segment-pairs, a pair-wise Adjacency matrix is constructed with the $i^{th}$ row and the $j^{th}$ column corresponding to the similarity score between $segment\text{-}pair_i$ and $segment\text{-}pair_j$. Then hierarchical weighted-single-linkage clustering is used.

To compute the pair-wise Adjacency matrix ($Adj_{combi}$), an Adjacency matrix based on contextual scores ($Adj_{context}$) and an Adjacency matrix based on word token matches ($Adj_{wm}$) between pairs of segment-pairs is first obtained. Since a segment-pair appears multiple times in a parallel corpus, a list of all words (along with their frequency of co-occurrence with the segment-pair) appearing within a window of two words prior (left context) to and two words following (right context) the source and target sides of the segment-pairs is first obtained. Positive pointwise mutual information (PPMI) [7] is then calculated from the frequency counts. Hence, a segment$pair$-context matrix with segment-pairs as the rows and context words as the columns is obtained. Cosine similarity is then used to find similarity between all pairs of segment-pairs resulting in $Adj_{context}$. The $i^{th}$ row and $j^{th}$ column of the $Adj_{context}$ represents the contextual similarity between $segment\text{-}pair_i$ and $segment\text{-}pair_j$. The fraction of the number of source and target words in common between $segment\text{-}pair_i$ and $segment\text{-}pair_j$ is used to find $Adj_{wm(i,j)}$. To compute a combined similarity score between $segment\text{-}pair_i$ and $segment\text{-}pair_j$, $Adj_{context(i,j)}$ and $Adj_{wm(i,j)}$ are linearly combined as,

$$Adj_{combi(i,j)} = c * Adj_{wm(i,j)} + (1 - c) * Adj_{context(i,j)} . \qquad (9)$$

Weights (c,1-c) are tuned with hill-climbing with the optimization function in (10).

The clustering begins with each segment-pair as a separate cluster. Two closest clusters are merged iteratively until all the segment-pairs belong to one cluster. The reason for adopting this approach is that hierarchical clustering provides a principled way to determine the number of clusters [13]. Clustering can be stopped when the algorithm tries to cluster two distant clusters. Fig. 5 shows the average distance $Dist_c(t)$ between the two closest clusters that are merged at each step $t$ with weight $c$ in (9). For this example, clustering can be stopped at the $4019^{th}$ iteration (with *number of clusters=number of data points - 4019*). A sample cluster extracted is given Fig. 6.

$$\hat{c} = \arg \max_c [max(Dist_c(t + 1) - Dist_c(t))] . \qquad (10)$$

extremely happy ↔ 非常 高兴

is happy ↔ 高兴

very glad ↔ 很 高兴

very glad ↔ 非常 高兴

very pleased ↔ 十分 高兴

very pleased ↔ 很 高兴

very pleased ↔ 非常 高兴

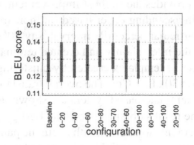

**Fig. 6.** A sample cluster from clusters generated

**Fig. 7.** BLEU scores with segment-pairs filtered at various percentile intervals of segment-pair frequencies

## 3  Experimental Setup and Results

English-French (Eng-Fre) and English-Chinese (Eng-Chi) language-pairs were used. We simulate sparsity by choosing smaller training data sets for both the language-pairs. For Eng-Chi, three sets of size 15k, 30k and 200k sentence pairs from the FBIS data [23] were selected as training data. Two sets of size 30k and 100k from the Hansard corpus [9] were selected for the experiments with Eng-Fre. To tune the EBMT system, a tuning set of 500 sentences was chosen for both the language-pairs. The test data consisting of 4000 sentences were selected randomly from the corpus. As the test data was extracted from the parallel corpus, only one reference file was used. The target half of the training data was used to build 4-gram language models with Kneser-Ney smoothing [17]. The data was segmented using the Stanford segmenter [28]. We had the Stanford parsed data [21] for both Chinese and English and so we obtained chunks and phrase labels from these parse trees using a set of rules. For Eng-Fre, chunking was performed using [26] on English and French independently.

The most important rules used in chunking Eng-Chi are as follows. A subtree was made a chunk if it included words and POS tags. If there is a subtree such as [NP (NN propaganda) (NN drive)], the subtree that qualifies to be a chunk is [NP propaganda drive] and not the unary rules ([NN propaganda]). The trees were flattened based on subtrees closer to the leaves, making sure that subtrees within complex embeddings are flattened correctly. When a PP contains a single NP, the NP was not separated. If a PP has more than one phrase, then the preposition is made one chunk and the other phrases are flattened as separate chunks. Verbs and conjunctions were separated as chunks with their POS tag as their chunk label.

We used 4-gram word-based BLEU to judge the quality. To test if the improvements are statistically significant, we used the Wilcoxon Signed-Rank Test [32]. For this, we divided the test sentences into 10 subfiles each consisting of 400 test sentences.

We used the EBMT system with no templates as our baseline system. Segment-pairs that occur frequently in the training corpus do not contribute much to the improvement in quality. This is because highly frequent segment-pairs appear in many contexts and hence the EBMT system will be able to extract long target fragments without the need for templates. So in order to find the right percentile interval where the template-based system provides the highest improvement, the segment-pairs from Sect. 2.3 were first sorted in ascending order based on their frequency of occurrence in the training data. For a particular percentile interval, say 20%-80%, we clustered segment-pairs that belong to the interval only and created templates with the resulting clusters. Fig. 7 shows the effect of templates created by clustering segment-pairs from various percentile intervals on 30k Eng-Chi. The overall translation scores obtained with the baseline system and the template-based system are shown in Table 1. Due to space constraints, we only show the results obtained from the best percentile intervals in our final results. Higher improvements were seen with segment-pairs from the mid-frequency region. Improvements were seen on all the subfiles and were found to be statistically significant.

**Table 1.** Translation Quality with templates created using clustered segment-pairs (G-EBMT) and the Baseline (EBMT)

| Language-Pair | Data Size | Baseline (EBMT) | Templates (G-EBMT) |
|---|---|---|---|
| Eng-Chi | 15k | 10.76 | 11.47 |
|  | 30k | 12.45 | 13.23 |
|  | 200k | 17.85 | 18.17 |
| Eng-Fre | 30k | 15.77 | 17.18 |
|  | 100k | 17.23 | 18.11 |

## 4   Conclusion and Future Work

In this paper, we showed how to use knowledge about source and target languages to extract syntactically coherent segment-pairs and also suggested a clustering technique based on semantic-similarity. We were able to achieve significant improvements over the baseline EBMT system in data sparse conditions. As future work, it would be interesting to explore other clustering strategies. One possibility is to cluster all segment-pairs that contain the same sequence of chunk labels on their source and target sides.

## References

1. Block, H.U.: Example-Based Incremental Synchronous Interpretation. In: Wahlster, W. (ed.) Vermobil: Foundations of Speech-to-Speech Translation. Springer, Heidelberg (2000)
2. Brown, R.D.: Automated dictionary extraction for "knowledge-free" example-based translation. In: Proceedings of the Seventh International Conference on Theoretical and Methodological Issues in Machine Translation, pp. 111–118 (1997)
3. Brown, R.D.: Example-Based Machine Translation in the PANGLOSS System. In: Proceedings of The International Conference on Computational Linguistics, pp. 169–174 (1998)
4. Brown, R.D.: Automated Generalization of Translation Examples. In: Proceedings of The International Conference on Computational Linguistics, pp. 125–131 (2000)
5. Brown, R.D.: Transfer-Rule Induction for Example-Based Translation. In: Proceedings of The Machine Translation Summit VIII Workshop on Example-Based Machine Translation, pp. 1–11 (2001)

6. Brown, R.D.: A Modified BWT for highly scalable Example-based translation. In: Proceedings of The Association for Machine Translation in the Americas, pp. 27–36 (2004)
7. Bullinaria, J., Levy, J.: Extracting semantic representations from word co-occurrence statistics: A computational study. In: Behavior Research Methods, pp. 510–526 (2007)
8. Chiang, D.: A hierarchical phrase-based model for statistical machine translation. In: Proceedings of the 43rd Annual Meeting on Association for Computational Linguistics, pp. 263–270 (2005)
9. Consortium, L.L.D.: Hansard corpus of parallel english and french. Linguistic Data Consortium (1997)
10. Gangadharaiah, R., Brown, R.D., Carbonell, J.G.: Spectral clustering for example based machine translation. In: HLT-NAACL (2006)
11. Gangadharaiah, R., Brown, R.D., Carbonell, J.G.: Automatic determination of number of clusters for creating templates in example-based machine translation. In: Proceedings of The Conference of the European Association for Machine Translation (2010)
12. Gough, N., Way, A.: Robust Large-Scale EBMT with Marker-Based Segmentation. In: Proceedings of The Conference on Theoretical and Methodological Issues in Machine Translation, pp. 95–104 (2004)
13. Goutte, C., Toft, P., Rostrup, E., Nielsen, F.A., Hansen, L.K.: On Clustering fMRI Time Series. NeuroImage, 298–310 (1998)
14. Guvenir, H.A., Cicekli, I.: Learning translation templates from examples. Information Systems, 353–363 (1998)
15. Harris, Z.: Distributional structure. Word 10(23), 146–162 (1954)
16. Kaji, H., Kida, Y., Morimoto, Y.: Learning Translation Templates from Bilingual Text. In: Proceedings of The International Conference on Computational Linguistics, pp. 672–678 (1992)
17. Kneser, R., Ney, H.: Improved backing-off for m-gram language modeling. In: Proceedings of the IEEE ICASSP, vol. 1, pp. 181–184 (1995)
18. Koehn, P.: Pharaoh: A Beam Search Decoder for Phrase-Based Statistical Machine Translation Models. In: Frederking, R.E., Taylor, K.B. (eds.) AMTA 2004. LNCS (LNAI), vol. 3265, pp. 115–124. Springer, Heidelberg (2004)
19. Koehn, P., Hoang, H., Birch, A., Callison-Burch, C., Federico, M., Bertoldi, N., Cowan, B., Shen, W., Moran, C., Zens, R., Dyer, C., Bojar, O., Constantin, A., Herbst, E.: Moses: Open Source Toolkit for Statistical Machine Translation. In: Annual Meeting of ACL, demonstration (2007)
20. Lafferty, J., McCallum, A., Pereira, F.: Conditional random fields: Probabilistic models for segmenting and labeling sequence data. In: Proceedings of The International Conference on Machine Learning, pp. 282–289 (2002)
21. Levy, R., Manning, C.D.: Is it harder to parse chinese, or the chinese treebank? In: Association for Computational Linguistics, pp. 439–446 (2003)
22. McTait, K.: Translation patterns, linguistic knowledge and complexity in ebmt. In: Proceedings of The Machine Translation Summit VIII Workshop on Example-Based Machine Translation, pp. 23–34 (2001)
23. NIST: Machine translation evaluation (2003)
24. Och, F.J., Ney, H.: A systematic comparison of various statistical alignment models. Computational Linguistics 29(1), 19–51 (2003)
25. Papineni, K., Roukos, S., Ward, T., Zhu, W.: Bleu: a method for automatic evaluation of machine translation. In: Association for Computational Linguistics, pp. 311–318 (2002)
26. Schmid, H.: Probabilistic part-of-speech tagging using decision trees. In: International Conference on New Methods in Language Processing, pp. 44–49 (1994)

27. Somers, H.L., McLean, I., Jones, D.: Experiments in multilingual example-based generation. In: International Conference on the Cognitive Science of Natural Language Processing (1994)
28. Tseng, H., Chang, P., Andrew, G., Jurafsky, D., Manning, C.: A conditional random field word segmenter. In: Fourth SIGHAN Workshop on Chinese Language Processing (2005)
29. Turney, P.D., Pantel, P.: From frequency to meaning: Vector space models of semantics. Journal of Artificial Intelligence Research, 141–188 (2010)
30. Veale, T., Way, A.: Gaijin: A bootstrapping, template-driven approach to example-based mt. In: International Conference, Recent Advances in Natural Language Processing, pp. 239–244 (1997)
31. Vogel, S.: Pesa phrase pair extraction as sentence splitting. In: Machine Translation Summit X (2005)
32. Wilcoxon, F.: Individual comparisons by ranking methods (1945)
33. Yamada, K., Knight, K.: A syntax-based statistical translation model. In: Association for Computational Linguistics, pp. 523–530 (2001)
34. Zens, R., Och, F.J., Ney, H.: Phrase-based statistical machine translation. In: Jarke, M., Koehler, J., Lakemeyer, G. (eds.) KI 2002. LNCS (LNAI), vol. 2479, pp. 18–32. Springer, Heidelberg (2002)

# A Multi-view Approach for Term Translation Spotting

Raphaël Rubino and Georges Linarès

Laboratoire Informatique d'Avignon
339, chemin des Meinajaries, BP 91228
84911 Avignon Cedex 9, France
{raphael.rubino,georges.linares}@univ-avignon.fr

**Abstract.** This paper presents a multi-view approach for term translation spotting, based on a bilingual lexicon and comparable corpora. We propose to study different levels of representation for a term: the context, the theme and the orthography. These three approaches are studied individually and combined in order to rank translation candidates. We focus our task on French-English medical terms. Experiments show a significant improvement of the classical context-based approach, with a F-score of 40.3 % for the first ranked translation candidates.

**Keywords:** Multilingualism, Comparable Corpora, Topic Model.

## 1 Introduction

Bilingual term spotting is a popular task which can be used for bilingual lexicon construction. This kind of resource is particularly useful in many Natural Language Processing (NLP) tasks, for example in cross-lingual information retrieval or Statistical Machine Translation (SMT). Some works in the literature are based on the use of bilingual parallel texts, which are often used in SMT for building translation tables [1,2]. However, the lack of parallel texts is still an issue, and the NLP community tends to use a forthcoming bilingual resource in order to build bilingual lexicons: bilingual comparable corpora.

One of the main approaches using non-parallel corpora is based on the assumption that a term and its translation share context similarities. It can be seen as a co-occurrence or a context-vector model, which depends on the lexical environment of terms [3,4]. This approach stands on the use of a bilingual lexicon, also known as bilingual seed-words. These words are used as anchor points in the source and the target language. This representation of the environment of a term has to be invariant from a language to another in order to spot correct translations. The efficiency of this approach depends on context-vectors accuracy. Authors have studied different measures between terms, variations on the context size, and similarity metrics between context-vectors [5,6,7].

In addition to context information, heuristics are often used to improve the general accuracy of the context-vector approach, like orthographic similarities

A. Gelbukh (Ed.): CICLing 2011, Part II, LNCS 6609, pp. 29–40, 2011.

between the source and the target terms [8]. Cognate-based techniques are popular in bilingual term spotting, in particular for specific domains. It can be explained by the large amount of transliteration even in unrelated languages. Also, related languages like Latin languages can share similarities between a term and its translation, like identical lemmas. We refer to this particularity as cross-languages cognates.

However, a *standard* context-vector approach combined with graphic features does not allow to handle polysemy and synonymy [9]. This limit can be overtaken by the introduction of semantic information. It is precisely the investigation presented in this paper: the combination of context, topic and graphic features for bilingual term spotting with comparable corpora. Each feature is a different view of a term. In a first step, we propose to study each feature individually. Then, we combine these features in order to increase the confidence on the spotted candidates. We focus on spotting English translations of French terms from the medical domain. We refer as a *term* in a terminological sense: a single or multi-word expression with a unique meaning in a given domain.

For the context-based approach, we want to tackle the context limitation issue, capturing information in a local and in a global context. We assume that some terms can be spotted using a close context, while other terms are characterized by a distant one.

For the topic feature, we want to handle polysemious terms and synonyms. We assume that a term and its translation share similarities among topics. The comparable corpora are modeled in a topic space, in order to represent context-vectors in different latent semantic themes. One of the most popular methods for semantic representation of a corpus is the so-called topic model. Topic models are widely used for statistical analysis of discrete data in large document collections. Data decomposition into latent components emerges as a useful technique in several tasks, such as information retrieval or text classification. This technique can be applied to many kinds of documents like scientific abstracts, news articles, social network data, etc. The Latent Dirichlet Allocation (LDA) [10] fits to our needs: a semantic based *bag-of-words* representation and unrelated dimensions (one dimension per topic).

Finally for the cognate approach, we investigate the efficiency of the classic Levenshtein distance [11] between source language and target language terms.

In order to increase the general precision of our system, a *vote* is used to combine the results of the three features. These three views are part of our multi-view approach for term-translation spotting. To the best of our knowledge, there is no study combining these particular features for this task.

The remainder of this paper is organized as follows: we describe the lexical based approach and its different variants in the next section. The topic model approach is introduced in section three. We also want to give details about the use of cognate terms, or orthographic features, in section four. Then we present the framework of the experiments in section five, followed by the results obtained with several configurations. Finally, results are discussed in the last section.

## 2   Context-Based Approach

Bilingual comparable corpora are a set of non-parallel texts that have common topics and are written independently in each language. In comparable corpora, a term and its translation share similarities through the vocabulary surrounding them. Based on this assumption, different techniques are well presented in the literature. One of the first studies is made on bilingual co-occurrence pattern matching [3]. The context-vector is introduced in [12], relying on a bilingual lexicon. With the same philosophy, other works are done with a thesaurus [13]. This approach is the basis of the work presented in this paper. It is also possible to induce a seed-words lexicon from monolingual sources, as described in [14].

Other studies were made on the association measures between a term and its context in order to build the most accurate context-vector. Some of the popular distance measures used in the literature are mutual information, log-likelihood and odds ratio [15,16,17]. Once the context-vectors are built, a similarity metric is used to rank the target terms according to the source term. For the bilingual term spotting task, similarity metrics between vectors have been studied, like the city-block metric or the cosine distance [12,18]. Among the large number of studies on association measures and similarity metrics, the used technique depends on the task: domain specific terms to translate, different sizes of corpus for each language, the amount of words and their translations in the seed-words lexicon, etc. Several combinations were studied in [7], and the most efficient configuration on their task was the odds ratio with the cosine distance. In our studies, we decide to implement a system based on this latter work, which stands as a baseline. The odds ratio ($odds = \frac{\tau_{11} \cdot \tau_{22}}{\tau_{12} \cdot \tau_{21}}$) is a coefficient of association strength which can be computed on a 2x2 contingency table. In our case, we want to compute the association strength between a candidate (in the source or in the target language) and words of the seed-words lexicon. The four elements in the contingency table are the possible observation (or absence) of two terms in a given window size. The most common practice is to use a sliding window around the term to translate. The size of the window limits the context. It can be fixed, 10 words for instance, or dynamic, like a sentence or a paragraph. One of the parameters we emphasize in this paper is precisely the size of the context. We build our implementation to be

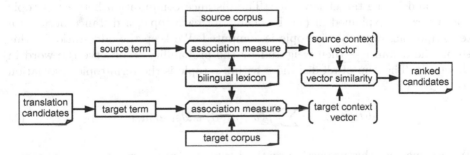

**Fig. 1.** Architecture of the context-based approach for term translation spotting

able to modify the size of the sliding window used to count words co-occurrence. The general architecture of the context-vector approach is described in Fig.1. We use different sizes of window because we assume that a term can be characterized by a close context and by a distant one.

## 3   Topic Model

The main idea in topic model is to produce a set of *artificial* dimensions from a collection of documents. It is based on the assumption that documents are mixtures of topics, and a topic is a probability distribution over words. One of the popular approaches used in automatic indexing and information retrieval is the Latent Semantic Analysis (LSA, or LSI for Latent Semantic Indexing) [19]. This method is based on a term-document matrix which describes the occurrences of terms in documents. This high dimensional sparse matrix is reduced by Singular Value Decomposition (SVD). In [20], the Probabilistic LSA (PLSA) is introduced to give a robust statistical foundation to the LSA. Based on the likelihood principle, a mixture decomposition is derived from a latent class model. With this technique, the order of the documents is taken into account. This model is introduced for bilingual term spotting in [9] to handle polysemy and synonymy, as a Bilingual PLSA approach. For more exchangeability, we decide to use the Latent Dirichlet Allocation (LDA), first introduced in [10]. The general principle of LDA stands on the computation of the multinomial probability $p(w_n|z_n,\beta)$ conditionned on the topic $z_n$, for $N$ words $w_n$ of a document $M$ in a collection.

A Multilingual LDA is introduced in [21] to mine topics from Wikipedia. They build a comparable corpus from aligned documents (articles) and use a modified LDA to model this corpus. However in this latter work, multilingual topic alignment stands on the use of links among documents. This kind of resources are not taken into account in our studies.

We want to obtain the distribution over topics of the bilingual lexicon used for the context-based approach. First, we filter the source language corpus with this bilingual lexicon. The resulting corpus contains only the vocabulary from the bilingual lexicon. Second, this reduced corpus is modeled in a latent topic space. The output model is then translated in the target language with the bilingual lexicon. Our aim is to select the most pertinent topics for candidates in the source and in the target language. The distance computation between a topic and a term is explained in (1). Basically, for each topic, a distance between a term and each word of the topic is computed. We keep the odds-ratio for this step. The distance is then weighted by the probability to observe the word in this topic. The sum of all distances within a topic is the term-topic association score (see Fig.2).

$$d(term, z) = \sum_n p(w_n|z_n, \beta) \; odds(term, w_n) \; . \tag{1}$$

We assume that this projection method leads to an important issue: the representation $p(w_n|z_n,\beta)$ of a word $w_n$ in the target language is not re-estimated.

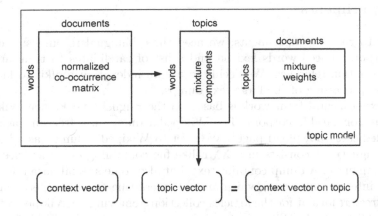

**Fig. 2.** Representation of the context of a term in different topics

This *naïve* projection of the topic space does not reflect the reality in the target language. Indeed, the topic alignment of two separately built latent spaces would be the perfect solution. However, the use of comparable corpora can limit the distortion effects of the words-over-topics distribution. Furthermore, we combine the weight of each word in a topic with a lexical distance measure to clear this hurdle, keeping in mind that this imprecise technique needs to be improved. An example of bilingual topic alignment is proposed in [22]. They introduce a multilingual topic model for unaligned text, designed for non-parallel corpora processing. This model can be used to find and match topics for, at least, two languages. They focus on building consistent topic spaces in English and German, based on the matched vocabulary in both languages.

## 4   Cognate Approach

As it is well described in [14], related languages like German and English share orthographic similarities which can be used for bilingual term spotting. Often call cognates, these particular terms can be compared by several techniques. The most popular one is the edit distance (or Levenshtein distance).

In order to handle the suffix or prefix transformation between languages, some researchers use transformation rules or limit the comparison to language roots. This approach works when the languages concerned by term spotting are related. Experiments with Latin languages often yields to good results.

A model introduced in [8] uses a Romanization system for Chinese (as the target language) and maps the source (English) and the target language letters of two terms. This technique allows comparing the spelling of unrelated languages.

In our experiments on French and English terms, we use the classic Levenshtein distance between two bilingual terms. We compute the distance between the four letters at the beginning of two terms and between the whole terms.

# 5   Experiments

In order to run our experiments, we need three bilingual resources: a comparable corpora, a seed-words lexicon and a list of candidates to translate with their translation reference. We decide to use an indexation toolkit to facilitate statistical processing of the large text dumps.

Our experimental framework is based on the English and French Wikipedia dumps as comparable corpora. The Wikipedia dumps are free and accessible on the dedicated download page[1]. We refer to Wikipedia dumps as all articles in one language. It consists in a XML-like file containing every articles in the selected language. A dump contains text, but also some special data and syntax (images, internal links, etc.) which are not interesting for our experiments. We use the *trectext* format for the articles collection, removing all Wikipedia tags. A stop-word list is used to filter the textual content of the articles. Table 1 contains the details about the comparable corpora. The context-vector approach leads to better results with a large amount of comparable texts. In this paper, we use the *Lemur Toolkit*[2] to index the Wikipedia dumps. After this step, counting co-occurrences of two terms in a fixed window can be done in one query. The candidates to translate are extracted from the Medical Subject Heading (MeSH) thesaurus[3] along with their translation references (one translation per source term) [23]. We use the same bilingual lexicon and the same candidates as in [7] to be able to compare our results to this baseline. The bilingual seed-words lexicon is taken from the Heymans Institute of Pharmacologys *Multilingual glossary of technical and popular medical terms*[4].

**Table 1.** Details about the bilingual resources used in our experiments

| corpus | documents | tokens | unique tokens |
|---|---|---|---|
| candidates | - | - | 3,000 |
| seed-words | - | - | 9,000 |
| Wikipedia FR | 872,111 | 118,019,979 | 3,994,040 |
| Wikipedia EN | 3,223,790 | 409,792,870 | 14,059,292 |

For the semantic part of our framework, we use an implementation of LDA based on the Gibbs Sampling technique[5]. We build different models with variations on the number of topics (from 20 to 200 topics). Each model is estimated with 2,000 iterations.

We want to study the different approaches separately. For each set of experiments, 3,000 translations have to be spotted. The result of each approach is a ranked list of the translation candidates, the first ranked candidate is the best

---

[1] http://download.wikimedia.org/backup-index.html
[2] http://www.lemurproject.org
[3] http://www.nlm.nih.gov/mesh/
[4] http://users.ugent.be/~rvdstich/eugloss/welcome.html
[5] http://gibbslda.sourceforge.net

translation according to the system. We observe the position of the translation reference and report the accuracy of each approach. Then we combine the results of the different views by a vote. We want to see if a correct translation is ranked first by the majority of the judges. The main advantage of this method is the ability to reach a very high precision with a lot of judges combined, but the recall may be low. We assume that the complementarity of the three different views can increase the recall, and the number of judges can maintain a high precision.

## 5.1   Context-Based Approach

First, we build one context-vector for each term. We make variations of the window size in order to capture different context information. Then, each target context-vector is ranked by cosine similarity to the source one. We compare source and target vectors built with the same parameters. We compute the recall scores on the first ranked target candidates for each window size. We also present the recall scores for terms spotted with one window size and not spotted by the others. Table 2 contains the results for each window size individually. The recall

**Table 2.** Recall for the context-based approach from rank 1 to 100 with different sliding window sizes. The *unique1* row shows the amount of correct translations spotted at rank 1 by the current window size and not by the other window sizes.

|         | win 10 | win 20 | win 30 | win 40 | win doc |
|---------|--------|--------|--------|--------|---------|
| 1       | 31.1   | 32.9   | **33.7** | 32.4 | 15.6    |
| 10      | 57.6   | 59.6   | 60.6   | 58.6   | 37.7    |
| 50      | 69.3   | 71.8   | 72.6   | 71.8   | 54.6    |
| 100     | 73.4   | 76.0   | 76.9   | 76.6   | 61.5    |
| unique1 | **4.5** | 1.7   | 1.5    | 1.7    | 3.1     |

scores are relatively close for the windows of a size between 20 and 40 words. The best configuration in our context-based approach is a 30 words window size, reaching a recall score of 33.7 % for the first ranked target language candidate. We show with these results that a term can be characterized by a close context (a small window size) or a global one. Some of the source terms to translate are locally linked to a vocabulary, because their translations are spotted using a 10 words window, and this characteristic is invariant in the target language. For these terms, a larger window introduces noise and is not efficient to spot the correct translation. This is the main motivation for the vote combination of the context-based approach with different window sizes, presented in Table 3.

In this experiment, the recall score for all the window sizes is low, because a small amount of candidates are ranked first by the 5 judges. However, the precision score is relatively high. The best F-score is reached by the combination of the 20 and 30 words windows. As expected, the higher precision is reached with the combination of all window sizes.

**Table 3.** Scores for the vote combination on the context-based approach, with 2 and 3 sizes combined, and all the window sizes, at rank 1. Scores for a 30 words window are also detailed.

|           | win30 | win20+win30 | win10+win30+win40 | all sizes |
|-----------|-------|-------------|-------------------|-----------|
| recall    | **33.7** | 27.8     | 19.2              | 6.2       |
| precision | 33.7  | 50.5        | 75.9              | **83.7**  |
| F-score   | 33.7  | **35.8**    | 30.6              | 11.5      |

## 5.2   Topic-Based Approach

With the topic model approach, we want to see if a semantic space can be used to filter the terms among the 3,000 English translations. For each English or French term of the candidate list, we measure the distance with each topic (semantic class) of the model. Basically, we want to see if a source term and its translation reference are at the same place in the topic space. This feature can be useful to filter target terms which are not on the same topics as a source term. For a source term, we check which target terms share the same topics, according to an ordered topic distance table. If a term to translate and its translation reference share the same *first* topics, it increases a recall score (the resulting recall is then divided by the number of candidates). A recall score of 100 means that each source terms and its translation reference share the same top ranked topic. We measure it for the three top ranked topics for each target term and present the results in Table 4. We also measure the precision of this approach. The precision decreases if there are target terms which share similar topics with the source term but are not the translation reference. A precision score of 0.0 indicates that all translation candidates are returned by the system for one source term.

Our semantic-based approach on bilingual term spotting can be seen as a validation step in our experiments. The translation candidates are ranked according to their similarity with a source term in the topic space. In a 50 dimensions topic space, 1/3 of the candidates to translate have their reference at the first rank. This score decreases when we increase the number of dimensions. This variation

**Table 4.** Scores for the topic-based approach. The three first topics of a target term are presented. Several models are tested: with 20, 50, 100 and 200 dimensions.

|   |           | 20   | 50   | 100  | 200  |
|---|-----------|------|------|------|------|
|   | recall    | 23.6 | 33.4 | 33.0 | 10.4 |
| 1 | precision | 0.12 | 0.06 | 0.07 | 0.18 |
|   | F-score   | 0.23 | 0.12 | 0.14 | 0.35 |
|   | recall    | 35.9 | 42.2 | 38.3 | 18.8 |
| 2 | precision | 0.1  | 0.06 | 0.07 | 0.15 |
|   | F-score   | 0.19 | 0.12 | 0.14 | 0.31 |
|   | recall    | 44.1 | 45.9 | 41.2 | 24.1 |
| 3 | precision | 0.08 | 0.06 | 0.07 | 0.14 |
|   | F-score   | 0.16 | 0.12 | 0.13 | 0.28 |

impacts the precision, which indicates that incorrect target terms are returned by the system. The models used in these experiments are not adapted to specific domains. All medical terms are close to a few topics, that is why a huge amount of source and target language terms are at the same place in the topic space.

## 5.3    Cognate Approach

The Levenshtein distance is used to rank the translation candidates according to a source language term. We measure the recall of this approach for several ranks, between the 4 letters at the beginning of two terms and between the full terms. Results are presented in Table 5. For each rank, if the translation reference is found, the recall score increases. We compute the precision score according to the amount of target terms at the current rank which are not the translation reference.

**Table 5.** Recall results with the noise scores for the cognate approach at different ranks: from 1 to 10. Two edit distances are tested: between the 4 beginning letters and between the full terms.

|         |           | 1    | 2    | 3    | 4    | 5    | 10   |
|---------|-----------|------|------|------|------|------|------|
|         | recall    | 34.0 | 39.0 | 45.9 | 65.4 | 100  | 100  |
| letters | precision | 15.9 | 3.9  | 0.5  | 0.1  | 0.03 | 0.03 |
|         | F score   | 21.7 | 7.2  | 1.0  | 0.2  | 0.07 | 0.07 |
|         | recall    | 50.7 | 54.8 | 59.6 | 67.4 | 77.3 | 99.3 |
| term    | precision | 83.5 | 29.6 | 5.6  | 1.4  | 0.5  | 0.2  |
|         | F-score   | 63.1 | 38.5 | 10.3 | 2.8  | 1.0  | 0.3  |

We can see that a classic Levenshtein distance allows to find 50.7 % of correct translations at the first rank with a precision score of 83.5 %. This result is not surprising in our case, because English and French domain specific terms are more likely to have common roots. Taking the ten first ranked candidates, the cognate approach yields to a 99 % recall, but the large number of wrong target terms spotted decreases the precision. However, this feature can be added to the context-based method in order to re-rank the translation candidates.

## 5.4    Combination

We present the combination of the three views in Table 6. Three combinations are studied: two *classic* combinations (context and topic or context and cognate), and the three approaches together. The context-based approach combined with the cognate approche is the baseline. Each feature, or view, is a judge in a vote combination. We propose two different vote configurations. The majority and the unanimity of the judges can determine which translation candidate is ranked first. If any of the vote configuration is fulfilled, the system is not able to spot a target language term among the candidates. We include the context-based

**Table 6.** Scores at rank 1 for the combination of our approaches. Unanimity is indicated in brackets if the judges are more than 2. The context feature is noted *cont*, the cognate feature is noted *cogn*.

|        |           | cont+topic | cont+cogn  | cont+topic+cogn |
|--------|-----------|------------|------------|-----------------|
|        | recall    | 13.9       | 19.0       | 24.2 (7.6)      |
| win30  | precision | 100        | 99.1       | 99.3 (100)      |
|        | F-score   | 24.4       | 31.9       | 39.0 (14.1)     |
|        | recall    | 20.8 (2.6) | 21.1 (4.0) | 26.9 (1.7)      |
| all    | precision | 76.2 (100) | 76.6 (97.6)| 80.4 (100)      |
|        | F-score   | 32.7 (5.0) | 33.1 (7.7) | 40.3 (3.4)      |

approach as five different judges with all window sizes or as one judge with one fixed window size. We show the results for both of them. A 30 words window is chosen for the fixed window size.

The context-vector approach combined with the Levenshtein distance yields to a recall of 19 % with a precision score of 99.1 %. Using all window sizes leads to a slight improvement of the recall score (21.1 %) but the precision decreases, according to the majority of the judges. Taking the unanimity in this configuration completely degrades the recall (4 %), showing that the overlap of all first ranked judges individually is very low. However, our system allows to spot only correct translations. There are no incorrect target language terms spotted when the precision reaches 100 %.

Two main aspects of the multi-view approach are outstanding with these results. The first one is the best F-score obtained with the combination of all views. This is the best configuration for an acceptable recall and a relatively high precision. The second one is the high precision reached by the combination of the three approaches with a fixed window size for the context feature. The recall is 2.7 % lower than with the combination of all judges.

As reported in Sect.5.2 with the best configuration, the precision of the topic-based approach is very low because of the topic model, which is not adapted to the medical domain. We are not including this precision in the combination results including the topic feature because we want to investigate the general coverage of our approach in this configuration. The semantic information is used to validate a translation candidate ranked first by the majority of the other approaches.

## 6   Discussion

We presented in this paper a multi-view approach for medical term translation spotting based on comparable corpora and a bilingual seed-words lexicon. The context-vector approach is combined with a topic-based model in order to handle polysemy and synonymy. We also include an orthographic feature to boost our results. The combination of the 3 approaches yields to a recall score of 26.9 % with a precision of 80.4 %. These results show the complementarity of the three features. Compared to a state-of-the-art approach combining the context and the cognate approches, our multi-view approach leads to a 12.6% absolute improvement of the F-score.

In future work, we plan to improve the topic model for the target language. Instead of translating the source language topic-model into the target language, the comparable corpora can be used to compute the probabilities of vocabulary over topics distribution. Another possible improvement on the topic representation is to select a subset of the comparable corpora in order to model the domain specific part only. This technique may lead to a finer grained topic model.

In the experiments presented in this paper, each word of the bilingual lexicon is used to build the context-vector of a source term. The context-based approach can be improved with a more precise study on ambiguous seed-words. This technique would reduce the size of the context-vector to the most discriminant dimensions for a term to translate.

## Acknowledgements

This research has been partly funded by the National Research Authority (ANR), AVISON project (ANR-007-014).

## References

1. Brown, P., Della Pietra, S., Della Pietra, V., Jelinek, F., Lafferty, J., Mercer, R., Roossin, P.: A Statistical Approach to Machine Translation. Computational Linguistics 16, 79–85 (1990)
2. Koehn, P.: Europarl: A Parallel Corpus for Statistical Machine Translation. In: MT Summit, vol. 5, Citeseer (2005)
3. Fung, P.: Compiling Bilingual Lexicon Entries from a Non-parallel English-Chinese Corpus. In: Proceedings of the 3rd Workshop on Very Large Corpora, pp. 173–183 (1995)
4. Rapp, R.: Identifying Word Translations in Non-parallel Texts. In: Proceedings of the 33rd ACL Conference, pp. 320–322. ACL (1995)
5. Chiao, Y., Zweigenbaum, P.: Looking for Candidate Translational Equivalents in Specialized, Comparable Corpora. In: Proceedings of the 19th Coling Conference, vol. 2, pp. 1–5. ACL (2002)
6. Rubino, R.: Exploring Context Variation and Lexicon Coverage in Projection-based Approach for Term Translation. In: Proceedings of the RANLP Student Research Workshop, Borovets, Bulgaria, pp. 66–70. ACL (2009)
7. Laroche, A., Langlais, P.: Revisiting Context-based Projection Methods for Term-translation Spotting in Comparable Corpora. In: Proceedings of the 23rd Coling Conference, Beijing, China, pp. 617–625 (2010)
8. Shao, L., Ng, H.: Mining New Word Translations from Comparable Corpora. In: Proceedings of the 20th ACL Conference, p. 618. ACL (2004)
9. Gaussier, E., Renders, J., Matveeva, I., Goutte, C., Dejean, H.: A Geometric View on Bilingual Lexicon Extraction from Comparable Corpora. In: Proceedings of the 42nd ACL Conference, p. 526. ACL (2004)
10. Blei, D.M., Ng, A.Y., Jordan, M.I.: Latent Dirichlet Allocation. The Journal of Machine Learning Research 3, 993–1022 (2003)

11. Levenshtein, V.: Binary Codes Capable of Correcting Deletions, Insertions, and Reversals. Soviet Physics Doklady 10, 707–710 (1966)
12. Rapp, R.: Automatic Identification of Word Translations from Unrelated English and German Corpora. In: Proceedings of the 37th ACL Conference, pp. 519–526. ACL (1999)
13. Déjean, H., Gaussier, E., Renders, J., Sadat, F.: Automatic Processing of Multilingual Medical Terminology: Applications to Thesaurus Enrichment and Cross-language Information Retrieval. Artificial Intelligence in Medicine 33, 111–124 (2005)
14. Koehn, P., Knight, K.: Learning a Translation Lexicon from Monolingual Corpora. In: Proceedings of the ACL Workshop on Unsupervised Lexical Acquisition, vol. 9, pp. 9–16. ACL (2002)
15. Church, K.W., Hanks, P.: Word Association Norms, Mutual Information, and Lexicography. Computational Linguistics 16(1), 22–29 (1990)
16. Dunning, T.: Accurate Methods for the Statistics of Surprise and Coincidence. Computational Linguistics 19, 61–74 (1993)
17. Evert, S.: The Statistics of Word Cooccurrences: Word Pairs and Collocations. Ph.D. Thesis, Institut für maschinelle Sprachverarbeitung, Universität Stuttgart (2004)
18. Fung, P., McKeown, K.: Finding Terminology Translations from Non-parallel Corpora. In: Proceedings of the 5th Workshop on Very Large Corpora, pp. 192–202 (1997)
19. Deerwester, S., Dumais, S., Furnas, G., Landauer, T., Harshman, R.: Indexing by Latent Semantic Analysis. Journal of the American Society for Information Science 41, 391–407 (1990)
20. Hofmann, T.: Probabilistic Latent Semantic Indexing. In: Proceedings of the 22nd ACM SIGIR Conference, pp. 50–57. ACM, New York (1999)
21. Ni, X., Sun, J., Hu, J., Chen, Z.: Mining Multilingual Topics from Wikipedia. In: Proceedings of the 18th International Conference on WWW, pp. 1155–1156. ACM, New York (2009)
22. Boyd-Graber, J., Blei, D.M.: Multilingual topic models for unaligned text. In: Proceedings of the 25th Conference on Uncertainty in Artificial Intelligence, pp. 75–82 (2009)
23. Langlais, P., Yvon, F., Zweigenbaum, P.: Translating medical words by analogy. In: Intelligent Data Analysis in Biomedicine and Pharmacology, Washington, DC, USA, pp. 51–56 (2008)

# ICE-TEA: In-Context Expansion and Translation of English Abbreviations

Waleed Ammar, Kareem Darwish, Ali El Kahki, and Khaled Hafez[1]

Cairo Microsoft Innovation Center, Microsoft, 306 Chorniche El-Nile, Maadi, Cairo, Egypt
{i-waamma,kareemd,t-aleka}@microsoft.com,
hafez.khaled@gmail.com

**Abstract.** The wide use of abbreviations in modern texts poses interesting challenges and opportunities in the field of NLP. In addition to their dynamic nature, abbreviations are highly polysemous with respect to regular words. Technologies that exhibit some level of language understanding may be adversely impacted by the presence of abbreviations. This paper addresses two related problems: (1) expansion of abbreviations given a context, and (2) translation of sentences with abbreviations. First, an efficient retrieval-based method for English abbreviation expansion is presented. Then, a hybrid system is used to pick among simple abbreviation-translation methods. The hybrid system achieves an improvement of 1.48 BLEU points over the baseline MT system, using sentences that contain abbreviations as a test set.

**Keywords:** statistical machine translation, word sense disambiguation, abbreviations.

## 1 Introduction

Abbreviations are widely used in modern texts of several languages, especially English. In a recent dump of English Wikipedia,[2] articles contain an average of 9.7 abbreviations per article, and more than 63% of the articles contain at least one abbreviation. At sentence level, over 27% of sentences, from news articles, were found to contain abbreviations. The ubiquitous use of abbreviations is worth some attention. Abbreviations can be acronyms, such as NASA, which are pronounced as words, or initialisms, such BBC, which are pronounced as a sequence of letters.

Often abbreviations have multiple common expansions, only one of which is valid for a particular context. For example, Wikipedia lists 17 and 15 valid expansions for IRA and IRS respectively. However, in the sentence: *"The bank reported to the IRS all withheld taxes for IRA accounts."* IRA conclusively refers to *"individual retirement account"* and IRS refers to *"internal revenue service"*. Zahariev (2004) states that 47.97% of abbreviations have multiple expansions (at WWWAAS[3])

---

[1] Author was an intern at Microsoft and is currently working at the IBM Technology Development Center in Cairo.
[2] http://dumps.wikimedia.org/enwiki/20100312/
[3] World-Wide Web Acronym and Abbreviation Server
http://acronyms.silmaril.ie/

A. Gelbukh (Ed.): CICLing 2011, Part II, LNCS 6609, pp. 41–54, 2011.

compared to 18.28% of terms with multiple senses (in WordNet), suggesting that abbreviations are highly polysemous with respect to regular words. Table 1 lists some popular abbreviations with multiple expansions.

**Table 1.** Some popular polysemous abbreviations

| Abb. | Expansion |
|------|-----------|
| TF | Term Frequency |
| TF | Task Force |
| IDF | Israel Defense Forces |
| IDF | Inverse Document Frequency |
| IDF | Intel Developers Forum |
| CIA | Central Intelligence Agency |
| CIA | Certified Internal Auditor |
| IRA | Irish Republican Army |
| IRA | Individual Retirement Account |
| AP | Advanced Placement |
| AP | Associated Press |
| AP | Access Point |
| ATM | Asynchronous Transfer Mode |
| ATM | Automated Teller Machine |
| ATM | Air Traffic Management |

Abbreviations pose interesting challenges and opportunities in Statistical Machine Translation (SMT) systems such as (Koehn et al., 2003; Quirk et al. 2005; Galley et al., 2006; Chiang, 2007). Some of the challenges include:

1. The proper abbreviation translation may not exist in parallel data that was used for training. Given the dynamic aspect of abbreviations, where tens of new abbreviations appearing every day (Molloy, 1997), parallel text used for training may be limited or out-of-date. Typically, available parallel text hardly covers one or two (if any) common translations of an abbreviation, overlooking less common translations.
2. Many abbreviations are polysemous. Even in cases when multiple translations are observed in parallel training text, sufficient context is often not available to enable a language model to promote the proper translation.

Intuitively, an SMT system may have a better chance at translating an expanded abbreviation than the abbreviation itself. If an abbreviation can be properly expanded prior to translation, the ambiguity is removed (availing problem 2), and the MT system may be able to produce a reasonable translation even if it does not exist in training data (availing problem 1).

The contributions of this paper are:

1. an efficient Information Retrieval (IR) based technique for abbreviation expansion,
2. the use of abbreviation expansion to enhance translation of sentences that contain abbreviations, and
3. a hybrid system that picks from among four different abbreviation translation methods.

In this work, abbreviation expansion is treated as a retrieval problem using a probabilistic retrieval model to compute the similarity between observed context and each of existing contexts of expansions that share the same abbreviation. As for abbreviation translation, the hybrid system picks from: direct in-context abbreviation translation, in-context and out-of-context translation of expansion, and transliteration. The paper demonstrates the effectiveness of the proposed methods on English to Arabic MT. Unlike English, abbreviations are rare in Arabic.

Abbreviation expansion is a special case of Word Sense Disambiguation (WSD). However, abbreviations have characteristics that necessitate handling them differently. Unlike normal words, abbreviations have well defined senses. Also, it is relatively easy to get training documents that contain abbreviations along with their expansions. Most research on WSD addresses these two aspects of disambiguation (i.e. definition of word senses and sense-annotated corpora), which is not a major concern for disambiguation of abbreviations. In addition, given their dynamic nature, many abbreviations have low chance to appear in parallel data compared to normal words. Consequently, special approaches to disambiguate and then translate abbreviations are needed.

The remainder of the paper is organized as follows: Sections 2 and 3 explain the proposed approaches for abbreviation expansion and abbreviation translation respectively; Section 4 describes the experimental setup and reports on results; Section 5 provides related work in the literature; Section 6 concludes the paper and proposes future work.

## 2  Abbreviation Expansion

### 2.1  Problem Statement

Given text T which contains an ambiguous abbreviation $\alpha$ and given a set of possible expansions E = {$e_1$, $e_2$, ... $e_n$}, abbreviation expansion is defined as the problem of selecting the proper expansion $e_k \in E$ of $\alpha$ given T.

### 2.2  Retrieval-Based Solution

The proposed approach is based on the assumption that contextual text T relates to documents which contain the correct expansion $e_k$ more than documents which contain other expansions $e_{i \neq k}$. For each abbreviation-expansion pair found in a document, the tuple {abbreviation, expansion, context} is constructed. Context refers to a set of sentences that contain the expansion for the abbreviation. The tuples are indexed offline using an IR engine. At query time, the index is queried using significant terms in text T as keywords, restricting results to those where abbreviation = α. The highest ranking expansion is assumed to be the proper expansion $e_k$.

Introducing possible expansions methods is beyond the scope of this paper; interested readers can refer to (Yeates, 1999; Hiroko and Takagi, 2005; Larkey et al., 2000; Xu and Huang, 2006; Zahariev, 2004). In addition, several resources on the web maintain up-to-date abbreviation definitions and serve them for free (e.g. The Internet Acronym Server[4], Acronym Finder[5] and Abbreviations[6]).

Given a database of abbreviations and their possible expansions, it is straight-forward to obtain training documents which contain a particular abbreviation expansion. Web search engines can be used for this purpose by specifying the abbreviation expansion as a phrase in addition to the abbreviation itself. However, since the authors did not have access to any database of abbreviation expansions, a method similar to that of Larkey et al. (2000) was used to identify abbreviations and their expansions in Wikipedia articles, creating a database of abbreviations and expansions (more details in section 4). The method relied on using heuristics to automatically identify abbreviations and their expansions in a large corpus. The corpus used herein was the English Wikipedia pages.

One of the advantages of using an IR engine is that, unlike binary discriminative classifiers, features (i.e. words in all contexts) assume consistent weights across classes (i.e expansions). Unlike most related work (e.g. Zahariev, 2004; Gaudan et al., 2005) where a classifier is built for each expansion requiring multiple classifiers to be used for each abbreviation, IR engine can ascertain the best expansion by quering one index.

For this work, retrieval was performed using Indri, a commonly used open source IR engine, which is based on inference networks and allows for fielded search (Metzler and Croft, 2004).

## 2.3 Learning Unseen Expansions

The proposed solution for abbreviation expansion cannot resolve abbreviations not defined in training documents. In order to keep the system up-to-date and complement shortages that may exist in training corpus, acquiring new tuples of abbreviation-expansion-context has to be an ongoing process. This is achieved by mining input text T to identify abbreviation definitions that may exist, in parallel to the normal processing described in the previous subsection (2.2). Texts which contain such tuples are incrementally indexed, and added to the training corpus for later use.

## 3 Abbreviation Translation

This section discusses several methods to translate a sentence S that contains an ambiguous abbreviation α. Given that different methods have advantages and disadvantages, a hybrid system that utilizes language modeling is used to pick from among the output of all methods.

---

[4] http://acronyms.silmaril.ie/
[5] http://www.acronymfinder.com/
[6] http://www.abbreviations.com/

## 3.1 Leave and Translate (LT)

This is the baseline. In this method, no special treatment is given to abbreviations. A syntactically informed phrasal SMT system, similar to that of Menezes and Quirk (2008) and Quirk et al. (2005) was used. This method performs well only with popular and unambiguous abbreviations (e.g. UNESCO, WWW), but it suffers from the problems mentioned in the introduction.

## 3.2 Expand and Translate in-Context (ETC)

In this method, abbreviations are expanded prior to translation. The rationale behind this method is that MT systems may have a better chance of translating an abbreviation expansion than translating the abbreviation itself. Usually, abbreviation expansions have reduced lexical ambiguity and improved lexical coverage as the constituents of an expansion are more likely to have relevant entries in the phrase-table compared to abbreviations. Also, expansion of abbreviations is informed by more context than language models which may only account for small windows of word n-grams. The proposed method works as follows:

1. Find the most likely expansion $e_k$ of the abbreviation $\alpha$ given its context.
2. Replace $\alpha$ in the sentence S with $e_k$, producing modified sentence S`.
3. Translate the modified sentence S` using baseline MT system.

For example, consider the following two sentences:

$S_1$: **_ATM_** is a networking protocol.
$S_2$: There's a nearby **_ATM_** in case you need to withdraw cash.

Using the LT method (as in subsection 3.1) to translate the English sentences to Arabic leads to identical translations of *ATM* for both sentences:

$LT(S_1)$: .جهاز الصراف الآلي بروتوكول شبكة اتصال
$LT(S_2)$: .يوجد جهاز الصراف الآلي قريبة في حال أردت سحب المال

In contrast, ETC first transforms the English source sentences to:

$S`_1$: **_Asynchronous transfer mode_** is a networking protocol.
$S`_2$: There's a nearby **_automatic teller machine_** in case you need to withdraw money.

Then, ETC translates the modified sentences, producing a much better translation for *ATM* in the first sentence:

$ETC(S_1)$: .وضع النقل غير المتزامن بروتوكول شبكة اتصال
$ETC(S_2)$: .يوجد جهاز الصراف الآلي قريبة في حال أردت سحب المال

A drawback of this method is that the MT decoder may inappropriately breakup phrases to match against the phrase-table. For example, the decoder may decide to translate *"nearby automatic"* and *"machine in case"* as phrases.

### 3.3  Expand and Translate Out-of-Context (ETOC)

To avoid the drawback described in 3.2, this method gains partial control over the segmentation of modified source sentences by translating the expansion in isolation, and then replacing the abbreviation in the source sentence prior to translation by the MT engine, as follows:

1. Find the most likely expansion $e_k$ for the abbreviation $\alpha$ given its context *(identical to ETC's step 1)*.
2. Translate the most likely expansion $e_k$ in isolation to target language phrase A.
3. Replace the abbreviation $\alpha$ in the source sentence with an OOV word, producing modified sentence S`.
4. Translate S`, *producing* T.
5. Replace the OOV word in T by A.

Building on the *ATM* example, the isolated translations of the expansions (step 2) produce:

$A_1$: وضع النقل غير المتزامن

$A_2$: جهاز الصراف الالي

Replacing the abbreviation $\alpha$ with translation A (step 3) produces:

S`$_1$: **OOV** *is a networking protocol.*
S`$_2$: *There's a nearby* **OOV** *in case you need to withdraw money.*

Translating the modified sentence S` (step 4) produces:

$T_1$: **OOV** بروتوكول شبكة اتصال.
$T_2$: يوجد **OOV** قريبة في حال أردت سحب المال.

Replacing the **OOV** word with the expansion translation **A** (step 5) produces:

$ETOC(S_1)$: **وضع النقل غير المتزامن** بروتوكول شبكة اتصال.
$ETOC(S_2)$: يوجد **جهاز الصراف الآلي** قريبة في حال أردت سحب المال.

One caveat that limits the usefulness of this method is that the introduction of out-of-vocab words confuses the translation model. In order to reduce dependence on any particular decoder and to enhance reproducibility of this work, authors preferred not to introduce changes to the decoder to address this issue.

### 3.4  Transliteration (TT)

Some abbreviations are sufficiently popular that people use the abbreviated form in several languages. For example, the most common Arabic translation of NASA and BBC are ناسا and بي بي سي respectively. In such cases, transliteration can be the preferred translation method. When a popular abbreviation is an acronym (e.g. NASA, AIDS), the phonetically equivalent word in Arabic is a borrowed word (NASA→ناسا, AIDS→إيدز). When a popular abbreviation is an initialism[7] (e.g. BBC, AFP), a letter-by-letter transliteration is usually the most common translation (e.g. AFP→أ ف ب, BBC→بي بي سي).

---

[7] Despite the difference between acronyms and initialisms, people often refer to both as acronyms.

In order to find the most common Arabic transliteration, Table 2 was used to produce possible transliterations of initialisms and acronyms. In short, English letters were replaced by their corresponding phonetic equivalents and then a language model (trained exclusively on target-language named-entities from publicly available ACE[8] and Bin-Ajeeba[9] corpora) was consulted to select the most likely transliteration. Phonetic transliteration of acronyms is left for future work.

**Table 2.** Arabic mappings of English letters

| English letter(s) | Mappings for acronyms | Mappings for initalisms | English letter(s) | Mappings for acronyms | Mappings for initialisms |
|---|---|---|---|---|---|
| A | أ|ا | إيه | N | ن | إن |
| B | ب | بي | O | و|أ | أو |
| C | س|ك | سي | P | ب | بي |
| D | د|ض | دي | Q | ك|ق | كيو |
| E | إ|أي | إي | R | ر | آر |
| F | ف | إف ، | S | س|ز | إس |
| G | ج | جي | T | ت | تي |
| H | ه | إتش | U | و|يو | يو |
| I | ي | أي | V | ف | في |
| J | ج | جيه | W | و | دبليو |
| K | ك | كي | X | س|ك | إكس |
| L | ل | إل | Y | ي | واي |
| M | م | إم | Z | ز|س | زد |

## 3.5 Hybrid (HYB)

None of the aforementioned methods is expected to consistently yield the best translation results. For example, if an abbreviation $\alpha$ appears (with the sense $e_k$) a sufficient number of times in the parallel training data and the general language model can properly pick the proper translation, then the LT method is likely to produce a correct translation. If the abbreviation is not present in the parallel training data, but its constituents do, methods ETC and ETOC are expected to produce better translations. If the abbreviation is used at the target language as a borrowed word, TT would be the method of choice.

The hybrid method translates the sentence using the four methods (LT, ETC, ETOC and TT) and selects the most fluent translation (as estimated by target language model probability).

---

[8] http://projects.ldc.upenn.edu/ace/
[9] http://www1.ccls.columbia.edu/~ybenajiba/

## 4  Experiments

### 4.1  Abbreviation Expansion

In this work, abbreviation expansion was examined for the English language. English Wikipedia articles were scanned for abbreviation-expansion pairs. An abbreviation-expansion pair was extracted when an expansion was followed by an abbreviation between brackets, where letters in the abbreviation matched sequences of initial and middle letters of words in the expanded form. Frequency of an abbreviation-expansion pair has to surpass a threshold (3 was used as the threshold) to qualify as a valid pair. As a by-product of this process, example documents containing the abbreviations and their expansions were automatically obtained.  In all, the constructed collection contained unique 10,664 abbreviations with 16,415 unique expansions, extracted from roughly 2.9 million Wikipedia articles. The number of expansions per abbreviation was 1.54 on average with a variance of 1.66.

Context documents were indexed using Indri fielded indexing, with the fields being the abbreviation, the expansion, and full text of the document.

The test set of abbreviations, expansions and contexts contained 500 English Wikipedia articles, randomly extracted and manually revised. Mentions of the abbreviation expansions were removed from the context.  For testing, the context query for an abbreviation was taken as the 10 words preceding, 10 words trailing the abbreviation (excluding stopwords).  If an abbreviation appeared more than once in the article, context words were aggregated for all occurrences. The 50 unique words with highest term frequency in the article were selected as the context query for the abbreviation.  The query was submitted to Indri constraining the results while restricting the abbreviation field to the abbreviation at hand. The expansion corresponding to the highest ranking result was chosen as the proper expansion. Accuracy was used as the figure of merit to evaluate abbreviation expansion. When calculating accuracy, the system received a 1 if the top returned expansion was correct and 0 otherwise.[10]

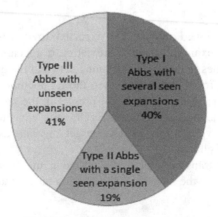

**Fig. 1.** Breakdown of test set types

---

[10] This is referred to as precision@1 in IR literature.

Figure 1 provides a breakdown of abbreviations in the test set as follows: Type I: 202 (out of 500) polysemous abbreviations had an average of 3.7 possible expansions each; Type II: 94 abbreviations had only 1 expansion (so the proposed approach did not have a choice but to select the correct expansion); and Type III: 204 abbreviation-expansion pairs were not previously observed (hence the proposed technique had no chance of finding them). Table 3 presents the results obtained for abbreviation expansion. The results reported here include accuracy when all test data is used (types I, II, and III), when excluding test items for which the correct expansion was never seen in training data (types I, II), and when excluding abbreviations which have a single possible expansion as well (type I). The baseline reported here is a process that selects an expansion randomly assuming uniform distribution for the possible expansions. Unfortunately, the authors did not have access to datasets reported on in the literature for comparative evaluation.

**Table 3.** Accuracy of abbreviation expansion

| Test set | Baseline | IR |
|---|---|---|
| All (500 test tuples) | 35% | 53% |
| Types I & II (296 tuples) | 44% | 90% |
| Type I only (202 tuples) | 27% | 86% |

When considering the abbreviations for which no expansions were seen in training, the proposed approach achieved 53% accuracy. Overcoming such low performance would require increasing the training data by learning new abbreviation-expansion pairs as explained in section 2. When excluding expansions that were missing from the training data, the system yielded 90% accuracy, which would be typical if the training set was to be expanded. It is noteworthy that when examining the mistakes that were made by the system, they included several examples such as "Singapore Art Museum" where the system's guess was "Seattle Art Museum"; such examples are probably harder to disambiguate than others. When abbreviations with a single expansion were excluded, the accuracy was 86%.

## 4.2 Abbreviation Translation

For abbreviation translation, the aforementioned translation methods were tested individually as well as the hybrid method with different combinations. All experiments were performed using a syntactically informed statistical MT system similar to (Menezes and Quirk, 2008). Performance was evaluated using BLEU score, as measured by NIST's mteval (v. 13). Parallel training data was comprised of LDC news corpora[11], UN data, as well as automatically mined parallel sentences from the web (Quirk et al., 2007). A total of 11M sentences were used in training, with an average English and Arabic sentence lengths of 23 and 27 tokens respectively.

---

[11] LDC2004T18, LDC2004T17, LDC2007T24, LDC2008T02, LDC2008T09 and LDC2009T09.

500K parallel sentences were held out for testing and tuning purposes. An automatic process extracted 756 sentence pairs from the held out parallel data such that a unique abbreviation exists in each source sentence. Out of those, 500 sentence pairs were used as a test set, the rest were added to the development set. The test set had average English and Arabic sentence lengths of 30 and 29 respectively. Note that the unique abbreviations condition imposed a limit on the size of the test set. Further, BLEU scores reported here are lower than typical Arabic-to-English scores due to the lack of multiple reference translations for the English-to-Arabic test set.

The test set used for abbreviation translation was different than the one used for abbreviation expansion (Section 4.1). In abbreviation expansion, each test sample must identify the proper expansion of the abbreviation, which is not available for abbreviations in the parallel test set. On the other hand, abbreviation translation requires each test sample to contain the proper translation of the sentence, which is not available for Wikipedia articles used to test abbreviation expansion.

SRILM toolkit was used to build an Arabic trigram language model with Good-Turing smoothing for the hybrid method. The language model was constructed using texts in Arabic Wikipedia, Arabic Giga Word collection[12], and the Arabic portion of training data.

Table 4 lists the results of using the aforementioned methods for translation of sentences that contain abbreviations. While individual methods, namely ETC and ETOC, showed a small improvement over the baseline (LT), the hybrid system effectively combined translations from all four methods to achieve a significant improvement of 1.48 BLEU points. Using the hybrid method to pick among different methods consistently gave better results than individual methods, suggesting that target language model was effective in combine several abbreviation translation methods.

**Table 4.** BLEU score for abbreviation expansion translation using different methods

| Method/Combination | BLEU |
|---|---|
| Baseline (LT) | 16.60 |
| ETC | 17.01 |
| ETOC | 16.98 |
| TT | 14.65 |
| Hybrid (LT, ETC) | 17.35 |
| Hybrid (LT, ETOC) | 17.70 |
| Hybrid (LT, ETC, TT) | 17.27 |
| Hybrid (LT, ETOC, TT) | 17.68 |
| Hybrid (LT, ETC, ETOC) | 18.04 |
| Hybrid (LT, ETC, ETOC, TT) | 18.08 |

---

[12] LDC2003T12.

However, individual methods contributed differently to this improvement. Implementing the hybrid method using different combinations helps analyze the contribution of each method. Combinations (LT, ETC) and (LT, ETOC) gave improvements of 0.75 and 1.10 BLEU points, respectively. This confirms the assumption that expanding abbreviations before translation is beneficial.

Adding transliteration (TT) to any combination seemed to either degrade or yield (almost) the same BLEU score. This is probably attributed to the type of abbreviations for which TT was designed. It was expected to produce meaningful results for popular abbreviations. Nevertheless, such popular abbreviations are also expected to appear frequently in parallel training data. Consequently, baseline MT system would be sufficient to find the proper translation of such abbreviations in the phrase table, refuting the need to use TT.

One factor that limited the gain of abbreviation expansion methods was the prevalence of sentences in the test set where abbreviations were not fully translated. For example, the reference translation for the sentence containing the abbreviation KAC (Kuwait Airways Corporation) only referred to KAC as "the corporation" (الشركة). While this translation is valid when the full name is written earlier in the text, the translation is not complete in isolation. One way to avoid this problem is to perhaps use multiple reference translations, or to manually create the references in isolation of the full context of the documents.

## 5 Related Work

The problem of abbreviation expansion can be viewed as a special case of word sense disambiguation (WSD) (Zahariev, 2004). Over the past sixty years, sophisticated approaches were developed to address WSD. Interested readers are referred to a recent comprehensive survey on WSD by Navigli (2009). Although polysemy (i.e. lexical ambiguity) in abbreviations is often greater than polysemy in regular words (Zahariev, 2004), the representation of word senses in abbreviations is less of a problem than in regular words. For instance, most people would distinguish [gold: noun] and [gold: adjective] as different senses, but some people will go further and argue that [gold: of the color of gold] and [gold: consisting of gold] should be two distinct senses as well. Fortunately, this problem almost does not exist for abbreviations, making it more feasible to find a satisfactory solution to the problem given available resources.

Several supervised and semi-supervised learning approaches were used to solve the abbreviation expansion problem. In general text, Zahariev (2004) used a support vector machine (SVM) classifier with a linear kernel. A model is trained for each abbreviation, with distinct expansions representing different classes. Terms occuring in the same document as the abbreviation were used as features. Training data were obtained by searching the web for PDF documents containing both an abbreviation and any of its expansions. Though effective, building SVM models for each expansion of every abbreviation was computationally intensive. SVM attempted to assign different weights to different features and these weights were different from one model to the next.

Solving this problem for the medical domain captured the interest of many researchers due to the widespread use of abbreviations in the biomedical domain. Pakhomov et al. (2002) approached the problem using maximum entropy classification, Gaudan et al. (2005) used an SVM classifier, and Stevenson et al. (2009) used a vector space model.

Roche and Prince (2008) ranked the expansions of a given abbreviation by calculating the cubic mutual information function and Dice's coefficient based on the number of web pages. Given an abbreviation, contextual text, and the set of possible expansions, their idea was to find the number of web pages containing the expansion and keywords from the context, then to divide this value by the number of pages containing individual key/expansion words. The evaluation was done for French as well as medical abbreviations.

Some research efforts targeted translation of abbreviations. Callison-Burch et al. (2006) looked at the broader problem of using paraphrases to improve lexical coverage in MT. Along the same line, Li and Yarowsky (2008a; 2008b) used an unsupervised technique to address translation of Chinese abbreviations. English named entities (NEs) were extracted from a monolingual corpus, and translated using a baseline MT system into Chinese. Then, Chinese abbreviation-expansion pairs were extracted from monolingual Chinese text, and matched with their English NE translations using the Chinese automatic translation obtained before as a bridge. Then, Chinese abbreviations and their corresponding English NE translations were added to the phrase table of the baseline MT system. While this approach effectively solved the first problem mentioned in the introduction (i.e. the proper phrase pair does not exist in the phrase table), it does not solve the second problem (i.e. the high polysemy of abbreviations) because the decoder was still responsible for disambiguating between multiple expansions (i.e. translations) of a polysemous abbreviation using the target language model. On the other hand, the proposed approach at hand addresses English abbreviations, solves both identified problems, and experiments with several methods of translating an abbreviation.

Some researchers (Chan et al., 2007; Carpuat and Wu, 2007) also studied the integration of WSD in SMT by introducing elaborate modifications to the decoders. In this work, although abbreviation expansion was a special case of WSD, a design decision was taken to simplify the integration by using the decoder as a black box, making it much easier to implement, replicate and scale to different SMT systems.

# 6 Conclusion and Future Work

A retrieval-based algorithm for abbreviation expansion was presented. Using a retrieval engine for abbreviation expansion availed the need to build separate classification models for different abbreviation. The described algorithm was both efficient and effective, yielding an accuracy of 90%.

Regarding translation, expanding abbreviations before translation was a simple but useful modification. A hybrid system that utilized a variety of abbreviation translation methods was presented. While individual methods showed small improvements, combining several methods achieved significant improvement of 1.48 BLEU points.

This work can be extended in three directions. One direction is to generalize the proposed IR-based disambiguation technique for words rather than abbreviations. The main difficulty here lies in the definition of word senses and developing sense-annotated corpora. The second direction is to enhance the proposed abbreviation translation approach. In particular, a proper way to condense translated abbreviation expansions is needed. For example, a professional translator would translate English "UN" into French "ONU", while the proposed approach would translate it to French "Organisation des Nations Unies". Also, using acronym phonetic transliteration may make the TT method more effective. The third direction is to make use of abbreviation expansion in other IR/NLP tasks that exhibit some sort of language understanding (e.g. query expansion and question answering).

## Acknowledgements

Authors would like to thank Arul Menezes and Mei-Yuh Hwang for their scientific and technical support, Hany Hassan and Khaled Ammar for their numerous valuable comments, and Amany Shehata for performing human evaluations.

## References

Agirre, E., Martinez, D.: Smoothing and word sense disambiguation. In: Vicedo, J.L., Martínez-Barco, P., Muñoz, R., Saiz Noeda, M. (eds.) EsTAL 2004. LNCS (LNAI), vol. 3230, pp. 360–371. Springer, Heidelberg (2004)

Callison-Burch, C., Koehn, P., Osborne, M.: Improved Statistical Machine Translation Using Paraphrases. In: NAACL 2006 (2006)

Carpuat, M., Wu, D.: Improving Statistical Machine Translation using Word Sense Disambiguation. In: Proceedings of EMNLP 2007, pp. 61–72 (2007)

Chan, Y., Ng, H., Chiang, D.: Word Sense Disambiguation Improves Statistical Machine Translation. In: Proceedings of ACL 2007, pp. 33–40 (2007)

Chiang, D.: Hierarchical Phrase-Based Translation. Computational Linguistics 33(2), 201–228 (2007)

Galley, M., Graehl, J., Knight, K., Marcu, D., DeNeefe, S., Wang, W., Thayer, I.: Scalable Inference and Training of Context-Rich Syntactic Translation Models. In: Proceedings of COLING/ACL 2006, pp. 961–968 (2006)

Gaudan, S., Kirsch, H., Rebholz-Schuhmann, D.: Resolving abbreviations to their senses in Medline. Bioinformatics 21(18), 3658–3664 (2005)

Hiroko, A., Takagi, T.: ALICE: An Algorithm to Extract Abbreviations from MEDLINE. Journal of the American Medical Informatics Association, 576–586 (2005)

Koehn, P., Och, F., Marcu, D.: Statistical phrase-based translation. In: Proceedings of NAACL 2003, pp. 48–54 (2003)

Larkey, L., Ogilvie, P., Price, A., Tamilio, B.: Acrophile: an automated acronym extractor and server. In: Intl. Conf. on Digital Libraries archive, 5th ACM Conf. on Digital libraries, pp. 205-214 (2000)

Li, Z., Yarowsky, D.: Unsupervised Translation Induction for Chinese Abbreviations using Monolingual Corpora. In: ACL 2008 (2008a)

Li, Z., Yarowsky, D.: Mining and modeling relations between formal and informal Chinese phrases from web corpora (2008b)

Menezes, A., Quirk, C.: Syntactic Models for Structural Word Insertion and Deletion. In: Proceedings of the 2008 Conference on Empirical Methods in Natural Language Processing, pp. 735–744, Honolulu (October 2008)

Metzler, D., Croft, W.B.: Combining the Language Model and Inference Network Approaches to Retrieval. Information Processing and Management Special Issue on Bayesian Networks and Information Retrieval 40(5), 735–750 (2004)

Molloy, M.: Acronym Finder (1997), from http://www.acronymfinder.com/ (retrieved February 8, 2010)

Navigli, R.: Word Sense Disambiguation: a Survey. ACM Computing Surveys 41(2) (2009)

Pakhomov, S.: Semi-Supervised Maximum Entropy Based Approach to Acronym and Abbreviation Normalization in Medical Texts. In: ACL 2002 (2002)

Quirk, C., Menezes, A., Cherry, C.: 2005. Dependency Treelet Translation: Syntactically Informed Phrasal SMT. In: ACL 2005 (2005)

Quirk, C., Udupa, R., Menezes, A.: Generative Models of Noisy Translations with Applications to Parallel Fragment Extraction. In: European Assoc. for MT (2007)

Roche, M., Prince, V.: AcroDef: A quality measure for discriminating expansions of ambiguous acronyms. In: Kokinov, B., Richardson, D.C., Roth-Berghofer, T.R., Vieu, L. (eds.) CONTEXT 2007. LNCS (LNAI), vol. 4635, pp. 411–424. Springer, Heidelberg (2007)

Roche, M., Prince, V.: Managing the Acronym/Expansion Identification Process for Text-Mining Applications. Int. Journal of Software and Informatics 2(2), 163–179 (2008)

Stevenson, M., Guo, Y., Al Amri, A., Gaizauskas, R.: Disambiguation of Biomedical Abbreviations. In: BioNLP Workshop, HLT 2009 (2009)

Xu, J., Huang, Y.: Using SVM to Extract Acronyms from Text. In: Soft Computing - A Fusion of Foundations, Methodologies and Applications, pp. 369–373 (2006)

Yeates, S.: Automatic Extraction of Acronyms from Text. In: New Zealand Computer Science Research Students Conference 1999, pp. 117–124 (1999)

Zahariev, M.: Automatic Sense Disambiguation for Acronyms. In: SIGIR 2004, pp. 586–587 (2004)

# Word Segmentation for Dialect Translation

Michael Paul, Andrew Finch, and Eiichiro Sumita

National Institute of Information and Communications Technology
MASTAR Project
Kyoto, Japan
michael.paul@nict.go.jp

**Abstract.** This paper proposes an unsupervised word segmentation algorithm that identifies word boundaries in continuous source language text in order to improve the translation quality of statistical machine translation (SMT) approaches for the translation of local dialects by exploiting linguistic information of the standard language. The method iteratively learns multiple segmentation schemes that are consistent with (1) the standard dialect segmentations and (2) the phrasal segmentations of an SMT system trained on the resegmented bitext of the local dialect. In a second step multiple segmentation schemes are integrated into a single SMT system by characterizing the source language side and merging identical translation pairs of differently segmented SMT models. Experimental results translating three Japanese local dialects (Kumamoto, Kyoto, Osaka) into three Indo-European languages (English, German, Russian) revealed that the proposed system outperforms SMT engines trained on character-based as well as standard dialect segmentation schemes for the majority of the investigated translation tasks and automatic evaluation metrics.

## 1 Introduction

Spoken languages distinguish regional speech patterns, the so-called *dialects*: a variety of a language that is characteristic of a particular group of the language's speakers. A *standard dialect* (or *standard language*) is a dialect that is recognized as the "correct" spoken and written form of the language. Dialects typically differ in terms of morphology, vocabulary and pronunciation. Various methods have been proposed to measure the relatedness between dialects using phonetic distance measures [1], string distance algorithms [2,3], or statistical models [4]. Concerning data-driven natural language processing (NLP) applications, research on dialect processing focuses on the analysis and generation of dialect morphology [5], parsing of dialect transcriptions [6], spoken dialect identification [7], and machine translation [8,9,10].

For most of the above applications, explicit knowledge about the relation between the standard dialect and the local dialect is used to create local dialect language resources. In terms of morphology, certain lemmata of word forms are shared between different dialects where the usage and order of inflectional affixes might change. The creation of rules that map between dialectic variations can

A. Gelbukh (Ed.): CICLing 2011, Part II, LNCS 6609, pp. 55–67, 2011.

help to reduce the costs for building tools to process the morphology of the local dialect [5]. Similarly for parsing, it is easier to manually create new resources that relate the local dialect to the standard dialect, than it is to create syntactically annotated corpora from scratch [6].

For machine translation (MT), linguistic resources and tools usually are available for the standard dialect, but not for the local dialects. Moreover, applying the linguistic tools of the standard dialect to local dialect resources is often insufficient. For example, the task of *word segmentation*, i.e., identifying word boundaries in continuous text, is one of the fundamental preprocessing steps of MT applications. In contrast to Indo-European languages like English, many Asian languages like Japanese do not use a whitespace character to separate meaningful word units. However, the application of a linguistically motivated standard dialect word segmentation tool to a local dialect corpus results in a poor segmentation quality due to different morphologies of verbs and adjectives, thus resulting in a lower translation quality for SMT systems that acquire the translation knowledge automatically from a parallel text corpus. For example, applying a Japanese segmentation tool to the standard dialect phrase "...*ITADAKEMASUKA*" ("could you please ...") might result in a segmentation like "*ITADAKE* (auxiliary verb) | *MASU* (inflectional form) | *KA* (question particle)", whereas the non-linguistic segmentation "*I | TA | DAKE | SHIMA | HEN | KA*" is obtained for the corresponding Kyoto dialect phrase "...*ITADAKESHIMAHENKA*". Therefore, a word segmentation method of the local dialect that is "consistent with the standard dialect" is one that takes into account meaningful standard dialect translation units and adjust its segmentation to inflectional variations of the local dialect.

Moreover, in the case of small translation units, e.g. single Japanese characters, it is likely that such tokens have been seen in the training corpus, thus these tokens can be translated by an SMT engine. However, the contextual information provided by these tokens might not be enough to obtain a good translation. For example, a Japanese-English SMT engine might translate the two successive characters "*HAKU*" ("white") and "*CHOU*" ("bird") as "*white bird*", while a human would translate "*HAKUCHOU*" as "*swan*". Therefore, the longer the translation unit, the more context can be exploited to find a meaningful translation. On the other hand, the longer the translation unit, the less likely it is that such a token will occur in the training data due to *data sparseness* of the language resources utilized to train the statistical translation models. Therefore, a word segmentation that is "consistent with SMT models" is one that identifies units that are small enough to be translatable but large enough to be meaningful in the context of the given input sentence, achieving a trade-off between the *coverage* and the *translation task complexity* of the statistical models in order to improve translation quality.

Various word segmentation approaches already have been proposed. Purely dictionary-based approaches like [11] addressed these problems by maximum matching heuristics. Recent research on unsupervised word segmentation focuses on approaches based on probabilistic methods [12,13]. However, the use of monolingual probabilistic models does not necessarily yield better MT performance [14]. Improvements have been reported for approaches taking into account not only monolingual, but also bilingual information, to derive a word segmentation suitable for SMT [15]. In addition, the usage of multiple word segmentation schemes for the source language also can help to improve translation quality [16,17,18]. The method proposed in this paper differs from previous approaches in the following ways:

- it works for any language pair in which the source language is unsegmented and the target language segmentation is known.

- it applies machine learning techniques to identify segmentation schemes that improve the translation quality for a given language pair.

- it exploits knowledge about the standard dialect word segmentation to derive the word segmentation of the local dialect

- it decodes directly from unsegmented text using segmentation information implicit in the phrase-table to generate the target and thus avoids issues of consistency between phrase-table and input representation.

- it uses segmentations at all iterative levels of the bootstrap process, rather than only those from the best single iteration, which allows for the consideration of segmentations from different levels of granularity.

The unsupervised iterative learning approach is applied to learn and integrate multiple word segmentation schemes that are consistent with the standard dialect segmentation and the utilized SMT models as described in Section 2. Experiments were carried out for the translation of three Japanese local dialects (Kumamoto, Kyoto, Osaka) into three Indo-European language (English, German, Russian). The utilized language resources and the outline of the experiments are summarized in Section 3. The results reveal that the integration of multiple segmentation schemes improves the translation quality of local dialect MT approaches and that the proposed system outperforms SMT engines trained on character-based as well as standard dialect segmentation schemes for the majority of the investigated translation tasks.

## 2   Word Segmentation

The proposed word segmentation method is a language-independent approach that treats the task of word segmentation as a *phrase-boundary tagging* task.

The unsupervised learning method uses a parallel text corpus consisting of initially unigram segmented source language character sequences and whitespace-separated target language words. The initial bitext is used to train a standard phrase-based SMT system ($SMT_{chr}$). The character-to-word alignment results of the SMT training procedure[1] are exploited to identify successive source language characters aligned to the same target language word in the respective

---

[1] For the experiments presented in Section 3, the GIZA++ toolkit was used.

bitext and to merge these characters into larger translation units, defining its granularity in the given bitext context. Unaligned source language characters are treated as a single translation unit.

In order to obtain a local dialect word segmentation that is consistent with the standard dialect, the source language corpora for both (standard and local) dialects are preprocessed using the linguistically motivated segmentation scheme. The initial source language corpus used to iteratively train the local dialect segmentation scheme is then created by keeping those word segments that are identical in both pre-segmented corpora and by characterizing all other non-matching tokens for each sentence of the local dialect corpus (see Section 2.1).

The translation units obtained are then used to learn the word segmentation that is most consistent with the phrase alignments of the given SMT system by aligning pre-segmented source language sentences to word units separated by a whitespace in the target language. A Maximum-Entropy (ME) model is applied to learn the most consistent word segmentation, to re-segment the original source language corpus, and to re-train a phrase-based SMT engine based on the learned segmentation scheme (see Section 2.2).

This process is repeated as long as an improvement in translation quality is achieved (see Section 2.3). Eventually, the concatenation of succeeding translation units will result in overfitting, i.e., the newly created token can only be translated in the context of rare training data examples. Therefore, a lower translation quality due to an increase of untranslatable source language phrases is to be expected.

However, in order to increase the *coverage* and to reduce the *translation task complexity* of the statistical models, the proposed method integrates multiple segmentation schemes into the statistical translation models of a single SMT engine so that longer translation units are preferred for translation, if available, and smaller translation units can be used otherwise (see Section 2.4).

## 2.1   Segmentation Data

The language resources required for the proposed method consist of a sentence-aligned corpus for the standard dialect, the local dialect, and the target language. In order to train two baseline systems, the source language data sets are initially segmented by (1) inserting white-spaces between successive characters (*character*) and (2) applying the linguistically-motivated word segmentation tool of the standard dialect (*standard*). The target language data sets are preprocessed by a tokenizer that simply separates punctuation marks from given word tokens.

Based on the *standard* segmentations of the standard dialect and the local dialect, an additional segmentation (*mapped*) of the local dialect that is consistent with standard dialect segmentation is obtained as follows. For each of the corpus sentences, the *standard* segmentation of both source language dialects are compared using maximum matching heuristics in order to identify segments that occur in both sentences. The respective local dialog sentences are re-segmented by keeping matching segments as is and characterizing all unmatched token parts. For example, given the Japanese sample of the introduction, the non-character

```
(1)     proc annotate-phrase-boundaries( Bitext ) ;
(2)   begin
(3)      for each (Src, Trg) in {Bitext} do
(4)          A ← align(Src, Trg) ;
(5)          for each i in {1, . . . , len(Src)-1} do
(6)              Trg_i ← get-target(Src[i], A) ;
(7)              Trg_{i+1} ← get-target(Src[i+1], A) ;
(8)              if null(Trg_i) or Trg_i ≠ Trg_{i+1} then
(9)                  (* aligned to none or different target *)
(10)                 Src_ME ← assign-tag(Src[i],' E') ;
(11)             else
(12)                 (* aligned to the same target *)
(13)                 Src_ME ← assign-tag(Src[i],' I') ;
(14)             fi ;
(15)             Corpus_ME ← add(Src_ME) ;
(16)          od ;
(17)          (* last source token *)
(18)          LastSrc_ME ← assign-tag(Src[len(Src)],' E') ;
(19)          Corpus_ME ← add(LastSrc_ME) ;
(20)      od ;
(21)      return( Corpus_ME ) ;
(22)  end ;
```

**Fig. 1.** ME Training Data Annotation

segments "*RESUTORAN*" and "*SAGASHI*" are identical in both dialects thus are kept in the *mapped* segmentation. However, the unmatched non-linguistic segmentation part "*I | TA | DAKE | SHIMA | HEN | KA*" is characterized resulting in a segmentation that is consistent with the standard dialect and that avoids noisy segmentation inputs for the iterative word segmentation learning method.

```
      character : RE|SU|TO|RA|N|O|SAGA|SHI|TE|I|TA|DA|KE|SHI|MA|HE|N|KA|.
standard dialect : RESUTORAN|O|SAGASHI|TE|ITADAKE|MASE|N|KA|.
   local dialect : RESUTORAN|O|SAGASHI|TE|I|TA|DAKE|SHIMA|HEN|KA|.
         mapped : RESUTORAN|O|SAGASHI|TE|I|TA|DA|KE|SHI|MA|HE|N|KA|.
```

The obtained translation units are then used to learn the word segmentation that is most consistent with the phrase alignments of the given SMT system. First, each character of the source language text is annotated with a word-boundary indicator where only two tags are used, i.e, "*e*" (end-of-word character tag) and "*i*" (in-word character tag). The annotations are derived from the SMT training corpus as described in Figure 1. An example for the boundary annotations (*tagged*) is given below for the *mapped* local dialect segmentation of the previous Japanese example.

```
tagged : RE_i SU_i TO_i RA_i N_e O_e SAGA_i SHI_e TE_e
         I_e TA_e DA_e KE_e SHI_e MA_e HE_e N_e KA_e ._e
```

Using these corpus annotations, a Maximum-Entropy (ME) model is used to learn the word segmentation consistent with the SMT translation model (see Section 2.2) and resegment the original source language corpus.

## 2.2   Maximum-Entropy Tagging Model

ME models provide a general purpose machine learning technique for classification and prediction. They are versatile tools that can handle large numbers of features, and have shown themselves to be highly effective in a broad range of NLP tasks including sentence boundary detection or part-of-speech tagging [19].

A *maximum entropy classifier* is an exponential model consisting of a number of binary feature functions and their weights [20]. The model is trained by adjusting the weights to maximize the entropy of the probabilistic model given constraints imposed by the training data. In our experiments, we use a *conditional maximum entropy* model, where the conditional probability of the outcome given the set of features is modeled [21]. The model has the following form:

$$p(t, c) = \gamma \prod_{k=0}^{K} \alpha_k^{f_k(c,t)} \cdot p_0$$

where:

$t$   is the tag being predicted;
$c$   is the context of $t$;
$\gamma$   is a normalization coefficient;
$K$   is the number of features in the model;
$f_k$   are binary feature functions;
$a_k$   is the weight of feature function $f_k$;
$p_0$   is the default model.

The feature set is given in Table 1. The *lexical context features* consist of target words annotated with a tag $t$. $w_0$ denotes the word being tagged and $w_{-2}, \ldots, w_{+2}$ the surrounding words. $t_0$ denotes the current tag, $t_{-1}$ the previous tag, etc. The *tag context features* supply information about the context of previous tag sequences. This conditional model can be used as a classifier. The model is trained iteratively, and we used the improved iterative scaling algorithm (IIS) [19] for the experiments presented in Section 3.

**Table 1.** Feature Set of ME Tagging Model

| Lexical Context Features | $< t_0, w_{-2} >$   $< t_0, w_{-1} >$ |
|---|---|
|  | $< t_0, w_0 >$ |
|  | $< t_0, w_{+1} >$   $< t_0, w_{+2} >$ |
| Tag Context Features | $< t_0, t_{-1} >$   $< t_0, t_{-1}, t_{-2} >$ |

## 2.3  Iterative Bootstrap Method

The iterative bootstrap method to learn the word segmentation that is consistent with an SMT engine is summarized in Figure 2. After the ME tagging model is learned from the initial character-to-word alignments of the respective bitext ((1)-(4)), the obtained ME tagger is applied to resegment the source language side of the unsegmented parallel text corpus ((5)). This results in a resegmented bitext that can be used to retrain and reevaluate another engine $SMT_1$ ((6)), achieving what is hoped to be a better translation performance than the initial SMT engine ($SMT_{chr}$).

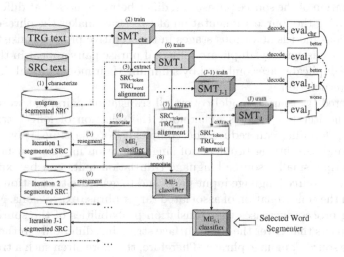

**Fig. 2.** Iterative Bootstrap Method

The unsupervised ME tagging method can also be applied to the token-to-word alignments extracted during the training of the $SMT_1$ engine to obtain an ME tagging model $ME_1$ capable of handling longer translation units ((7)-(8)). Such a bootstrap method iteratively creates a sequence of SMT engines $SMT_i$ ((9)-(J)), each of which reduces the translation complexity, because larger chunks can be translated in a single step leading to fewer word order or word disambiguation errors. However, at some point, the increased length of translation units learned from the training corpus will lead to overfitting, resulting in reduced translation performance when translating unseen sentences. Therefore, the bootstrap method stops when the $J^{th}$ resegmentation of the training corpus results in a lower automatic evaluation score for the unseen sentences than the one for the previous iteration. The ME tagging model $ME_{J-1}$ that achieved the highest automatic translation scores is then selected as the best single-iteration word segmenter.

## 2.4   Integration of Multiple Segmentation Schemes

The integration of multiple word segmentation schemes is carried out by merging the translation models of the SMT engines trained on the characterized and iteratively learned segmentation schemes. This process is performed by linearly interpolating the model probabilities of each of the models. In our experiments, equal weights were used; however, it might be interesting to investigate varying the weights according to iteration number, as the latter iterations may contain more useful segmentations.

In addition to the model interpolation, we also remove the internal segmentation of the source phrases by splitting them into characters. The advantages are twofold. Primarily it allows decoding directly from unsegmented text. Moreover, the segmentation of the source phrase can differ between models at differing iterations; removing the source segmentation at this stage makes the phrase pairs in the translations models at various stages in the iterative process consistent with one another. Consequently, duplicate bilingual phrase pairs appear in the phrase table. These duplicates are combined by summing their model probabilities prior to model interpolation.

The rescored translation model covers all translation pairs that were learned by any of the iterative models. Therefore, the selection of longer translation units during decoding can reduce the complexity of the translation task. On the other hand, overfitting problems of single-iteration models can be avoided because multiple smaller source language translation units can be exploited to cover the given source language input parts and to generate translation hypotheses based on the concatenation of associated target phrase expressions. Moreover, the merging process increases the translation probabilities of the source/target translation parts that cover the same surface string but differ only in the segmentation of the source language phrase. Therefore, the more often such a translation pair is learned by different iterative models, the more often the respective target language expression will be exploited by the SMT decoder.

The translation of unseen data using the merged translation model is carried out by (1) characterizing the input and (2) applying the SMT decoding in a standard way.

## 3   Experiments

The effects of using different word segmentations and integrating them into an SMT engine are investigated using the *Basic Travel Expressions Corpus* (BTEC), which is a collection of sentences that bilingual travel experts consider useful for people traveling abroad [22]. For the word segmentation experiments, we selected Japanese (ja), a language that does not naturally separate word units, and the local dialects from Kyoto (ky), Kumamoto (ku), and Osaka (os) area. For the target language, we investigated three Indo-European languages, i.e., English (en) and German (de) and Russian (ru).

**Table 2.** BTEC Language Resources

| Source (sentences) | train (160,000) | dev (1,000) | eval (1,000) | Target (sentences) | train (160,000) | dev (1,000) | eval (1,000) |
|---|---|---|---|---|---|---|---|
| ja  voc | 16,823 | 1,389 | 1,368 | en  voc | 15,364 | 1,254 | 1,283 |
| len | 8.5 | 8.1 | 8.1 | len | 7.5 | 7.1 | 7.2 |
| ky  voc | 17,384 | 1,461 | 1,445 | de  voc | 25,699 | 1,485 | 1,491 |
| len | 9.2 | 8.8 | 8.9 | len | 7.1 | 6.7 | 6.8 |
| ku  voc | 17,260 | 1,449 | 1,442 | ru  voc | 36,181 | 1,794 | 1,800 |
| len | 8.9 | 8.5 | 8.5 | len | 6.4 | 6.1 | 6.1 |
| os  voc | 17,341 | 1,470 | 1,459 | | | | |
| len | 8.8 | 8.5 | 8.3 | | | | |

Table 2 summarizes the characteristics of the BTEC corpus used for the training (*train*) of the SMT models, the tuning of model weights and the stop conditions of the iterative bootstrap method (*dev*), and the evaluation of translation quality (*eval*). Besides the number of sentences, the vocabulary size (*voc*) and the sentence length (*len*), i.e.,the average number of words per sentence, is listed. The given statistics are obtained using the widely accepted Japanese morphological analyzer toolkit CHASEN[2].

For the training of the SMT models, standard word alignment [23] and language modeling [24] tools were used. Minimum error rate training (MERT) was used to tune the decoder's parameters and performed on the *dev* set using the technique proposed in [23]. For the translation, an inhouse multi-stack phrase-based decoder [25] was used.

For the evaluation of translation quality, we applied two standard automatic metrics: the BLEU metric whose scores range between 0 (worst) and 1 (best) [26], and the TER metric whose scores are positive with 0 being the best possible [27]. Table 3 compares the translation results of our method (*proposed*) that integrates the linguistic motivated and iteratively learned segmentation schemes to two baseline systems that were trained on parallel text with the source language side segmented (1) character-wise (*character*) and (2) using the standard dialect segmentation scheme (*standard*).

The experimental results revealed that the proposed method outperforms both baseline systems for the majority of investigated translation tasks. The highest BLEU gains were achieved for the Kumamoto dialect (+1.1% / +3.9%), followed by the Osaka dialect (+0.9% / +4.9%), and the Kyoto dialect (+0.4% / +2.5%) relative to the *standard* / *character* SMT systems, respectively. In terms of the TER score, the proposed method worked equally well for the Kyoto and Osaka dialects (-1.7% / -4.7%) and achieved slightly lower gains for the Kumamoto dialect (-1.0% / -4.6%).

Table 4 summarizes the number of out-of-vocabulary (OOV) words for the *standard* and *proposed* translation experiments. The proposed method uses

---

[2] http://chasen.naist.jp/hiki/ChaSen/

**Table 3.** Automatic Evaluation Results

**\* → en**

| BLEU (%) | word segmentation | | | TER (%) | word segmentation | | |
|---|---|---|---|---|---|---|---|
| | character | standard | proposed | | character | standard | proposed |
| ku | 52.21 | 55.32 | **56.43** | ku | 38.32 | 35.39 | **34.36** |
| ky | 53.16 | 55.30 | **55.69** | ky | 37.44 | 35.60 | **33.80** |
| os | 50.46 | 52.40 | **53.35** | os | 39.51 | 37.94 | **36.60** |

**\* → de**

| BLEU (%) | word segmentation | | | TER (%) | word segmentation | | |
|---|---|---|---|---|---|---|---|
| | character | standard | proposed | | character | standard | proposed |
| ku | 46.49 | 49.75 | **50.40** | ku | 45.69 | 41.40 | **41.08** |
| ky | 46.87 | 50.01 | **50.34** | ky | 45.39 | 42.43 | **40.68** |
| os | 46.81 | 49.94 | **50.25** | os | 44.67 | 41.89 | **40.25** |

**\* → ru**

| BLEU (%) | word segmentation | | | TER (%) | word segmentation | | |
|---|---|---|---|---|---|---|---|
| | character | standard | proposed | | character | standard | proposed |
| ku | 44.72 | **49.28** | 47.89 | ku | 46.37 | 43.42 | **42.63** |
| ky | 44.96 | **49.10** | 48.46 | ky | 46.79 | 43.98 | **42.49** |
| os | 43.48 | 48.23 | **48.36** | os | 48.90 | 45.14 | **43.75** |

**Table 4.** OOV Statistics

**\* →en**

| | standard | proposed |
|---|---|---|
| ku | 0.8% | 0.1% |
| ky | 1.1% | 0.1% |
| os | 1.1% | 0.1% |

**\* →de**

| | standard | proposed |
|---|---|---|
| ku | 1.3% | 0.1% |
| ky | 1.3% | 0.1% |
| os | 1.4% | 0.1% |

**\* →ru**

| | standard | proposed |
|---|---|---|
| ku | 1.6% | 0.1% |
| ky | 1.7% | 0.2% |
| os | 1.8% | 0.2% |

segmentations at all iterative levels of the bootstrap process, thus the number of unknown words can be drastically reduced.

However, the obtained gains differ considerably between the target languages[3]. There are large differences in the morphology and grammatical characteristics of these languages. For example, in all three languages, words are modified to express different grammatical categories such as tense, person, number, gender, case, etc. However, Russian features the highest degree of inflection, followed by German and then English, which is indicated by the much larger vocabulary size of Russian in Table 2. Similarly, there are large differences in their grammars where the Russian grammar is much more complex than German and English.

---

[3] All three investigated target languages belong to the same language family of *Indo-European* languages all of which are in general extensively inflected. English and German are both *Germanic languages*, where Russian belongs to the group of *Slavic languages*.

The experimental results summarized in Table 3 indicate that the translation of Japanese dialects into English is the easiest task, followed by the German and Russian translation tasks. For English and German, positive results were obtained for all investigated translation tasks and evaluation metrics. For the translation experiments into Russian, however, the proposed method could not improve the performance of the system trained on the *standard* segmentation scheme in terms of BLEU scores.

One possible explanation for this result is the higher inflectional degree and grammar complexity of Russian. During the data preparation, we currently simply tokenize the target languages sets, but do not apply any morphological analyzer. Large word form variations due to the inflectional characteristics of the language cause a larger vocabulary thus increasing the *data sparseness* problem of SMT systems, i.e., phrases that have never been seen in the training data cannot be translated. As a consequence, the more source language words cannot be translated, the lower the automatic evaluation scores are likely to be.

As a counter-measure we plan to apply morphological tools to the target language training data sets, so that word forms and inflectional attributes of the source language can be better aligned to the word forms and inflectional attributes of the target language. This will allow for better alignments and more fine-grained levels of segmentation granularity during the segmentation process of the Iterative bootstrap learning approach and therefore will hopefully improve the translation performance of the proposed system further.

## 4  Conclusions

This paper proposes a new language-independent method to segment local dialect languages in an unsupervised manner. The method exploits explicit word segmentation knowledge about the standard dialect in order to automatically learn word segmentations of the local dialect that (1) take into account meaningful standard dialect translation units and (2) integrate segmentations from different levels of granularity.

The effectiveness of the proposed method was investigated for the translation of three Japanese local dialects (Kumamoto, Kyoto, Osaka) into three Indo-European languages (English, German, Russian). The automatic evaluation results showed consistent improvements of 2.5~3.9 BLEU points and 2.9~5.1 TER points compared to a baseline system that translates characterized input. Moreover, the proposed method improved the SMT models trained on the standard dialect segmentation by 0.3~1.1 BLEU points and 0.3~1.7 TER points for the translation into English and German.

The results of the translation experiments into Russian indicate that the proposed method could be improved by applying morphological tools to the target language training data sets in order to learn word form and inflectional attribute correspondences between the source and target languages on a more fine-grained level.

## Acknowledgment

This work is partly supported by the Grant-in-Aid for Scientific Research (C) Number 19500137.

## References

1. Nerbonne, J., Heeringa, W.: Measuring Dialect Distance Phonetically. In: Proc. of the ACL SIG in Computational Phonology, Madrid, Spain, pp. 11–18 (1997)
2. Heeringa, W., Kleiweg, P., Gosskens, C., Nerbonne, J.: Evaluation of String Distance Algorithms for Dialectology. In: Proc. of the Workshop on Linguistic Distances, Sydney, Australia, pp. 51–62 (2006)
3. Scherrer, Y.: Adaptive String Distance Measures for Bilingual Dialect Lexicon Induction. In: Proc. of the ACL Student Research Workshop, Prague, Czech Republic, pp. 55–60 (2007)
4. Chitturi, R., Hansen, J.: Dialect Classification for online podcasts fusing Acoustic and Language-based Structural and Semantic Information. In: Proc. of the ACL-HLT (Companion Volume), Columbus, USA, pp. 21–24 (2008)
5. Habash, N., Rambow, O., Kiraz, G.: Morphological Analysis and Generation for Arabic Dialects. In: Proc. of the ACL Workshop on Computational Approaches to Semitic Languages, Ann Arbor, USA, pp. 17–24 (2005)
6. Chiang, D., Diab, M., Habash, N., Rainbow, O., Shareef, S.: Parsing Arabic Dialects. In: Proc. of the EACL, Trento, Italy, pp. 369–376 (2006)
7. Biadsy, F., Hirschberg, J., Habash, N.: Spoken Arabic Dialect Identification Using Phonotactic Modeling. In: Proc. of the EACL, Athens, Greek, pp. 53–61 (2009)
8. Weber, D., Mann, W.: Prospects for Computer-Assisted Dialect Adaption. American Journal of Computational Linguistics 7(3), 165–177 (1981)
9. Zhang, X., Hom, K.H.: Dialect MT: A Case Study between Cantonese and Mandarin. In: Proc. of the ACL-COLING, Montreal, Canada, pp. 1460–1464 (1998)
10. Sawaf, H.: Arabic Dialect Handling in Hybrid Machine Translation. In: Proc. of the AMTA, Denver, USA (2010)
11. Cheng, K.S., Young, G., Wong, K.F.: A study on word-based and integrat-bit Chinese text compression algorithms. American Society of Information Science 50(3), 218–228 (1999)
12. Venkataraman, A.: A statistical model for word discovery in transcribed speech. Computational Linguistics 27(3), 351–372 (2001)
13. Goldwater, S., Griffith, T., Johnson, M.: Contextual Dependencies in Unsupervised Word Segmentation. In: Proc. of the ACL, Sydney, Australia, pp. 673–680 (2006)
14. Chang, P.C., Galley, M., Manning, C.: Optimizing Chinese Word Segmentation for Machine Translation Performance. In: Proc. of the 3rd Workshop on SMT, Columbus, USA, pp. 224–232 (2008)
15. Xu, J., Gao, J., Toutanova, K., Ney, H.: Bayesian Semi-Supervised Chinese Word Segmentation for SMT. In: Proc. of the COLING, Manchester, UK, pp. 1017–1024 (2008)
16. Zhang, R., Yasuda, K., Sumita, E.: Improved Statistical Machine Translation by Multiple Chinese Word Segmentation. In: Proc. of the 3rd Workshop on SMT, Columbus, USA, pp. 216–223 (2008)
17. Dyer, C.: Using a maximum entropy model to build segmentation lattices for MT. In: Proc. of HLT, Boulder, USA, pp. 406–414 (2009)

18. Ma, Y., Way, A.: Bilingually Motivated Domain-Adapted Word Segmentation for Statistical Machine Translation. In: Proc. of the 12th EACL, Athens, Greece, pp. 549–557 (2009)
19. Berger, A., Pietra, S.D., Pietra, V.D.: A maximum entropy approach to NLP. Computational Linguistics 22(1), 39–71 (1996)
20. Pietra, S.D., Pietra, V.D., Lafferty, J.: Inducing Features of Random Fields. IEEE Transactions on Pattern Analysis and Machine Intelligence 19(4), 380–393 (1997)
21. Ratnaparkhi, A.: A Maximum Entropy Model for Part-Of-Speech Tagging. In: Proc. of the EMNLP, Pennsylvania, USA, pp. 133–142 (1996)
22. Kikui, G., Yamamoto, S., Takezawa, T., Sumita, E.: Comparative study on corpora for speech translation. IEEE Transactions on Audio, Speech and Language 14(5), 1674–1682 (2006)
23. Och, F.J., Ney, H.: A Systematic Comparison of Statistical Alignment Models. Computational Linguistics 29(1), 19–51 (2003)
24. Stolcke, A.: SRILM an extensible language modeling toolkit. In: Proc. of ICSLP, Denver, USA, pp. 901–904 (2002)
25. Finch, A., Denoual, E., Okuma, H., Paul, M., Yamamoto, H., Yasuda, K., Zhang, R., Sumita, E.: The NICT/ATR Speech Translation System. In: Proc. of the IWSLT, Trento, Italy, pp. 103–110 (2007)
26. Papineni, K., Roukos, S., Ward, T., Zhu, W.-J.: BLEU: a Method for Automatic Evaluation of Machine Translation. In: Proc. of the 40th ACL, Philadelphia, USA, pp. 311–318 (2002)
27. Snover, M., Dorr, B., Schwartz, R., Micciulla, L., Makhoul, J.: A study of translation edit rate with targeted human annotation In: Proc. of the AMTA, Cambridge and USA, pp. 223–231 (2006)

# TEP: Tehran English-Persian Parallel Corpus

Mohammad Taher Pilevar[1], Heshaam Faili[1], and Abdol Hamid Pilevar[2]

[1] Natural Language Processing Laboratory,
University of Tehran, Iran
{t.pilevar,h.faili}@ut.ac.ir
[2] Faculty of Computer Engineering,
Bu Ali Sina University, Hamedan, Iran
pilevar@basu.ac.ir

**Abstract.** Parallel corpora are one of the key resources in natural language processing. In spite of their importance in many multi-lingual applications, no large-scale English-Persian corpus has been made available so far, given the difficulties in its creation and the intensive labors required. In this paper, the construction process of Tehran English-Persian parallel corpus (TEP) using movie subtitles, together with some of the difficulties we experienced during data extraction and sentence alignment are addressed. To the best of our knowledge, TEP has been the first freely released large-scale (in order of million words) English-Persian parallel corpus.

## 1  Parallel Corpora

Text corpus is a structured electronic source of data to be analyzed for natural language processing applications. A corpus may contain texts in a single language (monolingual corpus) or in multiple languages (multilingual corpus). Corpora are the main resources in corpus linguistics to study the language as expressed in samples or real world text. Parallel corpora are specially formatted multilingual corpora whose contents are aligned side-by-side in order to be used for comparison purpose.

While there are various resources such as newswires, books and websites that can be used to construct monolingual corpora, parallel corpora need more specific types of multilingual resources which are comparatively more difficult to obtain. As a result, large-scale parallel corpora are rarely available especially for lesser studied languages like Persian.

### 1.1  Properties of Parallel Corpora

Parallel corpora possess some properties that should be taken into account in their development [1]. The first feature is the structural distance between the text pair which indicates whether the translation is literal or free. Literal and free translations are two basic skills of human translation. A literal translation (also known as word-for-word translation) is a translation that closely follows the form of source language. It is admitted in the machine translation community that the training data of literal type better suits statistical machine translation (SMT) systems at their present level of intelligence [2].

A. Gelbukh (Ed.): CICLing 2011, Part II, LNCS 6609, pp. 68–79, 2011.

The second feature is the amount of noise available in the text pair. Noise is defined as the amount of omissions or the difference in segmentations of the text pair. Another important feature of a parallel corpus is its textual size. The value of a parallel corpus usually grows with its size and with the number of languages for which translations exist. Other features include typological distance, error rate and acceptable amount of manual checking.

## 1.2 Previous Parallel Corpora for Persian

Persian (locally called Farsi) is an Indo-Iranian branch of the Indo-European languages which uses a modified Arabic script and is spoken in Iran, Afghanistan, Tajikistan, by minorities in some of the countries in the south of the Persian Gulf, and some other countries. In total, it is spoken by approximately 134 million people around the world as first or second language[1]. It is written from right to left with some letters joined as in Arabic. Persian is a highly inflective language in which a great number of different word-forms are created by the attachment of affixes. Persian is a null-subject, or pro-drop language, so personal pronouns (e.g. I, he, she) are optional.

Until the release of TEP, there were quite a few parallel corpora for Persian language which were either small in size or unavailable for research purpose. Lack of such a resource hindered research in multilingual NLP applications such as statistical machine translation for Persian language.

Shiraz corpus is a bilingual parallel tagged corpus consisting of 3000 Persian sentences with the corresponding English sentences. The corpus is collected from Hamshahri newspaper online archive and all its sentences are manually translated at CRL3 of New Mexico State University [3].

In [4], in order to train a Persian-English speech to speech translation device, the authors have collected a corpus of medical-domain parallel cross-lingual transcripts which is composed of about 300K words.

The authors in [5] present a method to create Persian-English sentence-aligned corpus by mining Wikipedia. They used Wikipedia as a comparable corpus and extracted aligned sentences from it to generate a bilingual parallel corpus. They ran the method on 1600 page pairs which yielded about 12530 sentence pairs. The resulting corpus, however, has not yet been released.

In [6], Miangah reports an attempt to constitute an English-Persian parallel corpus composed of digital texts and web documents containing little or no noise. The corpus consists of total 3.5M English and Persian words aligned at sentence level (about 100K sentences, distributed over 50,021 entries). Although the corpus seems to have been offered in ELRA's website[2], we could not obtain a copy of it for academic purpose. Upon our inquiry, the developers expressed their unwillingness to release the corpus for it being still under development.

## 1.3 Using Movie Subtitles to Construct Parallel Corpora

The first resource that usually comes under consideration for construction of parallel corpora is literary translations. They are however less common for machine

---

[1] Languages of the World, 2005.
[2] http://catalog.elra.info

translation purpose, because they do not usually adopt literal translations and therefore involve many content omissions. This non-literal type of translation does not suit the word alignment process which is an essential step in the training of statistical machine translation systems. Translated books are not only unsuitable for the purpose, but also protected by copyright. Literal translations such as Hansards are commonly used in MT community as a resource to generate parallel corpora. For European languages, the Europarl corpus has become quite a standard one. Unfortunately, there exists no similar resource for Persian language.

To acquire a fairly large parallel corpus that could provide the necessary training data for experiments on statistical machine translation, we chose to mine movie subtitles; a resource which until recently has not been utilized by NLP tasks. There are various advantages in using movie subtitles [7], such as:

- They grow daily in amount: due to high demand, the online databases of movie subtitles are one of the fastest growing multilingual resources.

- They are publicly available and can be downloaded freely from a variety of subtitle websites.

- The subtitle files contain timing information which can be exploited to significantly improve the quality of the alignment. Fig. 1 shows a small part of a movie subtitle file.

- Translated subtitles are very similar to those in the original language – contrary to many other textual resources; the translator must adhere to the transcript and cannot skip, rewrite, or reorder paragraphs.

```
145                                         151
00:22:52,800 --> 00:22:58,800              00:22:51,717 <-- 00:22:57,717
This place is totaled.                     اینجا کاملا به هم ریخته
And we didn't wreck it. we're losing our touch bro!   و ما از این قضیه نمی ترسیدیم. ما جسمون رو از دست دادیم برادر

146                                         152
00:22:59,400 --> 00:23:04,100              00:23:03,008 <-- 00:22:58,300
The important thing is that no             نکته مهم اینه که کسی آسیب ندیده
one got hurt. Except for that guy.         البته به جز اون پسره

147                                         153
00:23:04,100 --> 00:23:08,100              00:23:07,008 <-- 00:23:03,008
And, and those three... and her.           و اون سه تا...و این دختره

148                                         154
00:23:09,900 --> 00:23:14,400              00:23:13,300 <-- 00:23:08,800
I told you to take them back, and          من بهت گفتم که اونها رو برگردون ولی تو نگهشون داشتی
you kept them! Now look what they've done.   حالا نگاه کن که چه کار کردن

149                                         155
00:23:14,400 --> 00:23:19,200              00:23:18,092 <-- 00:23:13,300
Okay granted, we do have some discipline issues.   ببین مسلما ما چندتا مسئله انضباطی داریم
Eating kids is not a discipline issue.     خوردن بچه ها مسئله نظم و انضباط نیست
```

**Fig. 1.** A manually aligned part of a movie subtitle pair

There are however disadvantages to using movie subtitles as a bilingual resource:

- Movie subtitles typically contain transcriptions of spontaneous speech and daily conversations which are informal in nature, and therefore the output of a machine translation system trained on them will be biased towards spoken language.

- After investigating the translated sentences in a statistical machine translation trained on an English-Persian corpus of movie subtitles [8], we observed that the average sentence length ratio of Persian to English is about 0.7 (which is not the case in human translation). This means that this resource is not well-suited for machine translation purpose.

- Punctuations are not usually included in movie subtitles, and therefore sentence limits are not available. This is especially problematic for a language like Persian whose sentences do not begin with a capital letter or a similar distinctive feature. For movie subtitles, the alignments are usually made between individual lines in subtitle files according to the timing information. However these individual lines are sometimes neither complete sentences nor complete phrases. This in turn leads to several problems. In 3.4.1, we will discuss some more problems faced while constructing parallel corpora from movie subtitles.

- In Persian, words are spoken in many ways, and therefore written in many different forms in an informal text like movie subtitle. Unifying these forms to avoid the scarcity is to be done manually and needs great effort.

Some of these problems can be tackled by applying rule-based correction methods. Building aligned bilingual corpora from movie subtitles were first presented in [9]. They proposed a semi-automatic method which needs human operator to synchronize some of the subtitles. Tiedemann created a multilingual parallel corpus of movie subtitles using roughly 23,000 pairs of aligned subtitles covering about 2,700 movies in 29 languages [10]. He proposed an alignment approach based on time overlaps. The authors in [7] proposed a methodology based on the Gale and Church's sentence alignment algorithm, which benefits from timing information in order to obtain more accurate results.

## 2 Persian Informal/Spoken Language

In most languages, people talk differently from the way they write. The language in its spoken (oral) form is usually much more dynamic and immediate than its written form. The written form of a language usually involves a higher level of formality, whereas the spoken form is characterized by many contractions and abbreviations. In formal written texts, longer and more difficult sentences tend to be used, because people can re-read the difficult parts if they lose track. The spoken form is shorter also due to semantic augmentation by visual cues that are not available in written text.

The size of the vocabulary in use is one of the most noticeable differences between oral and written forms of discourse. Written language uses synonyms instead of repeating the same word over and over again. This is, however, not the case in oral language which usually makes use of a more limited vocabulary. The level of difficulty in pronunciation may also affect the words chosen. Oral languages tend to use words of fewer syllables.

In addition to the aforementioned general differences between spoken and written forms of a language, Persian language introduces a variety of differences which further expand this gap. In addition to many informal words not appropriate to be used in formal language, there are remarkable variations in pronunciation of words.

As a case in point, the word "nan" ("a" is pronounced as the only vowel in the word "got" in English), which means bread, is changed into "noon" ("oo" as in "cool") in spoken language. This alteration between "aa" and "oo" is quite common but has no rule; so in many words speaker is not allowed to interchange "aa" and "oo" in colloquial language. Another common case is changing the last part of verbs. For example, the verb "mi:ravad" ("i:" as in "see" & "a" as in "cat"), which means she/he is going, changes into "mi:reh" ("i:" as in "see" & "eh" as in "pen").

A subtitle file reflects exact conversions of a movie in written form. A Persian subtitle file, therefore, involves all of described features of the spoken form of Persian language.

## 3   Development of the Corpus

### 3.1   Resources

Around 21000 subtitle files were obtained from Open-subtitles[3], a free online collection of movie subtitles in many languages. It included subtitles of multiple versions of the same movie or even multiple copies of the same version created by different subtitle makers. For each movie, a subtitle pair was extracted by examining the file size and timing information of available subtitle files. These information were used to confirm that the subtitle file pair belonged to the same version of a movie. Duplicates were then removed to make the resource unique and avoid redundancy. It resulted in about 1200 subtitle pairs. Each pair comprised of two textual files (in srt format), containing subtitles of the same version of a movie in both Persian and English languages.

### 3.2   Preprocessing

The movie subtitles database is entirely made up of user uploads and due to lack of a standard checking procedure, they need to be overviewed first. This overview includes checking if movies are tagged with the correct language, or if they are encoded in the same character encoding. We will talk more about this in 3.3.

Out of available subtitle formats, we selected those formatted using two most popular formats: SubRip files (usually with extension '.srt') and microDVD subtitle files (usually with extension '.sub'). We then converted files with these formats to a standard XML format.

### 3.3   Subtitle Alignment

Subtitle alignment is essentially similar to normal sentence alignment. Movie subtitles, however have an advantage as most of alignments are 1:1 and that they carry additional information that can help alignment.

We used the method proposed in [7] which is based on the algorithm proposed by Gale and Church with a small modification in order to take full advantage of timing information in movie subtitles. This method is a dynamic programming algorithm that

---

[3] www.opensubtitles.org

tries to find a minimal cost alignment satisfying some constraints. According to [11], the recursive definition of alignment cost is calculated by the following recurrence:

$$
C(i,j) = min \begin{cases} C(\text{i,j-1}) & + & d(0, t_j; 0, 0) \\ C(\text{i-1,j}) & + & d(s_i, 0; 0, 0) \\ C(\text{i-1,j-1}) & + & d(s_i, t_j; 0, 0) \\ C(\text{i-1,j-2}) & + & d(s_i, t_j; 0, t_{j-1}) \\ C(\text{i-1,j-1}) & + & d(s_i, t_j; s_{i-1}, 0) \\ C(\text{i-2,j-2}) & + & d(s_i, t_j; s_{i-1}, t_{j-1}) \end{cases}
\tag{1}
$$

where $C(i,j)$ is the alignment cost of a sentence in one language ($s_i$, $i=1...I$) with its translation in another language ($t_j$, $j=1..J$). $d(e,f)$ is the cost of aligning $e$ with $f$. Gale and Church defined $d(e,f)$ by means of relative normalized length of sentence in characters, namely $l(e)/l(S_e)$ and $l(f)/l(S_f)$ where $l(S_e)$ and $l(S_f)$ are the total lengths of the subtitle files of the first and second language, respectively. The authors in [7] defined a new cost function that also used the timing information. The specific cost function for subtitle alignment is as follows:

$$
d(e, f) = \lambda \left( \frac{dur(e)}{dur(S_e)} - \frac{dur(f)}{dur(S_f)} \right)^2 + (1 - \lambda) \left( \frac{l(e)}{l(S_e)} - \frac{l(f)}{l(S_f)} \right)^0
\tag{2}
$$

where the duration $dur(s)$ of subtitle $s$ is defined as:

$$
dur(s) = end(s) - begin(s)
\tag{3}
$$

And $\lambda$ is a language-dependent parameter whose value can be determined using grid-search and represents the relative importance of the timing information. We used the above algorithm for aligning subtitles using which we were able to produce highly accurate alignments.

### 3.4  Problems in Corpus Building

### 3.4.1  Problems with Subtitles
As mentioned earlier in 1.3, there are some disadvantages to making parallel corpora from movie subtitles. Apart from that, we experienced various impediments in extracting the parallel content from subtitle files. Some of these issues are listed below:

1. Noise: most of the subtitle files begin or end with advertising messages, comments about the subtitle creation/translation team or similar content which do not usually match between the file pair. An example is shown in Fig. 2. These non-matching contents of subtitle file pair introduce difficulty while aligning their sentences. There is no straightforward method to tackle this noise and hence, a manual process is required to chop off these contents from subtitle files. The user needs to spend considerable amount of time to remove this kind of noise in a text editing software.

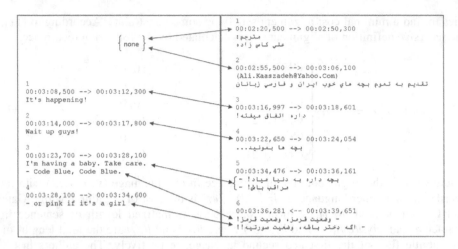

**Fig. 2.** An example for the available noise in subtitle pairs (advertisement at the beginning of Persian subtitle file)

2. The timing in subtitles is usually specified by frame numbers and not by real time, and therefore the absolute time values cannot be utilized for alignment and converting it to real time values is not always possible. Hence we use normalized subtitle duration instead [7]. This results in a significant reduction in the alignment error rate.

3. Another important problem with the subtitle files as parallel corpus construction resource is that their sentences usually do not end with full stops. As a result, the outputs of alignment phase are chunk pairs rather than sentence pairs. This is not obviously very desirable since it reduces the quality of further processing such as syntactic parsing. This is especially problematic for languages like Persian in which, the sentences do not begin with capital letters or a similar distinct notation and therefore no sentence splitting method is available.

4. Sometimes the frame rates of subtitle pairs do not match. This can however be easily sorted out using available subtitle editing software, but the process of identifying such mismatches is itself time taking.

5. Some of the subtitles are specifically intended for people who are deaf and hard-of-hearing. These are a transcription rather than a translation, and usually contain descriptions of important non-dialog audio such as "[sighs]" or "[door creaks]". These content sometimes do not match between subtitle pairs and introduce some difficulties while alignment. They are however easy to detect as they usually come within square brackets. This enables a simple regular expression to remove them completely.

6. Subtitle files do not have a standard for encoding. In order to overcome this, we converted all files to UTF-8 encoding.

7. While processing, we figured out that there are many subtitles with incorrect language tags. We used a simple Trigram Technique to detect the language of subtitle files. Therefore, each subtitle file was analyzed during preprocessing to

check whether the specified language tag correctly describes the language of that file. We also ignored those subtitles that contained multiple translations within the same file.

8. Another major drawback with the use of movie subtitles in corpus construction is that the resulting corpus cannot be easily annotated in an automatic manner. We made an effort to generate parse-trees of the sentences in TEP. However we soon realized that the spoken nature of the sentences in the corpus does not allow a reliable parsing. This was also problematic in the case of part of speech tagging. Especially in the case of Persian in which many words are different in spoken and written forms, no simple remedy exists to efficiently generate PoS tags using a PoS tagger trained on the available formal training texts.

### 3.4.2 Problem with Persian

In this section we report on some problems we ran into while developing English-Persian parallel corpus, most of which originated from specific features of Persian language. We will also discuss possible solutions to tackle some of these problems.

Persian uses code characters that are very similar to that of Arabic with some minor differences. The first difference is that the Persian alphabet adds 4 letters to that of Arabic. The other one is that the Persian employs some Arabic or ASCII characters beside the range of Unicode characters dedicated to it. Hence, the letters ک (kaf) and ی (ye) can be expressed by either the Persian Unicode encoding (U+06A9 and U+064A) or by the Arabic Unicode (U+0643 and U+06CC or U+0649) [12] and [13]. Therefore, to standardize the text, we replaced all Arabic characters with their equivalent Persian characters.

Another problematic issue while processing Persian texts is the internal word boundary in multi-token words that should be presented by a pseudo-space which is a zero-width non-joiner space. Amateur Persian typists tend to type a white-space instead of the pseudo-space in multi-token words. In such a case, a single multi-token word is broken up into multiple words which in turn introduce several problems while processing. For example words such as "می‌شود" and "پایان‌نامه" are sometimes mistakenly typed as "می شود" and "پایان نامه" which are both broken into two independently meaningful words when separated by a space. Obviously, such an issue affects statistical analysis of the text. In Persian, there exist many multi-token words, for which the insertion of pseudo-space is optional. For instance morphemes like the plural morpheme (ها), comparative suffix (تر، ترین) can be either joined or detached from words. This can result in distribution of the frequency of such words between different typing styles. However in a standard Persian corpus, these affixes are very limited in number and do not usually include ambiguities, and therefore a major part of such problems can be overcome.

As mentioned earlier in 2, there usually exist some differences in pronunciation of words between spoken (informal) and written (formal) Persian. As a case in point, the word آتش (pronounced as /ʌtæsh/) which means fire is changed into آتیش (pronounced as /ʌtɪsh/) in informal Persian. This alteration between "æ" and "ɪ" is quite common but has no rule. There are many more cases where a difference between spoken and formal Persian exists. This phenomenon is not observed in English, except for a few words like "them" which is sometimes written as "em" in spoken language. Table 1 shows examples of such words along with their frequencies in the TEP corpus.

This feature of Persian language has a negative effect on the quality of applications such as word frequency analysis or statistical machine translation when trained on a corpus of spoken language. Our effort in finding a set of rules to efficiently switch Persian words between their spoken and written forms did not result in any concrete way to merge these multiple styles into a unique form. Unlike the case of morphemes that can be automatically resolved, this multi-style issue of Persian cannot be overcome in a straightforward manner. Hence, we tried to manually transfer as many multi-style forms to a unique form as possible.

**Table 1.** Examples for Persian words having different written styles in formal and informal language (along with their frequencies in TEP corpus)

| Spoken form | Freq. | Formal form | Freq. |
|---|---|---|---|
| خوندم (khundam) | 194 | خواندم (khāndam) | 23 |
| آتیش (ātish) | 140 | آتش (ātash) | 972 |
| نمیتونم (nemitunam) | 2381 | نمی‌توانم (nemitavānam) | 56 |
| بهش (behesh) | 5674 | به او (be oo) | 2384 |

## 4  Statistics of TEP

Table 2 summarizes the statistics of the first release of TEP. Fig. 3 and 4 show the sentence length distributions of English sentences in characters and words respectively, whereas Fig. 5 and 6 illustrate that of Persian sentences. As observed in Table 2, the number of unique words in Persian side of the corpus is about 1.6 times more than that of English. This is due to the rich inflectional morphology of Persian language. We can also conclude that Persian sentences are on average constructed using fewer characters in comparison to their equivalent English sentences.

**Table 2.** Statistics of TEP

|  | English side | Persian side |
|---|---|---|
| Corpus size (in words) excluding punctuations | 3,787,486 | 3,709,406 |
| Corpus size (in characters excluding space) | 15,186,012 | 13,959,741 |
| Average sentence length (in words) | 6.829 | 6.688 |
| No. of unique words | 75,474 | 120,405 |
| Corpus size (in lines) | 554,621 | |

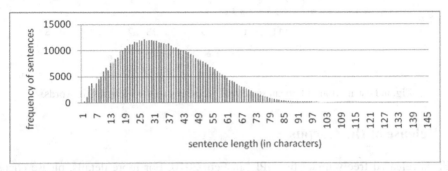

**Fig. 3.** Distribution of English sentences according to their lengths (in characters)

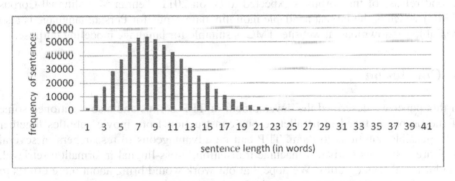

**Fig. 4.** Distribution of English sentences according to their lengths (in words)

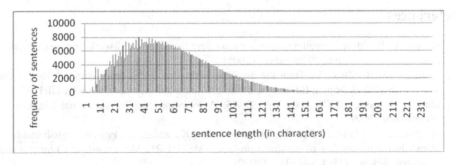

**Fig. 5.** Distribution of Persian sentences according to their lengths (in characters)

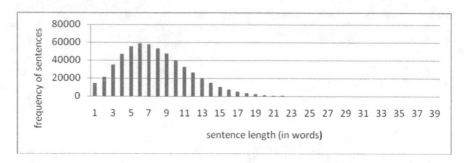

**Fig. 6.** Distribution of Persian sentences according to their lengths (in words)

## 5  Release of the Corpus

TEP is released freely under the GPL[4] in Feb. 2010. For more details, please check the website[5] of Natural Language Processing laboratory of University of Tehran. The second release of the corpus is expected to be on 2011. Tehran Monolingual Corpus (TMC) which is the largest available monolingual corpus for Persian language is also available for download at website. TMC is suitable for language modeling purpose.

## 6  Conclusion

In this paper we described the development of TEP corpus and also mentioned some of the problems faced in parallel corpus construction from movie subtitles together with possible solutions to them. TEP can be advantageous to researchers in several NLP areas such as statistical machine translation, cross-lingual information retrieval, and bilingual lexicography. We hope that our work would bring about more efforts to develop large-scale parallel corpora for Persian language.

## References

1. Rosen, A.: Building a parallel corpus for too many language. JRC workshop on exploiting parallel corpora in up to 20 languages (2005)
2. Han, X., Li, H., Zhao, T.: Train the machine with what it can learn-corpus selection for smt. In: 2nd Workshop on Building and Using Comparable Corpora: from Parallel to Non-parallel Corpora, BUCC 2009, pp. 27–33. Association for Computational Linguistics, Morristown (2009)
3. Amtrup, J.W., Mansouri Rad, H., Megerdoomian, K., Zajac, R.: Persian-english machine translation: An overview of the shiraz project. NMSU, CRL, Memoranda in Computer and Cognitive Science (MCCS-00-319) (2000)
4. Georgiou, P.G., Sethy, A., Shin, J., Narayanan, S.: An english-persian automatic speech translator: Recent developments in domain portability and user modeling. In: Proceedings of ISYC 2006, Ayia (2006)

---

[4] GNU General Public Licensing (www.gnu.org/licenses/gpl.html)
[5] http://ece.ut.ac.ir/nlp/

5. Mohammadi, M., GhasemAghaee, N.: Building bilingual parallel corpora based on wikipedia, pp. 264–268. IEEE Computer Society, Los Alamitos (2010)
6. Mosavi Miangah, T.: Constructing a large-scale english-persian parallel corpus. META 54(1), 181–188 (2009)
7. Itamar, E., Itai, A.: Using movie subtitles for creating a large-scale bilingual corpora. In: Proceedings of the Sixth International Language Resources and Evaluation (LREC 2008). European Language Resources Association (ELRA), Marrakech (2008)
8. Pilevar, M.T., Feili, H.: Persiansmt: A first attempt to english-persian statistical machine translation. In: Proceedings of 10th International Conference on statistical analysis of textual data, JADT 2009, pp. 1101–1112 (2009)
9. Mangeot, M., Giguet, E.: Multilingual aligned corpora from movie subtitles. Technical report, Condillac-LISTIC (2005)
10. Tiedemann, J.: Improved sentence alignment for movie subtitles. In: Proceedings of Int. Conf. on Recent Advances in Natural Language Processing (RANLP 2007), pp. 582–588 (2007)
11. Gale, W.A., Church, K.W.: A program for aligning sentences in bilingual corpora. In: Proceedings of the 29th Annual Meeting on Association for Computational Linguistics, ACL 1991, pp. 177–184. Association for Computational Linguistics, Morristown (1991)
12. Megerdoomian, K.: Persian computational morphology: A unification-based approach. Cognitive Science (2000)
13. Ghayoomi, M., Momtazi, S., Bijankhan, M.: A study of corpus development for persian. International Journal on Asian Language Processing 20(1), 17–33 (2010)

# Effective Use of Dependency Structure for Bilingual Lexicon Creation

Daniel Andrade[1], Takuya Matsuzaki[1], and Jun'ichi Tsujii[2,3]

[1] Department of Computer Science, University of Tokyo, Tokyo, Japan
[2] School of Computer Science, University of Manchester, Manchester, UK
[3] National Centre for Text Mining, Manchester, UK
{daniel.andrade,matuzaki,tsujii}@is.s.u-tokyo.ac.jp

**Abstract.** Existing dictionaries may be effectively enlarged by finding the translations of single words, using comparable corpora. The idea is based on the assumption that similar words have similar contexts across multiple languages. However, previous research suggests the use of a simple bag-of-words model to capture the lexical context, or assumes that sufficient context information can be captured by the successor and predecessor of the dependency tree. While the latter may be sufficient for a close language-pair, we observed that the method is insufficient if the languages differ significantly, as is the case for Japanese and English. Given a query word, our proposed method uses a statistical model to extract relevant words, which tend to co-occur in the same sentence; additionally our proposed method uses three statistical models to extract relevant predecessors, successors and siblings in the dependency tree. We then combine the information gained from the four statistical models, and compare this lexical-dependency information across English and Japanese to identify likely translation candidates. Experiments based on openly accessible comparable corpora verify that our proposed method can increase Top 1 accuracy statistically significantly by around 13 percent points to 53%, and Top 20 accuracy to 91%.

## 1   Introduction

Even for resource rich languages like Japanese and English, where there are comprehensive dictionaries already available, it is necessary to constantly update those existing dictionaries to include translations of new words. As such, it is helpful to assist human translators by automatically extracting plausible translation candidates from comparable corpora. The term comparable corpora refers to a pair of non-parallel corpora written in different languages, but which are roughly about a similar topic. The advantage of using comparable corpora is that they are abundantly available, and can also be automatically extracted from the internet, covering recent topics [1].

In this paper, we propose a new method for automatic bilingual dictionary creation, using comparable corpora. Our method focuses on the extraction and comparison of *lexical-dependency* context across unrelated languages, like Japanese

A. Gelbukh (Ed.): CICLing 2011, Part II, LNCS 6609, pp. 80–92, 2011.

and English, which, to our knowledge, has not yet been addressed. In order to take into account that the dependency structure in unrelated languages is not always comparable, we additionally include into our model a bag-of-words model on the sentence level. In total, we extract lexical-dependency context from four statistical models: three dependency models, which extract the successors, predecessors and siblings information, and a bag-of-words model on the sentence level. By combining the information appropriately in a combined probabilistic model, we are able to capture a richer context, and can show a significant improvement over previously reported methods.

In the next section, we review the most relevant previous researches, and then explain our proposed method in Section 3, followed by an empirical evaluation in Section 4. Finally, in Section 5, we will summarize our findings.

## 2 Previous Work

The basic idea behind using comparable corpora to find a new translation for a word $q$ (query), is to measure the context of $q$ and then compare the context with each possible translation candidate, using an existing dictionary [2][3]. We will call words for which we have a translation in the given dictionary, *pivot* words. First, using the source corpus, they calculate the degree of association of a query word $q$ with all pivot words $x$. The correlation is calculated using the co-occurrence frequency of $q$ and $x$ in a word-window, sentence, or document ([3], [1] and [4] respectively). In this way, they get a context vector for $q$, which contains the correlation, in each position, to a certain pivot word $x$. Using the target corpus, they then calculate a context vector for each possible translation candidate $c$, in the same way. Finally, they compare the context vector of $q$ with the context vector of each candidate $c$, and retrieve a ranked list of translation candidates.

Most of the previous work uses a bag-of-words model to count the co-occurrence frequency ([1], [2], [4] among others). Here, we summarize the work, which does not model the context as a plain bag-of-words model. In [5], instead of using all words in a word-window, only the verbs occuring in a verb-noun dependency are considered. In [6] it is suggested to use lexico-syntactic patterns, which were manually created. They first POS-tag the corpus, and then extract instances of lexico-syntactic patterns with regular expressions. Using their approach for English, we would, for example, extract the pattern instance <see, *subj*, man> from "A man sees a dog" using an appropriate regular expression. The pattern <see, *subj*, *> is then translated into the target language and used to score possible translation candidates for "man". For the experiments the researchers use related languages: Spanish and Galician, which allows them to use cognates, in addition to the bilingual dictionary. The work in [3] assumes that the word order of the source and target language is the same. They create a feature vector for each position in the word window, which are combined into one, long, feature vector. For example, if the size of the word-window is two, they create one feature vector for a pivot word $x$, which occurs before $q$, and one feature vector for $x$ occurring after $q$, and then append the two, to retrieve the final feature vector for $q$. In the actual experiments the two closely related languages: English and German, were used. The work in [7] improves the latter model, by creating a feature vector for the predecessor and successor in the dependency graph, instead of using the word order in the sentence. Their actual experiments uses the related languages: Spanish and English, and an unaligned, parallel corpus.

In summary, we can see that previous works, which use lexico-syntactic information, only applied their methods to related language pairs, and used either manually created regular expressions, or predecessor/successor information from the dependency tree. The problem is that lexico-syntactic information is more difficult to compare for an unrelated language pair, like Japanese and English. We therefore suggest the combination of the lexical-dependency information with sentence bag-of-words information. Furthermore, we claim that siblings in the dependency tree can contain relevant context information.

## 3   Proposed Method

In the first step, our proposed method extracts pivot words, which have a positive association to the query word in a sentence, and in various dependency positions. A pivot word $x$ is positively associated to the query in a certain context $C$, if the probability that we observe $x$, increases when we know that the query occurs in context $C$. Context $C$ can be a sentence, or predecessor, successor, sibling from the dependency tree. A formal definition will be provided in Section 3.2. In the same way, we also extract for all possible translation candidates, the pivot words, which are positively associated.

In the second step, we compare each translation candidate to the query word. Our proposed similarity measure uses the number of overlapping pivot words between the query and translation candidate. After having determined the number of overlapping pivots for several dependency positions, and for the sentence level, we combine this information in a probabilistic framework to retrieve a similarity score. The details of the similarity measure are described in Section 3.3.

For capturing the relevant context of a word $w$, we consider not only its successor and predecessors, but also the siblings in the dependency tree, to be important. The motivation behind this idea, is that siblings can contain, for example, adverbs, subject or objects, which are helpful clues for finding the translation of the word $w$. For example in a sentence like "The side door does not open with key", we can get "key" as a sibling of "door" (see Figure 1). This can be an important clue for the word "door": assuming we already know the translation of "key", then pivot word "key" can help us in finding the correct translation of "door".

Before we start describing our method in more detail, we will shortly discuss the necessary transformations of the dependency structures in Japanese and English in order to make a comparison for this very different language pair possible.

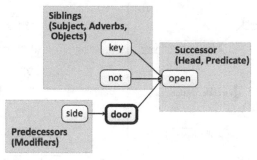

Fig. 1. Example of the dependency information used by our approach. Here, from the perspective of "door".

### 3.1 Comparing Japanese and English Dependency Structures

Dependency structures in Japanese and English cannot be compared directly with each other. The main two reasons are as follows:

- Traditionally, in Japanese, dependencies are defined between chunks of words, called bunsetsu, and not between words. A bunsetsu usually contains one or more content words, and ends with a function word. For example, this is the case for the bunsetsu: カード (card) キー (key) で (*particle*); see also Figure 2.
- Function words are difficult to compare in Japanese and in English. For example, the particle が (*ga*), which marks the subject in Japanese, cannot be related to an English word. Auxiliary verbs in English, like "can" are realized in a different position of the dependency tree in Japanese.

To make the dependency structures of English and Japanese comparable, we first introduce dependencies between words in a bunsetsu in Japanese; in the second step, we remove non-content words from the Japanese and English dependency structures.[1] An example for these transformations applied to a Japanese and to an English sentence can be seen in Figure 2. Looking at the English and Japanese sentences after transformations, we find that the relation "door" → "open" holds in Japanese as well as in English.

### 3.2 Extracting Pivots with Positive Association

In the following we extract pivot words $x$, which are positively associated with a query word $q$. The word $q$ is the source word for which we want to find an appropriate translation. First, we define the positive association between a pivot word $x$, and a word $q$, with respect to a sentence. Let $s$ be the set of words occuring in a given sentence. We define binary random variable $X_s$ and $Q_s$, in

---

[1] Working description can be found in "Supplement.pdf" available at
www.CICLing.org/2011/software/76

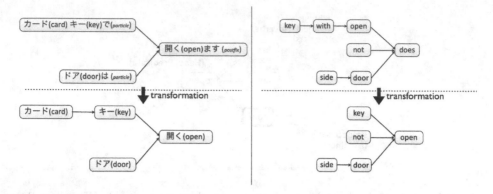

84     D. Andrade, T. Matsuzaki, and J. Tsujii

**Fig. 2.** Dependency structure transformations for the Japanese sentence ドアはカード キーで開きます (The door opens with the card key). And a similar English sentence "Side door does not open with key".

the following way: $Q_s :\Leftrightarrow q \in s$ and $X_s :\Leftrightarrow x \in s$. Our statistical model considers each sentence in the corpus as the result of a random experiment, where $Q_s$ and $X_s$ are sampled from a Bernoulli distribution. Therefore, the number of our random experiments $n$, is equal to the number of sentences. Furthermore, it follows that $Q_s$ and $X_s$ are true, if and only if word $q$ and word $x$ both occur in the same sentence. We say a pivot word $x$ is positively associated with a query word $q$ if:

$$\frac{p(X_s|Q_s)}{p(X_s)} > 1 .$$

In other words, we assume a positive association on the sentence level, if the probability of observing the pivot word $x$ in a sentence increases, when we know that the word $q$ occurs in the sentence.

Having defined the word occurrence as a Bernoulli trial, we then use the framework proposed in [4] to extract only the positive associations that are statistically significant. They suggest to define beta distributions over $p(X|Q)$ and $p(X)$. They then estimate the likelihood that the ratio of these two probabilities is larger than 1. This likelihood is then used to extract the sets of positively associated pivots. The corresponding parameters of the two beta distributions are defined as

$$\alpha'_{x|q} := f(x,q) + \alpha_{x|q} ,$$
$$\beta'_{x|q} := f(q) - f(x,q) + \beta_{x|q} , \text{ and}$$
$$\alpha'_x := f(q) + \alpha_x, \ \beta'_x = n - f(x) + \beta_x .$$

$\alpha_{x|q}, \beta_{x|q}, \alpha_x, \beta_x$ define the beta priors, which are estimated from the whole corpus, as described in [4]. Thus, we can see that the sufficient statistics are $f(x,q)$, $f(q)$, $f(x)$ and $n$. For our sentence model, $f(x,q)$ corresponds to the number of sentences in which word $x$ and $q$ occur together; $f(q)$, $f(x)$ are the number of sentences in which $q$ and $x$ occur, respectively. In the following, we will define how to calculate these sufficient statistics for each dependency model, by explicitly formulating the underlying statistical model.

The first model for capturing dependency information, is used to find the pivot words $x$, which are positively associated with $q$, in the situation where $x$ occurs as a predecessor of $q$. For a formal description of the model, we use the following definitions. First, we denote {w,v} a *dependency word pair*, if there is an edge in the dependency graph such that either $w \rightarrow v$ or $v \rightarrow w$ holds. For the pivot word $x$ and any word $v$, we define the random variable $X_{pred}$ such $X_{pred} :\Leftrightarrow x \rightarrow v$. Furthermore, for query word $q$ and any word $w$, we define $Q_{succ}$, such that $Q_{succ} :\Leftrightarrow w \rightarrow q$. We have the desired property, that if $X_{pred}$ and $Q_{succ}$ is true for a certain dependency word pair, then the pivot word $x$ is the predecessor of $q$. We can therefore define the positive association between $q$ and $x$ occurring as $q$'s predecessor, as:

$$\frac{p(X_{pred}|Q_{succ})}{p(X_{pred})} > 1.$$

In contrast to the bag-of-words model, we consider each dependency word pair {w, v}, in a sentence as a random experiment. This means that the total number of random experiments, $n$, equals the sum of the number of edges in the dependency trees of all sentences in the corpus. Formally[2]

$$n = \sum_s |\{(w_1, w_2)|w_1 \rightarrow w_2\}| = \sum_s (|s| - 1)$$

Another stochastic model, that is used to extract pivot words $x$, which are positively associated with $q$ such that $x$ occurs as a *successor* of $q$, is defined analogously.

Finally we extract positive associations between siblings in the dependency tree. Thus, we consider the number of parent nodes, which have two or more children as a random experiment. In other words, each set of children having the same parent are considered to be in a bag-of-words, where each such bag-of-words is considered to be the result of one random experiment. This modeling decision is similar to the bag-of-words modeling for a sentence, where each sentence represented a random experiment. Formally, we first divide the set of words $s$ into a set of subsets $\mathfrak{T}$ such that each subset $T$ is the set of children, which have the same parent.[3]

$$\mathfrak{T} = \{T | T \in \{\{b|b \in s, b \rightarrow a\}|a \in s\}, |T| \geq 2\}.$$

---

[2] The equality holds, because each word has exactly one out-going edge, except the root, which has no out-going edge.

[3] Note that $\mathfrak{T}$ does not cover $s$, and is therefore not a partition of $s$. However, the elements in $\mathfrak{T}$ are disjoint, since in our dependency tree, a node has, at most, one parent.

For each element $T$ in $\mathfrak{T}$, we define the random variable $X_{sib} :\Leftrightarrow x \in T$ and $Q_{sib} :\Leftrightarrow x \in T$. For example, the random variable $X_{sib}$ is true, exactly if the word $x$ occurs in a certain set of siblings $T$. And therefore $X_{sib}$ and $Q_{sib}$ are true, if and only if, they both occur in $T$. Analogously to before, we now define the positive association between $x$ and $q$, such that both are siblings as:

$$\frac{p(X_{sib}|Q_{sib})}{p(X_{sib})} > 1 \,.$$

In order to calculate the number of random experiments for this statistical model, we have to enumerate over all sets $T$, which is $|\mathfrak{T}|$, for all sentences $s$.

## 3.3   Comparing the Lexical-Dependency Information

In order to be able to compare the extracted pivot words for a query $q$ with the ones extracted for a translation candidate $c$, we translate, with the existing dictionary, each pivot word $x$ in the source language into a pivot word $x'$ in the target language. Note that for simplicity we assume a one-to-one correspondence between $x$ and $x'$; in case there is more than one translation listed in the dictionary we select the one which is closest in relative frequency (see Section 4 for details). We then use as similarity measure between $q$ and $c$ the degree of surprise that by chance the $m$ pivots, $x_1, ..., x_m$, which are positively associated with $q$, are also positively associated (p.a.) with $c$; i.e.

$$- log\, p(c \text{ p.a. with } x'_1, ..., x'_m) \tag{1}$$

In order to calculate $p(c \text{ p.a. with } x'_1, ..., x'_m)$ we use:

$$p(c \text{ p.a. with } x'_1, ..., x'_m) := p(s(c, x'_1), \ldots, s(c, x'_m)) \,, \tag{2}$$

where $s(c, x')$ means that the pivot word $x'$ is positively associated with candidate $c$ on the sentence level. For simplicity, it is assumed that having a positive association with pivot $x'_i$ is stochastically independent of having a positive association with pivot $x'_j$ ($x'_i \neq x'_j$); we can then write

$$p(c \text{ p.a. with } x'_1, ..., x'_m) = \prod_{i \in \{1, ..., m\}} p(s(c, x'_i)) \tag{3}$$

with

$$p(s(c, x'_i)) := \frac{|\{x' \in X'|s(c, x')\}|}{|X'|} \,, \tag{4}$$

where $X'$ is the set of all pivots in the target language. Note that the calculation of $p(s(c, x'_i))$ is independent of $x'_i$, since it is just the probability that *any* pivot $x'$, out of the set of all pivot words $X'$, is positively associated with candidate $c$.[4]

---

[4] That means, $p(s(c, x'_i))$ is the same for all $x'_i$, but in general different for different candidates $c$.

This definition leads to the desired bias that candidates that are positively associated with more pivots, have a higher probability of having a pivot in common with the query, by pure chance. Therefore, the resulting definition of similarity lowers the score of candidates, which occur very frequently, since words with a high frequency tend to have more associated pivots.

So far, we considered only the pivot words, which are positively associated on the sentence level. We now extend the similarity measure, by additionally considering $t_j(c, x')$, for $t_j$ being *predecessor*, *successor*, or *sibling*. However, note that some pivot words do not occur in a syntactic relationship, but nevertheless occur often in the same sentence as $q$ and $c$. We describe here the case in which a pivot word $x_j$ is at most in one syntactic relationship, associated with query word $q$ and candidate $c$. Let $x_1, ..., x_l, l \leq m$, be the pivots, which have the same syntactic relationship with $q$ and $c$. Let $x_{l+1}, ..., x_m$ be the pivots, which have either no relationship, or different syntactic relationship with $q$ and $c$, but are positively associated on the sentence level with both.[5] We can then include the dependency information into the similarity measure by extending equation (2) in the following way:

$$p(c \text{ p.a. with } x'_1, ..., x'_m) := p(s(c, x'_1), \quad , s(c, x'_m), t_1(c, x'_1), ..., l_l(c, x'_l)) \quad (5)$$

Note that we cannot simply factorize, as was done before, in equation (3), since the statistical independency clearly does *not* hold, as $t_j(c, x'_j) \Rightarrow s(c, x'_j)$.[5] However, we can group and factorize in the following way, using the same assumptions as before:

$$p(c \text{ p.a. with } x'_1, ..., x'_m) = \prod_{k \in \{l+1,...,m\}} p(s(c, x'_k)) \cdot \prod_{j \in \{1,...,l\}} p(t_j(c, x'_j), s(c, x'_j)),$$

which can be rewritten as

$$p(c \text{ p.a. with } x'_1, ..., x'_m) = \prod_{k \in \{1,...,m\}} p(s(c, x'_k)) \cdot \prod_{j \in \{1,...,l\}} p(t_j(c, x'_j)|s(c, x'_j)).$$

Thus, in the case where there is additional correspondence in the syntactic position, in which the pivot $x_j$ occurs with $q$ and $c$, we lower the probability by the factor $p(t_j(c, x'_j)|s(c, x'_j))$. As a consequence, the similarity for $c$, increases by $-log\, p(t_j(c, x'_j)|s(c, x'_j))$, due to the definition in equation (1). This probability $p(t_j(c, x'_j)|s(c, x'_j))$ could be easily estimated for each candidate $c$ individually; however, since we expect that this probability is roughly the same for all candidates, we can achieve a more reliable estimate by:

$$p(t_j(c, x'_j)|s(c, x'_j)) = \frac{\sum_{c^*} |\{x' \in X' |\, t_j(c^*, x') \wedge s(c^*, x')\}|}{\sum_{c^*} |\{x' \in X' |\, s(c^*, x')\}|},$$

which is the number of times we observe that any pivot word $x'$ is in positive association with any candidate $c$ on sentence level *and* with respect to relationship $t_j$, divided by the number of times any pivot word $x'$ is in positive association only on the sentence level.

---

[5] See also Remark1 and Remark2 in document "Supplement.pdf" available at www.CICLing.org/2011/software/76

## 4   Experiments

For our experiments, we use a collection of car complaints compiled by the Japanese Ministry of Land, Infrastructure, Transport and Tourism (MLIT)[6], and a different collection of car complaints, which is compiled by the USA National Highway Traffic Safety Administration (NHTSA)[7]. The corpora are non-parallel, but loosely comparable in terms of its content. The Japanese and English corpus contains 24090 sentences, and 47613 sentences, respectively.

The Japanese corpus was morphologically analyzed and dependency parsed using Juman and KNP[8]. The English corpus is POS-tagged and stemmed with Stepp Tagger [8] [9], and dependency parsed with the MST parser[10].

In the corpora, we consider all content words occurring more than 3 times. We used the Japanese-English dictionary JMDic[9], to automatically determine the pivots. Precisely speaking, a pivot is a pair of words consisting of one Japanese content word, and one English content word. In the case where there are several English translations in the dictionary, we take the translation that is closest in relative frequency to the Japanese word. In total 1796 pivot words were determined this way. The gold-standard is extracted from the dictionary, by using the Japanese and English noun pairs, which actually occur in the corpora. We then removed general nouns like 可能性 (possibility) to get a final list of 443 domain-specific terms (FULL).

Note that the evaluation with respect to such a gold-standard is a conservative approximation to the real performance, since several other correct or plausible translations, which are *not* listed in the dictionary, are counted as wrong translations. For the evaluation, we remove the corresponding <query, answer> pair from the pivot word list. The final gold-standard contains word pairs in the wide frequency range from 2284 to 3. Since it is known that the difficulty of finding a correct translation depends on the word's frequency [5], we also did two additional experiments, using only the high and low frequent word pairs from the gold-standard, respectively. We follow the suggestion from [5], and use the minimum frequency of the query and answer word, in order to find the 100 highest frequent word pairs (HIGH) and the 100 lowest frequent word pairs (LOW). Since most queries have only one translation, we evaluated the output of each method only by their accuracy.[10]

First of all, we compare the baseline method that uses only a bag-of-words model on the sentence level (SENTENCE), to several other methods that have been previously proposed. "Hypergeometric" refers to the method described in [4]; it uses the same method as SENTENCE to extract pivots that are positively

---

[6] http://www.mlit.go.jp/jidosha/carinf/rcl/defects.html

[7] http://www-odi.nhtsa.dot.gov/downloads/index.cfm

[8] http://www-lab25.kuee.kyoto-u.ac.jp/nl-resource/

[9] http://www.csse.monash.edu.au/~jwb/edict_doc.html

[10] To provide a more complete picture of the performance of each method, we show the accuracy at different ranks. The accuracy at rank $r$ is determined by counting how often at least one correct answer is listed in the top $r$ translation candidates suggested for a query, divided by the number of all queries in the gold-standard.

associated, but differ in the similarity method. In [4] they suggested the use of the hypergeometric distribution, whereas we suggested the use of Equation (2). A commonly used baseline is the method proposed in [2], that measures positive association with Tf-idf, and compares the context vectors using the cosine similarity ("Tf-idf"). Finally, the method proposed in [7] is identical to the method "Tf-idf", except that instead of a bag-of-words model on the sentence level, it uses the predecessor and successor information ("Tf-idf dependency").[11] The results are summarized in Figure 3. The baseline method SENTENCE improves on the method "Hypergeometric" throughout the whole gold-standard. "Tf-idf dependency" does *not* improve over "Tfi-idf", which uses the complete sentence as one bag-of-words. We note that [7] reported a small improvement of "Tf-idf dependency" over "Tfi-idf", which uses *not* a complete sentence, but a small word-window of size 4. We therefore suppose that a word-window of size 4 is too small. This is also confirmed by our next experiments.

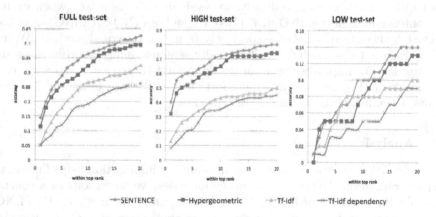

**Fig. 3.** Accuracy at different ranks for the baseline SENTENCE and several previously proposed methods

The next experiments evaluate our proposed method, that uses all dependency information (predecessors, successor and siblings) and combines the information with the sentence level information (PROPOSED). The results in Figure 4 show that our method improves over SENTENCE, that uses no dependency information.[12] We also investigated, in more detail, whether there was an improvement, and how much improvement we achieved by including the information

---

[11] Note that in the original method, they also include the predecessors of predecessor, and successor of the successor. However, since we 'hop over' function words we expect that the results are similar, since a majority of semantic relations can be reached this way. For example: verb ← preposition ← noun.

[12] The improvement of our method is also statistically significant with a p-value 0.01, using pairwise-comparison on the result for the complete gold-standard.

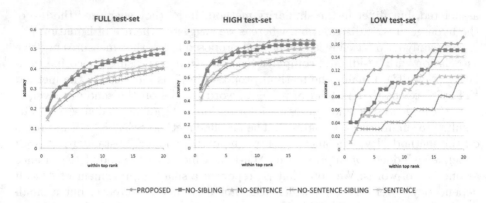

Fig. 4. Accuracy at different ranks for method PROPOSED and several baselines

from different contexts. Specifically, we evaluated the accuracy when excluding sentence information (NO-SENTENCE), siblings (NO-SIBLING), and both (NO-SENTENCE-SIBLING), respectively from the proposed method. We can see in Figure 4 that including additionally sentence information as suggested in Equation (5), improves the accuracy by around 10 percent. Furthermore, we can observe in Figure 4, a small but constant improvement of around 2 percent points, when including the sibling information.

## 4.1 Analysis

We showed that on average including dependency information helps finding the correct translation. To see why accuracy increases, we might look at a concrete example. If the query word is コンピュータ (computer), the method SENTENCE returns "wire" at first rank, and the correct translation at rank 45. However, the method PROPOSED returns the correct translation "computer" at first rank. Recall that the similarity score is based on how many overlapping pivot words we have for each context. If the context is set to a sentence, we get that the query word and "wire" have 13 pivot words in common, and is therefore preferred to "computer" which has only 6 in common. However, when looking at the dependency contexts, we find that "computer" has in total 12 pivot words in common with the query, the wrong translation "wire" has only 5. For example, "error" is a common sibling of the query and "computer", but not for "wire"; "failure" is a common successor of the query and "computer", but not for "wire".

Another insight we got is that including additional sentence information is crucial, since if we do not include it, the accuracy drops below the much simpler model, which uses only sentence information. We can actually make a more differentiated statement. When we compare the results for *HIGH* and *LOW* in Figure 4, we can see that NO-SENTENCE actually provides a higher accuracy than SENTENCE, for cases where we have enough observations (see results *HIGH*); otherwise, the reverse holds (see results *LOW*). This is explained in that we expect to capture most of the important pivot words by the three dependency groups, successor, predecessor and sibling, given that the query and answer occur often enough. However, if the frequency is too low, using dependency information does not give sufficient information any more and using sentence bag-of-words – although more noisy – is better in that case. For these reasons, it is advantageous to also incorporate sentence level information.

## 5  Summary

We proposed a new method for bilingual lexicon creation, which extracts positively associated pivots for a query and its translation candidates from the dependency tree, and from the bag-of-words in a sentence. We define four stochastic models: for the successor, predecessors, siblings, and for modeling co-occurrence on the sentence level. Since each model is a Bernoulli trial, we can use the beta distribution to extract only the pivot words, which are positively associated with statistical significance. In the second step, we estimate the probability that the query and a translation candidate have such pivot words in common in either any dependency position, or on the sentence level. This probability is then used to calculate a similarity score between the query and a translation candidate.

In comparison to previous methods, we engender a statistically significant increase in the accuracy. We also analyzed our method in more detail by excluding sentence and sibling information, and conclude that both of these two contextual information is important, and is one reason for the performance improvement, in comparison to the method "Tf-idf dependency" [7].

## Acknowledgment

This work was partially supported by Grant-in-Aid for Specially Promoted Research (MEXT, Japan). The first author is supported by the MEXT Scholarship and by an IBM PhD Scholarship Award.

## References

1. Utsuro, T., Horiuchi, T., Hino, K., Hamamoto, T., Nakayama, T.: Effect of cross-language IR in bilingual lexicon acquisition from comparable corpora. In: Proceedings of the Conference on European Chapter of the Association for Computational Linguistics. Association for Computational Linguistics, pp. 355–362 (2003)
2. Fung, P.: A statistical view on bilingual lexicon extraction: from parallel corpora to non-parallel corpora. In: Farwell, D., Gerber, L., Hovy, E. (eds.) AMTA 1998. LNCS (LNAI), vol. 1529, pp. 1–17. Springer, Heidelberg (1998)
3. Rapp, R.: Automatic identification of word translations from unrelated English and German corpora. In: Proceedings of the Annual Meeting of the Association for Computational Linguistics. Association for Computational Linguistics, pp. 519–526 (1999)

4. Andrade, D., Nasukawa, T., Tsujii, J.: Robust measurement and comparison of context similarity for finding translation pairs. In: Proceedings of the International Conference on Computational Linguistics, International Committee on Computational Linguistics, pp. 19–27 (2010)
5. Pekar, V., Mitkov, R., Blagoev, D., Mulloni, A.: Finding translations for low-frequency words in comparable corpora. Machine Translation 20, 247–266 (2006)
6. Otero, P., Campos, J.: Learning spanish-galician translation equivalents using a comparable corpus and a bilingual dictionary. In: Gelbukh, A. (ed.) CICLing 2008. LNCS, vol. 4919, pp. 423–433. Springer, Heidelberg (2008)
7. Garera, N., Callison-Burch, C., Yarowsky, D.: Improving translation lexicon induction from monolingual corpora via dependency contexts and part-of-speech equivalences. In: Proceedings of the Conference on Computational Natural Language Learning. Association for Computational Linguistics, pp. 129–137 (2009)
8. Tsuruoka, Y., Tateishi, Y., Kim, J., Ohta, T., McNaught, J., Ananiadou, S., Tsujii, J.: Developing a robust part-of-speech tagger for biomedical text. In: Bozanis, P., Houstis, E.N. (eds.) PCI 2005. LNCS, vol. 3746, pp. 382–392. Springer, Heidelberg (2005)
9. Okazaki, N., Tsuruoka, Y., Ananiadou, S., Tsujii, J.: A discriminative candidate generator for string transformations. In: Proceedings of the Conference on Empirical Methods in Natural Language Processing. Association for Computational Linguistics, pp. 447–456 (2008)
10. McDonald, R., Crammer, K., Pereira, F.: Online large-margin training of dependency parsers. In: Proceedings of the 43rd Annual Meeting on Association for Computational Linguistics. Association for Computational Linguistics, pp. 91–98 (2005)

# Online Learning via Dynamic Reranking
# for Computer Assisted Translation

Pascual Martínez-Gómez, Germán Sanchis-Trilles, and Francisco Casacuberta

Instituto Tecnológico de Informática
Universidad Politécnica de Valencia
{pmartinez,gsanchis,fcn}@dsic.upv.es

**Abstract.** New techniques for online adaptation in computer assisted translation are explored and compared to previously existing approaches. Under the online adaptation paradigm, the translation system needs to adapt itself to real-world changing scenarios, where training and tuning may only take place once, when the system is set-up for the first time. For this purpose, post-edit information, as described by a given quality measure, is used as valuable feedback within a dynamic reranking algorithm. Two possible approaches are presented and evaluated. The first one relies on the well-known perceptron algorithm, whereas the second one is a novel approach using the Ridge regression in order to compute the optimum scaling factors within a state-of-the-art SMT system. Experimental results show that such algorithms are able to improve translation quality by learning from the errors produced by the system on a sentence-by-sentence basis.

## 1  Introduction

Statistical Machine Translation (SMT) systems use mathematical models to describe the translation task and to estimate the probabilities involved in the process. [1] established the SMT grounds formulating the probability of translating a source sentence $\mathbf{x}$ into a target sentence $\hat{\mathbf{y}}$, as

$$\hat{\mathbf{y}} = \underset{\mathbf{y}}{\operatorname{argmax}} \ \Pr(\mathbf{y} \mid \mathbf{x}) \tag{1}$$

In order to capture context information, *phrase-based* (PB) models [2,3] were introduced, widely outperforming single word models [4]. PB models were employed throughout this paper. The basic idea of PB translation is to segment the source sentence $\mathbf{x}$ into *phrases* (i.e. word sequences), then to translate each source phrase $\tilde{x}_k \in \mathbf{x}$ into a target phrase $\tilde{y}_k$, and finally reorder them to compose the target sentence $\mathbf{y}$.

Recently, the direct modelling of the posterior probability $\Pr(\mathbf{y} \mid \mathbf{x})$ has been widely adopted. To this purpose, different authors [5,6] propose the use of the so-called log-linear models, where the decision rule is given by the expression

$$\hat{\mathbf{y}} = \underset{\mathbf{y}}{\operatorname{argmax}} \sum_{m=1}^{M} \lambda_m h_m(\mathbf{x}, \mathbf{y})$$

$$= \underset{\mathbf{y}}{\operatorname{argmax}} \ \boldsymbol{\lambda} \cdot \mathbf{h}(\mathbf{x}, \mathbf{y}) = \underset{\mathbf{y}}{\operatorname{argmax}} \ s(\mathbf{x}, \mathbf{y}) \tag{2}$$

where $h_m(\mathbf{x}, \mathbf{y})$ is a score function representing an important feature for the translation of $\mathbf{x}$ into $\mathbf{y}$, $M$ is the number of models (or features) and $\lambda_m$ are the weights

A. Gelbukh (Ed.): CICLing 2011, Part II, LNCS 6609, pp. 93–105, 2011.

of the log-linear combination. $s(\mathbf{x}, \mathbf{y})$ represents the score of a hypothesis $\mathbf{y}$ given an input sentence $\mathbf{x}$, and is not treated as a probability since the normalisation term has been omitted. Common feature functions $h_m(\mathbf{x}, \mathbf{y})$ include different translation models (TM), but also distortion models or even the target language model (LM). Typically, $\mathbf{h}(\cdot|\cdot)$ and $\lambda$ are estimated by means of training and development sets, respectively.

Adjusting both feature functions or log-linear weights leads to one important problem in SMT: whenever the text to be translated belongs to a different domain than the training or development corpora, translation quality diminishes significantly [4]. Hence, the *adaptation* problem is very common, where the objective is to improve systems trained on out-of-domain data by using very limited amounts of in-domain data.

Typically, the weights of the log-linear combination in Equation 2 are optimised by means of Minimum Error Rate Training (MERT) [7] in two basic steps. First, $N$ best hypotheses are extracted for each one of the sentences of a development set. Next, the optimum $\lambda$ is computed so that the best hypotheses in the *nbest* list, according to a reference translation and a given metric, are ranked higher within such *nbest* list. Then, these two steps are repeated until convergence, where no further changes in $\lambda$ are observed. However, such algorithm has an important drawback. Namely, it requires a considerable amount of time to translate the development (or adaptation) set several times, and in addition it has been shown to be quite unstable whenever the amount of adaptation data is small [8]. For these reasons, using MERT in an online environment, where adaptation data is arriving constantly, is usually not appropriate.

Adapting a system to changing tasks is specially interesting in the Computer Assisted Translation (CAT) [9] and Interactive Machine Translation (IMT) paradigms [10,11], where the collaboration of a human translator is essential to ensure high quality results. In these scenarios, the SMT system proposes a hypothesis to a human translator, who may amend the hypothesis to obtain an acceptable target sentence in a post-edition setup. The system is expected to learn dynamically from its own errors making the best use of every correction provided by the user by adapting the system *online*, i.e. without the need of an expensive complete retraining of the model parameters.

We analyse two online learning techniques to use such information to hopefully improve the quality of subsequent translations by adapting the scaling factors of the underlying log-linear model in a sentence-by-sentence basis.

In the next Section, existing online learning algorithms applied to SMT and CAT are briefly reviewed. In Section 3, common definitions and general terminology are established. In Section 4.1, we analyse how to apply the well-known perceptron algorithm in order to adapt the log-linear weights. Moreover, we propose in Section 4.2 a completely novel technique relying on the method of Ridge regression for learning the $\lambda$ of Eq. 2 discriminatively. Experiments are reported in Section 5, a short study on metric correlation is done in Section 6 and conclusions can be found in the last Section.

## 2   Related Work

In [12], an online learning application is presented for IMT, where the models involved in the translation process are incrementally updated by means of an incremental version of the Expectation-Maximisation algorithm, allowing for the inclusion of new phrase pairs into the system. The difference between such paper and the present one is that the techniques proposed here do not depend on how the translation model has been trained, since it only relies on a dynamic reranking algorithm which is applied to a *nbest* list, regardless of its origin. Furthermore, the present work deals with the problem of online learning as applied to the $\lambda$ scaling factors, not to the h features. Hence, the work in [12] and the present one can be seen as complementary.

The perceptron algorithm was used in [13] to obtain more robust estimations of $\lambda$, which is adapted in a batch setup, where the system only updates $\lambda$ when it has seen a certain amount of adaptation data. In Section 4.1, a similar algorithm is used to adapt the model parameters, although in the present work the perceptron algorithm has been applied in an online manner, i.e. in an experimental setup where new bilingual sentence pairs keep arriving and the system must update its parameters after each pair.

In [14] the authors propose the use of the Passive-Aggressive framework [15] for updating the feature functions h, combining both a memory-based MT system and a SMT system. Improvements obtained were very limited, since adapting h is a very sparse problem. For this reason, our intention is not to adapt the feature functions, but to adapt the log-linear weights $\lambda$, which is shown in [8] to be a good adaptation strategy. In [8], the authors propose the use of a Bayesian learning technique in order to adapt the scaling factors based on an adaptation set. In contrast, in the present work our purpose is to perform online adaptation, i.e. to adapt the system parameters after each new sample has been provided to the system. In this paradigm, the SMT system always proposes a target sentence to the user who accepts or amends the whole sentence. If the user post-edits the hypothesis, we obtain a reference along with the hypothesis and the online-learning module is activated.

The contributions of this paper are mainly two. First, we propose a new application of the perceptron algorithm for online learning in SMT. Second, we propose a new discriminative technique for incrementally learning the scaling factors $\lambda$, which relies on the concept of Ridge regression, and which proves to perform better than the perceptron algorithm in all analysed language pairs. Although applied here to phrase-based SMT, both strategies can be applied to rerank a *nbest* list, which implies that they do not depend on a specific training algorithm or a particular SMT system. Example data and software for reproducing the work described in this paper can be downloaded from http://www.CICLing.org/2011/software/50.

## 3  Online Learning in CAT

In general, in an online learning framework, the learning algorithm processes observations sequentially. After every input, the system makes a prediction and then receives a feedback. The information provided by this feedback can range from a simple opinion of how good the system's prediction was, to the true label of the input in completely supervised environments. The purpose of online learning algorithms is to modify its prediction mechanisms in order to improve the quality of future decisions. Specifically, in a CAT scenario, the SMT system receives a sentence in a source language and then outputs a sentence in a target language as a prediction based on its models. The user, typically a professional human translator, post-edits the system's hypothesis thus producing a reference translation $\mathbf{y}^\tau$. Such a reference can be used as a supervised feedback. Our intention is to learn from that interaction. Then, Eq. 2 is redefined as follows

$$\hat{\mathbf{y}} = \operatorname*{argmax}_{\mathbf{y}} \sum_{m=1}^{M} \lambda_m^t h_m^t(\mathbf{x}, \mathbf{y})$$

$$= \operatorname*{argmax}_{\mathbf{y}} \boldsymbol{\lambda}^t \cdot \mathbf{h}^t(\mathbf{x}, \mathbf{y}) \tag{3}$$

where both the feature functions $\mathbf{h}^t$ and the log-linear weights $\boldsymbol{\lambda}^t$ vary according to the samples $(\mathbf{x}_1, \mathbf{y}_1), \ldots, (\mathbf{x}_{t-1}, \mathbf{y}_{t-1})$ seen before time $t$. We can either apply online learning techniques to adapt $\mathbf{h}^t$, or $\boldsymbol{\lambda}^t$, or both at the same time. In this paper, however, we will only attempt to adapt $\boldsymbol{\lambda}^t$, since adapting $\mathbf{h}^t$ is a very sparse problem implying the adaptation of several million parameters, which is not easily feasible when considering an on-line, sentence-by-sentence scenario.

Let $\mathbf{y}$ be the hypothesis proposed by the system, and $\mathbf{y}^*$ the best hypothesis the system is able to produce in terms of translation quality (i.e. the most similar sentence with respect to $\mathbf{y}^\tau$). Then, our purpose is to adapt the model parameters ($\boldsymbol{\lambda}^t$ in this case) so that $\mathbf{y}^*$ is rewarded (i.e. achieves a higher score according to Eq. 2).

We define the difference in translation quality between the proposed hypothesis $\mathbf{y}$ and the best hypothesis $\mathbf{y}^*$ in terms of a given quality measure $\mu(\cdot)$:

$$l(\mathbf{y}) = |\mu(\mathbf{y}) - \mu(\mathbf{y}^*)|, \tag{4}$$

where the absolute value has been introduced in order to preserve generality, since in SMT some of the quality measures used, such as TER [16], represent an error rate (i.e. the lower the better), whereas others such as BLEU [17] measure precision (i.e. the higher the better). In addition, the difference in probability between $\mathbf{y}$ and $\mathbf{y}^*$ is proportional to $\phi(\mathbf{x})$, which is defined as

$$\phi(\mathbf{y}) = s(\mathbf{x}, \mathbf{y}^*) - s(\mathbf{x}, \mathbf{y}). \tag{5}$$

Ideally, we would like that increases or decreases in $l(\cdot)$ correspond to increases or decreases in $\phi(\cdot)$, respectively: if a candidate hypothesis $\mathbf{y}$ has a translation quality $\mu(\mathbf{y})$ which is very similar to the translation quality provided by $\mu(\mathbf{y}^*)$, we would like that such fact is reflected in the translation score $s$, i.e. $s(\mathbf{x}, \mathbf{y})$ is very similar to $s(\mathbf{x}, \mathbf{y}^*)$. Hence, the purpose of our online procedure should be to promote such correspondence after processing sample $t$.

A coarse-grained technique for tackling with the online learning problem in SMT implies adapting the log-linear weights $\boldsymbol{\lambda}$. The aim is to compute the optimum weight vector $\hat{\boldsymbol{\lambda}}_t$ for translating the sentence pair observed at time $t$ and then update $\boldsymbol{\lambda}$ as:

$$\boldsymbol{\lambda} = \boldsymbol{\lambda}_{t-1} + \alpha \hat{\boldsymbol{\lambda}}_t \tag{6}$$

for a certain learning rate $\alpha$.

The information that is usually taken into account to compute $\hat{\boldsymbol{\lambda}}_t$ is more general and imprecise than the information used when adapting feature functions, but the variation in the score of Eq. 2 can be higher since we will be modifying the scaling factors of the log-linear model. That is, when adapting the system to a different domain, we are going to adjust the importance of every single model to a new task in an online manner.

# 4  Online Learning Algorithms

## 4.1  Perceptron in CAT

The perceptron algorithm [18,19] is an error driven algorithm that estimates the weights of a linear combination of features by comparing the output $\mathbf{y}$ of the system with respect to the true label $\mathbf{y}^\tau$ of the corresponding input $\mathbf{x}$. It iterates through the set of samples a certain number of times (epochs), or until a desired convergence is achieved.

The implementation in this work follows the proposed application of a perceptron-like algorithm in [13]. However, for comparison reasons in our CAT framework, the perceptron algorithm will not visit a sample again after being processed once.

Using feature vector $\mathbf{h}(\mathbf{x},\mathbf{y})$ of the system's hypothesis $\mathbf{y}$ and feature vector $\mathbf{h}(\mathbf{x},\mathbf{y}^*)$ of the best hypothesis $\mathbf{y}^*$ from the $nbest(\mathbf{x})$ list, the update term is computed as follows:

$$\lambda_t = \lambda_{t-1} + \epsilon \cdot \text{sign}(\mathbf{h}(\mathbf{x},\mathbf{y}^*) - \mathbf{h}(\mathbf{x},\mathbf{y})) \tag{7}$$

where $\epsilon$ can be interpreted as the learning rate.

## 4.2  Discriminative Regression

The problem of finding $\hat{\lambda}_t$ such that higher values in $s(\mathbf{x},\mathbf{y})$ correspond to improvements in the translation quality $\mu(\mathbf{y})$ as described in Section 3 can be viewed as finding $\hat{\lambda}_t$ such that differences in scores $\phi(y)$ of two hypotheses approximate their difference in translation quality $l(y)$. So as to formalise this idea, let us first define some matrices.

Let $nbest(\mathbf{x})$ be the list of $N$ best hypotheses computed by our TM for sentence $\mathbf{x}$. Then, a matrix $H_\mathbf{x}$ of size $N \times M$, where $M$ is the number of features in Eq. 2, containing the feature functions $\mathbf{h}$ of every hypothesis can be defined such that

$$\mathbf{s_x} = H_\mathbf{x} \cdot \lambda_t \tag{8}$$

where $\mathbf{s_x}$ is a column vector of $N$ entries with the log-linear score combination of every hypothesis in the $nbest(\mathbf{x})$ list. Additionally, let $H_\mathbf{x}^*$ be a matrix with $N$ rows such that

$$H_\mathbf{x}^* = \begin{bmatrix} \mathbf{h}(\mathbf{x},\mathbf{y}^*) \\ \vdots \\ \mathbf{h}(\mathbf{x},\mathbf{y}^*) \end{bmatrix}, \tag{9}$$

and $R_\mathbf{x}$ the difference between $H_\mathbf{x}^*$ and $H_\mathbf{x}$:

$$R_\mathbf{x} = H_\mathbf{x}^* - H_\mathbf{x} \tag{10}$$

The key idea for scaling factor adaptation is to find a vector $\hat{\lambda}$ such that differences in scores are reflected as differences in the quality of the hypotheses. That is,

$$R_\mathbf{x} \cdot \hat{\lambda}_t \propto l_\mathbf{x} \tag{11}$$

where $l_x$ is a column vector of $N$ rows such that $l_x = [l(y_1) \dots l(y_n) \dots l(y_N)]'$, $\forall y_n \in nbest(x)$. The objective is to find $\hat{\lambda}$ such that

$$\hat{\lambda} = \operatorname*{argmin}_{\lambda} |R_x \cdot \lambda - l_x| \tag{12}$$

$$= \operatorname*{argmin}_{\lambda} ||R_x \cdot \lambda - l_x||^2 \tag{13}$$

where $|| \cdot ||^2$ is the Euclidean norm. Although Eqs. 12 and 13 are equivalent (i.e. the $\lambda$ that minimises the first one also minimises the second one), Eq. 13 allows for a direct implementation thanks to the Ridge regression[1], such that $\hat{\lambda}$ can be computed as the solution of the overdetermined system $R_x \cdot \hat{\lambda} = l_x$, given by the expression

$$\hat{\lambda} = (R'_x \cdot R_x + \beta I)^{-1} R'_x \cdot l_x \tag{14}$$

where a small $\beta$ is used as a regularisation term to ensure $R'_x \cdot R_x$ has an inverse.

# 5   Experiments

## 5.1   Experimental Setup

Given that a true CAT scenario is very expensive for experimentation purposes, since it requires a human translator to correct every hypothesis, in this paper we will be simulating such scenario by using the reference present in the test set. However, such reference will be fed one at a time, given that this would be the case in an online CAT process.

Translation quality will be assessed by means of the BLEU [17] and TER [16] scores. BLEU measures $n$-gram precision with a penalty for sentences that are too short, whereas TER is an error metric that computes the minimum number of edits required to modify the system hypotheses so that they match the references. Possible edits include insertion, deletion, substitution of single words and shifts of word sequences. For computing $y^*$ as described in Section 3, either BLEU or TER will be used, depending on the evaluation measure reported (i.e. when reporting TER, TER will be used for computing $y^*$). However, it must be noted that BLEU is not well defined at the sentence level, since it implements a geometrical average of $n$-grams which is zero whenever there is no common 4-gram between reference and hypothesis, even if the reference has only three words. Hence, $y^*$ is not always well defined when considering BLEU. Such samples will not be considered within the online procedure. Another consideration is that BLEU and TER might not be correlated, i.e. improvements in TER do not necessarily mean improvements in BLEU. This is analysed more in detail in Section 6.

As baseline system, we trained a SMT system on the Europarl training data, in the partition established in the Workshop on SMT of the NAACL 2009[2]. Specifically, we will train our initial SMT system by using the training and development data provided that year. The Europarl corpus [20] is built from the transcription of European Parliament speeches published on the web. Statistics are provided in Table 1.

---

[1] Also known as Tikhonov regularisation in statistics.
[2] http://www.statmt.org/wmt10/

**Table 1.** Characteristics of Europarl corpus. Dev. stands for Development, OoV for "Out of Vocabulary" words, K for thousands of elements and M for millions of elements.

|  |  | Es | En | Fr | En | De | En |
|---|---|---|---|---|---|---|---|
| Training | Sentences | 1.3M | | 1.2M | | 1.3M | |
| | Running words | 27.5M | 26.6M | 28.2M | 25.6M | 24.9M | 26.2M |
| | Vocabulary | 125.8K | 82.6K | 101.3K | 81.0K | 264.9K | 82.4K |
| Development | Sentences | 2000 | | 2000 | | 2000 | |
| | Running words | 60.6K | 58.7K | 67.3K | 48.7K | 55.1K | 58.7K |
| | OoV. words | 164 | 99 | 99 | 104 | 348 | 103 |

**Table 2.** Characteristics of NC test sets. OoV stands for "Out of Vocabulary" words w.r.t. the Europarl training set. Data statistics were again collected after tokenizing and lowercasing.

|  |  | Es | En | Fr | En | De | En |
|---|---|---|---|---|---|---|---|
| Test 08 | Sentences | 2051 | | 2051 | | 2051 | |
| | Running words | 52.6K | 49.9K | 55.4K | 49.9K | 55.4K | 49.9K |
| | OoV. words | 1029 | 958 | 998 | 963 | 2016 | 965 |
| Test 09 | Sentences | 2525 | | 2051 | | 2051 | |
| | Running words | 68.1K | 65.6K | 72.7K | 65.6K | 62.7K | 65.6K |
| | OoV. words | 1358 | 1229 | 1449 | 1247 | 2410 | 1247 |

The open-source MT toolkit Moses[3] [21] was used in its default setup, and the 14 weights of the log-linear combination were estimated using MERT [22] on the Europarl development set. Additionally, a 5-gram LM with interpolation and Kneser-Ney smoothing [23] was estimated using the SRILM [24] toolkit.

To test the adaptation performance of different online learning strategies, we also considered the use of two News Commentary (NC) test sets, from the 2008 and 2009 ACL shared tasks on SMT. Statistics of these test sets can be seen in Table 2.

Experiments were performed on the English–Spanish, English–German and English–French language pairs, in both directions and for NC test sets of 2008 and 2009. However, for space reasons, we only report results for the 2009 test set from the English–*foreign* pair, since this year's SMT shared task of the ACL focused on translating from English into other languages. Nevertheless, the results presented here were found to be coherent in all the experiments conducted, unless stated otherwise.

As for the different parameters adjustable in the algorithms described in Section 4, they were all set according to preliminary investigation as follows:

– Section 4.1: $\epsilon = 0.001$
– Section 4.2: $\alpha = 0.005$, $\beta = 0.01$

For Section 4.1, instead of using the true best hypothesis, the best hypothesis within a given $nbest(\mathbf{x})$ list was selected.

## 5.2   Experimental Results

The result of applying the different online learning algorithms described in Section 4 can be seen in Fig. 1. `percep.` stands for the technique described in Section 4.1,

---

[3] Available from http://www.statmt.org/moses/

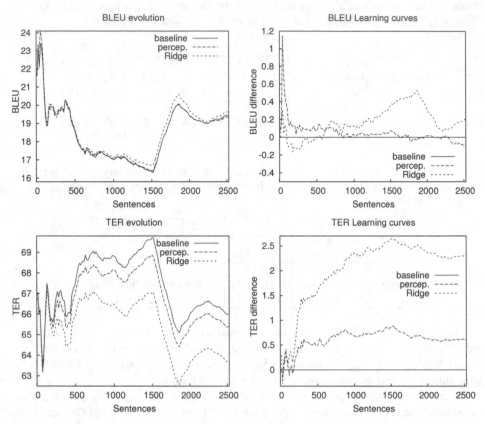

**Fig. 1.** BLEU/TER evolution and learning curves for English→French translation, considering all 2525 sentences within the NC 2009 test set. For clarity, only 1 every 15 points has been drawn. `percep.` stands for perceptron and `Ridge` for the technique described in Section 4.2.

and `Ridge` for the one described in Section 4.2. In the plots shown in this figure, the translation pair was English→French, the test set was the NC 2009 test set, and the size of the considered *nbest* list was 1000. The two plots on the left display the BLEU and TER scores averaged up to the considered $t$-th sentence. The reason for plotting the average BLEU/TER is that plotting individual sentence BLEU and TER scores would result in a very chaotic, unreadable plot given that differences in translation quality between two single sentences may be very big; in fact, such chaotic behaviour can still be seen in the first 100 sentences. The two plots on the right display the difference in translation quality between the two online learning techniques and the baseline.

The analysed online learning procedures perform better in terms of TER than in terms of BLEU (Fig. 1). Again, since BLEU is not well defined at the sentence level, learning methods that depend on BLEU being computed at the sentence level may be severely penalised. Although it appears that the learning curves peak at about 1500 sentences, this finding is not coherent throughout all experiments carried out, since such peak ranges from 300 to 2000 in other cases. This means that the particular shape of the

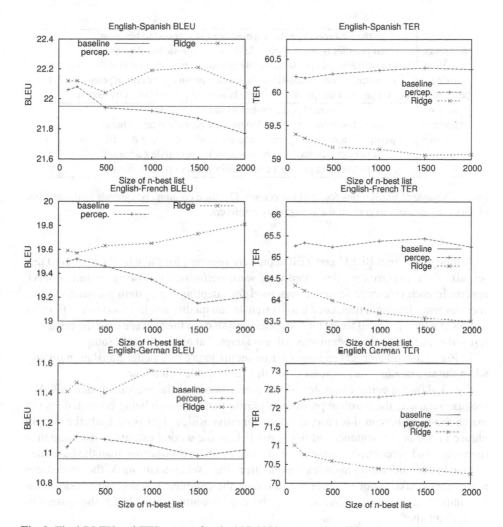

**Fig. 2.** Final BLEU and TER scores for the NC 2009 test set, for all language pairs considered. `percep.` stands for perceptron and `Ridge` for the technique described in Section 4.2.

learning curves depends strongly on the chosen test set, and that the information that can be extracted is only whether or not the implemented algorithms provide improvements.

In addition, it can be seen that the best performing method, both in terms of TER and in terms of BLEU, is the one described in Section 4.2. However, in order to assess these differences, further experiments were conducted. Furthermore, and in order to evidence the final improvement in translation quality that can be obtained after seeing a complete test set, the final translation quality obtained with varying sizes of $nbest(\mathbf{x})$ was measured. The results of such experiments can be seen in Fig. 2. Although the differences are sometimes scarce, they were found to be coherent in all the considered cases, i.e. for all language pairs, all translation directions, and all test sets.

| source | in the first round , half of the amount is planned to be spent . | |
|---|---|---|
| reference | au premier tour , la moitié de cette somme va être dépensée . | |
| baseline | dans la première phase , la moitié de la somme prévue pour être dépensé . | 8 |
| ridge | au premier tour , la moitié de la somme prévue pour être dépensé . | 4 |
| perceptron | dans un premier temps , la moitié de la somme prévue pour être dépensé . | 7 |
| source | it enables the probes to save a lot of fuel . | |
| reference | ainsi , les sondes peuvent économiser beaucoup de carburant . | |
| baseline | il permet à la probes de sauver une quantité importante de carburant . | 10 |
| ridge | il permet aux probes à économiser beaucoup de carburant . | 5 |
| perceptron | il permet à la probes à économiser beaucoup de carburant . | 6 |

**Fig. 3.** Example of translations found in the corpus. The third column corresponds to the number of necessary editions to convert the string into the reference.

Although the final BLEU and TER scores are reported for the whole considered test set, all of the experiments described here were performed following an online CAT approach: each reference sentence was used for adapting the system parameters after such sentence has been translated and its translation quality has been assessed. For this reason, the final reported translation score corresponds to the average over the complete test set, even though the system was still not adapted at all for the first samples.

In Fig. 2 it can be observed how Ridge seems to provide better translation quality when the size of $nbest(\mathbf{x})$ increases, which is a desirable behaviour.

Fig. 3 shows specific examples of the performance of the presented methods. For the first sentence, the baseline produces a phrase that, although being correct, does not match the reference; in this case, the discriminative Ridge regression finds the correct phrase in one of the sentences of the $nbest$ list. In the second example, discriminative regression and perceptron are able to find more accurate translations than the baseline.

One last consideration involves computation time. When adapting $\lambda$, the procedures implemented take about 100 seconds to rerank the complete test set. We consider this fact important, since in a CAT scenario the user is waiting actively for the system to produce a hypothesis.

## 6   Metric Correlation

From the experiments, it was observed that online learning strategies that optimise a certain quality measure do not necessarily optimise other measures.

To analyse such statement, 100.000 weight vectors $\lambda$ of the log-linear model were randomly generated and a static rerank of a fixed $nbest(\mathbf{x})$ list of hypotheses was performed for every sentence in a test set. For every weight vector configuration, BLEU (B) and TER (T) were evaluated for the test set of NC 2008 Spanish $\rightarrow$ English.

The correlation as defined by the covariance (cov) divided by the product of standard deviations ($\sigma$)

$$\rho_{B,T} = \frac{\mathrm{cov}(B,T)}{\sigma_B \sigma_T} \tag{15}$$

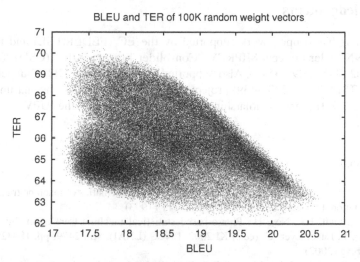

**Fig. 4.** Correlation between BLEU and TER of 100.000 configurations of $\lambda$. A slightly negative correlation can be appreciated, although not strong.

returned a value $\rho_{B,T} = -0.23798$. This result suggests that even if such correlation is not specially strong, one can expect to optimise TER (as an error metric) only to certain extent when optimising BLEU (as a precision metric), and vice-versa. A plot of the translation quality yielded by the random weight configurations is presented in Fig.4. It can be observed that it is relatively frequent to obtain low BLEU scores after optimising TER (high density area in the bottom left part of the graph). On the other hand, if BLEU is optimised, TER scores are reasonably good (right side of the plot).

## 7  Conclusions and Future Work

In this paper, two different online learning algorithms have been applied to SMT. The first one is a well-known algorithm, namely the perceptron algorithm, whereas the second one is completely novel and relies on the concept of discriminative regression. Both of these strategies have been applied to adapt the log-linear weights of a state-of-the-art SMT system, providing interesting improvements.

From the experiments conducted, it emerges that discriminative regression, as implemented for SMT in this paper, provides a larger gain than the perceptron algorithm, and is able to provide improvements from the very beginning and in a consistent manner, in all language pairs analysed.

Although BLEU is probably the most popular quality measure used in MT, it has been shown that its use within online, sentence-by-sentence learning strategies may not be very adequate. In order to cope with the discrepancies between optimising BLEU and TER, we plan to analyse the effect of combining both measures, and also consider other measures such as NIST which are also well defined at the sentence level.

We plan to analyse the application of these learning algorithms to feature functions to study how the behaviour of such techniques evolves in much sparser problems.

## Acknowledgements

This paper is based upon work supported by the EC (FEDER/FSE) and the Spanish MICINN under projects MIPRCV "Consolider Ingenio 2010" (CSD2007-00018) and iTrans2 (TIN2009-14511). Also supported by the Spanish MITyC under the erudito.com (TSI-020110-2009-439) project, by the Generalitat Valenciana under grant Prometeo/2009/014 and scholarship GV/2010/067 and by the UPV under grant 20091027.

## References

1. Brown, P., Pietra, S.D., Pietra, V.D., Mercer, R.: The mathematics of machine translation. In: Computational Linguistics, vol. 19, pp. 263–311 (1993)
2. Zens, R., Och, F.J., Ney, H.: Phrase-based statistical machine translation. In: Jarke, M., Koehler, J., Lakemeyer, G. (eds.) KI 2002. LNCS (LNAI), vol. 2479, pp. 18–32. Springer, Heidelberg (2002)
3. Koehn, P., Och, F.J., Marcu, D.: Statistical phrase-based translation. In: Proc. HLT/NAACL 2003, pp. 48–54 (2003)
4. Callison-Burch, C., Fordyce, C., Koehn, P., Monz, C., Schroeder, J.: (meta-) evaluation of machine translation. In: Proc. of the Workshop on SMT. ACL, pp. 136–158 (2007)
5. Papineni, K., Roukos, S., Ward, T.: Maximum likelihood and discriminative training of direct translation models. In: Proc. of ICASSP 1988, pp. 189–192 (1998)
6. Och, F., Ney, H.: Discriminative training and maximum entropy models for statistical machine translation. In: Proc. of the ACL 2002, pp. 295–302 (2002)
7. Och, F., Zens, R., Ney, H.: Efficient search for interactive statistical machine translation. In: Proc. of EACL 2003, pp. 387–393 (2003)
8. Sanchis-Trilles, G., Casacuberta, F.: Log-linear weight optimisation via bayesian adaptation in statistical machine translation. In: Proceedings of COLING 2010, Beijing, China (2010)
9. Callison-Burch, C., Bannard, C., Schroeder, J.: Improving statistical translation through editing. In: Proc. of 9th EAMT Workshop Broadening Horizons of Machine Translation and its Applications, Malta (2004)
10. Barrachina, S., et al.: Statistical approaches to computer-assisted translation. Computational Linguistics 35, 3–28 (2009)
11. Casacuberta, F., et al.: Human interaction for high quality machine translation. Communications of the ACM 52, 135–138 (2009)
12. Ortiz-Martínez, D., García-Varea, I., Casacuberta, F.: Online learning for interactive statistical machine translation. In: Proceedings of NAACL HLT, Los Angeles (2010)
13. España-Bonet, C., Màrquez, L.: Robust estimation of feature weights in statistical machine translation. In: 14th Annual Conference of the EAMT (2010)
14. Reverberi, G., Szedmak, S., Cesa-Bianchi, N., et al.: Deliverable of package 4: Online learning algorithms for computer-assisted translation (2008)
15. Crammer, K., Dekel, O., Keshet, J., Shalev-Shwartz, S., Singer, Y.: Online passive-aggressive algorithms. Journal of Machine Learning Research 7, 551–585 (2006)
16. Snover, M., Dorr, B., Schwartz, R., Micciulla, L., Makhoul, J.: A study of translation edit rate with targeted human annotation. In: Proc. of AMTA, Cambridge, MA, USA (2006)
17. Papineni, K., Roukos, S., Ward, T., Zhu, W.: Bleu: A method for automatic evaluation of machine translation. In: Proc. of ACL 2002 (2002)
18. Rosenblatt, F.: The perceptron: A probabilistic model for information storage and organization in the brain. Psychological Review 65, 386–408 (1958)

19. Collins, M.: Discriminative training methods for hidden markov models: Theory and experiments with perceptron algorithms. In: EMNLP 2002, Philadelphia, PA, USA, pp. 1–8 (2002)
20. Koehn, P.: Europarl: A parallel corpus for statistical machine translation. In: Proc. of the MT Summit X, pp. 79–86 (2005)
21. Koehn, P., et al.: Moses: Open source toolkit for statistical machine translation. In: Proc. of the ACL Demo and Poster Sessions, Prague, Czech Republic, pp. 177–180 (2007)
22. Och, F.: Minimum error rate training for statistical machine translation. In: Proc. of ACL 2003, pp. 160–167 (2003)
23. Kneser, R., Ney, H.: Improved backing-off for m-gram language modeling. In: IEEE Int. Conf. on Acoustics, Speech and Signal Processing II, pp. 181–184 (1995)
24. Stolcke, A.: SRILM – an extensible language modeling toolkit. In: Proc. of ICSLP 2002, pp. 901–904 (2002)

# Learning Relation Extraction Grammars with Minimal Human Intervention: Strategy, Results, Insights and Plans

Hans Uszkoreit

German Research Center for Artificial Intelligence (DFKI) and Saarland University
Stuhlsatzenhausweg 3, 66123 Saarbrücken, Germany
uszkoreit@dfki.de

**Abstract.** The paper describes the operation and evolution of a linguistically oriented framework for the minimally supervised learning of relation extraction grammars from textual data. Cornerstones of the approach are the acquisition of extraction rules from parsing results, the utilization of closed-world semantic seeds and a filtering of rules and instances by confidence estimation. By a systematic walk through the major challenges for this approach the obtained results and insights are summarized. Open problems are addressed and strategies for solving these are outlined.

**Keywords:** relation extraction, information extraction, minimally supervised learning, bootstrapping approaches to IE.

## 1 Introduction

While we still cannot build software systems that translate all or most sentences of a human language into some representation of their meanings, we are currently investigating methods for extracting relevant information from large volumes of texts. Some of the scientists working on information extraction view these methods as feasible substitutes for real text understanding, others see them as systematic steps toward a more comprehensive semantic interpretation. All agree on the commercial viability of effective information extraction applications, systems that detect references to interesting entities and to relevant relations between them, such as complex connections, properties, events and opinions.

One of the most intriguing but at the same time most challenging approaches to information extraction is the bootstrapping paradigm that starts from a very small set of semantic examples, called the *seed,* for discovering patterns or rules, which in turn are employed for finding additional instances of the targeted information type. These new instances will then be used as examples for the next round of finding linguistic patterns and the game repeats until no more instances can be detected. Since the seed can be rather small, containing between one and a handful of examples, this training scheme is usually called *minimally supervised learning*.

A. Gelbukh (Ed.): CICLing 2011, Part II, LNCS 6609, pp. 106–126, 2011.

The first information extraction systems based on such an approach were quite impressive; with a tiny set of prepared sample data thousands of new instances could be automatically extracted from texts. But at the same time they were also limited because they only extracted binary relations and did not exploit any tools for linguistic analysis [1,2].

In subsequent developments, the method was enriched by more sophisticated linguistic analysis techniques. But when it was first applied to benchmark data for language technology research, such as the widely known MUC management succession data, the results could be called encouraging at best. As we learned from a careful analysis of these results, the deficiencies of the proposed systems could not be fully explained by the remaining lack of linguistic sophistication but was in large parts also due to the nature of the data [3,4].

As we could demonstrate, a truly successful application of the paradigm requires learning data that exhibit certain graph-theoretical properties.

We built a system based on the initial architecture of [1,2], that incorporates much more linguistic processing than its predecessors. When we applied this system to data showing the needed types and degrees of redundancy, we could obtain much better results than all earlier bootstrapping systems for relation extraction [5]. However, intellectual error analysis disclosed that we were still not able to exploit the full potential of the learning data. The errors were partially caused by shortcomings of the linguistic analysis and partially by problems inherent to the approach. Thus we decided to further analyze the problems and then search for appropriate remedies.

From the gained insights we derived a medium-term research strategy. Parts of this strategy are chains of additional experiments with fine-grained diagnostics. The results of these experiments determine the evolution of the approach and shape our constantly changing complex research plan.

By taking advantage of the creative freedom associated with the special text sort of an invited keynote paper I will pursue four goals in parallel:

– provide an overview of the approach and its evolution,
– explain and illustrate the adopted research strategy,
– summarize the obtained results and their relevance for the strategy and
– outline planned steps and open problems.

This makes this paper quite different from our other publications, which concentrate on specific issues and solutions, i.e. on individual steps in the evolution of the general approach. The next section will provide an overview of our approach. The core of the paper is *Section Three* where I will follow a problem-driven approach for summarizing challenges, solutions, the evolution of the approach, intermediate results and open issues. In *Section Four* I will try to draw some conclusions from the experience with the general paradigm and our evolving approach and then close with a brief outlook on future research.

# 2   Approach

Relation extraction grammars or sophisticated relation detection classifiers map certain patterns in the linguistic form, the text, to some partial and lean representations of meaning, usually tuples of entities denoting a relation instance or the equivalent tagging of these tuples in the text. After extensive experience with the shortcomings of manual development of large relation extraction grammars such as high costs and coverage limitations, the hopes of the field have increasingly focused on automatic learning. The most widespread paradigm for the automatic training of RE systems is supervised learning, i.e. the learning from data that have been annotated in just the way the system is supposed to interpret data automatically after the acquisition process [6]. This method has led to impressive results. However the annotation is costly and often requires true specialists. Often inter-annotator agreement is rather low. Existing unsupervised methods, on the other hand, are not suited for most RE tasks, since the data only represent the form and give no indication of the targeted meaning types and representations.

The minimally supervised bootstrapping approach offers a way of reducing the human preparation to a minimum. With a few examples but without any annotation of the data, the system can learn entire RE grammars. Some minimally supervised systems start with rules or patterns, others with semantic examples. Our architecture permits different types of seeds. In order to better understand the operation of the system and some central insight gained, assume that we start with a single semantic instance $e_1$ as seed. From the seed $e_1$ we use IR methods to find candidate mentions $m_i$, i.e. sentences containing the named entities in $e_1$. From these mentions we derive our first rules $r_j$. This concludes our first cycle. In the second cycle, we use the rules to find more mentions and from these new mentions we extract new instances. This was our first double cycle.

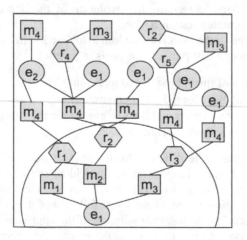

**Fig. 1.** The first learning cycle

This learning can only work if some instances are reported by several patterns and if some patterns are used for more than one instance. Actually, for any relation and any data set, there exists a hidden bipartite graph linking instances and patterns used for describing them in the data. (It does not matter whether we define this graph as directed or undirected.) Figure 1b shows such a graph. Since the graph in this example is not connected, we cannot reach all events from a single seed. If we work with one seed instance, we can at most find eight additional instances, but in some cases only two.

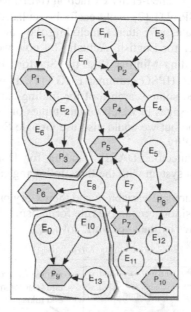

**Fig. 2.** A bipartite learning graph

In the bipartite graph we have two degree distributions: 1) events per patterns and 2) patterns per event. We hypothesize that the choice of the linguistic pattern in a mentioning does not strongly depend on the event instance but rather on the author or the publication media. Following a conjecture by Ken Church, we claim that the frequency distribution of patterns follows a Zipf-type skewed heavy-tailed distribution. This is in principle confirmed by our data.

The distribution of patterns per event depends on the number of mentions per event. This distribution differs depending on the domain and text base. In our news report data sets in prize domain, particularly in the Nobel Prize domain, we find a skewed distribution. In the MUC 6 management succession data, however, we find a boring distribution: nearly all events are just mentioned a single time, since the data all stem from the New York Times.

If both distributions are skewed, as in the prize domain, we get a scale-free graph, i.e., $P(k) \sim k - \gamma$. In this case, the graph displays the so-called small world property. We can reach nearly all events from any seed in a few steps, even if the graph grows. (In the talk, we will illustrate scale-free graphs in the world and in IE by examples.) The reason is simple: in order to learn the numerous less frequent patterns in the heavy tail of the distribution, we need "event hubs". But we need the less frequent patterns in order to get to many events mentioned only once. This insight concerning the importance on certain data properties needs to be taken into account for the design of an appropriate research strategy.

Next I would like to introduce our concrete architecture for learning extraction rules. Prerequisites are tools that can identify the relevant entity types and some method for finding and generalizing over the relevant patterns. For the first prerequisite we employ an existing named-entity extraction (NEE) tool that annotates the text by NE tags. Our system SProUT [7] could easily be replaced by other NEE tools. As a tool for finding candidates of patterns including the named entities, we utilize a widely-used generic parser for English. Actually by now we have tested the system with several parsers including MiniPar [8], the Stanford Parser [9], and the PET HPSG parser [10] with the HPSG grammar ERG [11]. If we work with larger volumes of data we also need some search tool for finding candidates of relation mentions, i.e., sentences or passages in which certain tuples of entities or certain patterns occur. Here we use Lucene but we could also use Google Search or any other search engine of similar functionality.

We call our own architecture DARE standing for Domain Adaptive Relation Extraction.[1] It was the first system with the three following capabilities:

1. Detecting $n$-ary relations of variable n and their $k$-ary projections where $0 < k \leq n$.
2. Employing a compositional recursive rule formalism
3. Annotating the entities by their semantic roles

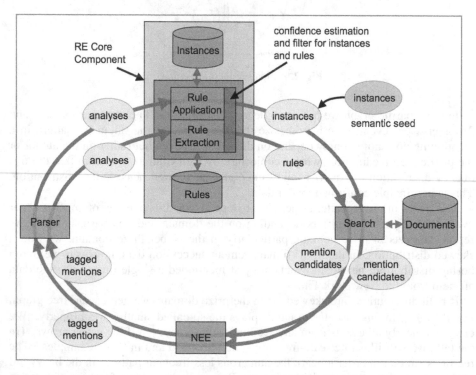

**Fig. 3.** Architecture of the system with the double learning cycle

---

[1] A public web demo of DARE displaying for each learning step all rules and instances can be accessed at http://dare.dfki.de.

The architecture in Fig. 3 shows the typical endless double circle of minimally supervised bootstrapping. The system is initialized with a semantic seed containing at least one sample instance, a tuple of named entities with their respective NE types. The first cycle will then find patterns, the second cycle instances and so forth. Usually, the bootstrapping terminates after 4-10 double-cycles when no more instances can be found. The architecture is generic enough to also permit an initialization with rules or even tagged mentions as seed.

Let us first focus on the rule learning mechanism. The following example relation comes from the prize award domain. The relation contains four arguments representing an event in which a person or an organization won a particular prize in a specific area and in a certain year:

(1) &lt;recipient, award, area, year&gt;

(2) is an example relation instance of (1), referring to an event mentioned in the sentence (3).

(2) &lt;Mohamed ElBaradei, Nobel, Peace, 2005&gt;

(3) Mohamed ElBaradei, won the 2005 Nobel Prize for Peace on Friday for his efforts to limit the spread of atomic weapons.

DARE learns three rules from the tree in (4), i.e., (5), (6) and (7).

(4)

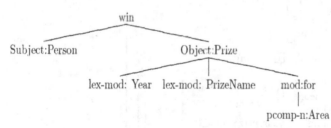

(5) extracts the semantic argument *area* from a prepositional phrase, while (6) extracts three arguments *year*, *prize* and *area* from the complex noun phrase and calls the rule (5) for the argument *area*.

(5)

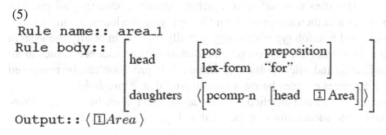

(6)

```
Rule name:: year_prize_area_1
Rule body::
```

$$
\begin{bmatrix}
\text{head} & \begin{bmatrix} \text{pos} & \text{noun} \\ \text{lex-form} & \text{``prize''} \end{bmatrix} \\
\text{daughters} & \left\langle \begin{bmatrix} \text{lex-mod} & [\text{head} \quad \boxed{1}\,Year] \end{bmatrix}, \right. \\
& \quad \begin{bmatrix} \text{lex-mod} & [\text{head} \quad \boxed{2}\,Prize] \end{bmatrix}, \\
& \quad \left. \begin{bmatrix} \text{mod} & \begin{bmatrix} \text{head} & \begin{bmatrix} \text{pos} & \text{preposition} \\ \text{lex-form} & \text{``for''} \end{bmatrix} \\ \text{rule} & area\_1:: \langle \boxed{3}\,Area \rangle \end{bmatrix} \end{bmatrix} \right\rangle
\end{bmatrix}
$$

Output:: $\langle \boxed{1}Year, \boxed{2}Prize, \boxed{3}Area \rangle$

(7) is the rule that extracts all four arguments from the verb phrase dominated by the verb "win" and calls (6) to handle the arguments embedded in the linguistic argument "object".

(7)

```
Rule name:: recipient_prize_area_year_1
Rule body::
```

$$
\begin{bmatrix}
\text{head} & \begin{bmatrix} \text{pos} & \text{verb} \\ \text{mode} & \text{active} \\ \text{lex-form} & \text{``win''} \end{bmatrix} \\
\text{daughters} & \left\langle \begin{bmatrix} \text{subject} & [\text{head} \quad \boxed{1}\,Person] \end{bmatrix}, \right. \\
& \quad \left. \begin{bmatrix} \text{object} & \begin{bmatrix} \text{head} & \begin{bmatrix} \text{pos} & \text{noun} \\ \text{lex-form} & \text{``prize''} \end{bmatrix} \\ \text{rule} & \begin{array}{l} year\_prize\_area\_1:: \\ \langle \boxed{4}Year, \boxed{2}Prize, \boxed{3}Area \rangle \end{array} \end{bmatrix} \end{bmatrix} \right\rangle
\end{bmatrix}
$$

Output:: $\langle \boxed{1}Recipient, \boxed{2}Prize, \boxed{3}Area, \boxed{4}Year \rangle$

In its core, the system uses non-statistical machine learning, acquiring all patterns found for certain tuples and then all tuples found by applying the learned patterns. For special types of data and relation types this may actually suffice. In reality the system is combined with a confidence estimation scheme that assigns confidence values to all learned tuples (instances) and rules (patterns). In this way, precision can be preserved by filtering out all instances or rules below a certain confidence threshold.

The DARE system has been applied to two application domains: Nobel Prize awards and management succession events. Table 1 gives an overview of the test data sets.

**Table 1.** Overview of Test Data Sets

| Data Set Name | Doc Number | Data Amount |
|---|---|---|
| Nobel Prize Corpus[2] | 3300 | 20 MB |
| MUC-6 | 199 | 1 MB |

**Table 2.** Performance of DARE for Nobel Prize Corpus with one example seed

| Data Set Name | Number of Seed | Precision | Recall |
|---|---|---|---|
| Nobel Prize Corpus | 1 | 80,59% | 69% |

**Table 3.** Performance of DARE for MUC-6 with different size of seed

| Data Set Name | Number of Seed | Precision | Recall |
|---|---|---|---|
| MUC-6 | 1 | 15,1% | 21,8% |
| | 20 | 48,4% | 34,2% |
| | 55 | 62,0% | 48,0% |

The comparison of the data properties of the Nobel Prize corpus and the MUC-6 corpus explains the performance gap between these two corpora. In Nobel Prize corpus, the connectivity between the patterns and instance mentions is close to small world property. Thus, with only one example, the performance is very satisfied. MUC-6 corpus has a different data property, since the patterns and instances mostly are not connected to each other. Therefore, the MUC-6 task is not suitable for DARE.

In addition to the connectivity of the corpus data, when we tested the system on a new task and domain, we observed that for this domain the interference of overlapping outside relations became a burden for the precision of the system. Various approaches to confidence estimation of learned rules have been proposed as well as methods for identifying "so-called" negative rules for increasing the precision value (e.g., [12,13,14]). For improving precision we added a novel method of exploiting negative evidence in addition to positive one. We first added negative examples, which we used for filtering and for the acquisition of negative patterns [15].

We then refined the way of providing at the same time positive and negative examples. We provide a closed-world fragment of the relation as extended seed. The seed consists of positive instances that are closed over some selected entities [16]. If one provides all Nobel-Peace-Prize winners of the nineties than the set is closed over the combination of prize and decade. This means implicitly the seed contains huge numbers of negative examples, i.e. all tuples of <person, Nobel-Peace-Price, year> where year is between 1990 and 1999. With 600.000 people in our data base, this makes 6 Mio. tuples minus the 10+ Nobel-Peace-Prize instances of the nineties being the number of implicit negative examples. By these large numbers the chances for finding and filtering out wrong instances and patterns are considerably increased.

Thus we extend the validation method by an evaluation of extracted instances against some limited closed-world knowledge, while also allowing cases in which

---

[2] The Nobel Prize corpus is freely available for download at http://dare.dfki.de

knowledge for informed decisions is not available. In our work, closed-world knowledge for a target relation is the total set of positive relation instances for entire relations or for some selected subsets of individuals. For most real world applications, closed world knowledge can only be obtained for relatively small subsets of individuals participating in the relevant relations. We store the closed-world knowledge in a relational database, which we dub "closed world knowledge database" (abbr. *cwDB*). Thus, a *cwDB* for a target relation should fill the following condition:

A *cwDB* must contain all correct relation instances (*insts*) for an instantiation value (*argValue*) of a selected relation argument *cwArg* in the target relation.

An example of a *cwDB* is the set of all prize winners of a specific prize area such as Peace, where PRIZE AREA is the selected *cwArg* and *argValue* is *Peace*. Note that the merger of two *cwDBs*, for example with PRIZE AREAs *Peace* and *Literature*, is again a *cwDB* (with two *argValues* in this case). Given a *cwDB* of a target relation and its *argValue* of its selected argument cwArg, the validation of an extracted instance (inst) against the *cwDB* is defined as follows.

$$
\begin{aligned}
inst\ correct\ &\Leftrightarrow\ &&inst \in cwDB\\
inst\ wrong\ &\Leftrightarrow\ &&inst \notin cwDB\ \wedge\\
&&&cwArg(inst) = argValue\\
inst\ unknown\ &\Leftrightarrow\ &&(\ \ inst \notin cwDB\ \wedge\\
&&&cwArg(inst) \neq argValue\ \ )\\
&\vee\ &&(\ \ inst \notin cwDB\ \wedge\\
&&&cwArg(inst)\ \text{is}\ unspecified\ \ )
\end{aligned}
$$

The confidence value of the extracted instances is estimated based on their validation against the *cwDB* or the confidence value of their ancestor seed instances from which their extraction rules stem. Furthermore, the specificity of the instances (percentage of the filled arguments) and the learning depth (iteration step of bootstrapping) are parameters too. Given the scoring of instance inst, the confidence estimation of a rule is the average score of all insts extracted by this rule:

$$
\mathbf{confidence}(rule) = \frac{\sum_{inst \in \mathbb{I}} \mathbf{score}(inst)}{|\mathbb{I}|}
$$

where $\mathbb{I} = \mathbf{getExtractedInstances}(rule)$

Rules extracting wrong instances are lowered in rank. Our first experiments conducted on the same data "Nobel Prize award data" demonstrate: 1) limited closed-world knowledge is very useful and effective for improving rule confidence estimation and precision of relation extraction; 2) integration of "soft" constraints boosts the confidence value of the good and relevant rules, but without strongly decreasing the recall value.

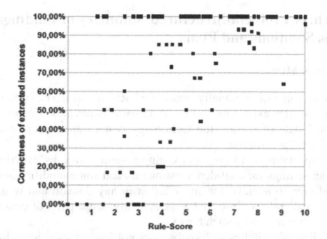

**Fig. 4.** Rule confidence value (Rule-Score) and their extraction performance with cwDB

Figure 4 shows the correlation between the confidence score of the rules and their extraction precision. Although the development curve here is not as smooth as the ideal case, the higher scored rules have better precision values than most of the lower scored rules. However, we can observe that some very good rules are scored low, located in the left upper corner. The reason is that many of their extracted instances are unknown, even if their extracted instances are mostly correct. Figure 5 shows the IE performance development with respect to rule score threshold. Given all the rules with a score of 4 or higher, DARE achieves the best modified F-Measure 92,21% with an improvement of precision of about 11 percentage points compared to the DARE baseline system after the integration of the closed-world knowledge.

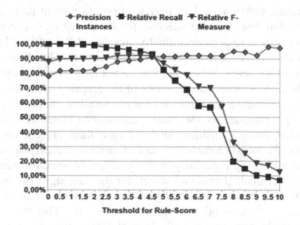

**Fig. 5.** Evaluation of Extraction Performance of DARE with cwdb

# 3 Research: A Problem-Structured Summary of Findings, Insights, Solutions and Plans

## 3.1 The Success Metrics

The commercial value of minimally supervised RE systems, i.e. systems that automatically adapt to new extraction tasks and domains is obvious. However, according to the published state of the art, the technology is not quite mature enough yet for immediate practical application.

The naïve expectation is to tune an existing system a little by experimenting with some parameter settings for confidence estimation and some modifications to the rule extraction and then to achieve for any relation in any domain and in any text base precision and recall values close to 100%. Actually, large parts of contemporary research in our field is driven by such hopes.

If the twiddling of switches and knobs does not lead to significant improvements during one annual publication cycle then researchers may loose interest since there are still so many unexplored gadgets in the ever growing inventory of machine learning techniques.

The reality is that we need to understand the underlying mechanisms of this type of learning much better and we may also have to investigate the relevant aspects of language structure and language use before we can predict for which relations, domains, data collections, parsers and NEE systems the approach can yield satisfactory results and, moreover, to determine what could be done to systematically increase the application potential of the paradigm. The means for obtaining such deeper understanding is a systematic loop of diagnostic experimentation and system modification.

One has to be careful in defining the targets of such a systematic system evolution. When it comes to selecting the metrics for measuring the success of the approach we must not readily accept the sporting rules for performance comparison defined for some shared tasks. Let me provide three examples to underline this word of caution.

Accepting external benchmark data may lead onto a garden path. It is important that your training and test data are suitable for the approach. Several publications have reported on attempts to apply minimally supervised RE to the MUC 6 management succession data referred to in the previous section. The discouraging results could easily be misinterpreted. When the data do not have certain properties one cannot even study the non-obvious ramifications of the learning approach. Therefore we held on to our own data sets even if this did not please all of our reviewers.

Accepting the F1 measure as the main performance criterion also seems to be somewhat deceptive. From a business angle of view there are not so many applications for which the F1 measure correlates to commercial success. However, there are large classes of applications, especially in the intelligence area, for which recall optimization is mission critical. If analysts search for the first signs of some new development, may it be of financial, technological or military nature, a recall of 0,99 or higher would more than outweigh a low precision of 0,01. This would simply mean that the analysts would have to manually look at 100 documents in order to get the one that really matters. In this way an RE system with an F1 score of 0,02 can serve as the basis for a true killer application, whereas a system achieving an F1 of 0.5 with both precision and recall values of 0.5 may not be usable at all.

On the other side of the scale are applications that sample large numbers of events in order to automatically compute statistical surveys and analyze trends. Here it does not matter so much if the recall is small, as long as the minimal sample size needed for significant findings is reached and the factors limiting the recall are statistically independent from the factors that matter for the statistical survey. Thus we may safely conclude that measures independently lifting recall at some expense for precision or vice versa may indeed be very important for building powerful applications. They may also shed more light onto the hidden mechanisms behind the learning approach.

The third example is the provision of similar learning and application data, i.e., the partitioning of a homogeneous data set into training and test data. It may well be that for many learning tasks any application data do not possess the properties required for effective minimally supervised learning. This does not mean that are no data suitable for training the system. It could just be that the training data are quite different from the application data.

Finally, let me point out the important difference between applications that are instance-oriented in contrast to the ones that are mention-oriented. Most information gathering applications including intelligence tools want to use textual data for learning about the world. If a market research application tries to find out whether the authors of certain statements actually own or have owned the type of car they are commenting on, texts stating that the person has sold such a car are important because they entail that the person had owned the relevant type of car. Consider in comparison applications that observe which events are reported by which media, then the texts need to be explicitly on the events. In this case, the relevant relation actually holds between the reporting media and the targeted events, which in turn are relations as well.

In the following I will concentrate on applications that are instance-oriented because mention-oriented applications can easily be redefined as instance-oriented by making the reporting part of the relation.

Starting from the assumption that recall and precision are our ultimate performance measures we have identified the different problems for these measures grouped by their causes.

## 3.2  Insufficient Recall

### 3.2.1  Problems Caused by the Data

Most learning methods need a minimum of data. The amount of data needed and the usefulness of the data for maximizing recall largely depend on the density and frequency distribution of the occurrences of patterns to be learned. For our learning method, the main properties are the distribution of mentions to instances and the distribution of patterns to instances, since we would like a fully or at least highly connected learning graph. In addition the grammatical and orthographical consistency of the data also influences the recall.

*Lack of data*
Depending on the domain, text sort and relevant relations we may not find a sufficient volume of data. If we consider professional biographical information, we will find

many more texts on artists, politicians and scientists than on shepherds, flexographers and proofreaders. Even generic relations that may be connected with many people can be rarely mentioned in texts. There will be far fewer mentions of people cancelling a household insurance than of people getting divorced.

If there is a lack of (available) data we always have the option to create such data. This can be expensive. However, it is even harder to create data with the important redundancies, i.e., the distributional properties needed for minimally supervised boot-strapping. Thus it may be cheaper in this case to produce labeled data for supervised learning. In our architecture, supervised learning can be naturally combined with minimally supervised learning. In our learning cycle we can insert labeled data, i.e. tagged mentions, easily before parsing and rule extraction. They will be treated as if they had been found by searching for seed instances. They would receive the highest possible confidence value.

However before resorting to artificial data production, we can try to exploit data from other domains or similar relations. We ran successful experiments in learning the extraction of prize winning and other award events from massively available news on Nobel Prize awards [17]. We call these learning data from other domains and relations auxiliary data. When we learned the "marry" relation, we found not surprisingly that marriages are reported very often for pop stars but much more rarely for business leaders [18]. However, most patterns used in the mentions of marriages of business leaders also occur in reports on pop celebrity marriages.

*Error-ridden or inconsistent data*
Rules learned from misspelled input or from input not conforming to grammatical and orthographic standards will hardly carry over to other mentions. Applying our current architecture with the existing NEE and parsing components to types of user-generated input exhibiting large proportions on non-conformity and other properties of sponta-neous language would not work. Relaxing the matching in rule application would be a burden for precision. The most straightforward remedy would be the deployment of language checking and normalization technology. Such systems may have to be spe-cially designed for certain types of non-standard language, but they would be needed for other language technology applications as well such as machine translation or information retrieval.

*Lack of redundancy and skewed distributions in the data*
Without having a skewed distribution of reports to instances leading to nearly scale-free learning graphs the graphs are more likely to be non-connected and even if they are connected the learning paths can become very long leading to increased noise. Even large volumes of available data may not exhibit the desired properties. In these cases one should first look for appropriate learning data with more redundancy such as adding texts from additional sources covering the same subject area and time. We successfully tried this method when we exploited the popularity of Nobel Prizes in the media for learning rules that we then also applied to extracting winning events of less popular prizes, e.g., Albert Lasker Award, Pritzker Prize, Turner Prize and Prix_de_Rome Prize.

If no better auxiliary data can be found, missing patterns may have to be injected by hand-crafted or hand collected data. Sometimes just a few more patterns might suffice for building bridges to unconnected islands or continents in the learning graph.

*Lack of redundancy/variants concerning patterns*
Theoretically it may also happen that the variance of patterns or their distribution with respect to mentions does not suffice for learning. This could for instance be the case when large parts of the authors use one or more conventionalized or prescribed patterns but the goal is to reliably learn the unknown rare patterns, which are not used for any instances with multiple mentions. We have not encountered such a situation and would not know how to solve the problem with the existing machinery.

### 3.2.2 Problems Caused by Linguistic Processing
Despite considerable progress in named-entity detection and generic parsing, our existing analysis methods are still far from being reliable.

*NEE Errors*
Depending on the types of named entities and on the variance in the data, NEE can vary between rather reliable and extremely error prone. Person names, locations, dates and times are much better covered than organizations, products, technologies and other large classed of NEEs.

*Parsing Errors*
The parsers we first employed (MiniPar and Stanford Parser) are great tools but they charge a certain price for their robustness. Parsing errors mainly effect precision, more about this in 3.3.2. But wrong parses may also reduce recall when a structure is wrongly selected that does not allow the appropriate RE rule to apply. When we tested our system with the PET/ERG parser because of its better precision, we also had to deal with its lower coverage and with its imperfect statistical parse selection scheme. Lower coverage is less of a problem because we can always keep the Stanford Parser as a fallback for sentences that PET/ERG cannot parse. But the parse selection errors reduced our recall. Thus we utilized our confidence estimation for re-ranking the parses. It turned out that for the sentences relevant for our RE task, we could improve both recall of our RE performance and the accuracy of the PET/ERG parser. In this way, we used a little bit of semantics in syntactic disambiguation [19].

### 3.2.3 Problems Caused by the Approach
The problems considered up to this point were extrinsic factors since they depend on available data and other processing tools. Unfortunately, there are also problems intrinsic to the minimally supervised bootstrapping approach, at least to its initial instantiation.

*Rule Under-Generalization*
When the DARE rule extraction copies those parts from the dependency tree that connect the relation arguments to build a rule from this partial graph, the question arises which sibling nodes of the arguments to include into the rule. Some of these siblings are needed, others are free adjuncts that make the rule too specific. (9) is derived from the HPSG parsing tree of (8). It is a rule for extracting marriage

relationships. (9) is too specific because it keeps the adjective "first" modifying the lexical head "wedding". The chance that (9) will match another sentence in the corpus is relatively small. But if we delete the node "first", the result constitutes a good rule for the marriage relation.

(8) Mira Sorvino's first wedding to actor Christopher Backus - a private civil cere-
    mony at the Santa Barbara Courthouse June 11 - was lovely.

(9)

```
┌ rule_7743                                                                              ┐
│ PATTERN ┌ pattern                                                                   ┐ │
│         │ HEAD          ("first_n1", )                                              │ │
│         │ SP-HD_N_C ┌ sp-hd_n_c                                                   ┐ │ │
│         │           │ HEAD          ("apostrophe_s_2_lex", )                      │ │ │
│         │           │ SP-HD_HC_C ┌ sp-hd_hc_c                    ┐                │ │ │
│         │           └            └ HEAD ⓪ <person> ┘            ┘                │ │ │
│         │ HDN-AJ_RC_C ┌ hdn-aj_rc_c                                             ┐ │ │
│         │             │ HEAD          ("wed_v1", )                             │ │ │
│         │             │ HD-AJ_VMOD_C ┌ hd-aj_vmod_c                          ┐ │ │ │
│         │             │              │ HEAD          ("to", )                │ │ │ │
│         │             │              │ HD-CMP_U_C ┌ hd-cmp_u_c            ┐ │ │ │ │
│         └             └              └            └ HEAD ① <person> ┘ ┘ ┘ ┘ │
│ OUTPUT ┌ relation   ┐                                                               │
│        │ person2 ①  │                                                               │
│        └ person1 ⓪  ┘                                                               │
└                                                                                        ┘
```

If the pattern is frequent, it will most likely also show up some other place without the free adjunct. The over-specific rule does not hurt but it may not add more in-stances unless some mention contains exactly the same free modification. We are currently working on two possible remedies. One is a rule generalization algorithm that generalizes over sets of rules that only differ with respect to one sibling phrase of the arguments. If such rule simplification is applied to freely it can easily lead to overly permissible rules. The other solution is the exploitation of the richer informa-tion in the PET/ERG parses for detecting truly optional adjuncts.

*Non-local mentions*
The first systems employing minimally supervised learning for RE only learned pat-terns of a very constrained syntactic structure [3, 4, 20, 21].

When DARE was first tested it constituted quite a leap forward since it detected any patterns containing a fixed minimum number of relation arguments as long as it was fully contained within one sentence. But very often the mention of a relation instance extends over more than one sentence. A simple form of such non-locality is mentions in which a named entity is introduced one or more sentences before the relation pattern occurs. In these cases the place of the argument in the pattern is filled by an anaphoric element, which can be a pronoun or a co-referring non-pronominal phrase.

(10) The scientist won the 2005 Nobel Prize for Peace on Friday for his efforts to
     limit the spread of atomic weapons.

More than one argument can be realized by anaphoric elements. The relation can also be broken up in two or more parts, which can be scattered across several sentences. The parts are usually connected by anaphoric elements again.

Since reliable general anaphora resolution is an unsolved problem, we sought a solution in which only certain anaphoric relations had to be resolved, namely the ones needed for getting the complete instance tupel. Similar to the case of parse re-ranking we also thought of a way for utilizing our own confidence estimation for filtering hypothesized anaphoric references. Using this approach we were able to improve our recall with some minor damage to our precision, namely, the recall has improved by 3.5 % and the F1 measure gained 1.9 %.

The novel strategy opens new ways of bringing semantic domain knowledge as reflected in the data into anaphora resolution. We even achieved a slight improvement of the F1 score [22]. However, just a boost of recall by itself would have been a useful result because of the application relevance discussed in 3.1.

### 3.3  Insufficient Precision

The better our means are for controlling precision the more can we employ additional methods for boosting recall. The central mechanisms for increasing precision are the use of negative seed data, confidence estimation and deeper parsing methods.

### 3.3.1  Problems Caused by the Data

*Unreliable data*
A common but by far not a predominant source of precision errors is wrong data. Many such errors can be prevented by the employed statistical confidence estimation and filtering through thresholds. The use of negative data also contributes to this filtering. The confidence estimation is set up in a way that it takes into account the trustworthiness of sources. If an application can provide such information, precision will further increase.

### 3.3.2  Problems Caused by Linguistic Processing

*NEE Errors*
Nothing will be said here about errors resulting from the erroneous detection of named entities. This is not a major error source and NEE is likely to further improve.

*Parsing Errors*
Extensive experiments in [23] showed that the largest portion of erroneously extracted relation instances, namely 38:2%, are due to errors of the employed dependency parser system. [24] compares the performance of the dependency parsing results of MiniPar, Stanford Parser with the syntactic and the semantic analysis of the deep HPSG parser with the English ERG grammar. It turns out that the syntactic analysis of the deep parser delivers the best precision, while the semantic analysis of the deep parser yields in almost the same recall as the Stanford parser.

### 3.3.3  Problems Caused by the Approach
As we could already show for recall deficiencies, also some observed classes of precision errors are caused by inherent problems of the approach.

*Embeddings into negation and other non truth-value preserving modalities*
When mentions of a relation instance occur in contexts that do not preserve the truth value of the embedded propositions wrong instances are detected. (11) provides an ironic context, while (12) mentions the fictitious Nobel Prize winner in a TV program.

(11) It's also possible [that O.J. Simpson will find the real killer, that Bill Clinton will enter a monastery and that Rudy Giuliani will win the Nobel Peace Prize.]

(12) In NBC "West Wing," [we get President Josiah Bartlett, a Nobel Prize Winner]

Although many of these instances can be filtered out by our confidence estimation, a general and truly effective solution requires a sufficient knowledge of the patterns signaling such contexts. The use of negative seed data can help in learning such contexts. However, the current rule extraction algorithm will not consider linguistic patterns outside the minimal relation mention. To this end, we need to learn overt and less overt negations and patterns signaling possibility, demand, assumption, speculation, expectation, thinking, communication etc. Whether these can be learned through minimal supervision bootstrapping or whether they need to injected by labeled data or rules we do not yet know.

*Confusion with other Relations*
A side effect of learning from semantic seed is the confusion among relations with overlapping extensions. Consider the relation:

(13) management_succession <Person_in, person_out, position, organization>

Let us assume that we have a positive seed: <Jeremy Knoll, Paolo Bronte, CEO, Moonitec>.

There are many situations that may involve the two persons, the job title and the company, such as the one reported by the following sentence:

(14) Paolo Bronte, CEO of Moonitec, met with Jeremy Knoll, an independent NYC-based management consultant.

We have found numerous confusions of this type. When searching for mentions of marriages of two actors in a certain year, we found reports on the two co-starring in a movie. When aiming for mentions of a scientist winning a certain prize in a certain year, we found reports on her being nominated for exactly the same prize in exactly the same year. Some of these confusions are more surprising than others but others are quite expected. Among the latter are the nomination events instead of award winning events and events of meeting instead of getting married.

When we were searching for current and former spouses of celebrities, we also found many reports on divorces. These reports were significant since every divorced spouse could safely be added to the list of former spouses. Thus we have three types of overlap: non-overlap, partial overlap and inclusion (with mutual inclusion meaning complete extensional overlap) of two relations.

If the inclusion goes in the right direction, the exploitation of entailment is a welcome side effect of the approach, for the mentions of the included relation always yield valid instances. In case of other overlaps, the confusion can lead to wrong results. Often it helps to add another argument for learning. If for instance a wedding date is added to the marriage relation or a hiring date to the management succession relation, then most cases of confusion can be avoided. However, the additional argument usually drastically reduces the number of mentions so that there is a price to pay.

Another method we tried was to take special care in the selection of optimal seeds. We defined a property of seeds, which we called "distinctiveness". We found that in relations between people, the ideal seed involves one individual who is extremely popular so that many mentions will be found and another individual who does not participate in any other relations that are mentioned in reports. We could show that for the marriage relation, we could improve precision by 30% simply by picking the best seeds. However the selection of optimal seeds is a non-trivial intellectual challenge either requiring good quantitative data on occurrences of candidates in the documents or a good deal of general knowledge coupled with intuition.

The extension of our approach by closed world seeds, i.e., by implicit negative examples, turned out to be the most effective measure against the confusion effect. After we added negative data, precision could be boosted by 10%. We also found that after this increase in precision through the use of negative seeds, the selection of the seed did not have any influence on precision anymore.

*Rule Over-Generalization*
In some rare cases overly general rules have been extracted. However, we do not attribute these errors to the rule extraction algorithms since all cases of such rules could be traced back to incorrect parsing results.

## 4 Conclusion and Outlook

On and off and parallel to other projects we have worked for several years on our minimally supervised bootstrapping approach to relation extraction. After these years we feel that we are beginning to really understand the problems and mechanisms connected with the application of the paradigm to human language. We have gained better insights into the advantages and limitations of the approach.

On the other hand we also feel that research on intelligent combinations of semantics-oriented learning with generic lexical, morphological and syntactic language processing has just begun. This line of research is very tempting not only because it creates the prerequisites for new powerful applications but also because it opens new avenues for a much more systematic investigation of methods for semantic interpretation.

Among the many directions for further research I would like to single out three themes from which I personally expect important results in the near future:

- learning of negative and other modal contexts
- learning of patterns that occur in many relations such as patterns signaling the agent, time, location, purpose or cause of events
- application of minimally supervised RE to opinion and sentiment analysis

Another promising theme for future research is the combination of different learning methods in a single RE system. Up to now we have not used and validated the opportunities for combining minimally supervised and supervised learning that our approach allows. In our application centered projects we are encountering extraction tasks for which no data exist that possess the required properties for minimally supervised learning. But often these need to be combined with others for which appropriate learning data exist. And then there will always be borderline cases. A modular generic framework such as DARE allowing experimentation with combinations of different linguistic processing tools, different confidence estimation methods, different types of seeds and different IE paradigms ranging from minimally supervised bootstrapping via supervised learning all the way to the manual production of rules offers ample opportunities for further systematic exploration of the vast search space.

## Acknowledgements

The work presented here is based on joint ideas by Feiyu Xu and myself. It was carried out in a research group that we jointly led. I am also grateful to Li Hong, Peter Adolphs, Sebastian Krause and Yi Zhang who have made important contributions to the results described and cited in this paper. I would also like to express my gratitude to the reviewers of our cited publications and to the audiences of our conference presentations, invited talks and keynote lectures for their valuable comments.

The research was partially supported through a research grant to the project TAKE, funded by the German Ministry for Education and Research (BMBF, FKZ: 01IW08003) and by a grant to the project KomParse by the ProFIT programme of the Federal State of Berlin, co-financed by the European Union through the European Regional Development Fund (EFRE).

## References

1. Brin, S.: Extracting patterns and relations from the world wide web. In: Atzeni, P., Mendelzon, A.O., Mecca, G. (eds.) WebDB 1998. LNCS, vol. 1590, pp. 172–183. Springer, Heidelberg (1999)
2. Agichtein, E., Gravano, L.: Snowball: Extracting relations from large plain-text collections. In: Proceedings of the Fifth ACM International Conference on Digital Libraries (2000)
3. Yangarber, R.: Scenario Customization for Information Extraction. Dissertation, Department of Computer Science, Graduate School of Arts and Science, New York University, New York, USA (2001)

4. Sudo, K., Sekine, S., Grishman, R.: An improved extraction pattern representation model for automatic IE pattern acquisition. In: Proceedings of the 41st Annual Meeting of the Association for Computational Linguistics, pp. 224–231 (2003)

5. Xu, F., Uszkoreit, H., Li, H.: A seed-driven bottom-up machine learning framework for extracting relations of various complexity. In: Proceedings of ACL 2007, 45th Annual Meeting of the Association for Computational Linguistics, Prague, Czech Republic (2007)

6. Muslea, I.: Extraction patterns for information extraction tasks: A survey. In: AAAI Workshop on Machine Learning for Information Extraction, Orlando, Florida (1999)

7. Drozdzynski, W., Krieger, H.-U., Piskorski, J., Schäfer, U., Xu, F.: Shallow processing with unification and typed feature structures — foundations and applications. Künstliche Intelligenz 1, 17–23 (2004)

8. Lin, D.: Dependency-based evaluation of MINIPAR. In: Abeillé, A. (ed.) Treebanks - Building and Using Parsed Corpora. Kluwer Academic Publishers, Dordrecht (2003)

9. de Marneffe, M., Manning, C.D.: The stanford typed dependencies representation. In: Coling 2008: Proceedings of the Workshop on Cross-Framework and Cross-Domain Parser Evaluation, Manchester, UK (2008)

10. Callmeier, U.: PET – a platform for experi- mentation with efficient HPSG processing techniques. Natural Language Engineering 6(1), 99–107 (2000)

11. Flickinger, D.: On building a more efficient grammar by exploiting types. Natural Language Engineering 6(1), 15–28 (2000)

12. Yangarber, R.: Counter-training in discovery of semantic patterns. In: Proceedings of the 41st Annual Meeting of the Association for Computational Linguistics, Sapporo, Japan (2003)

13. Etzioni, O., Cafarella, M., Downey, D., Popescu, A., Shaked, T., Soderland, S., Weld, D., Yates, A.: Unsupervised named-entity extraction from the web: An experimental study. Artificial Intelligence 165(1), 91–134 (2005)

14. Bunescu, R.C., Mooney, R.J.: Learning to extract relations from the web using minimal supervision. In: Proceedings of the 45th Annual Meeting of the Association for Computational Linguistics (2007)

15. Uszkoreit, H., Xu, F., Li, H.: Analysis and Improvement of Minimally Supervised Machine Learning for Relation Extraction. In: Horacek, H., Métais, E., Muñoz, R., Wolska, M. (eds.) NLDB 2009. LNCS, vol. 5723, pp. 8–23. Springer, Heidelberg (2010)

16. Xu, F., Uszkoreit, H., Krause, S., Li, H.: Boosting Relation Extraction with Limited Closed-World Knowledge. In: Poster Volume of the Proceedings of the 23rd International Conference on Computational Linguistics, Beijing, China (2010)

17. Xu, F., Uszkoreit, H., Li, H., Felger, N.: Adaptation of relation extraction rules to new domains. In: Proceedings of the Poster Session of the Sixth International Conference on Language Resources and Evaluation, LREC 2008, Marrekech, Morocco (2008)

18. Li, H., Xu, F., Uszkoreit, H.: Minimally Supervised Learning of Relation Extraction Rules from Various Domains. DFKI research report (2010)

19. Xu, F., Li, H., Zhang, Y., Krause, S., Uszkoreit, H.: Minimally Supervised Domain Adaptive Parse Reranking (2011) (forthcoming)

20. Sudo, K., Sekine, S., Grishman, R.: An improved extraction pattern representation model for automatic IE pattern acquisition. In: Proceedings of the 41st Annual Meeting of the Association for Computational Linguistics, pp. 224–231 (2003)

21. Greenwood, M.A., Stevenson, M.: Improving semi-supervised acquisition of relation extraction patterns. In: Proceedings of the Workshop on Information Extraction Beyond The Document, pp. 29–35. Association for Computational Linguistics, Sydney (2006)

22. Xu, F., Uszkoreit, H., Li, H.: Task driven coreference resolution for relation extraction. In: Proceedings of the European Conference for Artificial Inteligence ECAI 2008, Patras, Greece (2008)
23. Xu, F.: Bootstrapping Relation Extraction with Semantic Seeds. PhD thesis, Saarland University, Saarbrücken, Germany (2007)
24. Adolphs, P., Li, H., Uszkoreit, H., Xu, F.: Deep Linguistic Knowledge for Relation Extraction (2011) (forthcoming)
25. Kozareva, Z., Hovy, E.: Learning arguments and supertypes of semantic relations using recursive patterns. In: Proceedings of the 48th Annual Meeting of the Association for Computational Linguistics, Uppsala, Sweden, pp. 1482–1491 (July 2010)

# Using Graph Based Method to Improve Bootstrapping Relation Extraction

Haibo Li, Danushka Bollegala, Yutaka Matsuo, and Mitsuru Ishizuka

Graduate School of Information Science and Technology
University of Tokyo
7-3-1 Hongo, Bunkyo-ku, Tokyo 113-8656, Japan
lihaibo@mi.ci.i.u-tokyo.ac.jp, danushka@iba.t.u-tokyo.ac.jp,
matsuo@biz-model.t.u-tokyo.ac.jp, ishizuka@i.u-tokyo.ac.jp

**Abstract.** Many bootstrapping relation extraction systems processing large corpus or working on the Web have been proposed in the literature. These systems usually return a large amount of extracted relationship instances as an out-of-ordered set. However, the returned result set often contains many irrelevant or weakly related instances. Ordering the extracted examples by their relevance to the given seeds is helpful to filter out irrelevant instances. Furthermore, ranking the extracted examples makes the selection of most similar instance easier. In this paper, we use a graph based method to rank the returned relation instances of a bootstrapping relation extraction system. We compare the used algorithm to the existing methods, relevant score based methods and frequency based methods, the results indicate that the proposed algorithm can improve the performance of the bootstrapping relation extraction systems.

## 1 Introduction

For many real world applications, background knowledge is intensively required. The acquisition of relational domain knowledge is still an important problem. Relation extraction systems extract structured relations from unstructured sources such as documents or web pages. These structured relations are as useful as knowledge. Acquiring relational facts *Acquirer–Acquiree* relation or *Person–Birthplace* relation with a small number of annotated data could have an important impact on applications such as business analysis research or automatic ontology construction.

Currently, research in relation extraction focuses mainly on pattern learning and matching techniques for extracting relational entity pairs from large corpora or the Web. The Web forms a fertile source of data for relation extraction, but users of relation extraction system are typically required to provide a large amount of annotated text to identify the interesting relation. This requirement is not feasible in real world applications. Therefore, many systems have been proposed to address the task of Web-based relation extraction, which usually only need a small number of seed entity pairs of relations. These systems typically build on the paradigm of bootstrapping of entity pairs and patterns as proposed by Brin[1].

However, an entity pair often has more than one type of semantic relations in real world. Consequently, a bootstrapping-based extraction system might introduce some

A. Gelbukh (Ed.): CICLing 2011, Part II, LNCS 6609, pp. 127–138, 2011.

irrelevant noises into further iteration. For example, given the entity pair (*Bill Gates, Microsoft*) for *CEO-of-Company* relation, two context patterns: "*was the CEO of*" and "*has retired from*" can be easily extracted from the Web. These context patterns express two different relationships and only the former is relevant to the target relation. If the irrelevant context pattern is used for further iteration, more irrelevant context patterns will be introduced into the extracted results. Although many filtering functions have been proposed in the literature, the extraction precision still is not satisfactory[2]. A significant number of noise or weakly relevant relationship instances are returned. Therefore, we use a graph based multi-view learning algorithm to rank all the extracted entity pairs by their relevance to the given seeds.

A semantic relation between two entities can be represented from two different aspects or views: the entity pair itself and the surrounding context. For example, the *Person–Birthplace* relation can be expressed as a set of entity pairs, such as (*Albert Einstein, Ulm*) and (*Jesus, Bethlehem*). From a lexical pattern view, the *Person–Birthplace* relation can also be represented with some context patterns, such as "*A was born in B*", "*B, the birth place of A*". Meanwhile, for a bootstrapping relation extraction system, an entity pair by context pattern co-occurrence matrix can be constructed easily. Then we construct two weighted complete graphs for all entity pairs and context patterns respectively. Concretely, we use each entity pair as vertex to construct a complete graph $G_e$, the edges are weighted with certain similarity between the entity pairs. We construct a context pattern graph $G_c$ similarly. Since $G_e$ and $G_c$ is composed of entity pairs and context patterns in each view respectively, the two graphs are termed intra-view graph. We also construct a bipartite graph $G_b$ composed with entity pairs and context patterns. Each edge on $G_b$ links an entity pair and a context pattern. These edges are weighted with the relevance of the linked entity pair and context pattern. Vertices of graph $G_b$ are composed of entity pair view and context pattern view, so $G_b$ is an inter-view graph. We combine these three graphs together to accurately compute the ranking score of each entity pair.

The multi-view learning algorithm is based on inter-view and intra-view consistency assumptions. Given an entity pair $e$, if $e$ strongly links to other high ranking score entity pairs on $G_e$, then $e$ is likely to achieve a high ranking score. Meanwhile, on the graph $G_b$, if $e$ links frequently to the context patterns whose ranking scores are high, then $e$ is likely to achieve a high ranking score. The intra-view consistency assumption means that nearby entity pairs on $G_e$ are likely to have similar ranking score. In addition, the context pattern graph provides us with similarity between extracted context pattern and given context pattern seed. The inter-view consistency assumption makes the similarity between context patterns useful to compute the relational similarity between entity pairs. Because if an entity pair $e$ frequently co-occur with a context pattern $c$ which is very similar to the given context pattern seeds, then ranking score of entity pair $e$ should be high.

The remainder of the paper is organized as follows. The following section presents a discussion of related works. Subsequently, a bootstrapping-based relation extraction system is introduced. Using this relation extraction system, we extract some entity pairs for ranking. Thereafter, the multi-view ranking algorithm and some empirical results are presented. Finally, we conclude this paper.

## 2   Related Work

Bootstrapping-based relation extraction [1,3,4,5,6] leverage large amounts of data on the Web efficiently. The method is initialized using a seed set; it extracts relative facts or relations. Sergey Brin propose DIPRE system [1] to extract *author–book* relation form the Web; The Snowball system[3] extracts entity pairs including a predefined relation from a corpus. KnowItAll[4] and Espresso[5] is also bootstrapping-based system, but a different type of pattern evaluation method is used. They compute co-occurrence of context patterns and entity pairs to filter context patterns. The SatSnowball [6] extends the Snowball using statistical methods and extracts entity pairs and keywords around the entities. Furthermore, both DIPRE [1] and SatSnowball use a general form to represent extracted patterns. Although these general form patterns improve the coverage of extracted patterns, they decrease the precision. Moreover, general form patterns cannot be used directly as a query for a Web search engine, which is an efficient tool to retrieve texts on the Web. A crucial issue of these system is to filter noise out of the instances for further iteration. Sebastian Blohm et. al. systematical evaluated the impact of different filtering functions [2].

Open Information Extraction (Open IE)[7] is a domain independent information extraction paradigm which uses some generalized patterns to extract all potential relations between name entities. The generalized patterns are extracted from a dependency parsed corpus. Although Open IE is different from the bootstrapping-based method, the ranking of extracted entity pairs is also useful to retrieve these entity pairs.

Many previous reports have described that the proper use of unlabeled data can complement a traditional labeled dataset to improve the performance of a supervised algorithm. For example, a named entity classification algorithm proposed in [8], which is based on co-training framework, can reduce the need for supervision to a handful of seed rules. Label propagation [9] is a graph-based semi-supervised learning model in which the entire dataset is presented as a weighted graph; then the label score is propagated on this graph. Zhou et al. proposed a label propagation algorithm working on spectral graph [10]. In this paper, our Multi-View Ranking algorithm combines the label propagation approach with the multiple views idea from co-training.

## 3   A Framework for Bootstrapping Based Relation Extraction

This section provides an overview of the bootstrapping relation extraction framework which is used to extract entity pairs for ranking. The main components are explained in upcoming sections. Figure 1 portrays the framework architecture. In the framework, each sentence containing target relation is represented as a tuple, $(e, c)$, where $e = (e_a, e_b)$ is an entity pair and $c$ is a context pattern. The *input* of framework is a sentence set $S^0 = \{(e_i, c_j)|i = 1, 2, \cdots, n; j = 1, 2, \cdots, m; \}$ which is composed of the entity pair $e_i$ and the context pattern $c_j$. The *output* of the framework is a set of entity pairs and a set of context patterns. The system distends $S^0$ to construct a potential target entity pair set $E$ and a context pattern set $C$.

The *Extraction* part uses a dual extraction model. $E^0 = \{e_i|(e_i, c.) \in S^0\}$ and $C^0 = \{c_j|(e., c_j) \in S^0\}$ represent respectively the entity pairs and context patterns

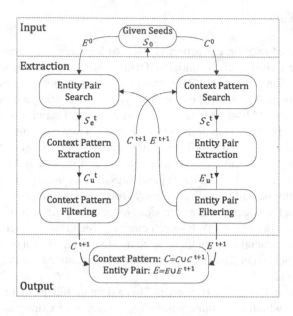

**Fig. 1.** Framework of a bootstrapping relation extraction system

in $S^0$. In $t$-th extracting iteration, we respectively submit some queries generated from entity pair set $E^t$ and context pattern set $C^t$ (at the beginning $t = 0$) to a search engine. The context pattern set $C^t$ is used in the Context Search part and the entity pairs in $E^t$ are used in the Entity Pair Search part. We collect some top ranking web pages returned by the search engine.

In the $t$-th entity pair search step, we use an entity pair $e$ to generate some queries for a search engine. These queries are designed to retrieve the web pages in which the two entities occur. We collect some top ranking pages containing the entity pair. Then, these web pages are split into sentences. We extract the sentences in which the two entities appear simultaneously. The context patterns used in our extraction system are the contexts between an entity pair. We select and submit the lexical context patterns composed of less than five words to the search engine. We claw some top ranking pages which contain the context pattern. Then, these web pages are split into sentences. Then, the sentences captaining the context pattern are selected for further steps. In this way, the sentences set $S_e^t$ and $S_c^t$ is constructed respectively.

In the Entity Pair Extraction step, a Named Entity Recognition tool is used to label the named entities in each sentence. Then all entity pair candidates are extracted and added to the candidate set $E_u^t$. The system selects a subset of entity pair, $E^{t+1} \subseteq E_u^t$, for $t+1$ round of extraction. Simultaneously, these selected entity pairs in $E^{t+1}$ are added to the output. Similarly, the set of context pattern $C_u^t$ is extracted and some context patterns, $C^{t+1} \subseteq C_u^t$, are selected for $t+1$ round of entity pair expansion. The corresponding context patterns in $C^{t+1}$ are also added to the output.

# 4   Entity Pair and Context Pattern Filtering Function

Because of the many-to-many relation between the entity pairs and the context patterns, extracted entity pairs and context patterns are not all applicable to the next round of extraction. For an entity pair, some context patterns that represent different types of relation may be extracted. For example, using the entity pair (*Albert Einstein, Ulm*), we can extract two types of context pattern:"*A was born in B*" and "*A's stay in B*". The two context patterns have totally different semantic relation. Therefore, the context pattern and the entity pair filtering is necessary.

In order to select most promising context patterns for further iteration, we measure each context pattern $c \in C_u^t$ using the entity pair set $E^t$ as follow:

$$S(c) = \frac{1}{|E^t|} \sum_{e \in E^t} \frac{|e_a, c, e_b|}{|e_a, *, e_b|}$$

We write $|e_a, c, e_b|$ to denote the number of sentence $s \in S_c^t$ in which both entity pair $e$ and context pattern $c$ appear. $|e_a, *, e_b|$ denotes the number of sentence $s \in S_c^t$ containing entity pair $(e_a, e_b)$. The top $n$ patterns are selected for the next iteration. If the extracted pattern is less than $n$, we use all extracted patterns for further iteration.

Similarly, we measure the entity pair $e \in E_u^t$ using the context pattern set $C^t$ as follow:

$$S(e) = \frac{1}{|C^t|} \sum_{c \in C^t} \frac{|e_a, c, e_b|}{|*, c, *|}$$

where $|*, c, *|$ is the number of sentence $s \in S_c^t$ containing context pattern $c$. We select the top $m$ entity pairs for the next iteration. If the number of entity pair in $E_u^t$ is less than $m$, we use all entity pairs in $E_u^t$ for the next iteration.

# 5   Semi-supervised Multi-view Ranking

In this section, we illustrate the graph based multi-view algorithm which ranks all the extracted entity pairs by their relevance to the given entity pair seeds.

## 5.1   Intra-view and Inter-view Graphs Generation

Before applying the multi-view ranking algorithm, we need to construct a co-occurrence matrix $M$ of the extracted entity pairs in $E$ and the extracted context patterns in $C$. $m_{ij}$ is the co-occurrence frequency of entity pair $e_i = (e_{ia}, e_{ib}) \in E$ and context pattern $c_j \in C$. For entity pair $e_i$ and context pattern $c_j$, we submit a query, like "$e_{ia}\ c_j\ e_{ib}$", to the search engine. Then the number of hits returned by the search engine is set as $m_{ij}$. For two entity pairs $e_h$ and $e_i$, the corresponding rows $M_h.$ and $M_i.$ are the vector form of entity pairs. The context patterns in $C$ are treated as features to represent entity pairs. Consequently, we use function $Sim_e(M_h., M_i.)$ to calculate the similarity between $e_h$ and $e_i$. Similarly, we can compute the similarity of context pattern $c_j$ and $c_k$ using the function $Sim_c(M._j, M._k)$.

---

**Algorithm 1.** Multi-View Ranking

---

Given:
  – Intra-view similarity matrices $T^e$, $T^c$
  – Inter-view correlation matrices $T^b$, $[T^b]^\top$
  – Ranking score vector of two views $Y = \begin{bmatrix} Y^e \\ Y^c \end{bmatrix}$
1. Generate Matrix $T$.

$$T = \begin{bmatrix} T^e & T^b \\ [T^b]^\top & T^c \end{bmatrix}$$

2. Normalize matrix $T$: $W = \frac{1}{\lambda}T$, where $\lambda = \max_i \sum_j T_{ij}$.
3. Propagate on Matrix $W$.
    **repeat**
      $Y_{t+1} \leftarrow WY_t + Y_0$
    **until** converge
4. Sort entity pairs in $E$ by corresponding score in $Y^e$.

---

Following [9], we use both labeled and unlabeled nodes to create fully connected graph. We construct two intra-view graphs $G_e = < V_e, L_e >$, $G_c = < V_c, L_c >$. Here, $V_e = E$ and $V_c = C$ respectively represent the data points in *entity pair* view and *context pattern* view; the weighted edges in $L_e$ and $L_c$ correspond to similarities between data points in different views. Taking $G_e$ as example, a $|E| \times |E|$ matrix $T^e$ is constructed to represent graph $G_e$. The edge between entity pair $e_h$ and $e_i$ is weighted as Eq.1. Parameter $\sigma$ is the average similarity of all node pairs in $G_e$. Using the same method, we construct weighted graph $G_c$ and get a $|C| \times |C|$ matrix $T^c$.

$$T_{hi}^e = \exp\left(\frac{-Sim_e(M_{h\cdot}, M_{i\cdot})}{2\sigma^2}\right) \tag{1}$$

We also construct a bipartite graph $G_b = < V_b, L_b >$. The vertex set $V_b$ is composed of entity pairs and context patterns, the edges of $G_b$ connect an entity pair and a context pattern. The function $Sim_b(e_i, c_j)$ is designed to measure the correlation of the entity pair $e_i$ and the context pattern $c_j$. We construct a $|E| \times |C|$ matrix $T^b$ to express this graph.

$$T_{ij}^b = \exp\left(\frac{-Sim_b(e_i, c_j)}{2\sigma^2}\right)$$

Above matrices $T^e$, $T^c$ and $T^b$ are used as the input of the multi-view ranking algorithm.

## 5.2    Multi-view Ranking Algorithm

Let $Y^e$ be a $|E|$ dimensional ranking score column vector, where $[Y^e]_i$ denotes the similarity of $e_i$ to the given seeds. The given seeds are evaluated as 1 in $Y_0^e$, other elements are initialized as zero. Similarly, $Y^c$ is a $|C|$ ranking score column vector, whose i-*th* row represents the ranking score of context pattern $c_i$. In addition, $Y_0^c$ is initialized similarly as $Y_0^e$. Let $Y$ be a $(|E| + |C|)$-dimensional column vector.

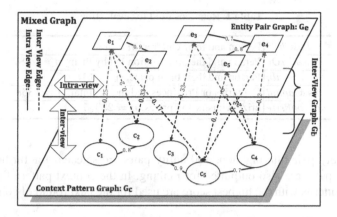

**Fig. 2.** Illustration to multi-view ranking algorithm. This mixed graph contains two intra-view graphs and an inter-view graph.

Algorithm 1 presents the multi-view ranking algorithm. In the first step of the algorithm, we use three graphs mentioned above to generate a matrix $T$. Putting the three graphs together makes the algorithm more concision. Matrix $T$ can represent a mixed graph, as Figure 2 shows, in which both entity pairs and context patterns are vertices. On this mixed graph $G = <V, L>$, we have $V = E \cup C$ and $L = L_e \cup L_c \cup L_b$. The ranking score of each entity pair is decided by both linked entity pairs and linked context patterns on graph $G$. On the context pattern graph $G_c$, the context patterns that are similar to give seeds get high ranking score. Each context pattern's ranking score is propagated to linked entity pairs through inter-view graph $G_b$. The weight of edges in graph $G_b$ control the context patterns' influence to the linked entity pairs.

In the second step, the matrix $M$ is normalized symmetrically, which is necessary for the convergence of the following iteration.

In the third step, the label score is propagated on the mixed graph generated in the first step. Finally, entity pairs $e_i = (e_{ia}, e_{ib})$ whose relevance scores in $[Y^e]_i$ are the highest $l$ are returned.

## 6  Experiments

### 6.1  Relation Extraction System

In order to generate entity pairs for ranking experiment, we built a relation extraction system using the framework in Section 3. In this relation extraction system, we index 4556821 wikipedia[1] pages and built a local search engine using Lucene[2] toolkit. In entity pair extraction step of the relation extraction system, we construct a dictionary based named entity recognizer. The used entity dictionary is composed with all extracted named entities of YAGO project[3].

---

[1] http://www.wikipedia.org/
[2] http://lucene.apache.org/
[3] http://www.mpi-inf.mpg.de/yago-naga/yago/

**Table 1.** Relations used for evaluation

| | |
|---|---|
| *hasChild* | Person and their children, $n = 4454$ |
| *isLeaderOf* | Person and the organization led by them, $n = 2887$ |
| *bornIn* | Person and their birth place, $n = 36189$ |
| *hasCapital* | Countries or Provinces and their capital, $n = 1368$ |
| *hasWonPrize* | prize winners and prizes, $n = 23075$ |

In the entity pair filtering step, 100 entity pairs are selected for further iteration. These entity pairs are also outputted for ranking. In the context pattern filtering step, 50 context patterns with the highest score are used for the next round of context pattern searching.

We repeat the bootstrapping process 5 times for every relation. At the beginning, 50 entity pairs and 10 context patterns are inputted as seeds.

### 6.2    Datasets and Evaluation Measures

We construct a large relation set using the result of YAGO project. We select 5 types of relation extracted by YAGO project for our experiments. Some facts about these selected relations are given in Table 1.

In order to evaluate the ranking performance, we use the extracted entity pairs of YAGO project as golden standard. For each relation type, we use $rel(i)$ to measure the relevance of a given entity pair $e_i$. For a target relation type, $rel(i)$ is set to 1 when $e_i$ appears in the corresponding golden standard, otherwise $rel(i)$ is set to 0. For two lists showing the same instances with different order, we suppose that the list in which highly relevant instances appear earlier (have higher ranks) is more useful. Recently, some experiments show that the Mean Average Precision (MAP) measure is sensitive to the query set size or may even provide misleading results. Comparing with MAP, Normalized Discounted Cumulative Gain ($nDCG$) appeared more robust to query set size and more relaible[11]. Therefore, we adopt $nDCG$ to measure ranking quality. $nDCG$ emphasizes highly relevant instances appearing early in the result list. The $nDCG$ measure is built on $DCG$ metric which is defined as follow:

$$DCG@p = \sum_{i=1}^{p} \frac{2^{rel_i} - 1}{\log_2(1 + i)}$$

For a extracted entity pair set, an ideal list can be produced by sorting entity pairs of the list by $rel(i)$. $IDCG@p$ is used to annotate the $DCG$ at position $p$ of this ideal list. Then, we have $nDCG@p$ as follow:

$$nDCG@p = \frac{DCG@p}{IDCG@p}$$

We compare the ranking results of 5 relations using the $nDCG@p$ measure.

## 6.3   Baseline Methods

We compare our algorithm against following methods:

**VSM:** This method is a vector based method which is proposed by Turney et al.[12]. Since the co-occurrence matrix $M$ of entity pair and context pattern is built as mentioned in previous section, the entity pair can be presented with context patterns using the rows of $M$. The similarity between the entity pairs can be computed as the cosine of the two corresponding vectors. We compute the similarity of candidate entity pair and given seed entity pair as below:

$$VSM(e) = \frac{\sum_{i=1}^{|S|} cos(s_i, e)}{|S|}$$

where $S$ is the set of entity pair seeds. Then the entity pairs which have the highest similarity score are selected.

**CON:** This is the measure proposed by Agichtein et.al[3]. The patterns are measured by the *confidence*, by which the context patterns that tend to generate wrong entity pairs are filtered. In this experiment, we use the instance confidence to measure the quantity of context pattern and entity pair.

$$CON(e) = \frac{e_{positive}}{e_{positive} + e_{negative}}$$

in which $e_{positive}$ is the number of positive matches of entity pair or context pattern. Taking entity pair as example, if entity pair $e$ matches the pattern $c$ which can be found in previous iteration, then this match is considered as a *positive* match. Otherwise, the match is *negative*.

**LRA:** The Latent Relational Analysis(LRA) is proposed by Turney [12]. For a matrix $M$, supposing the rows represent the entity pairs and the columns represent context patterns. Then Singular value decomposition(SVD) is performed on the matrix, in which the matrix toolkit [4] is used. The relation similarity of entity pair can be measured by the cosine of the angle between the two vector in matrix $U_k \Sigma_k$. Similarly, the relevance of context pattern can be measured using the vector in matrix $\Sigma_k V_k^T$. In our experiment, $k$ is set as 10. LRA is the current state-of-the-art relation similarity measure.

**PMI:** The *Espresso* information extraction system[5] uses the pointwise mutual information (PMI) to measure the relation between context pattern and entity pair:

$$PMI(e_i, c_j) = \frac{|e_{ia}, c_j, e_{ib}|}{|e_{ia}, *, e_{ib}||*, c_j, *|}$$

The ranking score of entity pair $e_i$ is set as $\sum_{j=1}^{|C|} PMI(e_i, c_j)$.

---

[4] http://code.google.com/p/matrix-toolkits-java/

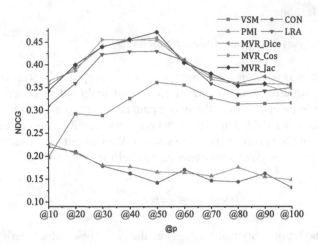

**Fig. 3.** Average $nDCG@p$ of five relations

In this experiment, we take Dice coefficient as inter-view measure to weight the graph $G_b$ and test the sensitivity of multi-view ranking to the intra-view similarity measure. We test three frequently used similarity measures in the naturel language processing community. These measures are used to weight the edges of graph $G_e$ and $G_c$. Following definitions only take graph $G_e$ as example, context patterns' similarity are calculated using column vector $M_{.j}$ and $M_{.k}$.

**Dice:** Dice coefficient is a usually used measure in Natural Language Processing community. In this experiment we want to test the sensitivity of label propagation algorithm to the similarity measures. The Dice coefficient is used to weight the graph $G_e$ which is defined as:

$$Sim_e(M_{h.}, M_{i.}) = 2\frac{|M_{h.} \bigcap M_{i.}|}{|M_{h.}| + |M_{i.}|}$$

**Cos:** In this method, the cosine similarity is used to weight the graph $G_e$ and $G_c$. Given co-occurrence matrix of context pattern and entity pair $M$, we can construct the graph $G_e$ with the following cosine similarity measure:

$$Sim_e(M_{h.}, M_{i.}) = Cosine(M_{h.}, M_{i.})$$

**Jac:** In this setting, we compute the *Jaccard* score between each entity pairs $e_h$ and $e_i$, using following equation:

$$Sim_e(M_{h.}, M_{i.}) = \frac{|M_{h.} \cap M_{i.}|}{|M_{h.} \cup M_{i.}|}$$

**Table 2.** Average $nDCG@p$ of Multi-View Ranking and Baselines, $(p = \{10, 20, ..., 100\})$

| Method \\ Relation | VSM | CON | PMI | LRA | MVR-Dice | MVR-Cos | MVR-Jac |
|---|---|---|---|---|---|---|---|
| hasChild | 0.2691 | 0.2541 | 0.2492 | 0.3995 | **0.4247** | 0.4120 | 0.4083 |
| isLeaderOf | 0.1977 | 0.1431 | 0.1562 | 0.3703 | **0.4004** | 0.3791 | 0.3927 |
| bornIn | 0.4052 | 0.1450 | 0.1588 | 0.3964 | 0.3878 | **0.4122** | 0.4049 |
| hasCapital | **0.4424** | 0.1380 | 0.1538 | 0.3758 | 0.3960 | 0.3896 | 0.4018 |
| hasWonPrize | 0.2292 | 0.1495 | 0.1617 | 0.3132 | 0.3655 | **0.3656** | 0.3587 |
| Average | 0.3087 | 0.1659 | 0.1759 | 0.3710 | **0.3949** | 0.3917 | 0.3932 |

### 6.4  Experimental Results

These seven methods described above presented for comparison in table 2 and Figure 3.

Figure 3 shows the average $nDCG@p$ score of five relationships. We observe from Figure 3 that multi-view ranking algorithm(MVR) outperforms other methods. Comparing vector space mode (VSM) with measure based CON and PMI, we can observe that VSM works better than CON and PMI. Comparing multi-view ranking based method (MVR) with LRA, the multi-view ranking algorithms outperform LRA in most cases except $p = 60$ and $p = 100$.

Table 2 shows the performance of these algorithms on different relations. The results show that MVR algorithm outperforms other methods except the hasCapital relation. Furthermore, the multi-view ranking based algorithms get the highest $nDCG$ score in average of five relations. A close look into the contexts extracted from the relation extraction system reveal that context patterns of hasChild relation contain less noise. Then when we rank the extracted entity pairs of hasChild relation, most algorithms achieve better performance than other relations.

## 7  Conclusions

We propose a graph based multi-view learning algorithm to rank the returned relation instances of a bootstrapping-based relation extraction system. We compare the MVR algorithm to the existing methods, relevant score based methods and frequency based methods, the results indicate that the MVR can improve the performance of the relation extraction systems.

## References

1. Brin, S.: Extracting patterns and relations from the world wide web. In: WebDB Workshop at EDBT 1998, pp. 172–183 (1998)
2. Blohm, S., Cimiano, P., Stemle, E.: Harvesting relations from the web: quantifiying the impact of filtering functions. In: Proceedings of the 22nd National Conference on Artificial Intelligence, pp. 1316–1321 (2007)

3. Agichtein, E., Gravano, L.: Snowball: Extracting relations from large plain-text collections. In: Proceedings of the Fifth ACM International Conference on Digital Libraries (2000)
4. Etzioni, O., Cafarella, M., Downey, D., Popescu, A.M., Shaked, T., Soderland, S., Weld, D.S., Yates, A.: Unsupervised named-entity extraction from the web: an experimental study. Artificial Intelligence 165, 91–134 (2005)
5. Pantel, P., Pennacchiotti, M.: Espresso: Leveraging generic patterns for automatically harvesting semantic relations (2006)
6. Zhu, J., Nie, Z., Liu, X., Zhang, B., Wen, J.R.: Statsnowball: a statistical approach to extracting entity relationships. In: Proceedings of the 18th International World Wide Web Conferece, pp. 101–110 (2009)
7. Banko, M., Cafarella, M.J., Soderl, S., Broadhead, M., Etzioni, O.: Open information extraction from the web. In: IJCAI, pp. 2670–2676 (2007)
8. Collins, M., Singer, Y.: Unsupervised models for named entity classification. In: Proc. Joint SIGDAT Conference on Empirical Methods in Natural Language Processing and Very Large Corpora, pp. 100–110 (1999)
9. Zhu, X., Ghahramani, Z.: Learning from labeled and unlabeled data with label propagation. Technical report, Technical Report CMU-CALD-02-107 (2002)
10. Zhou, D., Bousquet, O., Lal, T.N., Weston, J., Schölkopf, B.: Learning with local and global consistency. In: Advances in Neural Information Processing Systems, vol. 16, pp. 321–328. MIT Press, Cambridge (2004)
11. Radlinski, F., Craswell, N.: Comparing the sensitivity of information retrieval metrics. In: Proceeding of the 33rd international ACM SIGIR Conference on Research and Development in Information Retrieval 2010, pp. 667–674 (2010)
12. Turney, P.D.: Similarity of semantic relations. Computational Linguistics 32, 379–416 (2006)

# A Hybrid Approach for the Extraction of Semantic Relations from MEDLINE Abstracts

Asma Ben Abacha and Pierre Zweigenbaum

LIMSI, CNRS, F-91403 Orsay, France
{asma.benabacha,pz}@limsi.fr

**Abstract.** With the continuous digitisation of medical knowledge, information extraction tools become more and more important for practitioners of the medical domain. In this paper we tackle semantic relationships extraction from medical texts. We focus on the relations that may occur between diseases and treatments. We propose an approach relying on two different techniques to extract the target relations: (i) relation patterns based on human expertise and (ii) machine learning based on SVM classification. The presented approach takes advantage of the two techniques, relying more on manual patterns when few relation examples are available and more on feature values when a sufficient number of examples are available. Our approach obtains an overall 94.07% F-measure for the extraction of cure, prevent and side effect relations.

## 1 Introduction

Relation extraction is a long-standing research topic in Natural Language Processing, and has been used to help, among others, knowledge acquisition [1], information extraction [2], and question answering [3]. It has also received much attention in the medical [4] and biomedical domains [5]. With a large amount of information, health care professionals need fast and precise search tools such as question-answering systems [6]. Such systems need to correctly interpret (i) the questions and (ii) the texts from which answers will be extracted, hence the need for information extraction approaches such as [4,7,8]. The complexity of the task lies both in the linguistic issues known in open-domain tasks and in domain-specific features of the (bio)medical domain.

We propose here a hybrid approach to the detection of semantic relations in abstracts or full-text articles indexed by MEDLINE. This approach combines (i) a pattern-based method and (ii) a statistical learning method based on an SVM classifier which uses, among others, semantic resources. Their combination is based on a confidence score associated to the results of each method. We focus on extracting relations between a *disease* and a *treatment*. The obtained results are good and show the interest of combining both types of methods to disambiguate the multiple relations that can exist between two medical entities.

A. Gelbukh (Ed.): CICLing 2011, Part II, LNCS 6609, pp. 139–150, 2011.

In section 2 we present some related works in semantic relation extraction. We then describe the proposed hybrid approach and its components in section 3. In section 4 we present our experiments and results[1].

## 2   Previous Work

### 2.1   Open-Domain Relation Extraction

Open-domain extraction of semantic relations between entities in text corpora has used approaches based on the co-occurrence statistics of specific terms [9] and machine-learning approaches (e.g., [10]), as well as more linguistic approaches based on patterns or extraction rules [1] or hybrid approaches which combine these two techniques [11]. Zhou et al. [12] use lexical, syntactic and semantic knowledge in relation extraction with an SVM classifier. The proposed feature-based approach outperforms tree kernel-based approaches, achieving 55.5% F-measure in relation detection and classification on the ACE2003 training data. Their results also suggest that for relation extraction the base phrase chunking information is very effective while additional information from full parsing gives limited enhancement.

### 2.2   Relation Extraction in the Biomedical Domain

Similar approaches exist in the medical domain. Stapley and Benoit [13] tackle the detection of relations between genes based on word co-occurrence statistics. Other approaches use high-precision methods based on manually-written rules or patterns. Cimino and Barnett [14] use patterns to detect relations in MED-LINE article titles. They rely on the MeSH descriptors associated with these articles in MEDLINE and on the co-occurrence of target terms in the same title to generate semantic relation extraction rules. Khoo et al. [15] address the extraction of causal relations in medical article abstracts by matching graphical patterns to syntactic dependency trees. Embarek and Ferret [8] propose to extract four relations (*detect*, *treats*, *sign_of*, *cures*) between five types of medical entities, based on patterns which are automatically built using an edit distance between sentences and a multilevel phrase matching algorithm. The SemRep system [4] identifies semantic relations ('predications') in the biomedical literature using manually developed rules. Schneider et al. [16] use syntactic patterns over parsed text, surface patterns and automatically learned 'transparent words' to detect protein-protein interactions. They obtain 80.5% precision and 21.0% 'loose recall' on the IntAct corpus.

In parallel, other works use machine-learning techniques to detect which semantic relationship links two occurrences of medical entities in a sentence. Xiao et al. [17] address the extraction of protein-protein interactions with a supervised learning method. Based on lexical, syntactic and semantic features, they obtain a recall of 93.9% and a precision of 88.0%. Roberts et al. [18] focus on

---

[1] This work has been partially supported by OSEO under the Quæro program.

semantic relations in medical texts (e.g. *has finding*, *has indication*) and propose a method to identify such relations based on supervised learning with an SVM classifier. Grouin et al. [19] used a hybrid approach for relation extraction from clinical texts under the i2b2/VA 2010 challenge. A pattern-based approach is applied first, then for sentences where no relation was found, a machine learning approach is automatically applied. They obtained 70.9% of F-measure.

## 2.3   Extracting Relations between Treatment and Disease Entities

Some works focus on the semantic relations that link a *disease* and a *treatment*. This is motivated by the importance of these two types of medical entities and their high frequency in both the biomedical literature and clinical texts. Various types of methods have been used for this task. Lee et al. [20] apply manually built patterns to oncology medical abstracts to identify *treats* relations between a *drug* and a *disease*. They obtain a recall of 84.1% and a precision of 48.1% on their test set of sentences. With the same type of method, Ben Abacha and Zweigenbaum [21] use patterns to extract relations of the same type between a *treatment* and a *disease*. Their patterns are built semi-automatically by (*i*) automatically compiling a set of sentences that contain both a *treatment* and a *disease*, then (*ii*) manually selecting the sentences that contain a treatment relation and finally (*iii*) building patterns from these sentences. They obtain 60.5% recall and 75.7% precision.

Beside pattern-based methods, machine learning is also used to identify relations between a *treatment* and a *disease*. Rosario and Hearst [22] tackle the disambiguation of seven relation types. They compare five generative models and a neural network model and find that the latter obtains the best results. Frunza and Inkpen [25] used the corpus of Rosario and Hearst [22] and tackled the extraction of three relations types (cure, prevent and side effect) between treatment and disease entities. They used Weka [23] and tested six different models to learn the target relations. Their results show that probabilistic and linear models give the best results.

## 2.4   Summary

Several methods have been proposed to extract semantic relations. Some approaches focus on recall while others favor extraction precision. A common point in semantic relation extraction is the need for domain knowledge to describe specific relations. Linguistic methods provide for deep analysis of the context of occurrence of each medical entity and relation, but some relations are still hard to detect due to the high variability of expression of these relations and to the sometimes complex structure of some sentences. Besides, learning-based approaches can only obtain a high accuracy when enough training examples are available for a given relation. In this context, we propose an approach which combines linguistic, statistic and knowledge-based methods to determine the relationships between two given medical entities. We focus on the relations between a *treatment* and a *disease*.

# 3   Material and Methods

In this section, we describe the train and test corpora. We then describe our hybrid approach for relation extraction.

## 3.1   Corpus Description

We use Rosario and Hearst [22]' corpus, which was also used by Frunza and Inkpen [24] for relation extraction. This corpus is extracted from MEDLINE 2001 and annotated with 8 semantic relations between diseases (DIS) and treatments (TREAT). These relations (cf. table 1) are: *Cure*, *Only DIS* (TREAT is not mentioned in the sentence), *Only TREAT* (DIS is not mentioned), *Prevent*, *Vague* (unclear relation), *Side Effect* and *No Cure*. The relations *Only DIS* and *Only TREAT* are not adapted to our objective as only one entity was annotated in the sentence. The small number of examples for the relations *Vague* and *No Cure* does not allow efficient learning-driven extraction. Consequently we chose only three relation types: *Cure*, *Prevent* and *Side Effect*.

**Table 1.** Initial Corpus

| Relation | Definition |
|---|---|
| nb (train, test) | *Example* |
| **Cure** | TREAT cures DIS |
| 810 (648, 162) | *Intravenous immune globulin for recurrent spontaneous abortion* |
| **Only DIS** | TREAT not mentioned |
| 616 (492, 124) | *Social ties and susceptibility to the common cold* |
| **Only TREAT** | DIS not mentioned |
| 166 (132, 34) | *Flucticasone propionate is safe in recommended doses* |
| **Prevent** | TREAT prevents the DIS |
| 63 (50, 13) | *Statins for prevention of stroke* |
| **Vague** | Very unclear relationship |
| 36 (28, 8) | *Phenylbutazone and leukemia* |
| **Side Effect** | DIS is a result of a TREAT |
| 29 (24, 5) | *Malignant mesodermal mixed tumor of the uterus following irradiation* |
| **No Cure** | TREAT does not cure DIS |
| 4 (3, 1) | *Evidence for double resistance to permethrin and malathion in head lice* |
| **Total relevant: 1724 (1377, 347)** | |
| **Irrelevant** | TREAT and DIS not present |
| 1771 | *Patients were followed up for 6 months* |
| (1416, 355) | |
| **Total: 3495 (2793, 702)** | |

We split the initial corpus into equally-sized train and test corpora for each target relation. However, each sentence in the corpus is annotated with only one relation and potentially many TREAT/DIS entities. It is often the case that one sentence contains more than one relation between different TREAT/DIS couples. We chose to duplicate such sentences to have multiple sentences annotated each by only one relation and one unique TREAT/DIS couple.

This method also allowed us to exploit sentences initially annotated as <to_see> sentences (containing more than one relation) by duplicating them into (many) sentences annotated each with one relation. For instance, the following corpus sentence: "<DIS_SIDE_EFF> Progressive multifocal leukoencephalopathy </DIS_SIDE_EFF> following <TREAT> oral fludarabine </TREAT> treatment of <DIS> chronic lymphocytic leukemia </DIS>" was rewritten into two sentences annotated respectively with the relation "side_effect" and "cure" and their corresponding TREAT/DIS entities.

The final number of sentences in the extended corpus is presented in Table 2 for each relation. The varied numbers of availabe examples for each relation allows testing and evaluating the contribution of different kinds of approaches.

**Table 2.** Number of train and test sentences for each relation

| Relation | Training Corpus | Test Corpus |
|---|---|---|
| Cure | 524 | 530 |
| Prevent | 43 | 33 |
| Side Effect | 44 | 28 |

### 3.2    Pattern-Based Approach

This approach is based on manually-constructed lexical patterns for each target relation [21]. Basically, patterns are regular expressions describing a set of target sentences containing medical entities at specific positions in a more or less specific lexical context. Table 3 presents the numbers of patterns used for each relation and simplified pattern examples. Each pattern consists in a sequence of words, semantic tags (i.e. DIS and TREAT) and generic markers representing a length-limited character sequence (e.g. [^\.]{0,75} indicates a sequence of 0 to 75 characters not including dots).

**Table 3.** Examples of relation patterns

| Relation | Patterns Number | Examples |
|---|---|---|
| Cure | 60 | DIS [^\.]{0.75} relieved by [^\.]{0.75} TREAT |
| Prevent | 23 | TREAT [^\.]{0,80} for prophylaxis (against\|of) [^\.]{0,40} DIS |
| Side Effect | 51 | [^(no\|without)] + DIS [^\.]{0,80} following ((the)? administration of)?[^\.]{0,80} TREAT |

Patterns were constructed manually from the training corpus and from other MEDLINE corpora used in [21]. The precision of pattern-extracted relations varies according to the lexical context specified by the matching pattern. We therefore qualify a pattern which describes a more precise lexical context as "more specific" (e.g. involves more words than other "less specific" patterns).

We take into account such specificity by associating a weight to each pattern. This weight is used (i) to extract relations by choosing the most specific matching patterns in case of multiple extraction candidates and (ii) in the hybrid extraction approach where it is a contributing factor in the selection of the relation to be extracted.

In order to compute the pattern weights, we organise them in a hierarchical manner. Patterns deriving from other patterns are considered as more specific. We define and annotate the *generalise* relationship between patterns to describe their hierarchical structure. Pattern weights are then computed automatically by decrementing the "leaf" patterns' weights (initially set to 1) according to the *generalise* relation. For instance, starting from the patterns set $E1$, the annotations set $E2$ is generated automatically using the coefficient $C$ ($C$ is an integer parameter).

$$E1 = \left\{ \begin{array}{l} < pattern1, \; specificity, \; \boldsymbol{P} > \\ < pattern2, \; generalises, \; pattern1 > \\ < pattern3, \; generalises, \; pattern2 > \end{array} \right\}$$

$$E2 = \left\{ \begin{array}{l} < pattern2, \; specificity, \; \boldsymbol{P/C} > \\ < pattern3, \; specificity, \; \boldsymbol{P/C/C} > \end{array} \right\}$$

If a pattern generalises many different patterns, we choose the minimum level of specificity of its derived patterns. Patterns are then used from the most specific to the most general in the relation extraction process. Table 4 presents some pattern examples.

**Table 4.** Examples of Weighted Patterns

| Pattern | Relation | Pattern Weight | Example |
|---------|----------|----------------|---------|
| TREAT for DIS | Cure | 0.50 | *Intralesional corticosteroid therapy* for *primary cutaneous B cell lymphoma.* |
| TREAT for the treatment of DIS | Cure | 0.75 | *Cognitive-behavioral group therapy is an effective intervention* for the treatment of *geriatric depression.* |

As indicated earlier, the pattern weights are used to compute a confidence index for the extracted relation. This index also takes into account the number of noun phrases between the considered source and target medical entities.

The intuition behind this second factor is to consider a relation extraction as "stronger" if there are only verbs and/or prepositions between the involved medical entities compared to the case where one or many noun phrases separate the entities in the sentence.

The confidence index $I$ of a relation occurrence $R$ extracted by a pattern $P$ from a sentence $S$, between two medical entities $E1$ and $E2$ is defined as:

$$I(R) = \frac{W(P)}{c^{NN(S,E1,E2)}} \tag{1}$$

- $W(P)$: Weight of P
- $NN(S, E1, E2)$: Noun-phrases number between E1 and E2 in S

## 3.3  Supervised Learning Method

This second method is based on a supervised learning technique. Given several predefined categories (here, the target relations, or the absence of relation), such a technique relies on a set of examples of these relations to take a decision in front of new examples. Each example is represented by a set of features. We use a linear classifier (SVM [25], using the LIBSVM library [26]) because of its known performance in text categorization.

The problem is modeled as follows: given two entities E1 and E2 in a sentence, determine the relationship which links them (or the absence of any relationship). We compiled three types of features to describe the data: ($i$) lexical features, ($ii$) morphosyntactic features, and ($iii$) semantic features.

**Lexical Features.** Covers word-related features: (1) the words of the source entity E1, (2) the words of the target entity E2, (3) the words between E1 and E2 (and their number), (4) the words before E1, (5) the words after E2, and (6) the lemmas of these words.

**Morphosyntactic Features.** Includes: (1) the parts-of-speech of all the words, (2) the verbs between E1 and E2, (3) the verbs before E1, and (4) the verbs after E2. We used TreeTagger [27] to perform POS tagging of the corpus sentences.

**Semantic Features.** Consists of those features which use external semantic resources. In the medical domain, the largest semantic resource is the UMLS [28]. The first class of semantic features relies on the UMLS Metathesaurus and includes: (1) the concept associated to E1, (2) the concept associated to E2, (3) the concepts found between E1 and E2. The second class relies on the UMLS Semantic Network and includes: (1) the semantic type of E1, (2) the semantic type of E2, (3) the semantic types of the medical entities between E1 and E2, and (4) the semantic relationships that can possibly link E1 and E2.

The present work focuses on the possible relations between *disease* and *treatment* medical entity types. The last feature is hence not discriminant in this particular case since the set of possible relations is the same for all entity pairs.

We also take into account the verbs types as verbs are often the first clues for relation identification. However, as far as we know, there are no verb-related semantic resources for the medical domain, therefore, we chose domain-independent ones: ($i$) the VerbNet semantic classes [29], and ($ii$) Levin's semantic classes [30], to type the verbs found between entities E1 and E2, before E1, and after E2.

### 3.4    Hybrid Method

This method combines the two preceding methods to compute a global result according to the confidence indices associated to the results of each method.

Relation extraction then depends on the influence, or weight, granted to each method. We rely on the number of training examples of a relation to compute the influence of the supervised learning approach on the extraction procedure. This weight, noted $\mu_S(R)$ for a given relation, obtains the following values for the target relations: 0.897 for *cure*, 0.056 for *prevent*, and 0.047 for *side effect*.

The influence of the pattern-based approach is computed with two different weights: a global weight $\mu_P(R)$, which is the complement of $\mu_S$ for a given relation $R$: $\mu_P(R) + \mu_S(R) = 1$, and a finer-grained weight for each extracted relation occurrence, which takes into account the confidence index computed for this relation occurrence (see Section 3.2). A pattern-extracted relation only has influence when ($i$) its confidence index is greater than a given threshold $I_{min}$ and ($ii$) its global weight is greater than or equal to the weight of the supervised learning method for the same relation: $\mu_P(R) \geq \mu_S(R)$.

## 4    Experiments

The training corpus described in section 3.1 was used (i) to design sentence patterns for relation extraction with the pattern-based method and (ii) as training corpus for our statistical approach. All methods were then tested on the test corpora. In this section, we first summarize the different experimental settings we used and present the obtained results with the classical measures of *precision*, *recall* and *F-measure*. Precision is the number of correct relations extracted divided by the number of returned relations. Recall is the number of correct relations extracted divided by the number of reference relations. F-measure is the harmonic mean of recall and precision.

### 4.1    Settings

Table 5 presents the 5 experimented settings. Multi-class machine learning (ML1) uses only one model for all relation types. It provides a multiple classification of sentences into three classes (one class per relation type). Mono-class machine learning (ML2) uses 3 different models, each associated to only one target relation. It provides a binary classification for each relation type (positive and negative).

**Table 5.** Experimental Settings

| | |
|---|---|
| **Pat** | Pattern-based Method |
| **ML1** | Multi-class Machine Learning |
| **ML2** | Mono-class Machine Learning |
| **H1** | Pat + ML1 |
| **H2** | Pat + ML2 |

## 4.2 Results

Table 6 presents the results for each relation type. Table 7 presents the overall recall, precision and F-measure values computed on all extracted relations.

**Table 6.** Precision $P$, Recall $R$ and F-measure $F$ of each relation for each setting

| Config. | Cure | | | Prevent | | | Side effect | | |
|---|---|---|---|---|---|---|---|---|---|
| | P | R | F | P | R | F | P | R | F |
| Pat | 95.55 | 32.45 | 48.44 | 89.47 | 51.51 | 65.37 | 65.21 | 53.57 | 58.63 |
| ML1 | 90.44 | **100** | 94.98 | 15.15 | 15.15 | 15.15 | 0 | 0 | 0 |
| ML2 | **99.43** | 91.97 | 95.55 | 90 | 27.27 | 41.86 | **100** | 7.14 | 13.33 |
| H1 | 95.07 | 98.30 | 96.66 | **90** | 54.54 | 67.92 | 65.21 | 53.57 | 58.82 |
| H2 | 95.42 | 98.30 | **96.84** | **90** | 54.54 | 67.92 | 68.00 | **60.71** | **64.15** |

**Table 7.** Precision $P$, Recall $R$ and F-measure $F$ on all relations

| Setting | Precision (%) | Recall (%) | F-measure (%) |
|---|---|---|---|
| Pat | 91.89 | 34.51 | 50.17 |
| ML1 | 90.52 | 90.52 | 90.52 |
| ML2 | 91.96 | 91.03 | 91.49 |
| H1 | 93.73 | 93.73 | 93.73 |
| H2 | 94.07 | 94.07 | **94.07** |

The Hybrid methods effectively outperform the pattern-based and machine learning approaches in terms of F-measure. The contribution of both hybrid approaches on the "prevent" and "side effect" relations is important w.r.t the results obtained for these relations by the machine learning approach. In the same way, their contribution is important on the "cure" relation w.r.t the pattern-based technique.

## 4.3 Discussion

Several semantic relation extraction approaches detect only whether a relation type $T$ occurs in a given sentence or not. In our approach we tackle the extraction of medical relationships between specific medical entities in the sentences. We can therefore extract many relations in one sentence (e.g. from the sentence: *"E1 treats E2 but increases the risk of E3"* we can extract cure(E1,E2) and side_effect(E1,E3) ).

Frunza and Inkpen [24] obtained good results on the same corpus as that used in our experiments. However, their objective was relation detection in a one-relation-per-sentence assumption rather than precise relation extraction with the identification of the relation's source and target (i.e. multiple-relations-per-sentence assumption). In the same context [24] considered the *Only DIS* and *Only TREAT* annotated sentences as negative examples when in our case negative sentences are sentences where source and target entities exist but are not linked with a semantic relation (i.e. cure, prevent and side effect).

A second aspect of our work is the contribution of hybrid approaches w.r.t pattern-based approaches and machine learning approaches for relation extraction. In our experiments we observed that pattern-based methods offer good precision values but can be weak when faced with heterogeneous vocabulary and sentences complexity. On the other hand, machine-learning techniques can be very efficient but need enough training examples to obtain good results.

Our method differs from Grouin et al. [19] approach since we rank our lexical patterns according to their specificity level and to sentence-level criteria (e.g. number of noun phrases between the entities). We also automatically compute weights for both pattern-based and machine learning-based approaches. These weights are then applied to integrate the output of each method.

The combination of pattern-based and machine learning approaches allows us to take advantage of both techniques. The proposed hybrid approach obtained good results on the "cure" relation extraction because an important number of training examples was available in the corpus (524 sentences in the training corpus). For the other two relations (*Prevent* and *Side Effect*), few training examples are available (respectively 43 and 44) and machine learning performance was largely diminished (cf. Table 6). However this lack was compensated by (i) the use of manually-constructed patterns and (ii) the weighting of both approaches which takes into account the number of training examples available.

## 5  Conclusions

In this paper we presented a hybrid approach for relation extraction between medical entities from biomedical texts. We particularly investigated the extraction of semantic relations between treatments and diseases. The proposed approach relies on (i) a pattern-based technique and (ii) a supervised learning method with an SVM classifier using a set of lexical, morphosyntactic and semantic features. We experimented this approach and compared it to the pattern-based and machine learning approaches applied separately. The obtained results show that the hybrid approach significantly outperforms the two mentioned techniques and provides a good alternative to enhance machine learning performance if few training examples are available.

We plan to evaluate our approach with other relation types and different corpora. It would also be interesting to integrate automatic medical entity recognition and test other learning models to compare them with the SVM model.

# References

1. Hearst, M.: Automatic acquisition of hyponyms from large text corpora. In: Proceedings of the 14th International Conference on Computational Linguistics (COLING 1992), pp. 539–545 (1992)
2. Agichtein, E., Gravano, L.: Snowball: Extracting relations from large plain-text collections. In: Proceedings of the 5th ACM International Conference on Digital Libraries, pp. 85–94 (2000)
3. Fleischman, M., Hovy, E., Echihabi, A.: Offline strategies for online question answering: Answering questions before they are asked. In: Proceedings of the 41st Annual Meeting of the Association for Computational Linguistics, Sapporo, Japan, pp. 1–7. Association for Computational Linguistics (2003)
4. Rindflesch, T.C., Bean, C.A., Sneiderman, C.A.: Argument identification for arterial branching predications asserted in cardiac catheterization reports. In: AMIA Annu Symp Proc., pp. 704–708 (2000)
5. Blaschke, C., Andrade, M.A., Ouzounis, C., Valencia, A.: Automatic extraction of biological information from scientific text: protein-protein interactions. In: ISMB 1999, pp. 60–67 (1999)
6. Zweigenbaum, P.: Question answering in biomedicine. In: de Rijke, M., Webber, B. (eds.) Proceedings Workshop on Natural Language Processing for Question Answering, EACL 2003, Budapest, pp. 1–4. ACL (2003)
7. Shadow, G., MacDonald, C.: Extracting structured information from free text pathology reports. In: AMIA Annu Symp Proc., Washington, DC (2003)
8. Embarek, M., Ferret, O.: Learning patterns for building resources about semantic relations in the medical domain. In: LREC 2008 (May 2008)
9. Hindle, D.: Noun classification from predicate argument structures. In: Proc. 28th Annual Meeting of the Association for Computational Linguistics (ACL 1990), Berkeley, USA (1990)
10. Wang, T., Li, Y., Bontcheva, K., Cunningham, H., Wang, J.: Automatic extraction of hierarchical relations from text. In: Sure, Y., Domingue, J. (eds.) ESWC 2006. LNCS, vol. 4011, pp. 215–229. Springer, Heidelberg (2006)
11. Suchanek, F.M., Ifrim, G., Weikum, G.: Combining linguistic and statistical analysis to extract relations from Web documents. In: KDD 2006: Proceedings of the 12th ACM SIGKDD International Conference on Knowledge Discovery and Data Mining (April 2006)
12. Zhou, G., Su, J., Zhang, J., Zhang, M.: Combining various knowledge in relation extraction. In: Proceedings of the 43th Annual Meeting of the Association for Computational Linguistics (2005)
13. Stapley, B., Benoit, G.: Biobibliometrics: information retrieval and visualization from co-occurrences of gene names in MEDLINE abstracts. In: Proceedings of the Pacific Symposium on Biocomputing, Hawaii, USA, pp. 529–540 (2000)
14. Cimino, J.J., Barnett, G.O.: Automatic knowledge acquisition from MEDLINE. Methods Inf. Med. 32(2), 120–130 (1993)
15. Khoo, C.S.G., Chan, S., Niu, Y.: Extracting causal knowledge from a medical database using graphical patterns. In: Proc. 38th Annual Meeting of the Association for Computational Linguistics (ACL 2000), pp. 336–343 (2000)

16. Schneider, G., Kaljurand, K., Rinaldi, F.: Detecting protein-protein interactions in biomedical texts using a parser and linguistic resources. In: Gelbukh, A. (ed.) CICLing 2009. LNCS, vol. 5449, pp. 406–417. Springer, Heidelberg (2009)
17. Xiao, J., Su, J., Zhou, G., Tan, C.: Protein-protein interaction extraction: a supervised learning approach. In: Proceedings of the 1st International Symposium on Semantic Mining in Biomedicine (SMBM) (2005)
18. Roberts, A., Gaizauskas, R., Hepple, M.: Extracting clinical relationships from patient narratives. In: BioNLP 2008 (2008)
19. Grouin, C., BenAbacha, A., Bernhard, D., Cartoni, B., Deléger, L., Grau, B., Ligozat, A.L., Minard, A.L., Rosset, S., Zweigenbaum, P.: CARAMBA: Concept, assertion, and relation annotation using machine-learning based approaches. In: Uzuner, Ö., et al. (eds.) i2b2 Medication Extraction Challenge Workshop (2010)
20. Lee, C., Khoo, C., Na, J.: Automatic identification of treatment relations for medical ontology learning: An exploratory study. In: McIlwaine, I. (ed.) Knowledge Organization and the Global Information Society: Proceedings of the Eighth International ISKO Conference (2004)
21. Ben Abacha, A., Zweigenbaum, P.: Automatic extraction of semantic relations between medical entities: Application to the treatment relation. In: Collier, N., Hahn, U. (eds.) Proceedings of the Fourth International Symposium on Semantic Mining in Biomedicine (SMBM), Hinxton, Cambridgeshire, UK, pp. 4–11 (2010)
22. Rosario, B., Hearst, M.A.: Classifying semantic relations in bioscience text. In: Proceedings of the 42nd Annual Meeting of the Association for Computational Linguistics (ACL 2004), Barcelona (July 2004)
23. Hall, M., Frank, E., Holmes, G., Pfahringer, B., Reutemann, P., Witten, I.H.: The WEKA data mining software: An update. SIGKDD Explorations 11(1) (2009)
24. Frunza, O., Inkpen, D.: Extraction of disease-treatment semantic relations from biomedical sentences. In: Proceedings of the 2010 Workshop on Biomedical Natural Language Processing, Uppsala, Sweden, pp. 91–98. Association for Computational Linguistics (2010)
25. Joachims, T.: Text categorization with support vector machines: Learning with many relevant features. In: Nédellec, C., Rouveirol, C. (eds.) ECML 1998. LNCS, vol. 1398. Springer, Heidelberg (1998)
26. Chang, C.C., Lin, C.J.: LIBSVM: a library for support vector machines (2001), Software available at http://www.csie.ntu.edu.tw/~cjlin/libsvm
27. Schmid, H.: Probabilistic part-of-speech tagging using decision trees. In: Proceedings of the International Conference on New Methods in Language Processing, Manchester, UK, pp. 44–49 (1994)
28. Lindberg, D.A., Humphreys, B., McCray, A.T.: The Unified Medical Language System. Methods of Information in Medicine 32(4), 281–291 (1993)
29. Kipper, K., Dang, H.T., Palmer, M.: Class-based construction of a verb lexicon. In: AAAI/IAAI, pp. 691–696
30. Levin, B.: English verb classes and alternation: A preliminary investigation. The University of Chicago Press, Chicago (1993)

# An Active Learning Process for Extraction and Standardisation of Medical Measurements by a Trainable FSA

Jon Patrick and Mojtaba Sabbagh

Health Information Technology Research Laboratory, School of Information Technology,
The University of Sydney, NSW, Australia
jonpat@it.usyd.edu.au, mojtaba.sabbagh@sydney.edu.au

**Abstract.** Medical scores and measurements are a very important part of clinical notes as clinical staff infer a patient's state by analysing them, especially their variation over time. We have devised an active learning process for rapid training of an engine for detecting regular patterns of scores, measurements and people and places in clinical texts. There are two objectives to this task. Firstly, to find a comprehensive collection of validated patterns in a time efficient manner, and second to transform the captured examples into canonical forms. The first step of the process was to train an FSA from seed patterns and then use the FSA to extract further examples of patterns from the corpus.

The next step was to identify partial true positives (PTP) from the newly extracted examples. A manual annotator reviewed the extractions to identify the partial true positives (PTPs) and added the corrected form of these examples to the training set as new patterns. This cycle was continued until no new PTPs were detected. The process showed itself to be effective in requiring 5 cycles to create 371 true positives from 200 texts. We believe this gives 95% coverage of the TPs in the corpus.

**Keywords:** Finite State Automata, Medical Measurements, Active Learning.

## 1 Introduction

Our work specializes in processing corpora of medical texts [2][8]. These corpora usually contain many years of patient records from hospital clinical departments.

Clinical notes are a distinctly different genre of text with characteristics such as: 30% non-word tokens, idiosyncratic spellings, abbreviations and acronyms, and poor grammatical structure. The capacity to process the notes accurately is a direct function of learning something about them, e.g. the correct expansion of an abbreviation, and then reusing that immediately to understand subsequent text. In real-life situations the turnover of staff in a local hospital is so great that the language in use in the notes has its own dynamism that makes the continual accumulation of knowledge about the notes fundamental to a successful implementation of any practical medical language processing (MLP) technology that will survive the test of time.

A. Gelbukh (Ed.): CICLing 2011, Part II, LNCS 6609, pp. 151–162, 2011.

The primary task in NLP is tokenization and all further processing is affected by the manner of tokenizing the text. In this work we deal with clinical text containing very complex patterns, such as medical measurements, chemical formula and tokens containing numbers and other non-alphabetic symbols (we refer to these tokens as complex tokens hereafter). A simple-minded approach to tokenization is likely to dismember these complex expressions and thereby make their correct recognition difficult. A good tokenization method should preserve the structure of the complex tokens and would subsequently increase the accuracy of processing modules later in a pipeline of information extraction.

From a wider perspective, there are many applications that need to process language. They can be as simple as identifying a short span of text that occurs locally in a sentence up to detecting the global structure of a document. We propose three levels of processing for identifying concepts in medical text: *Direct matching, Pattern matching, Statistical matching*.

Direct matching is principally defined by dictionary-based approaches for identifying concepts. We use this method for proof-reading, resolving unknown words [1] and also by adding a dynamic programming search by using a good concept matcher that has been built in our Laboratory [8]. Pattern matching is using rule-based strategies to recognise content, and statistical matching is typically using machine-learning methods. The task of a language processing system is to exploit all of these methods in an integrated pipeline that optimises each of their strengths and minimises their weaknesses.

Pattern matching is the subject of this article. Finite State Automata/Machine (FSA) are used to match pattern examples in text. A module has been developed that can be used for any project needs identifying regular language, and it is discussed in more detail in the next sections. Statistical matching is out of the scope of this work, but we have completed a number of projects to identify concepts that do not occur in a pre-determined structure and the relationships between them.

This paper consists of 9 sections. After the Introduction, we review related literature and then briefly describe two methods for performing this type of work. Section four is about creating and using finite state automata (FSA). More features of the FSA are described in section five. We discuss active learning workflow in section 6. Section 7 and 8 describe the completed work and conclusions.

## 2  Related Work

One of the applications of FSA is partial parsing. In many NLP projects, there is no need to have a complete parse graph of the sentence as they may be interested in grammatical chunks, such as NP, VP, etc. Here we can assume chunks belong to regular language and set up an FSA to detect chunks of interest. Abney [6] used cascaded FSAs to parse free text. He was interested in speed and efficiency of the FSAs and if they could be more accurate without sacrificing computational time.

Zhao et al [10] used the same approach along with adding a few heuristics called modality rules to improve their results. By these rules they infer the final outcome from the results come from many FSA running together instead of just relying on the longest match string. They improved the F-measure from 86.60 to 90.07 by these rules.

Carrasco et al [3] devised an algorithm for optimizing an incremental tree FSA. This is an algorithm for optimizing our FSA as it is built. The FSA generated from our examples is incremental and without loops, so this method is of potential value.

CSSR is an algorithm for generating probabilistic FSA [7], [9]. It creates a weighted FSA from training data. It is the same as our method but their algorithm builds a probabilistic FSA. We can compare their method with a Hidden Markov Model (HMM). HMM has a fixed structure and changes the weight of edges according to training data. In the CSSR algorithm, the model changes its structure to satisfy new training data. CSSR has been used to identify Name Entities in text with a score of F=89.01%. Setting the algorithm parameters such as Lmax that represent the maximum length of the patterns to be analysed is important. CSSR performs poorly under conditions of data sparseness.

## 3  Method

Two methods for extracting measurements from clinical text were tested, using REGEXP and using a trained FSA. We started with **Regular Expressions (RExp)** where a few rules could extract a reasonable amount of patterns but the number of measurement patterns was greater than at first expected. As more rules were developed to capture missed items, the rules became very complex in a manner that made it difficult to update them and any change had the risk of losing previously recognised patterns or introducing new false positives. Another problem is that the person who is doing the task should have an exhaustive knowledge about RExp and spend a considerable amount of time adjusting the rules. Finally, it is difficult to identify even a small opportunity for automating this process.

The second approach was using a trainable **FSA**. As RExp translates to an FSA computationally, we decided to build the FSA directly from patterns of data. In the RExp method we look at the patterns and infer some rules then Rexp was compiled to an FSA by Python, Perl etc. Effectively we developed our own module to create a shortcut that directly compiled text patterns to an FSA.

There is an argument here, if the expressions of interest in this study follow a regular language then finite state automata as a computational model is suitable for them. Some of these expressions are presented in table 1 showing a regularity in them. For example, most of them start with a header and are then followed by digits, also but there are a lot of exceptions available. These exceptions move the actual language of these expressions to unrestricted language thus other more elaborate computational models such as PDA (push down automata) are not able to cover the language correctly. Our solution (FSA) has at least the advantage of simplicity compared to other models especially the PDA, as there is no chance of taking advantage of stack because the expressions are more complex than just saving a part in a stack to help us to detect other parts. In other words, we try to use a big enough collection of FSAs to approximate the unrestricted language of expressions and our effort has been to make adding a new expression to the collection as easy as possible until filling the gap between the real language of expressions and the collection of FSA with the least effort.

## 4  Finite State Automata

Patterns and Examples are important concepts in this work and are defined here. In the corpus there are about 50 different types of measurements such as *Blood pressure (BP), Heart Rate (HR), Respiration Rate (RR), pH, Glascow Coma Scale (GCS), Temperature etc.* Each type was assigned a tag ( shown in parenthesis). Each measurement can be written in many different ways and so a standardised form was defined by a **pattern**. Hence there are many **examples** of the one pattern in the corpus.

One investigation has been performed on 30,000 examples of the Glasgow Coma Score (GCS). Depending on the level of generalization, between 2500 to 5000 different patterns can be recognized and interestingly 60% of them are hapaxes. The patterns used only once make the identification process complex. If they are ignored there is a loss in accuracy and adding each of them means one exception has been inserted and the more exceptions lead to more complexity. To tackle this problem we use a different generalization method to reduce the number of these exceptions.

### 4.1  Building an FSA

FSAs are built from training patterns, we put patterns in a text file with a simple format. The first column is the pattern type, for instance BP. The second column is the pattern itself that is a span of the text, the pattern is exactly reproduced there. Table 1 shows a sample of training patterns.

**Table 1.** A subset of training Types (columns 1, 3) with Patterns (columns 2, 4)

| Types | Pattern | Types | Pattern |
|-------|---------|-------|---------|
| MAP | MAP 62-75 | RR | RR 16-20 |
| BP | BP 140/65(84) | Measurement | 4mg/kg |
| HR | HR 72 | Measurement | 7mg/hr |
| RR | RR 13 | DRNAME | Dr. <:[A-Z][a-zA-Z]*<: [A-Z][a-zA-Z]*:> |

To build an FSA, firstly each pattern is tokenised then a light generalization is applied. We generalize patterns by converting all digits to a symbol and all whitespaces (tab, newline, space) to a single space. A branch of states is built for each pattern. The building algorithm begins from the start state. For the first pattern, no edges are radiating from the start state, so it creates a new state and an edge with a label of the current token connecting the two states together. Fig. 1 shows how the FSA is built with the application of 7 patterns. If the first part of one pattern is common with another pattern, no new states are made until a difference is detected. From that point the building algorithm makes new states for the rest of the pattern. States 19 to 22 and 23 show two patterns with the same start sequence.

Studying the FSA shows there is a substantial opportunity for optimization but at this stage this is not a priority. The FSAs built from patterns have 3 features. Firstly they have one start state, second the number of final states is the same as unique patterns in the training data, and third, there is no cycle in the automata. In each final state a span of the text has the same structure as the pattern used to make this branch of the FSA.

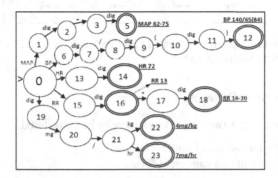

**Fig. 1.** Automaton built using patterns from Table 1. All numbers are generalized to the symbol *dig*.

## 4.2 Executing the FSA

Execution of the FSA over a text, identifies spans of the text, for which there is an example in the set of training patterns. The model of execution is to consider the FSA as a large tree with some active states. On each iteration the FSA processes one token. At the end of each iteration new active states are replaced with previous active states. The initial state should always be active so it is added to the list of active states at the beginning of each iteration because an example can start at any part of the text.

Apart from the initial state, a state becomes active if we can get to that state from an active state by a transition on the current token. These new active states are used in the next iteration to process the next token. If the new active state is a final state, then the algorithm saves all visited tokens from the initial state to the current final state as one matched region. Finally the algorithm returns the list of matched regions as its output.

## 4.3 General Edges

Sometimes, the token being processed is not of interest but rather its generalised format is important. For example, the pattern *Dr. John Smith* represents doctor name in a text. The FSA token processing is generalised so that after detecting *Dr.* it expects an alphabetical string starting with a capital letter and then another string with the same format. At this point the FSA instead of comparing two strings to find a match compares the format of the input token. To implement such an edge in the automata we defined generalised edges in the FSA. When the FSA gets to this edge in one state it looks at the RExp stored in the next state. (The RExp comes from a Patterns file) If the current token matches with this RExp, there is a transition to that state. Fig. 2 shows the structure of an FSA having a generalised edge.

General edges make the FSA non-deterministic. It means, at a specific node with a particular token, there are more than one token to transmit.

**Fig. 2.** Generalized token edges in a generated FSA. To hop from state 2 to 3 the current token should have the format [A-Z][a-zA-Z]*

### 4.4 Canonical Form of Patterns

On planning to use an FSA two major tasks have to be considered.

1. Detecting expressions of interest in text.
2. Converting them to a canonical form.

This process decreases the complexity of the text as it in effect ring-fences short token sequences that have a coordinated meaning but are more difficult to interpret when dealt with as isolated tokens. This allows the sequence to be treated as a whole lexical unit for other types of processing such as part-of-speech tagging, grammar parsing, and entity recognition. There are two steps to defining a canonical form for an expression of interest in a short token sequence: Firstly *marking up the patterns* and secondly *Defining a canonical template*.

A canonical form is actually a new format of the current examples in the text, and it often has the consequences of requiring the tokens of the text example to be reorganised for conformity. Hence before reordering the **components** of an example, the components need to be determined explicitly. A mark up process is used to identify the components. Before providing the details of canonicalisation an example of analysing a GCS is presented:

The original span in the text: GCS=M4+E2+V2=8

After annotating this text it becomes:

GCS=M<?vone:4/>+E<?vtwo:2/>+V<?vthree:2/>=<?vfour:8/>

However the Canonical template is defined in this form:

GCS: E=%(vtwo)s, V=%(vthree)s,M=%(vone)s, T=%(vfour)s

This example illustrates that a component is tagged by <?component-name:component-value/> symbols. The combination of <? shows the beginning of one component and /> indicates the completion of it. The component name and component value are separated by a colon (:) symbol. Now the FSA can recognise a component inside the example and marks the state matched to the component value by the component name.

When the FSA gets to the final state, it saves the extent of the span recognised in the text and executes a function that extracts the components from the matched region. This function traces the FSA backwards from the final state to the start and saves each value captured in each marked state into a (Python) dictionary data structure. The keys of the dictionary are component names and the values are component values thus capturing everything about the matched extent. The final action is to fill the canonical template with this data. It is a simple process to find the location of each component and replace the component name by its value. The canonical form of the above example will be:

GCS: E=2, V=2, M=4, T=8

Converting a token sequence to a canonical form is an important task but the drawback is each pattern has to be annotated and also the canonical form for each is provided manually. An effective way of applying this process computationally has yet to be found. This process has been tested and works effectively. The only issue that may occur is the duplication of the names of components in each pattern. A fixed set of names has to be used for all components of patterns to make sure no conflict occurs. As some parts of the patterns are common and will compile to the common nodes, they should have the same name. This means the first component of each pattern should be for instance *vone* (value one), the second *vtwo* and so on.

## 5   More Features

The FSA has two other features to increase its power of generalization. After testing the FSA on 100 clinical notes, 60% of the patterns have just one example, so these patterns have to be manually added to the training patterns, so a solution to make current patterns as general as possible until they match to a wide range of patterns was required. However adding new patterns may increase false positives (FPs) so it was necessary to find a tradeoff between FPs and the amount of manual work required.

### 5.1   Cascaded FSAs

The idea of cascaded FSAs comes from the fact that each pattern of measurement is composed of basic patterns. The basic patterns such as numbers, ranges, complex numbers (like 140/7) etc. can be factored out of the complex patterns. The effect of this strategy is the efficient use of patterns.

There are two lists of training patterns. One for building basic FSAs and another for complex FSAs. We define patterns like *BP range, HR range, RR range* instead of *BP 100-120, HR 60-100, RR 15-20* in complex pattern training set, now if a new range pattern is discovered and added to the basic pattern list, in effect 3 complex patterns have been added. In other words, the effect of adding one pattern in the basic FSA training data is equal to the number of times that basic type is used in complex training patterns.

This begs the question of: What should be the definition of the complex FSA? Processing patterns manually in a complex FSA is not a workable strategy in the long term. Our aim is to do the minimal amount of manual processing on patterns as possible, so we keep the format of complex patterns the same as basic patterns (Table 1) and when compiling complex FSAs, run basic FSAs over the complex pattern and convert the basic pattern examples to the appropriate tags first, then make the FSA from the processed patterns.

The same approach occurs when the text is passed to the complex FSA to identify complex examples. There are three levels of computation here (Fig. 3). On the bottom of Figure 3 is the text. A simple tokenization and generalization process promotes the text to level two. At level two the basic FSA finds basic patterns such as *Range, Digit,* and *Complex Digit* and replaces them with their type tags to arrive at the level three representation. Finally the complex FSA finds complex patterns and annotates the input text.

## 5.2 Irrelevant Words

A large percentage of the measurement patterns have just one example in the text. They can't be ignored yet processing all of them is difficult. After investigating these examples, it was evident that the problem of these patterns came about because some words or punctuation inserted inside the pattern structure. These are identified as Irrelevant Words (IWs).

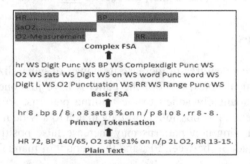

HR...........                    BP ...........................
SaO2...........................
O2-Measurement            RR........
**Complex FSA**
⬆
hr WS Digit Punc WS BP WS Complexdigit Punc WS
O2 WS sats WS Digit WS on WS word Punc word WS
Digit L WS O2 Punctuation WS RR WS Range Punc WS
**Basic FSA**
⬆
hr 8 , bp 8 / 8 , o 8 sats 8 % on n / p 8 l o 8 , rr 8 - 8 .
**Primary Tokenisation**
⬆
HR 72, BP 140/65, O2 sats 91% on n/p 2L O2, RR 13-15.
**Plain Text**

**Fig. 3.** Applying cascaded FSA on the text. There are three levels of processing: Primary Tokenisation and generalization, Basic FSA, Complex FSA.

While building an FSA, once an IW is found inside a pattern, the algorithm doesn't make a node in the FSA, it transitions back to the same state so as to consume the irrelevant token as a null transition. It will stay at this state until a key word arrives and thus enables it to change states. This approach reduces the number of states and also enables more generalized patterns.

The drawback of this approach is that it increases the number of FPs, where the FSA might be stuck at one state and consume the rest of document. Our solution to this problem is to decrease the generalisation and so prevent the automata from being trapped in the one state. The system has a definition of the maximum number of times that the null (irrelevant) transition at each state can be taken (**m**). This number is computed from training data and its default is zero. Building the FSA algorithm updates m when facing a pattern that has a number of irrelevant tokens which is more than the number stored in that state. In the execution, the FSA doesn't use null transitions more than m times for an active state. Table 2 compares the performance of FSAs when these features are added.

# 6  Active Learning Process

A workflow was devised to find the patterns of interest in a text. A seed set of training patterns was collected at first by a simple REGexp search and then a trainable FSA was built using these patterns. The FSA uses the examples of patterns as its training set and then is used to annotate a set of unseen documents. Then a discovery task is performed manually to check all annotations to find computed matches matching to parts of a larger entity or incorrect annotations. The examination of all new annotations is a time costly activity. The discovery task is stopped when a specific number of new patterns (for example, 20 new patterns) are recognised. These newly discovered patterns are added into the training data and a new cycle of annotation commenced.

To speed up the process, a mechanism for identifying FPs computationally based on partial matching is used. The aim is to find annotated examples where the token on the right side of the annotated region actually belongs to a valid example of interest, but as the training data does not have the whole pattern of the example it has only been partially captured. We call such partially identified examples Partial True Positives (PTP). The objective is to create a statistical model that generates a high probability when the righthand tokens belong to a valid example and low if not. The most straightforward strategy is to use n-grams to build the model.

## 6.1  N-gram Model

N-gram is a simple model and quite useful for learning local contexts. A bank of around 3000 pattern examples was used to build a 3-gram model with the SRILM language modeling toolbox [2]. The *probabilities* and *back off* were used to generate a model for each n-gram to compute the probability of the righthand token.

## 6.2  Generating the Probability of the Righthand Token

After annotating a document, each annotation is checked by computing the probability of it belonging to one or more tokens on its righthand side. For instance, using the GCS example, previously used in Section 4.5, the FSA annotated the string "GCS: V 1" as GCS which by itself is a PTP. The righthand tokens of the string is "M 6 E 4 = 11". The trigram that the PTP detector computes is the probability of ('V', '1', 'M'). 'V' and '1' come from annotated region and 'M' is found in the righthand context.

To improve the recall for a given PTP a selection of trigram probabilities is computed with the maximum subsequently accepted. There is an option to specify the number of tokens in the annotated portion to participate in building the trigram. For example if 2 is specified, then just the two closest tokens on the righthand of the annotated region are used, in the case of 3 then three pairs are made, that is ('V', '1'), (':', 'V'), (':', '1'). Also for the righthand region the number of tokens can be selected. In the case of a parameter value=1 only the first token is chosen. If parameter values were set at 3 for the annotated region and 2 for the righthand window, the algorithm would compute the probability of the following trigrams and use their maximum: ('V', '1', 'M'), (':', 'V', 'M'), (':', '1', 'M'), ('V', '1', '6'), (':', 'V', '6'), (':', '1', '6').

## 6.3 Post Processing Rules

After computing the probability for each annotated area, they are sorted according to the probability. There are some annotations that are not PTPs, but the model generates a high probability for them. For instance, "." or "," as the first token of the righthand window followed by a white space such as " . SpO2 91", is an example. To refine the PTP list, a very simple rule was devised that changed the probability of these types of patterns to a low value thus demoting their sort position leaving the real PTPs at the top of list. Now the annotator can readily find undiscovered patterns and add them into the valid training data.

# 7  Experiments

The FSA workflow was applied to 200 notes from the RPAH-ICU corpus. The training set for all notes is the same (289 patterns). Table 2 compares the FSA performance when these features were added.

Table 2. Comparing the Accuracy of Three Types of FSA

| Method | Patterns | Positive | PTP | TP |
|---|---|---|---|---|
| Simple FSA | 289 | 1000 | 71 | 929 |
| Cascaded FSA | 289 | 1032 | 44 | 988 |
| Full FSA | 289 | 1146 | 63 | 1083 |

The Full FSA contains the cascaded FSA and removal of Irrelevant Words processes which produced the best results overall. It uses the process of ignoring Irrelevant Words giving an increase of True Positives (TP) of 9% over Cascaded FSA. The cascaded automata was better than the Simple FSA by 6% in TPs.

The last cycle of the active learning process was to detect PTPs. A pair of tokens was chosen from an annotated region (lefthand side) and one from the righthand window. The algorithm was used to make various numbers of trigrams for each annotation and to determine the maximum number of PTPs with the largest probability for the various combinations generated. The output of the active learning process is the list of all examples sorted by probability of being a PTP. Different combinations of the number of tokens from the lefthand side and righthand side were tested. The best combination occurs when 3 tokens are chosen from the lefthand side and 2 tokens from the righthand side. In this case the number of PTP in the top 20 were exactly 20. If we increase the number of tokens in each side the number of PTP in the top 20 drops.

This workflow of successive discovery and reuse was applied to over 200 clinical notes to find valid measurement strings. In prior work we had annotated 100 notes in which it took approximately 10 hours to find all measurement instances. Table 3 shows how active learning can speed up the process so that we could annotate all measurements in this set of notes within less than 2 hours. Table 3 presents the number of PTPs in the top 20 of the list and the number of new patterns discovered after reviewing all the annotation instances. Besides, the amount of the time spent to find these new patterns, the number of annotated examples checked to find these new examples are presented.

**Table 3.** Review of 200 documents and progress rates for finding new measurement patterns

| Cycle | No. of Patterns | PTP in Top 20 | New Patterns | Time Spent | Examined |
|-------|------|------|------|------|------|
| 1 | 289 | 20 | 19 | 5 mins | 26 |
| 2 | 308 | 16 | 20 | 6 mins | 42 |
| 3 | 328 | 10 | 20 | 8 mins | 70 |
| 4 | 348 | 0 | 23 | 60 mins | 525 |
| 5 | 371 | 1 | 8 | 10 mins | 697 |

The First three cycles indicate that the PTP detector has done a good job and gathered many PTP examples on the top of the list and the annotator is easily able to detect them and add their expansions to the training data. The 4th cycle reveals that when the major PTPs were removed, the system cannot rank the remaining valid patterns at a high probability. This situation occurs because those patterns are very rare in the training data. As shown in cycle 4 the 23 discovered patterns are scattered over 525 examples and so revealing them took about 1 hour. In the fifth cycle the number of PTPs were very rare and one could easily look at the examples and pass them. There were no real PTPs when the log probability was less -5.

In this task the role of post processing is important. 505 examples out of 1413 were marked by these rules as not PTPs. This means we can ignore one third of annotated examples without missing any real PTPs.

# 8  Computational Complexity

Two algorithms are used in this process, firstly building the FSA and secondly executing the FSA. Building the FSA is fast. The runtime is k*m where "k" is the average length of patterns and "m" is the number of patterns in the training pattern set. We can assume k as a constant thus computational complexity for building the FSA is $O(m)$ that is a linear algorithm. The computational complexity of running the FSA is an order of m*n. "m" is the number of patterns in training data and "n" is the number of tokens in the input text. As the number of patterns (m) is very small compared to n, then the Order of the algorithm is $O(n)$. This means the growth of computational time depends linearly on the number of input text tokens.

# 9  Conclusion

In this study a pattern-matching engine consisting of a trainable FSA was created to capture non-lexical expressions of interest in clinical text. These expressions have a large amount of variety in terms of both many types and a multitude of various patterns for each type. Rule based FSAs such as [4], [3], [5] are not readily applicable to this task as each time a new pattern is discovered there needs to be a manual determination as to which rule in the FSA should be changed or a new rule added to cover the new pattern.

A dynamically trainable FSA engine was built to learn from text examples to resolve this problem, and processing functions were added to the automaton to increase the power of generalization of the machine. Although this strategy helps the FSA to identify more examples by learning a new pattern, it may however lead to more PTP. Table 2 shows the comparison of the effects of the generalisation functions and the cascaded FSA, which decreases the PTPs compared to the simple FSA.

Active learning has proven useful for improving the speed of capturing examples. At first we wanted to find a threshold for the probability of a detected pattern being a PTP and if the probability of an example is less than this threshold then we don't show it as PTP. However finding this threshold was not easy and there is always a chance that an unusual pattern makes a PTP although the probability of it is very low. In our experience if the log probability of an example were less than -5, it would not be a PTP. We decided to sort all annotated examples by their probability and so as to populate the top of the list with PTPs. This strategy allowed the annotator to search down the list and add the full examples pointed to by the PTPs to the training set.

# References

1. Patrick, J., Sabbagh, M., Jain, S.: Spelling correction in Clinical Notes with Emphasis on First Suggestion Accuracy. In: BioTxtM 2010, MALTA (2010)
2. Stolcke, A.: SRILM – An Extensible Language Modeling Toolkit. In: Proceedings ESCA Eurospeech, Denver, pp. 901–904 (2002)
3. Fael, C.C., Daciuk, J., Forcada, M.L.: Incremental Construction of Minimal Tree Automata. J. Algorithmica 55, 95–110 (2009)
4. Karttunen, L.: Applications of finite-state transducers in natural language processing. In: Yu, S., Păun, A. (eds.) CIAA 2000. LNCS, vol. 2088, p. 34. Springer, Heidelberg (2001)
5. Koskenniemi, K.: Two-level Model for Morphological Analysis. In: Proceedings of IJCAI 1983, pp. 683–685 (1983a)
6. Steven, A.: Partial parsing via finite-state cascades. J. Natural Language Engineering 2, 334–337 (1996)
7. Padro, M., Padro, L.: ME-CSSR: an extesion of CSSR using maximum entropy. In: Proceeding of the 2007 Conference on Finite-State Methods for NLP (FSMNLP), Germany (2007)
8. Patrick, J., Li, M.: High accuracy information extraction of medication information from clinical notes: 2009 i2b2 medication extraction challenge. JAMIA 17, 524–527 (2010)
9. Shalizi, C.R., Shalizi, K.L., Crutchfield, J.P.: An algorithm for pattern Learning Research. J. Journal of Machine Learning Research (2003)
10. Zhao, Y., Zeng, J., Yang, Y.: Chinese Partial Parsing with Modal Finite-State Approach. In: International Symposium on Computer Science and Computational Technology, vol. 2 (2008)

# Topic Chains for Understanding a News Corpus

Dongwoo Kim and Alice Oh

KAIST
Computer Science Department
Daejeon, Korea
dw.kim@kaist.ac.kr, alice.oh@kaist.edu

**Abstract.** The Web is a great resource and archive of news articles for the world. We present a framework, based on probabilistic topic modeling, for uncovering the meaningful structure and trends of important topics and issues hidden within the news archives on the Web. Central in the framework is a *topic chain*, a temporal organization of similar topics. We experimented with various topic similarity metrics and present our insights on how best to construct topic chains. We discuss how to interpret the topic chains to understand the news corpus by looking at long-term topics, temporary issues, and shifts of focus in the topic chains. We applied our framework to nine months of Korean Web news corpus and present our findings.

## 1 Introduction

The Web is a convenient and enormous source for learning about what is happening in the world. One can go to the Web site of any major news outlet or a portal site to get a quick overview of the important issues of the moment. However, it is difficult to use the Web to understand what has been happening over an extended period of time. We propose a computational framework based on probabilistic topic modeling to analyze a corpus of online news articles to produce results that show how the topics and issues emerge, evolve, and disappear within the corpus.

The problem of understanding a corpus of news articles over an extended period of time is challenging because one has to discover an unknown set of topics and issues from a large corpus of disparate sources, find and cluster similar topics, discover any short-term issues, and identify and display how the topics change over time. A narrower but similar problem has been studied in the TDT (topic detection and tracking) field [1] where the goal is to identify new events and track how they change over time. The events, however, are defined as happenings at certain places at certain times, and so they compose a small subset of general news topics and issues. For example, an earthquake in Haiti is an event, but the prolonged decline of real estate sales is not. The latter makes up a large portion of news, but the TDT community would only cover the former, whereas our research covers both. The probabilistic topic modeling community offers solutions such as Dynamic Topic Models [2] and Topics Over Time [3] for discovering topics

A. Gelbukh (Ed.): CICLing 2011, Part II, LNCS 6609, pp. 163–176, 2011.

and looking at how they change over time, but those models do not capture how topics newly emerge and disappear because they assume the same set of topics exist from the beginning through the end of the time-series data.

We propose *topic chains*, a framework for analyzing a sequential corpus, composed of similar topics appearing within a specified sliding window. Topic chains present a temporal and similarity-based organization of topics found by latent dirichlet allocation (LDA) [4]. Topic chains can be used to identify general topics, such as *labor unions* or the *stock market*, which occur in long topic chains. Short-term issues, such as the *death of Michael Jackson*, can be seen in short topic chains. Some short-term issues can be embedded within a long topic chain because they are related to general topics. One example of such issue is the recall of Toyota cars which is related to the general topic of the automobile industry. Those issues embedded within general topics can be identified by looking at *focus shifts* shown by words that change significantly within the topic chain.

Our contributions can be summarized as follows:

- We compare six frequently used similarity metrics using log likelihood of data for finding similar topics. We show that the two most frequently used metrics, cosine similarity and KL divergence, do not give the best results.
- We define and construct *topic chains* using the best similarity metric we found. We then illustrate how to further analyze topic chains to identify general topics and short-term issues.
- Overall, we propose a framework for understanding how topics and issues emerge, evolve, and disappear through time in a corpus of online news articles. This framework includes a set of analyses for a sequential corpus that other similar tools do not provide.

## 2   Related Work

This work can be positioned with respect to three related research areas: topic and event detection and tracking, probabilistic topic modeling, and temporal news mining.

Topic detection and tracking (TDT) is a well-studied task, summarized in Allan's book [1], and followed up by a line of research around *event threading* [5,6,7]. Both TDT and event threading solve a narrowly defined problem of looking for articles related to one or more *events*, where an *event* is defined as something that happens at a certain place at a certain time in the real world. We solve a much broader problem of discovering all topics and issues that occur in the corpus, whether or not they are directly related to concrete events in the world. Also, our definition of *issues* is more general than the definition of events by the TDT task. For example, the H1N1 influenza *issue* of 2009 is a series of related events such as deaths, vaccinations, and travel warnings, as well as non-events such as the safety of the vaccine, spatiotemporal course of the pandemic, and susceptibility of populations. We borrow two central aspects of the TDT task which are the discovery of new events and the evolution of events over time. We substitute our general definition of topics for their events such that our framework discovers new topics and how they evolve over time.

Probabilistic topic models [8] such as the frequently used latent dirichlet allocation (LDA) [4] discover all topics, regardless of event-like characteristics, that are highly represented in a corpus, and extensions to LDA, [9,2,10,3] consider the temporal aspect of the corpus as well. In [9], Wang et al. worked with asynchronous text streams to find common topics from documents with different timestamps. They found highly discriminative topics from asynchronous data and synchronized the documents according to topics. With dynamic topic models (DTM) [2], Blei and Lafferty analyzed how topics evolve over time in a sequential corpus, and they demonstrated how topics in the journal *Science* changed from 1881 to 1999. One limitation with DTM is that it only models the changes of word distributions within the topics and assumes the set of topics stays constant throughout the corpus, so it does not model how topics appear and disappear over time. The same limitation exists for the topic trend detection in [10]. With Topics over Time (TOT) [3], Wang and McCallum jointly model topics and timestamps to analyze when in the sequential data the topics occur. This model can discover when new topics appear and then disappear, but in this model, the topics stay the same over time. In our framework, different but similar topics form a topic chain so we can observe how the topics evolve over time.

Previous work on temporal news mining include [11,12,13]. Leskovec et al. [11] look at the news cycle by tracking how *memes* travel widely through the media sites and blogs. While this approach is very interesting, it does not capture the broad and overall picture of what topics and issues emerge and spread through the media sites. Shahat and Guestrin's work [12] looks at how two news articles can be connected through a series of articles in between them to form a coherent chain of articles. This is an effective solution to get a big picture of the story that connects two news articles. Mei and Zhai's work [13] is probably the closest to our work, but they work with data that is filtered for specific topics, such as the Asia Tsunami. They extend this work in [14] to include the spatial dimension. Our work aims to present an overall picture of topics and issues including how to identify general topics as well as temporal issues.

## 3   Overall Framework

Suppose there is a corpus of twelve months of news articles from major online newspapers that a user wishes to understand. A good way to do that is to break down the problem into finding the following details about the corpus:

- *Topic*: a *topic* is a major subject discussed in the corpus. Examples are "winter olympics", "healthcare reform", "the stock market".
- *Long-Term Topic*: if a *topic* lasts for a long time, we say it is a *long-term topic*. Examples are "the stock market", "Afghanistan war", "education".

- *Temporary issue*: if a *topic* lasts for a short time, we say it is a *temporary issue*. Examples are "the winter olympics", "earthquake in Haiti", "death of Michael Jackson".
- *Focus Shift*: a topic chain exhibits different focuses for each individual topic in the chain. An example of a focus shift is "Greece, moratorium" to "Europe, recession" in the "economy" *long-term topic*.

We propose a framework to analyze the corpus to find the *topics, long-term topics, temporary issues,* and *focus shifts*. In this section, we explain the parts that compose the overall framework.

1. **Discovering Topics**: We discover the topics in the corpus with latent dirichlet allocation (LDA) [4], the most widely used method of probabilistic topic modeling. LDA models topics as multinomial distributions of words.
2. **Measuring Topic Similarity**: We compare several methods for measuring topic similarity so that we can use the best method to find similar topics. We look at six popular similarity metrics and compare them in terms of log likelihood of data.
3. **Constructing Topic Chains**: A topic chain is a sequence of similar topics through time. Using the topic similarity metric, we look for similar topics within a sliding time window and add links between two similar topics to construct topic chains.
4. **Long-Term Topics and Temporary Issues**: After constructing the topic chains, we can identify *long-term topics* such as the stock market, *temporary issues* such as the Olympics. We can also identify *focus shifts* in *long-term topics*.

## 4   Topics

The first step in our analysis is finding topics in the corpus. Because we are looking at news data which are sequential by nature, we divide the corpus into several time slices, and for each time slice, we find a set of topics that are most salient in the documents within the time slice. We first describe the topic model we used for finding the topics, then we describe our dataset and the topics found in it.

### 4.1   Latent Dirichlet Allocation

LDA [4] is a widely used method for probability topic modeling. LDA is a generative model that models a document using a mixture of topics. In the generative process, for each document $d$, a multinomial distribution $\theta_d$ over topics is randomly sampled from a Dirichlet with parameter $\alpha$, and then to generate each word, a topic $z_n$ is chosen from this topic distribution, and a word, $w_n$, is generated by randomly sampling from a topic-specific multinomial distribution $\phi_{z_n}$. A topic-specific multinomial distribution $\phi_{z_n}$ is also randomly sampled from a

**Table 1.** Four topics discovered by LDA for the news dataset. Topics are randomly chosen and are represented by top ten probability words. Topic 1 is about "soccer game", topic 2 is about "market" and "business", topic 3 is about "smart phones", and topic 4 is about "research". Each topic is a multinomial distribution over words.

| Topic 1 | | Topic 2 | | Topic 3 | | Topic 4 | |
|---|---|---|---|---|---|---|---|
| Top words | Probability | Top words | Probability | Top words | Probability | Top words | Probability |
| game | 0.030 | growth | 0.035 | Apple | 0.024 | research | 0.078 |
| player | 0.026 | business | 0.034 | smartphone | 0.018 | professor | 0.042 |
| league | 0.025 | recovery | 0.031 | internet | 0.017 | science | 0.018 |
| coach | 0.023 | crisis | 0.026 | iphone | 0.016 | doctorate | 0.017 |
| soccer | 0.016 | prospect | 0.024 | mobile phone | 0.013 | discovery | 0.016 |
| season | 0.012 | policy | 0.023 | Google | 0.012 | analysis | 0.012 |
| leader | 0.011 | investment | 0.020 | computer | 0.011 | technology | 0.010 |
| competition | 0.011 | strategy | 0.018 | usage | 0.010 | universe | 0.010 |
| advance | 0.007 | market | 0.016 | advertise | 0.010 | plant | 0.009 |
| pro | 0.007 | consume | 0.015 | information | 0.008 | experiment | 0.009 |

Dirichlet with parameter $\beta$. From the generative process, we obtain the likelihood of a document:

$$p(\mathbf{w}, \mathbf{z}, \theta_d, \Phi | \alpha, \beta)$$
$$= \prod_{n=1}^{N} p(w_n | \phi_{z_n}) p(z_n | \theta_d) \cdot p(\theta_d | \alpha) \cdot p(\Phi | \beta).$$

The Dirichlet parameters $\alpha$ and $\beta$ are vectors that represent the average of the respective distributions. In many applications, it is sufficient to assume that such vectors are uniform and to fix them at a value pre-defined by the user, and these values act as smoothing coefficients.

## 4.2   Corpus

We collected over 130K news documents from the Web editions of three major Korean newspapers[1] between 2009-07-01 and 2010-04-10. Each news outlet covers a wide range of topics such as politics, economy, sports, entertainment, and culture, and show their own perspectives on cultural and social phenomena.

For the topic modeling task, we refined each document using a Korean morpheme analyzer and part-of-speech (POS) tagger provided by ETRI[2]. In the Korean language, each word can be broken down into morphemes. The morphemes are the smallest meaningful units, and each morpheme has a POS tag associated with it. Most of the morphemes do not carry semantic meaning but are instead used as syntactic markers, and almost every verb, adverb, and adjective can be broken down into morphemes with a noun token and one or more syntactic markers.

After preprocessing the documents as described, we divided the corpus into 28 time slices, ten days each. The average number of documents in each time

---

[1] http://www.yonhapnews.co.kr/, http://www.donga.com/, http://www.hani.co.kr/
[2] http://www.etri.re.kr

slice is 4,715, and the average number of unique words in each group is 13,611. We extracted 50 topics with LDA for every time slice using Gibbs sampling. To reduce the effort of estimating hyperparameters, we used symmetric Dirichlet priors. More specifically, for $\alpha$ and $\beta$, we adopted the commonly used values of 0.1 and 0.01 respectively. We set the number of topics to be 50 for one time slice, so the total number of topics is 1,400 for the entire corpus. We randomly chose 4 topics from the corpus and show them in Table 1, each topic represented with the words that have the highest probabilities in that topic.

# 5   Topic Similarity

To construct topic chains, we need to measure the similarity between a pair of topics. In previous topic modeling research where topic similarity must be measured, cosine similarity [15] and Kullback-Leibler (KL) divergence [16] are frequently used without any formal validation. There exist, however, several well-known metrics that can be used to measure topic similarity, so we compared them to see which metric would be best for our purpose. We considered six metrics and evaluated each metric using the negative log likelihood of corpus.

## 5.1   Six Metrics of Topic Similarity

A topic, $\phi_i$, is a multinomial distribution over the vocabulary, but it can also be viewed as a ranked list of words, or a $|W|$ dimensional vector, where each dimension $i$ is a probability of $w_i$ in that topic. A topic can also be represented by a set of topic words–words with a probability over a threshold. These various perspectives allow the following metrics for measuring similarity between topic $\phi_i$ and topic $\phi_j$:

- **Cosine Similarity** measures the similarity between two vectors by finding the cosine of the angle between them.
- **Jaccard's Coefficient** measures the similarity and diversity of two sets. It is defined as the size of the intersection divided by the size of the union of two sets.
- **Kendall's $\tau$ Coefficient** measures the association between two ranked lists.
- **Discounted Cumulative Gain(DCG)** measures the effectiveness of the ranked results of a web search algorithm.
- **Kullback-Leibler Divergence** is a non-symmetric measure of the difference between two probability distributions $p$ and $q$.
- **Jensen-Shannon Divergence** is the symmetric variation of KL divergence.

Each metric considers a different aspect of the relationship between two topics. Kendall's $\tau$ and DCG consider the ranks of words within a topic. KL divergence and JS divergence consider the divergence of multinomial topic probabilities, and lower divergence would indicate higher similarity between two topics. Cosine similarity measures the angle of two vectors, and Jaccard's coefficient looks at

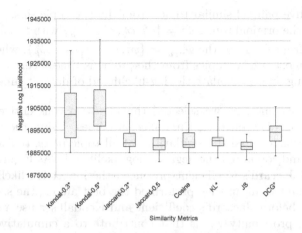

**Fig. 1.** Comparison of negative log likelihood for six similarity metrics using a boxplot. Negative log likelihood was computed for the corpus using the set of topics where five topics were substituted with the five most similar topics from another time slice, identified by each of the six similarity metrics. A better similarity metric gives a lower negative log likelihood. JS divergence and Jaccard's coefficient with 0.5 cumulative probability mass achieve better performances than the other metrics. An asterisk (*) next to a metric indicates statistically significant differences between the metric and JS divergence using t-test, $p < 0.01$.

the association between two sets. Jaccard's coefficient must use a partial set of words because it looks at the intersection and the union of the two sets of words that represent the topics. We use the top probability words that contribute to the cumulative probability mass, which is a parameter that must be set. We also use a partial set of top probability words for Kendall's $\tau$. This is because Kendall's $\tau$ is equally affected by the differences among high probability words and the differences among low probability words, but the words that have low probabilities in both topics should not contribute to the similarity score as much.

## 5.2    Comparing the Metrics

We compared the six metrics with the negative log likelihood of the corpus which measures how well the model explains the corpus. Starting from a set of topics extracted for a time slice, we substitute five topics with the topics from another time slice that are found to be most similar according to each of the six metrics to form six modified sets of topics. By comparing the negative log likelihoods using the modified sets of topics, we can see which metrics found the most similar topics. The process is as follows:

1. Train LDA for two consecutive time slices to get two sets of topics $\Phi^{t-1} = \{\phi_1^{t-1}, \phi_2^{t-1}, \phi_3^{t-1}, ...\phi_k^{t-1}\}$ and $\Phi^t = \{\phi_1^t, \phi_2^t, \phi_3^t, ...\phi_k^t\}$.
2. Compute the similarity score between $\phi_i^{t-1}$ and $\phi_j^t$ for every $i, j$.

3. Select top five pairs of similar topics from the two topic sets.
4. Substitute the original topics $\Phi^t = \{\phi_1^t, \phi_2^t, \phi_3^t, ... \phi_k^t\}$ with the five most similar topics from $t-1$. So the $\Phi_{new}^t = \{\phi_1^t, \phi_i^{(t-1)}, \phi_3^t, ... \phi_k^t\}$, where $i$ is a one of the five most similar topics from the previous time slice.
5. Finally, using $\Phi_{new}^t$, calculate the log-likelihood of data at time $t$.

To evaluate the metrics, we selected the first two consecutive time slices, and then trained LDA on each time slice 30 times. Using these 30 pairs of LDA results, we calculated the similarities of all topic pairs, replaced the most similar topics, and computed the negative log likelihoods. As Figure 1 shows, JS divergence and Jaccard's Coefficient produced the lowest log likelihood scores, which we interpret to mean they performed the best among the six metrics.

As we noted before, Jaccard's coefficient and Kendall's $\tau$ use a subset of the vocabulary–top probability words that contribute to a cumulative probability mass. The average size of the set of words with probability mass 0.5 is 39.56, and 0.3 is 13.58. The results show that Jaccard's coefficient can find similar topics at probability mass of 0.5, using only the top 40 words. Kendall's $\tau$ does not show good performance compared to Jaccard's coefficient although they use the same set of words. This result indicates that the ranking of top probability words does not matter much in judging topic similarity. DCG does not perform well for this topic similarity task even though it is a good metric of comparing ranked results in information retrieval (IR). This is because the typical results of IR include relevance scores, but the topics found by LDA do not have analogous scores to be used in place of relevance scores.

We further tested Jaccard's Coefficient with various probability masses. However, selecting a proper probability mass can be corpus-dependent. Hence, we conclude that JS divergence is best in terms of performance and generality, so we use JS divergence as the topic similarity metric in the rest of the paper.

## 6    Topics and Issues

Using the similarity discussed in the previous section, we construct topic chains to understand the topic trends in the main stream news. In this section, we discuss the construction of topic chains and associated parameters, interpretation of long topic chains, and the characteristics of short topic chains.

### 6.1    Constructing Topic Chains

We construct topic chains by finding similar topics within a certain time window. We use two parameters, similarity cut and sliding window, and follow this process:

1. Calculate the similarity between topic $\phi_i^t$ and topic $\phi_j^{t-1}$ for all topics at time $t-1$.

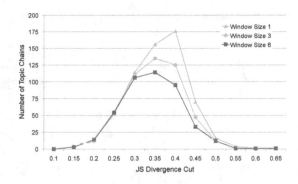

**Fig. 2.** Number of topic chains with different similarity cuts using JS divergence. The number of topic chains is significantly changed at JS divergence of 0.4. We chose JS divergence of 0.4 to construct topic chains.

2. If there are one or more topics such that $sim(\phi_i^t, \phi_j^{t-1})$ is greater than the similarity cut, we make links between all such topic pairs, and move to the next topic $\phi_{i+1}^t$.
3. If there are no similar topic pairs, we calculate similarity between $\phi_i^t$ and $\Phi_{t-2}$
4. Repeat, going back one more time slice, until one or more similar topics are found, or the time gap between the two time slices exceeds the sliding window size.

The two parameters, similarity cut and the window size, play important roles in determining the characteristics of the topic chains constructed. We discuss each of them below.

**Similarity Cut.** There is no standard similarity cut at which we can say two topics are similar, so we construct several topic chains, varying the similarity cut and looking at the effect on the resulting topic chains. Figure 2 shows how the number of topic chains changes with similarity cut using JS divergence. We define the size of a topic chain to be the number of topics in that chain, and we count topic chains whose size is greater than one. We also experimented with various sizes of the sliding window. If we set the JS divergence cut to a large value, then all topic nodes would be disconnected, and the total number of topic chains of size greater than one would be 0. Conversely, if we set the JS divergence cut to 0, then all topic nodes would be connected, and the number of topic chains would be 1. As Figure 2 shows, the number of topic chains changes significantly at 0.4. To see the relationship between JS divergence values and the similarity of two topics in a qualitative way, we can look at pairs of topics and the JS divergence values. From the qualitative analysis and the analysis of the number of topic chains, we decided that 0.4 is an appropriate threshold of JS divergence for constructing topic chains.

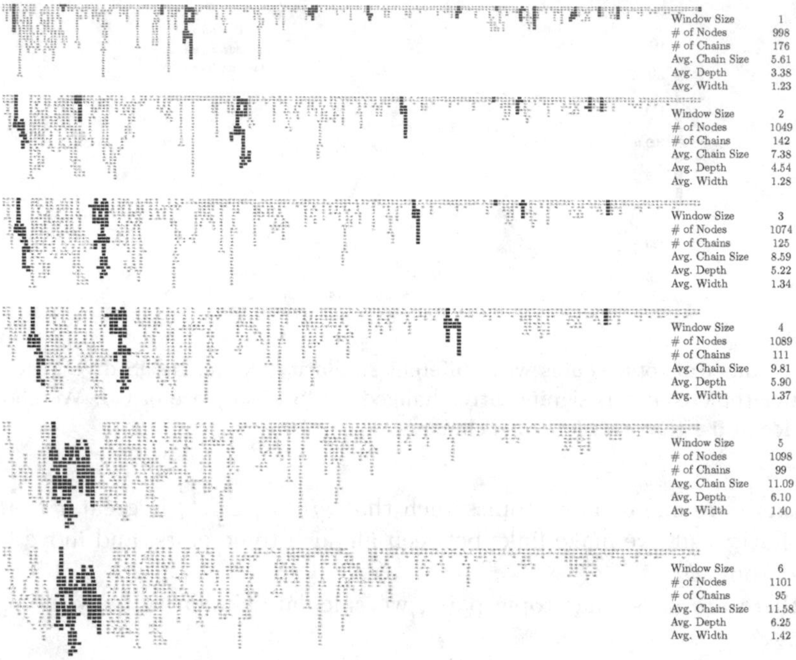
Window Size 1
# of Nodes 998
# of Chains 176
Avg. Chain Size 5.61
Avg. Depth 3.38
Avg. Width 1.23
Window Size 2
# of Nodes 1049
# of Chains 142
Avg. Chain Size 7.38
Avg. Depth 4.54
Avg. Width 1.28
Window Size 3
# of Nodes 1074
# of Chains 125
Avg. Chain Size 8.59
Avg. Depth 5.22
Avg. Width 1.34
Window Size 4
# of Nodes 1089
# of Chains 111
Avg. Chain Size 9.81
Avg. Depth 5.90
Avg. Width 1.37
Window Size 5
# of Nodes 1098
# of Chains 99
Avg. Chain Size 11.09
Avg. Depth 6.10
Avg. Width 1.40
Window Size 6
# of Nodes 1101
# of Chains 95
Avg. Chain Size 11.58
Avg. Depth 6.25
Avg. Width 1.42

**Fig. 3.** Six sets of topic chains constructed with sliding windows of sizes one to six. For each set of topic chains, every topic chain starts at the same vertical position. Within each topic chain, topics are temporally ordered, the oldest (first) topic at the top and going down to the most recent topic at the bottom. The number of nodes indicates the total number of topics that are connected with one or more similar topics. Blue nodes are those that belong to the largest topic chain in the last set of topic chains, constructed with the sliding window of size six. Blue nodes start out in the first set of topic chains as small topic chains, and they agglomerate as the size of sliding window increases. The full-size figure is available at *http://uilab.kaist.ac.kr/research/topic-chain/*

**Sliding Window Size.** The size of the sliding window is also an important factor for constructing the topic chains. If we use a sliding window of size one, it means that we only consider the previous time slice to find the similar topics for the topics of the current time slice. However, this Markov assumption is not generally helpful, as similar topics can appear over non-consecutive time slices, so proper consideration of the sliding time window is needed to capture these topic trends.

We vary the size of the sliding window from one to six and observe the changes in the resulting topic chains. Figure 3 shows the topic chains of size greater than one for the various window sizes with their descriptive statistics. First, the number of nodes indicates the total number of topics that belong in topic chains. This number excludes singleton topics and shows how many topics, out of 1,400 total, are matched with one or more similar topics within the time window. The

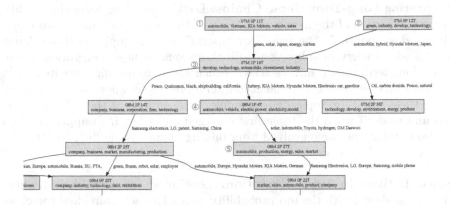

**Fig. 4.** Detecting focus shifts using difference of a word probability along the topic chain. Each rectangle represents a topic node, and contains top probability words. Edges connect two similar topics within a sliding window of size six, and the words next to the edge are the named entities whose probabilities are changed the most between two topics. xxM yyP zzT represent month, period, and topic number, respectively.

number of nodes increases at a faster rate from window size one to four and at a slower rate from window size four to six, and through that, we can see that similar topics do not necessarily appear in consecutive time slices.

Other graph characteristics also change with the size of the sliding window as shown in Figure 3. The total number of topic chains decreases as we increase the window size. This means many of the distributed small topic chains merge as the size of the sliding window increases. This is further evidenced by increases of both the average chain size and the average chain depth. The width of topic chain is the maximum number of topics of the same time slice in that chain. Unlike other increasing characteristics, the average width of the topic chain remains stable throughout the size of the sliding window. This is expected because topics of the same time slice represent different aspects of mainstream news.

Figure 3 illustrates how similar topics agglomerate as we vary the size of the sliding window. We painted in blue nodes of the largest topic chain at a sliding window of size six. We also painted the same nodes at the other sizes of the sliding window. At the window size of one, there are fourteen separate topic chains painted in blue. These chains join together to form larger chains as we increase the size of the sliding window.

## 6.2   Focus Shifts

When we construct topic chains, we find that there are long topic chains and short or singleton topic chains. Long topic chains tend to cover very general topics such as politics, economy, and sports, and we call them *long-term topics*.

**Interpreting Long-Term Topic Chains.** Looking at a long-term chain is like looking at a section of the newspaper. Many of the long-term topic chains could be labelled as "politics", "business", or "sports", and the topics in those chains reflect a wide variety of subjects within those general news categories. There are also long-term topics, such as H1N1, which are more specific news items but last for a long time. Our topic chains contain more helpful information for interpreting these long-term topics. For example, you can look at the "H1N1" topic chain and read off when the topic first emerged and when it disappeared. You can also see that the topic evolved from talking about "swine flu", to "travel", to "vaccinations" and "deaths".

**Named Entities in Topic Evolution.** Looking at the topic chains, where each node is shown with the top probability words for each individual topic, we can see the general evolution of the topic chain, but it is difficult to interpret the evolution to see what happened. This is because the words that represent the individual topics may be too general and occur in many topics throughout the topic chain. For example, words like *season, home run, game*, and *coach* are always top probability words in a topic chain about baseball. Those frequently occurring top words tell us what the general topic trend is, but it may be more interesting to see how the focus shifts for each topic within the chain.

To identify the words that can help to understand the focus of the topic chain at each time slice, we hypothesized that the words tagged as named entities–people, places and organizations–would be good discriminating words of the different focuses within the topic chain. We illustrate these named entities with the most changes in probabilities in Figure 4. Each rectangle represents a topic with the top five probability words. An edge connects two similar topics, and the words next to the edge are the named entities that change the most between the two topics. For example, topics 1 and 3 are both about the automobile industry, but the named entities *green, solar, Japan*, and *energy*, show that the focus is on green energy for topic 3. We can indeed find a related news article from the time period of topic 3 with the headline "Toyota makes eco-friendly solar car". Also we can see the evolution of the topic from 2 to 3. Topic 2 represents the general green (environmental-friendly) industry. By incorporating the focus words *automobile, hybrid, Hyndai Motors,* and *Toyota* this topic evolves into the topic of environmental-friendly automobiles, topic 3. From topic 3 to 4, the electric car and its battery problems received attention from news, and from 4 to 5, other alternative sources of energy, *solar* and *hydrogen* became the focus.

## 6.3   Short Topic Chains and Singleton Topics

We discussed long topic chains in the previous section, but short topic chains–chains of two or three topics, or singleton topics–are important for two reasons.

First, most of the short topic chains are about *temporary issues*. If a topic lasts over a long period of time, it would become part of a long topic chain. That means singleton topics and short topic chains are likely to be about temporary

**Table 2.** Examples of single node topic chains. First to sixth topics are temporary issues. First issue refers to the missile launch from North Korea, second issue is related to the death of Michael Jackson, fifth issue is related to the romance of Korean top actor and actress, and sixth issue is talking about Arbor Day on April 5. These are typical cases of temporary issues. However, the last example is not a coherent topic.

| Date | Topic |
|------|-------|
| OP 07M 2009Y | North Korea, missile, launch, range, UN Security Council, ship, navy, East sea, ballistic missile |
| OP 07M 2009Y | Jackson, family, funeral, cherish the memory of, Michael Jackson, son, LA, publish, report, death |
| OP 10M 2009Y | melamine, dry milk, region, environment, investigation, food, pollution, mercury, produce |
| 2P 12M 2009Y | flight, airport, passenger, airplane, search, terror, time, security, explosion, aircraft |
| OP 01M 2010Y | Hyesoo Kim, actor, 2010, ski, Haejin Ryu, once, 4, soul, colleague, lover |
| OP 04M 2010Y | tree, recover, park, culture, movement, development, environment, ecology, forest, designation |
| OP 02M 2010Y | Obama, Republicans, Jeju island, game, Jeju, golf, White house, Woods, gamers, budget |

issues, and we can see that is true for the examples of singleton topics and short chains listed in Table 2. Topics such as the death of Michael Jackson, reinforcing airport security at the end of the year, and romance between top actors do not last for a long time and can be considered as temporary issues.

Second, some of the singleton topics are useless. When we extract topics with LDA, the results do not consist of only meaningful topics. Sometimes LDA extracts topics that are not understandable as coherent topics. For example, the last topic in Table 2 is not a coherent topic. Constructing topic chains leaves bad results of LDA to be isolated as singleton topics. Conversely, topics that form long topic chains tend to be coherent. Evaluation of topics found by LDA is an on-going challenging research problem[17], so our topic chain framework may offer one solution of evaluating topics found in a sequential corpus. We will explore this in future work.

# 7   Discussions

In this paper, we proposed a framework for analyzing a corpus of news articles over a contiguous time period. Our framework discovers topics from the corpus, constructs topic chains using a topic similarity metric, identifies long-term topics and temporary issues, and detects focus shifts within each topic chain. An important contribution in this work is a comparison of various topic similarity metrics. We looked at six commonly used metrics and compared them using the negative log likelihood of corpus.

A secondary use of the topic chains is as an analysis tool to evaluate the quality of topics by a topic model. Most of the work on probabilistic topic modeling typically assume that the latent space is semantically meaningful, and so the topics are not systematically evaluated. In this work, we found that most of the topics that belong to long topic chains are semantically meaningful, whereas singleton topics are less coherent. Further analysis of the relationship among topics in the sequential corpus may find an effective way to analyze semantic meaningfulness of the topics.

# References

1. Allan, J.: Introduction to topic detection and tracking. In: Topic Detection and Tracking, Event-based Information Organization, pp. 1–16 (2002)
2. Blei, D., Lafferty, J.: Dynamic topic models. In: Proceedings of the 23rd International Conference on Machine Learning (2006)
3. Wang, X., McCallum, A.: Topics over time: a non-markov continuous-time model of topical trends. In: Proceedings of the 12th ACM SIGKDD International Conference on Knowledge Discovery and Data Mining (2006)
4. Blei, D., Ng, A., Jordan, M.: Latent dirichlet allocation. The Journal of Machine Learning Research, 993–1022 (2003)
5. Nallapati, R., Feng, A., Peng, F., Allan, J.: Event threading within news topics. In: Proceedings of the Thirteenth ACM International Conference on Information and Knowledge Management (2004)
6. Feng, A., Allan, J.: Finding and linking incidents in news. In: Proceedings of the Sixteenth ACM Conference on Information and Knowledge Management (2007)
7. Allan, J., Gupta, R., Khandelwal, V.: Temporal summaries of new topics. In: Proceedings of the 24th Annual International Conference on ACM SIGIR Research and Development in Information Retrieval (2001)
8. Blei, D., Lafferty, J.: Topic models. In: Text Mining: Theory and Applications, pp. 71–93 (2009)
9. Wang, X., Zhang, K., Jin, X., Shen, D.: Mining common topics from multiple asynchronous text streams. In: Proceedings of the Second ACM International Conference on Web Search and Data Mining (2009)
10. Bolelli, L., Ertekin, Ş., Giles, C.: Topic and trend detection in text collections using latent dirichlet allocation. In: Boughanem, M., Berrut, C., Mothe, J., Soule-Dupuy, C. (eds.) ECIR 2009. LNCS, vol. 5478, pp. 776–780. Springer, Heidelberg (2009)
11. Leskovec, J., Backstrom, L., Kleinberg, J.: Meme-tracking and the dynamics of the news cycle. In: Proceedings of the 15th ACM SIGKDD International Conference on Knowledge Discovery and Data Mining (2009)
12. Shahaf, D., Guestrin, C.: Connecting the dots between news articles. In: Proceedings of the 16th ACM SIGKDD International Conference on Knowledge Discovery and Data Mining (2010)
13. Mei, Q., Zhai, C.: Discovering evolutionary theme patterns from text: an exploration of temporal text mining. In: Proceedings of the Eleventh ACM SIGKDD International Conference on Knowledge Discovery and Data Mining (2005)
14. Mei, Q., Liu, C., Su, H., Zhai, C.: A probabilistic approach to spatiotemporal theme pattern mining on weblogs. In: Proceedings of the 15th International Conference on World Wide Web (2006)
15. He, Q., Chen, B., Pei, J., Qiu, B., Mitra, P.: Detecting topic evolution in scientific literature: how can citations help? In: Proceeding of the 18th ACM Conference on Information and Knowledge Management (2009)
16. Newman, D., Asuncion, A., Smyth, P.: Distributed algorithms for topic models. The Journal of Machine Learning Research 10, 1801–1828 (2009)
17. Chang, J., Boyd-graber, J., Gerrish, S., Wang, C., Blei, D.M.: Reading tea leaves: How humans interpret topic models. In: Advances in Neural Information Processing Systems (2010)

# From Italian Text to TimeML Document via Dependency Parsing

Livio Robaldo[1], Tommaso Caselli[2], Irene Russo[2], and Matteo Grella[3]

[1] Department of Computer Science, University of Turin
[2] Istituto di Linguistica Computazionale, CNR, Pisa
[3] Parsit s.r.l.
http://www.parsit.it/
robaldo@di.unito.it, tommaso.caselli@ilc.cnr.it,
irene.russo@ilc.cnr.it, matteo.grella@parsit.it

**Abstract.** This paper describes the first prototype for building TimeML xml documents starting from raw text for Italian. First, the text is parsed with the TULE parser, a dependency parser developed at the University of Turin. The parsed text is then used as input to the TimeML rule-based module we have implemented, henceforth called as 'The converter'. So far, the converter identifies and classifies events in the sentence. The results are rather satisfactory, and this leads us to support the use of dependency syntactic relations for the development of higher level semantic tools.

## 1 Introduction

The access to information through content has become the new frontier in NLP. Innovative annotation schemes such as TimeML [12] have push forward this aspect by creating benchmark corpora. The TimeBank corpus [13] has renewed the interest in temporal processing and in its use for complex NLP task such as Open-Domain Question-Answering [16], Summarization and Information Extraction.

The task of temporal processing can be split into different subtasks. First, the basic ontological entities involved, i.e. events and temporal expressions, must be recognized and treated on their own. Then, temporal relations between them can be computed. This paper describes an implemented event detector and classifier, which represents the first step of an ongoing research collaboration on the development of a TimeML-compliant tool for Italian.

In TimeML, an event is defined as something that holds true, obtains/happens, or occurs. Natural language (NL) offers a variety of means to realize events, namely verbs, complex VPs (such as light verb constructions or idioms), nouns (including nominalizations, second order nominals and type-coercions), predicative constructions, prepositional phrases or adjectival phrases. Two innovative aspects introduced by TimeML with respect to event detection and classification are represented by:

A. Gelbukh (Ed.): CICLing 2011, Part II, LNCS 6609, pp. 177–187, 2011.

a) the lenghth of the text span to be annotated.
b) Their classification, which is based on language independent criteria relevant for characterizing the nature of an event as being irrealis, factual, possible, reported, intensional and so forth.

Seven classes have been identified in TimeML:

1. *Reporting*: the action of a person, declaring something, narrating or informing about an event (e.g. 'say', 'tell', etc.).
2. *Perception*: events which involve the physical perception of another event (e.g. 'see', 'hear', etc.).
3. *I_action*: events which give rise to an intensional relation with their event argument (e.g. 'try', etc.).
4. *I_state*: events which give rise to an intensional state with their event argument (e.g. 'love', 'want', etc.).
5. *State*: temporally bound circumstances in which something obtains or holds true (e.g. 'peace', 'be in love', etc.).
6. *Occurrence*: events which describe things that happen in the world (e.g. 'happen', 'come', etc.).
7. *Aspectual*: events which describe an aspectual predication of another event (e.g. 'start', 'finish', etc.).

Event detection and classification has been usually tackled in recent years by applying different data-driven approaches. We have instead adopted a rule-based approach on the output of a dependency parser. The identification is built upon morpho-syntactic information and co-occurrences with specific keywords.

The remaining of the paper is organized as follows: in section 2 related works on this task are reviewed. Section 3 describes the formalism of the dependency parser we have used as input for detecting and classifying events according to the TimeML specifications adapted to Italian. Sections 4-5 describes our approach for event identification and detection. Section 6 reports the evaluation of the event detector and classification component on two set of data: the Italian training data released for the Task-13 (TempEval-2) at the SemEval 2010 workshop and the Wikipedia entries of the Turin University Treebank (TUT). Finally, section 7 presents the conclusion and future extensions of the algorithm.

## 2    Related Works

To the best of our knowledge, there is only one officially released corpus annotated with the TimeML specifications, the TimeBank [13]. However, the SemEval task, TempEval-2 [14], has provided an enlarged set of TimeML annotated documents for languages other than English, namely Italian, French, Spanish, Chinese and Korean. All approaches regarding TimeML event extraction have been evaluated either on the TimeBank corpus or on the TempEval-2 data.

The EVITA system [17] is the first system developed. It employs a hybrid approach by combining both linguistic and machine learning techniques. The results are evaluated against the TimeBank and the system reports 74.03% precision, 87.31% recall, and 80.12% on $F_{\beta=1}$ for event detection and an accuracy of 86.12% on classification.

One of the most recent system tested on the TimeBank corpus is [9]. It employs a statistical algorithm based on conditional random fields augmented with a variety of morpho-syntactic and semantic features. The system achieves 83.43% precision, 79.54% recall and 81.40% on $F_{\beta=1}$ for event detection and 68.84% precision, 60.15% recall and 64.20% on $F_{\beta=1}$ for event classification.

A further set of systems are those which partecipated to the TempEval-2 competition at SemEval. As for the event detection task, 5 systems took place: 4 for English [6], [7], [18] and 1 for English and Spanish [10]. The results are shown in Table 1.

As for Italian, no previous work on this task has been done.

**Table 1.** Results of TempEval-2 participants for event detection and classification

| System Name | Event Detection | | | Event Classification |
|---|---|---|---|---|
| | Precision | Recall | F-measure | |
| Edinburgh (EN) | 0.75 | 0.85 | 0.80 | 0.66 |
| JU_CSE (EN) | 0.48 | 0.56 | 0.52 | 0.53 |
| TipSem (EN) | 0.81 | 0.86 | 0.83 | 0.79 |
| TipSem-B (EN) | 0.83 | 0.81 | 0.82 | 0.79 |
| TRIOS (EN) | 0.80 | 0.74 | 0.77 | 0.77 |
| TRIPS (EN) | 0.55 | 0.88 | 0.68 | 0.67 |
| TipSem (ES) | 0.90 | 0.86 | 0.88 | 0.66 |
| TipSem-B (ES) | 0.92 | 0.85 | 0.88 | 0.66 |

## 3 The TULE Parser

The converter from Italian free text to TimeML documents takes as input the syntactic trees of the sentences in the text built by the TULE parser. TULE stands for 'Turin University Linguistic Environment' [8], and it outputs *Dependency* syntactic trees. The TULE software is free[1], and, according to [3], it is currently one of the parsers with the best attested performance for Italian.

A Dependency Grammar (DG) is a formalism that allows to describe NL syntax in terms of oriented relations between words, called 'dependency relations' or 'grammatical functions'. In particular, a DG analysis represents a NL sentence by means of a hierarchical arrangement of words linked via dependency relations.

Dependency relations are *oriented*; therefore, for each pair of linked words, we can identify a head (the origin of the link) and a dependent (the destination of the link). The dependent plays the role of "completion" of the head, i.e. it provides a sort of "parameter" to the latter, instantiating, in this way, the head

---

[1] It may be downloaded at *http://www.tule.di.unito.it*, under the section 'Download'.

meaning on the dependent meaning. Dependency relations are usually labeled in order to make explicit the function played by a dependent with respect to the head. Moreover, it is important to note that, in a dependency relation, not only the dependent, but the whole subtree having the dependent as root contribute to the "completion" of the head.

TULE has been used in a number of projects, including the development of TUT, a Treebank[2] for Italian [4]. The dependency arcs are labelled according to a dependency scheme that encodes the surface relations between words. The scheme is based on a twofold distinction between Functional and Non-functional dependents. The latter are dependents not having domain-based semantic import, e.g. *aux* (auxiliaries), *contin* (continuations, in idioms), *coord* (arcs related with conjunctions), *visitor* (e.g. in raising structures), and some particles void of semantic contents (as the Italian "accorger-si" - *remark* –, where the "si" reflexive pronoun is lexicalized into a pseudo-reflexive verb). The former are further split into Arguments and Modifiers, corresponding to the standard distinction between syntactic actants and syntactic circumstantials [11]. Arguments are mandatory argument as verbal complement like *subj* (subject), *obj* (object), *indcompl* (indirect complement), etc. Modifiers are optional and are further subclassified as *rmod* (restrictive modifiers) and *apposition*. An example is shown in fig.1.

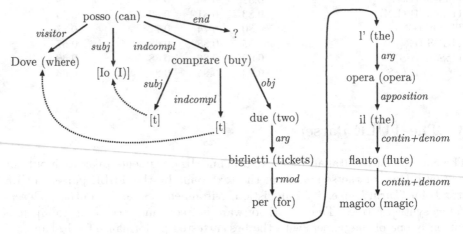

**Fig. 1.** TULE dependency tree for: "Dove posso comprare due biglietti per lopera Il Flauto Magico?" *(Where can [I] buy two tickets for the opera The Magic Flute?)*

Fig.1 also includes two traces; empty nodes include links (called *coref*) to their referents; these links, however, are not standard dependency arcs. Fig.1 also includes a compound name ("Il Flauto Magico", i.e. "The Magic Flute"), treated as an idiom (*contin+denom*) and acting as a denomination apposition of the noun "opera".

---

As it should be clear, Dependency Grammars are particularly suitable for the task of producing TimeML documents. The definition of the TimeML formalism is based upon the concept of *minimal chunk*. It is only the *head* of the Event Phrase that is annotated, and not the whole phrase. In a dependency tree, it is immediate to identify such an head, as it corresponds to the node in root position.

# 4   Event Detection via TULE Dependency Trees

According to the TimeML guidelines [12], events may be conveyed by four parts of speech only: verbs, nouns, adjectives, and prepositions. We implemented a Java™ module, that explores TULE dependency trees and, for each node belonging to any of these four classes, runs a set of ad-hoc if-then rules in order to decide if it must be annotated as Event. Below we report a brief description of the rules. These rules are very simple as they inspect the nearest nodes of the one under examination, and check if they belong to certain pre-built static lists of words and locutions, and/or if they satisfy some simple constraints. These lists have been created from the La Repubblica Corpus [1], on the basis of annotated documents used for the experimental annotation in the adaptation phase of the TimeML specifications to Italian [5].

However, the triviality of the rules stems straightforwardly from the syntactic structure, that already puts at disposal the relevant links involved in the words' meaning. In other words, the rules simply inspect the governor and the (nearest) modifiers of the node under examination, and check if they contain certain keywords.

## 4.1   Verbs

Verbs are the easiest words to process. Most of them, including modals and causative verbs, denote an event. The only exceptions are auxiliary verbs, that must not be annotated as Event. In TULE format, it is rather easy to identify auxiliary verbs, as they are linked to the main verb with an arc labelled as '*aux*'. The converter simply checks if the verb has such an arc. In case it does not, it is annotated.

## 4.2   Nouns

On the other hand, distinguishing between nouns that convey an event from those that do not involves more complex rules. The fact that a noun has an eventive meaning is strongly dependent on the context. For instance, the Italian word "assemblea" (*meeting*) is ambiguous between two senses: Human group, as (1.a), and Event, as in (1.b). Only the latter must be annotated as Event.

(1)   a. L'assemblea ha approvato il bilancio.
         (*The board meeting approved the budget*)
      b. L'**assemblea** è stata rinviata.
         (*The board meeting has been posponed*)

Accordingly, the converter checks nouns' senses. This is done with respect to MultiWordNet[3] (MWN). We automatically map each MWN synset into WordNet supersenses. This is possibile because MWN is strictly aligned with Princeton WordNet 1.6.

In case all senses associated with a noun correspond to certain MWN senses (e.g. 'act', 'event', 'phenomenon', etc.) the noun is annoted, in case none of them corresponds to one of them, the noun is *not* annotated, while in case only some of them do correspond, some further tests are performed. First, it is checked if the noun may denote a location, substance, shape or quantity *and* it has at least one modifier in the dependency tree. In case it does, it is *not* annotated as Event. Otherwise, we check if it is the subject or the object of a verb belonging to certain pre-built lists of verbs. Those verbs, among which "continuare" (*to continue*), "eseguire" (*to execute*), etc. have been identified as verbs that convey events in their syntactic arguments. Obviously, a lot of nouns denoting events are not captured by this simple rule. An example is shown in (2).

(2) I soldati di Napoleone erano stanchi dopo una lunga **campagna**.
    (*Napoleon's soldiers were tired after a long campaign*)

The noun "campagna" in Italian could mean 'military campaign', which is an event and so must be annotated, and 'countryside', which is a location and so must not be annotated. In the case of (2), it has the first sense, but the rule explained above is unable to detect it.

In order to harvest such names, we added some further rules: we check that either the noun includes, among its dependents, certain temporal modifiers as "nuovo" (*new*), "domani" *tomorrow*, etc., or that it is governed by certain "temporal" prepositions as "dopo" (*after*), "finché' (*until*), etc. In the case of (2), the second additional rule correctly identifies the event denoted by the noun, as it detects that the governor of "campagna" is the "temporal" preposition "dopo".

Two more cases must be covered: functional nouns, i.e. nouns associated with a value on a scale, and nouns which are part of locutions or common expressions in Italian. An example of the former is shown in (3.a) while an example of the latter, i.e. "prendere parte" (*to take part*), is shown in (3.b).

(3)  a. L'**utile** è di 30 milioni.
        (*the gain is of 30 millions*)
     b. Vi presero **parte** con la loro presenza.
        (*They took part in there with their presence*)

Two ad-hoc lists have been built to detect those cases: a list of verbs selecting functional names and a list of locutions. The converter checks if the governor of the noun belongs to one of the two lists.

Obviously, this is a very rough word-sense disambiguation. The definition of a more effective word-sense disambiguation module is seen as the object of future work.

---

[3] See [2] and the collection of references at http://multiwordnet.fbk.eu.

Finally, it must be pointed out that the converter is able to manage coordinations, as it applies the rules above to every noun in a coordination, and coreferences, as it considers the governors of every trace referring to the noun. This allows to identify events expressed by nouns modified by reduced relative sentences. For instance:

(4) I **risultati** ottenuti nel primo quadrimestre ...
    (*The results obtained in the first quarter ...* )

In TULE format, (4) is represented as in fig.2. The object of the verb "ottenuti" has been dropped, and a trace referring to "risultati" has been inserted to its place. "Ottenere risultati" (*to obtain results*) is considered as a common expression in Italian, and so it belongs to the list of locutions. Obviously, it can be detected just in case the noun is checked with respect to the governor of its trace.

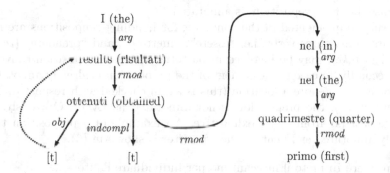

**Fig. 2.** Syntactic dependency tree associated with the phrase "I risultati ottenuti nel primo quadrimestre" (*The results obtained in the first quarter*)

### 4.3   Adjectives

Similarly to what is done with nouns, the converter loads and uses some lists of verbs and locutions to identify which adjectives must be annotated as Event. An example is the verb "rendere" (*to render, to make*): whenever an adjective occurs as its predicative complement, as in (5), the adjective denotes an event.

(5) ... che devono essere rese piú **esplicite** ...
     (*... that must be made more explicit ...* )

In TULE format, predicative complements are linked to their associated verbs with an arc labelled as *predcompl*, as in fig.3.

   As for nouns, it is easy to detect the occurrence of such patterns by checking the dependency links entering/exiting an adjective.

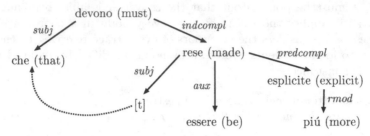

**Fig. 3.** Syntactic dependency tree associated with the phrase "che devono essere rese piú esplicite" *(that must be made more explicit)*

### 4.4 Prepositions

Finally, there are few cases where also a prepositional phrase may denote an event. According to the concept of minimal chunk, however, only the head of the phrase, i.e. the preposition, is annotated.

The rules implemented in the converter for handling prepositions are rather rigid. First, only three verbs, i.e. "essere", "mettere", and "prendere" (*to be, to put,* and *to take*), may be involved in patterns where a preposition conveys an event. Secondly, in case the argument of the preposition is already annotated as Event or it may denote a location (this is again checked with respect to MWN sense inventory), the preposition is not annotated as Event. Obviously, these rules are still far from being exhaustive. An example of a preposition that is correctly annotated as Event by the converter is shown in (6).

(6) ... mettere **in** moto il meccanismo per individuare l'autore ...
    (... *put in motion the mechanism for identifying the author* ...)

## 5   Event Classification

Similarly to the event detection, event classification has been implemented by means of a further set of if-then rules developed on a mapping between the TimeML classes and the semantic types of the PAROLE/SIMPLE/CLIPS lexicon[4]. The lexicon represents the largest computational lexical knowledge base for Italian, containing over 45 thousand lemmas and more that 57 thousand word senses, or semantic units. At the semantic layer of information, lexical units are structured in terms of a semantic type system and are characterized and interconnected by means of a rich set of semantic features and relations. The ontology is a multidimensional type system based on both hierarchical and non-hierarchical conceptual relations. The Event top node, whose sub-hierarchy mainly reflects the traditional event types, i.e. state, process and transition, has seven subtypes (Perception, Aspectual, State, Act, Psychological Event, Change, Cause Change) and has been used to develop the mapping between the TimeML classes and the semantic types of the resource.

---

[4] See [15] for details on the structure of the lexical resource.

Semantic information plays a primary role in the assignment of the TimeML classes. However, the semantic characteristics of an event are not always necessary and sufficient conditions for its classification. Other levels of linguistic information, like syntactic dependencies, verb form realization (finite *vs.* non finite forms) and the argument structure, may influence the class assignment or work as discriminating cues. The mapping provides each event denoting expression with a default semantic template in the resource, i.e. the ontology node. Once the event denoting expression is associated with its corresponding semantic type, a set of rules which keep into account the co-textual (i.e. syntactic) information obtained from the TULE parser apply and assign the corresponding TimeML class.

# 6   Evaluation

The evaluation of the converter has been performed on the Italian training set of the TempEval-2 task and on a subset of the Turin University Treebank (TUT). The former is a subset of the Italian TimeBank corpus, currently under development. It contains 52 articles from the ISST (Italian Syntactic Semantic Treebank). It contains 26,000 tokens, with 5,357 tokens tagged as events. The data set has been annotated by two annotators and validated by a an expert. The average precision and recall of the data is 0.89, with a K-value of 0.87. The TUT data is composed by 10 files which have not been previously annotated according to the TimeML specifications and used as a second evaluation test in order to identify possible discrepancies due to the parser errors. The annotation with the TimeML specifications has been perfomed as a correction of the output of the system. We have identified a total of 1,751 event tokens.

The results obtained for event detection are reported in Table 2.

**Table 2.** Results of the TULE converter for the task of event detection

| Data set | Precision | Recall | F-measure |
| --- | --- | --- | --- |
| SemEval Data | 0.7290 | 0.6792 | 0.7032 |
| TUT Data | 0.9252 | 0.9081 | 0.9165 |

As the data show, the TULE converter performs better on the TUT data set than on the Italian TempEval data set. The reason is that the former is manually annotated while the latter is automatically parsed. From this, we conclude that the main source of error can be identified in the performance of the parsing phase, not in the converter or in the use of the dependency grammars.

As for the event classification task, we have used a subset of 17 articles from the SemEval data set. This subset comprehends 8,617 tokens with a total of 1,423 tokens annotated as events. The annotation has been conducted by two annotators and validated by an expert, obtaining a K-value of 0.83. The TULE converter achieves an accuracy of 70.44% for the class assignment. The results were computed by considering only the events detected by the converter.

# 7    Conclusion and Future Work

The paper describes the implementation of a TimeML converter via dependency parsing. The results obtained for the subtasks of event detection and classification are satisfying and support the use of a dependecy parser to facilitate the development of these kinds of tools. In other words, from the fact that this trivial set of if-then rules is already sufficient for obtaining satisfying results, we conclude that Dependency Grammars represent a good choice for the task.

With respect to previous works, the main difference is represented by the actual realization of the TULE converter, which is a rule-based system, while the prototypes built for dealing with other languages are mainly based on machine learning techniques. Through a theoretical comparison of the results of our system with those of the systems realized for English and Spanish, we can observe that the performance of our converter can be improved. In particular, we are planning to integrate the converter with a machine learning algorithm to exploit the annotated data and induce more effective word-sense disambiguation rules for the identification of event realized by POS other than verbs. Moreover, the TULE converter will be further developed by implemeting all other TimeML tags and links.

# References

1. Baroni, M., Bernardini, S., Comastri, F., Piccioni, L., Volpi, A., Aston, G., Mazzoleni, M.: Introducing the "la Repubblica" corpus: A large, annotated, TEI(XML)-compliant corpus of newspaper italian. In: Proceedings of the Fourth International conference on Language Resources and Evaluation, LREC 2004 (2004)
2. Bentivogli, L., Pianta, E.: Exploiting parallel texts in the creation of multilingual semantically annotated resources: the multisemcor corpus. Natural Language Engineering 11(3), 247–261 (2005)
3. Bosco, C., Montemagni, A., Mazzei, A., Lombardo, V., Dell'Orletta, F., Lenci, A., Lesmo, L., Attardi, G., Simi, M., Lavelli, A., Hall, J., Nilsson, J., Nivre, J.: Comparing italian parsers on a common treebank: the evalita experience. In: Proc. of the 6th Int. Conf. on Language Resources and Evaluation (LREC 2010) (2010)
4. Bosco, C.: A grammatical relation system for treebank annotation. PhD thesis, University of Turin, Italy (2004)
5. Caselli, T.: TimeML Annotation Scheme for Italian - Version 1.3.1 (2010)
6. Grover, C., Tobin, R., Alex, B., Byrne, K.: Edinburgh-ltg: Tempeval-2 system description. In: Proc. of the 5th International Workshop on Semantic Evaluation, Uppsala, Sweden, pp. 333–336. Association for Computational Linguistics (July 2010)
7. Kumar Kolya, A., Ekbal, A., Bandyopadhyay, S.: Ju_cse_temp: A first step towards evaluating events, time expressions and temporal relations. In: Proc. of the 5th International Workshop on Semantic Evaluation, Uppsala, Sweden. ACL (2010)
8. Lesmo, L., Lombardo, V.: Transformed subcategorization frames in chunk parsing. In: Proc. of the 3rd Int. Conf. on Language Resources and Evaluation (LREC 2002), Las Palmas, pp. 512–519 (2002)

9. Llorens, H., Saquete, E., Navarro-Colorado, B.: Timeml events recognition and classification: Learning crf models with semantic roles. In: Proceedings of the 23rd International Conference on Computational Linguistics (Coling 2010), Beijing, China, pp. 725–733. Coling 2010 Organizing Committee (2010)
10. Llorens, H., Saquete, E., Navarro, B.: Tipsem (english and spanish): Evaluating crfs and semantic roles in tempeval-2. In: Proc. of the 5th International Workshop on Semantic Evaluation, Uppsala, Sweden, pp. 284–291. ACL (2010)
11. Mel'cuk, I.: Actants in semantics and syntax. Linguistics (42), 247–291 (2004)
12. Pustejovsky, J., Castao, J., Saurì, R., Ingria, R., Gaizauskas, R., Setzer, A., Katz, G.: TimeML: Robust specification of event and temporal expressions in text. In: Fifth International Workshop on Computational Semantics (IWCS-5) (2005)
13. Pustejovsky, J., Hanks, P., Saurì, R., See, A., Gaizauskas, R., Setzer, A., Radev, D., Sundheim, B., Day, D., Ferro, L., Lazo, M.: The TIMEBANK corpus. In: Corpus Linguistics 2003 (2003)
14. Pustejovsky, J., Verhagen, M.: Semeval-2010 task 13: Evaluating events, time expressions, and temporal relations (tempeval-2). In: Proceedings of the Workshop on Semantic Evaluations: Recent Achievements and Future Directions (SEW 2009), Boulder, Colorado, pp. 112–116. Association for Computational Linguistics (June 2009)
15. Ruimy, N., Monachini, M., Gola, E., Calzolari, N., Fiorentino, M.D., Ulivieri, M., Rossi, S.: A computational semantic lexicon of italian: SIMPLE. In: Linguistica Computazionale XVIII-XIX, Pisa, pp. 821–864 (2003)
16. Saquete, E., Gonzàlez, J.L.V., Martìnez-Barco, P., Munoz, R., Llorens, H.: Enhancing qa systems with complex temporal question processing capabilities. J. Artif. Intell. Res (JAIR) 35, 775–811 (2009)
17. Saurì, R., Knippen, R., Verhagen, M., Pustejovsky, J.: Evita: A robust event recognizer for qa systems. In: Proceedings of Human Language Technology Conference and Conference on Empirical Methods in Natural Language Processing (HLT/EMNLP 2005), pp. 700–707 (2005)
18. UzZaman, N., Allen, J.: Trips and trios system for tempeval-2: Extracting temporal information from text. In: Proc. of the 5th International Workshop on Semantic Evaluation, Uppsala, Sweden, pp. 276–283. ACL (2010)

# Self-adjusting Bootstrapping

Shoji Fujiwara[1] and Satoshi Sekine[2]

[1] Nikkei Digital Media, Inc.
shoji.fujiwara@nex.nikkei.co.jp
[2] Computer Science Department
New York University
sekine@cs.nyu.edu

**Abstract.** Bootstrapping has been used as a very efficient method to extract a group of items similar to a given set of seeds. However, the bootstrapping method intrinsically has several parameters whose optimal values differ from task to task, and from target to target. In this paper, first, we will demonstrate that this is really the case and serious problem. Then, we propose *self-adjusting bootstrapping*, where the original seed is segmented into the real seed and validation data. We initially bootstrap starting with the real seed, trying alternative parameter settings, and use the validation data to identify the optimal settings. This is done repeatedly with alternative segmentations in typical cross-validation fashion. Then the final bootstrapping is performed using the best parameter setting and the entire original seed set in order to create the final output. We conducted experiments to collect sets of company names in different categories. Self-adjusting bootstrapping substantially outperformed a baseline using a uniform parameter setting.

## 1 Introduction

Bootstrapping has been used in many information extraction tasks, such as harvesting names (Strzalkowski and Wang 96) (Collins and Singer 99), relations (Brin 98) (Agichtein and Gravano 00) (Ravichandran and Hovy 02) (Sun 09), and events (Yangarber et al. 00). Recently, there are more work on bootstrapping mostly using query logs (Pasca 07) (Pantel and Pennacchiotti 06) (Sekine and Suzuki 07). Given seeds of the desired names or relations (which we will hereafter call "items"), it gathers more items using a large un-annotated corpus. First, the most salient contexts of the seed items are found, then those contexts are used to find more items of the same kind. This process can be repeated to get more contexts and items. It is recognized as a very efficient method to extract a group of items similar to a given set of seeds, when there is enough data in the matrix of items and contexts. However, there is an essential problem in the bootstrapping method, namely parameter tuning. The bootstrapping method intrinsically has several parameters, such as the number of contexts to be used at each iteration, the number of items to be extracted at each iteration, and the scoring functions to calculate the similarity between contexts and between items. In

A. Gelbukh (Ed.): CICLing 2011, Part II, LNCS 6609, pp. 188–201, 2011.

previous work, these parameters have been chosen seemingly at random or sometimes empirically to optimize the performance of the target task. However, we have observed that the optimal parameter setting will be different for each different task, and even more problematically, may be different for different target sets within the same task. For example, as we will demonstrate in this paper, the optimal parameter settings to gather company names of different categories, such as banks, food-related companies, and electronic manufacturers, are quite different. This is intuitively understandable, because the characteristics of the features, i.e. the context of the company names of different categories, are quite different. For some company names in a particular category, a small number of contexts are very useful and the excess contexts serve only to add noise. However, finding company names in some other categories may not have strong contexts and need a lot of contexts.

In this paper we will propose a method to solve this problem. The basic idea is to segment the seed set into two sets: the real seed data and the validation data. We will use the real seed data for bootstrapping and will use the validation data to measure how well it works with a given parameter setting. This process will be done several times in a typical cross-validation arrangement so that all of the seed data will at some point be used for validation. For example, if one-third of the data is used for validation, then the experiment will be repeated three times. We will run the bootstrapping using many different parameter settings in order to find the optimal one by maximizing the results. In the following sections, we will explain the idea in detail, and we will report on its success in harvesting company names in 10 categories.

## 2  Bootstrapping

In this section, the basic bootstrapping algorithm will be briefly given using a simple example; then we will state the problem.

### 2.1  Example

In principle, the bootstrapping method works by gathering a set of items of the same kind using a small set of sample items (seeds) and an un-annotated corpus. The basic assumption is that similar items are likely to appear in similar contexts (Harris 54), such as the words surrounding the items or the dependency path between the items. We can define the context arbitrarily, but researchers in the field have been working to find the most effective set of contexts in order to gather the set of items correctly and efficiently. The set of items can be anything that may have a set of similar features, such as:

- A set of names in the same category, such as company names in the same category, country names, names of Roman Gods, name of mountains.
- A set of named entity pairs, such as companies which have subsidiary relations, country and capital pairs, or a person name and birth year.
- A set of expressions which express the same kind of relations.

We will describe a simple example. Assume we are interested in gathering the names of current and past presidents of nations across the world. Let's start with the following seeds:

Seeds of presidents =
{"Clinton", "Bush", "Putin", "Chirac"}

First, the contexts of these words are found in the large corpus. For simplicity's sake, the context here is defined as at most two words before the item or after the item. For example, the contexts in Table 1 may be found from the seed word "Clinton" (the tag #EOS# refers to a sentence boundary).

**Table 1.** Contexts for "Clinton"

| Freq. | Left context | item | Right cont. |
|---|---|---|---|
| 1932 | #EOS# | Clinton | , who |
| 1654 | President Bill | Clinton | #EOS# |
| 1476 | Hilary Rodham | Clinton | #EOS# |
| 1365 | by President | Clinton | #EOS# |
| 1288 | of President | Clinton | #EOS# |
| 712 | with President | Clinton | #EOS# |

There are specific contexts for "Clinton", such as the second and third lines in Table 1, but in addition, the table contains a number of general contexts for presidents. In particular, because we use more than one seed, there is a chance that good contexts for president names can be found collectively. In order to select good contexts to extract more items correctly, we will score the contexts. Many different scoring functions have been used, including relative frequency, mutual information, Jaccard coefficient, and Dice coefficient. By employing one such scoring function, we may bring the contexts for the four seed items to the top rank of the list, as follows:

**Table 2.** Examples of top scored contexts

| President | * | said yesterday |
|---|---|---|
| of President | * | in |
| President | * | the |
| President | * | who |

These top contexts are now used to gather more items, by searching for the matched items in the large corpus. Note that the number of patterns used to find the new items is arbitrary. It has been defined experimentally or completely intuitively in previous experiments. Now, matching the contexts in the corpus, we can extract candidates for president names. Next, we will score the candidates so that the items that are most likely to be presidents are found in the top of the list. Again, several statistical methods can be used to score the candidates. As a result, the following items might be extracted:

$$\text{Newly found items} =$$
$$\{\text{"Boris Yeltsin", "Jiang Zemin",}$$
$$\text{"Bill Clinton", "Saddam Hussein"}\}$$

Now, we have more items than we had at the initial stage, adding four new names to our four seed names. We can iterate the above procedures using the larger set of items.

### 2.2 Problems

Observant readers may have already found a serious problem in the description in the previous subsection. "Arbitrary" decisions had to be made on at least four occasions:

1.  The number of contexts to be used for item extraction at each iteration
2.  The scoring function to be used to rank the contexts
3.  The number of items to be extracted at each iteration
4.  The scoring function to be used to rank the candidate items

In previous work, these parameters were set arbitrarily or empirically. Some researchers set the first parameter to five, and some use a particular statistical function for the two scoring functions. Bootstrapping has been reported many times with varying degrees of success. It has worked very nicely for some tasks, while for others, it simply does not work at all. We believe one of the reasons for these outcomes stems from the parameter setting problem, which we try to solve in this paper. In section 5.2, we will show that, in fact, for different categories, the parameter settings that work best for bootstrapping are different.

## 3 Basic Idea

The basic idea of our approach to solve the problem can be summarized as the following.

*   There will be two stages. In the first stage, the best parameter setting is found using a part of the seed set for bootstrapping and the remainder for evaluating the result. In the second stage, the bootstrapping produces the final outputs using the entire seed set.
*   We segment the seed set into the real seed items and the validation items. For example, we use two-thirds (2/3) of the items for bootstrapping and then evaluate the result using the remaining data (validation data). Or we can use all but one of the seeds as the real seed data and the one remaining item for validation. If the validation items appear at the top of the list, it is regarded as an example of good bootstrapping. This process will be done multiple times in cross-validation, taking different subsets of the seeds as the real seeds and validation data.

- The bootstrapping will be conducted with different parameter settings. For example, we will prepare six functions to score the contexts and the candidate items. In addition, the number of contexts to be used for finding candidate items, and the number of items to be extracted at each iteration, can be varied (5, 10 and 20). The bootstrapping processes with the segmented seed sets runs as many times as there are combinations of parameters (actually 6x6x3x3=324). Then we will find which combination of parameters works the best for the task. Six standard scoring functions are used to calculate the score of contexts and items based on frequencies and co-occurrence frequencies (Manning and Schutze 99). These are, namely, Relative Frequency (RF), Relative Frequency using log (RFl), Mutual Information (MI), Jaccard Coefficient (JC), Dice Coefficient (DC) and Overlap Coefficient (OC), described in the next section. The scoring functions for the candidate items are defined in the same manner.
- The best parameter setting which maximizes the validation result is chosen for the final bootstrapping. The final bootstrapping will be run using the entire seed sets and the selected parameter setting.

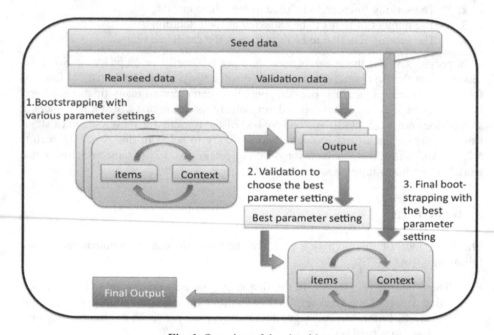

**Fig. 1.** Overview of the algorithm

## 4   Algorithm

### 4.1  Seed Segmentation

The segmentation of the seed set can be done in many ways, but we tried two ways to segment the seeds into the real seed and the validation data. One of them, mainly reported in this paper, is to randomly select two-thirds of the seeds to be the real seeds, and the rest (one-third) to be the validation data. The other method is to use all but one item as the real seeds and the one selected to be used for the validation data. The latter approach can be expected to achieve better results, but the computation cost is relatively large. We conducted the first experiment three times, and the second experiment a number of times equal to the number of seeds, so that all items in the seed set would serve as validation data once.

## 4.2 Context Search

The algorithm searches in the corpus for the contexts of each item. In this experiment, the context is defined as at most two words before and at most two words after the item, excluding the case with a single word before and after. More precisely, the *'s in the following three patterns are regarded as the contexts. We don't treat the previous context and following context separately; we take them together as a context. Here * and "*item*" represent tokens:

$$* * item * *$$
$$* \ item * *$$
$$* * item *$$

We used an ngram search tool (citation anonymous) to find the matching contexts quickly. We don't use single frequency contexts for each item. We use a sampling method when the number of matched contexts is too large. We observed almost no relative effect due to the sampling.

## 4.3 Context Scoring and Selection

As we have mentioned, the contexts are scored using six different functions, defined as follows:

$freq_i()$: instance freq. (number of instances)
$freq_t()$: type frequency (number of kinds)
$F$: frequency of all tokens
$Set$: set of items which match with c
$ALL$: set of all items

- Relative Frequency (RF)

$$score_{RF}(c) = \frac{\sum_{i \in Set} freq_i(i)}{\sum_{i \in ALL} freq_i(i)}$$

- Relative Frequency – log (RFl)

$$score_{RFl}(c) = \frac{\sum_{i \in Set} freq_i(i)}{\sum_{i \in ALL} \log(freq_i(i))}$$

- Mutual Information (MI)

$$score_{MI}(c) = \sum_{i \in Set} \log(\frac{freq_i(c,i) * F}{freq_i(c) freq_i(i)})$$

- Jaccard Coefficient (JC)

$$score_{JC}(c) =$$

$$\sum_{i \in Set} \frac{freq_i(c,i)}{freq_i(c) + freq_i(i) - freq_i(c,i)}$$

- Dice Coefficient (DC)

$$score_{DC}(c) = \sum_{i \in Set} \frac{2 * freq_i(c,i)}{freq_i(c) + freq_i(i)}$$

- Overlap Coefficient (OC)

$$score_{OC}(c) =$$

$$\sum_{i \in Set} \frac{freq_i(c,i)}{\min(freq_i(c), freq_i(i))}$$

The scores are calculated based on the accumulated set of items; the score should tell how reliable the contexts are for the items we already know. Once the score is calculated for each context, the highest scoring contexts are selected. Another parameter that needs to be tuned is the number of contexts to be selected. In our experiment, we used three values - 5, 10 and 20. Then, as per the bootstrapping method, the selected contexts will be used to find more items.

## 4.4 Item Scoring and Extraction

The selected contexts are used in order to find more items. Each item is limited to one or two tokens. More specifically, we search the large corpus for the following regular expression:

*pre-context (.+ ){1,2} post-context*

The scoring functions for item candidates are the same as the six scoring functions for the contexts. They are based on how reliable the items are based on the currently known contexts for the target. The number of items to be extracted is also varied among 5, 10 or 20. We run iterations until the number of items extracted reaches 100 in total.

## 4.5 Evaluation Using Validation Data

Once candidate items are extracted for bootstrapping with different parameter settings, the next task is to find which parameter setting is the best using the validation data. In our experiments, we used two evaluation functions. $E_1(p)$ is the summation of the reciprocal rank for the items in the validation data which appear in the 100 items accumulated in the bootstrapping. $E_2(p)$ is the summation of the weighted reciprocal rank. The weighting is done based on the iteration in which the item was found. The assumption is that the earlier an item is found in the iterations, the better. Here, *rank(i)* is a function which returns the rank of item *i* and *iteration(i)* is a function which returns at which iteration the item was extracted.

$$E_1(p) = \sum_{i \in \text{validation set}} \frac{1}{rank(i)}$$

$$E_2(p) = \sum_{i \in \text{validation set}} \frac{1}{rank(i)} \frac{1}{iteration(i)}$$

This formula was inspired by MRR (mean reciprocal rank), which is used in the IR and QA communities. It gives larger weight to the validation data that is ranked higher in the list of candidate items and it gives larger weight if the item was found an earlier iteration in the second evaluation function. In addition, it gives a better score if more validation data appears in the list. The parameter setting that gives the best evaluation value is selected as the parameter tuning result for the task.

### 4.6  Running the Bootstrapping with the Selected Parameter Setting

The selected parameter setting is used in the final bootstrapping. In this bootstrapping the entire seed set is used in the experiment.

## 5  Experiments

In this section, we report on the experiments and their results. First the experimental setting is explained in section 5.1. Then, the validation results using different parameter settings will be explained in section 5.2. We show that the best bootstrapping for company names in different categories requires completely different parameter settings. In section 5.3, we will show how the final bootstrapping works using the entire seed set with the best parameter tuning. In section 5.4, we will report the experiment using all but one for the real training and the one for the validation.

### 5.1  Corpus and Target Categories

We conducted experiments on company names. The Nikkei company categories are widely used in the financial world. There are more than 10 categories and each of the 3,835 listed trading companies, along with some other big companies, are placed in a category. The number of companies in each category ranges from about 50 to 400. The list consists only of trading companies and other big companies; mid-size and small companies in the same category are not listed. Each company belongs to only one category even if the company conducts business in multiple categories. So, there is a need to augment the list with smaller companies, as well as big companies that conduct business in multiple fields.

The un-annotated corpus used in the experiment is the Nikkei group newspapers (the most authoritative Japanese financial newspapers). The Nikkei group consists of four newspapers, The Nikkei Newspaper, The Nikkei Business Daily, The Nikkei Marketing Journal and The Nikkei Financial Daily. The corpus extends from 2000 to 2008, includes 493 million tokens and is 1.7GB in total. The sentences are tokenized using the automatic morphological analyzer JUMAN with minor amendments, such as the treatment of special symbols.

For our experiment, we selected 10 categories, in order to observe the effectiveness of our technique. The categories we used for the experiment are shown in Table 3.

**Table 3.** Nikkei company categories used in the experiment

| Real estate | Brokerage firm |
|---|---|
| Food | Automobile |
| Trading | Electrical Apparatus |
| Bank | Medicine |
| Construction | Retailing |

## 5.2  Validation Results

Table 4 shows a random sample of the validation results as $E_1(p)$ values (using the 2/3-1/3 segmentation of the seed set) for the companies in three categories, "Real estate", "Construction" and "Medicine". For each category of company names, we run the bootstrapping using different combinations of parameter values. As we can see in the table, the best parameter settings are different for each of the three categories. Note that because of the space limitation, only a small sample of the results are shown.

**Table 4.** Sample Validation results

| Field | S.F. Cont. | S.F. item | Num. of contexts/items | | |
|---|---|---|---|---|---|
| | | | 5/5 | 10/5 | 20/10 |
| Real estate | RF | IRF | 2.33 | 2.67 | 3.48 |
| | IRF | MI | 2.67 | 2.2 | 1.41 |
| | MI | DC | 2.7 | 2.87 | 1.11 |
| | JC | RF | 0.83 | 0.4 | 0.96 |
| | DC | OC | 3.12 | 5.33 | 2.48 |
| | OC | JC | 2.28 | 2.62 | 1.1 |
| Con-struc-tion | RF | MI | 1.17 | 0.87 | 1.7 |
| | OC | OC | 2 | 1.53 | 1.13 |
| | IRF | RF | 0.78 | 1.25 | 0.14 |
| | JC | DC | 2.03 | 3.07 | 0.87 |
| | DC | IRF | 2.58 | 2.4 | 0.21 |
| | MI | JC | 5.5 | 4.9 | 2.65 |
| Medi-cine | JC | DC | 0.2 | 1 | 0.5 |
| | RF | OC | 2.45 | 3 | 3.59 |
| | IRF | MI | 0.5 | 1.5 | 0.49 |
| | MI | JC | 3.28 | 0.2 | 0.5 |
| | OC | RF | 1.5 | 1.4 | 0.5 |
| | DC | IRF | 0.2 | 0.17 | 0 |

Table 5 shows the best parameter settings for all 10 categories in the 2/3-1/3 experiments. "E" indicates the evaluation functions used in the calculation of the validation result. The table shows that all six functions are used to achieve the best validation result, as well as different numbers of contexts and items. This result clearly shows that different parameter settings work best for different tasks in bootstrapping.

**Table 5.** Best parameter settings

| Field | E | Scoring func. | | Number of | |
|---|---|---|---|---|---|
| | | cont. | item | cont. | item |
| Real estate | $E_1$ | DC | OC | 10 | 5 |
| | $E_2$ | OC | OC | 10 | 5 |
| Food | $E_1$ | RF | MI | 20 | 5 |
| | $E_2$ | OC | MI | 5 | 5 |
| Trading | $E_1$ | DC | IRF | 5 | 20 |
| | $E_2$ | RF | RF | 20 | 5 |
| Bank | $E_1$ | DC | MI | 5 | 5 |
| | $E_2$ | RF | MI | 20 | 5 |
| Construction | $E_1$ | MI | JC | 5 | 5 |
| | $E_2$ | MI | JC | 5 | 5 |
| Brokerage firm | $E_1$ | OC | MI | 5 | 5 |
| | $E_2$ | JC | RF | 20 | 5 |
| Automobile | $E_1$ | JC | OC | 5 | 5 |
| | $E_2$ | JC | OC | 10 | 20 |
| Electrical Apparatus | $E_1$ | RF | RF | 10 | 5 |
| | $E_2$ | MI | MI | 10 | 5 |
| Medicine | $E_1$ | RF | OC | 20 | 10 |
| | $E_2$ | RF | OC | 20 | 10 |
| Retailing | $E_1$ | MI | MI | 10 | 5 |
| | $E_2$ | MI | MI | 10 | 5 |

## 5.3 Final Bootstrapping

Having found the best parameter setting for each category, the final bootstrapping runs with the best parameter setting on the entire seed set. Since our objective was to find more names, we decided to evaluate the results based on the number of new company names in the category found among at most 100 extracted items. So if the number of items to extract per iteration is 5, then 20 iterations will be conducted. Note that sometimes bootstrapping is not able to find 100 item candidates due to a lack of qualified contexts or items. We made two runs for each category using two sets of seeds randomly selected from the company names in each category. This is to avoid the effect of seed selection on the experiment. Also, we set the number of seeds for all categories to 50 in order to balance the results across the categories.

We expected to extract the largest number of items in the final bootstrapping when the best parameter setting was used. In table 6, we show the results for the 2/3-1/3 experiment. "The number of items" indicates the number of new companies found in all the categories combined, and "average rank" is the micro average of the ranks in terms of the number of items found in each category among 324 results. 324 is the number of combinations of the parameters (6x6x3x3). The baseline result is achieved by a single parameter setting that maximizes the performance for all categories combined at the final bootstrapping. So, the baseline is actually unrealistic and unfairly good, because it was chosen by looking at the final results.

**Table 6.** Results of 2/3-1/3 experiment

| Method | Num. of items | Average Rank |
|--------|---------------|--------------|
| Baseline | 319/985 | 55.3 |
| $E_1$ | 345/898 | 49.9 |
| $E_2$ | 384/866 | 34.2 |

Observing Table 6, the results are quite good, in particular, compared to a baseline that is unrealistically good. The baseline method, i.e. using one single parameter setting, which was DC/MI/20/20, extracts only 319 correct company names in total in all the categories out of 985 extracted names, while our proposed methods extract 345 and 384 company names in total, using $E_1$ and $E_2$ metrics respectively. These good results are achieved even though the total numbers of the extracted items are smaller. The average ranks of the final results also indicate that our proposed methods outperform the baseline. Comparing between two evaluation methods, $E_1$ and $E_2$, $E_2$ performs better. It means that the factor 1/iteration in the evaluation of validation data is helpful to select good contexts. Note that we can make another weaker baseline, which is to choose the parameter setting randomly. As we have 324 combinations of parameters, the average rank of the runs with all possible settings is about 162. Compared to this weak baseline, our proposed methods are far better.

### 5.4 All-but-one and One Experiment

In this subsection, we describe the experiment in which all-but-one seeds are used for the real seed set and the remaining one is used for the validation. This cross-validation experiment would be repeated the same number of times as the number of original seeds, so that all the seeds will become the validation data once. The experiment takes a very long time, i.e. we have to do the cross-validation experiment 50 times instead of 3 times. We conducted the experiment only for the electrical apparatus category. The results are shown in Table 7.

**Table 7.** All-but-one & one experiment

| | $E_1$ | | $E_2$ | |
|---|---|---|---|---|
| | # of items | Rank | # of items | Rank |
| 2/3-1/3 | 32/64 | 39 | 34/100 | 35 |
| ABO-1 | 36/100 | 30 | 29/60 | 43 |

In the table, results of two experiments, 2/3-1/3 and All-but-one and one (ABO-1) are shown. Two evaluation scorings ($E_1$ and $E_2$) are used and for each experiment, two values are shown. One is the number of correct items out of the number of extracted items and the second is the rank of the correctly extracted items. As we can see in the table, the number of correctly extracted items depends on the number of extracted items and the rank is very similar. Although we are planning to conduct similar experiments for the other categories, the experimental results indicate that the alternative segmentation methods of seeds don't have a major effect on the experiment.

# 6 Related Work

(Pantel 08) reported that the performance of bootstrapping varies widely for different target sets, supporting this claim with a very convincing graph. The problem was widely recognized, but to the best of the authors' knowledge there have been no serious attempts to address the problem. We can't judge without the detail of their experiments, but the results in this paper may suggest the explanation for the observation of (Pantel 08).

(Goldberg and Zhu 09) proposed that the cross validation method be used to find the optimal parameters based on a semi-supervised learning (SSL) paradigm. In addition, they tried to find the SSL algorithm that generates the optimal parameters. However, the tasks they reported are all binary classification problems using relatively small un-labeled data sets (100 or 1,000 items). Our task of harvesting names using bootstrapping is quite different because it is not a data classification problem where the validation data can always be useful. Another key difference is that we use an open-ended amount of un-labeled data.

# 7 Discussion

## 7.1 Why Different Parameter Settings Are Needed for Different Tasks

As we mentioned, the characteristics of the different categories are quite different. For some company names in a particular category, a small number of contexts are very useful. However, some other categories need a lot of contexts. Table 8 shows the number of contexts that are actually used in order to get correct new company names in the "bank" and "trading company" categories. These are the results from the best parameter settings for the two categories.

**Table 8.** Comparison of number of contexts

|  | Parameter | Num. of cont. | Num. of items |
|---|---|---|---|
| Bank | RF/MI/20/5 | 127 | 73/98 |
| Trading | DC/IRF/5/20 | 18 | 40/77 |

For the Bank category, 127 different contexts are needed to extract 73 items. In contrast, 18 different contexts suffice for the trading company category. Looking at the contexts, it is observed that the 18 contexts for the trading company category are very strong contexts for that kind of company, such as "trading company such as *" (note that the experiment was done in Japanese, the pattern has the company name in the middle of the pattern). However, the contexts for the bank category are a mix of such contexts and contexts involving lists of banks, such as "Bank-A, * and Bank-B". It may be due in part to the fact that most Japanese bank names have suffixes indicating bank, similar to "Citibank". However most trading companies do not have such suffixes. So the newspaper is more likely to use the phrase "trading companies such as *" rather than "banks such as *". It is obvious that "Citibank" is a bank without explicitly prefixing it with "banks such as *".

## 7.2 Coverage of Parameters

The candidate values of the parameters (scoring function, number of contexts and items) have to be defined in advance. It is possible that functions that are not in the list may perform better than any of the listed functions, or that a number of the contexts or items to be used or extracted at each iteration other than 5, 10 or 20 may work more accurately. We believe the selection of possible parameter values is not a process that can be done automatically, and some insight and heuristics about the task and the bootstrapping method are needed.

## 7.3 Meta Parameters

The main objective of our method is to avoid the parameter tuning which is necessary for bootstrapping. However, we have introduced several new parameters in our procedure, such as how to segment the seed set. It is natural to ask how to find the best values for these parameters. We conducted experiments using two different splits of the data. One used 2/3 and 1/3 segmentation between real seeds and validation data, while the other used only one seed for validation and the rest as real seeds. We observed only a small difference between the two methods. We believe that once there is enough seed data, they would work similarly. However, there is need for further investigation in this area.

# 8  Conclusion

We proposed self-adjusting bootstrapping. It segments the seed set into the real seed set and the validation data. The bootstrapping initially runs with the real seed set, trying various parameter settings in order to find the best parameter setting that optimizes the result as measured using the validation data. We conducted experiments using company names in different categories, and showed that different parameter settings worked best for company names in different categories. We showed that self-adjusting bootstrapping worked much better than a baseline using uniform parameter settings across all categories.

# References

Agichtein, E., Gravano, L.: Snowball: Extracting Relations from Large Plain-Text Collections. In: Proc. 5th ACM International Conference on Digital Libraries (ACM DL) (2000)

Brin, S.: Extracting Patterns and Relations from the World Wide Web. In: Proc. Conference of Extending Database Technology, Workshop on the Web and Databases (1998)

Collins, M., Singer, Y.: Unsupervised Models for Named Entity Classification. In: Proc. of the Joint SIGDAT Conference on Empirical Methods in Natural Language Processing and Very Large Corpora (1999)

Goldberg, A.B., Zhu, X.: Keepin' it real: Semi-supervised learning with realistic tuning. In: NAACL 2009, Workshop on Semi-supervised Learning for NLP (2009)

Manning, C.D., Schutze, H.: Foundations of Statistical Natural Language Processing. The MIT Press, Cambridge (1999)

Pantel, P.: Of search and Semantics. In: NSF Symposium on Semantic Knowledge Discovery, Organization and Use (2008)

Ravichandran, D., Hovy, E.: Learning surface text patterns for a question answering system. In: Proceedings of ACL 2002, Philadelphia, PA, pp. 41–47 (2002)

Sun, A.: A Two-Stage Bootstrapping Algorithm for Relation Extraction. In: Proceedings of Recent Advances in Natural Language Processing 2009, Borovets, Bulgaria (2009)

Strzalkowski, T., Wang, J.: A Self-Learning Universal Concept Spotter. In: COLING 1996 (1996)

Yangarber, R., Grishman, R., Tapanainen, P., Huttunen, S.: Automatic Acquisition of Domain Knowledge for Information Extraction. In: COLING 2000 (2000)

Paşca, M.: Organizing and Searching the World Wide Web of Fact — Step Two: Harnessing the Wisdom of the Crowds. In: Proceedings of the 16th International World Wide Web Conference (WWW 2007), pp. 101–110 (2007)

Pantel, P., Pennacchiotti, M.: Espresso: Leveraging Generic Patterns for Automatically Harvesting Semantic Relations. In: Proceedings of the 21st International Conference on Computational Linguistics and the 44th Annual Meeting of the Association for Computational Linguistics, pp. 113–120 (2006)

Sekine, S., Suzuki, H.: Acquiring Ontological Knowledge from Query Logs. In: Proceedings of the 16th International World Wide Web Conference (WWW 2007), pp. 101–110 (2007)

# Story Link Detection Based on Event Words

Letian Wang and Fang Li

Department of Computer Science & Engineering
Shanghai Jiao Tong University
Shanghai, China
{koh,fli}@sjtu.edu.cn

**Abstract.** In this paper, we propose an event words based method for story link detection. Different from previous studies, we use time and places to label nouns and named entities, the featured nouns/named entities are called event words. In our approach, a document is represented by five dimensions including nouns/named entities, time featured nouns/named entities, place featured nouns/named entities, time&place featured nouns/named entities and publication date. Experimental results show that, our method gain a significant improvement over baseline and event words plays a vital role in this improvement. Especially when using publication date, we can reach the highest 92% on precision.

**Keywords:** story link detection, event words, multidimensional model, nouns/named entities, featured nouns/named entities.

## 1 Introduction

Story link detection, which was first defined in the Topic Detection and Tracking (TDT) [1,2,12,14,16] competition program, is the task of determining whether two stories, such as news articles and/or radio broadcasts, are about the same event, or linked. Story link detection is important for many applications. For example, there are three reports whose titles are:

- Midterm election polls open in United States
- US presidential vote is underway
- Voting in parliamentary election starts in Japan

The content of three news stories above are very similar, because they are all about the election, they have many common words in the text such as "election", "vote", "candidate" and so on. But actually they are different because they are not the same event. The first one is related to the election in U.S.A. in 2006 while the second one is about the election in U.S.A in 2008 and the last one refers to the election in Japan in 2007. The task of story link detection is to find out if the two stories are about the same event even though they may have the same content.

According to TDT, two stories are linked if the events in the stories happened at some specific time and place. In this paper, we give a more explicit definition:

A. Gelbukh (Ed.): CICLing 2011, Part II, LNCS 6609, pp. 202–211, 2011.

**Definition 1.** *Two stories are linked if they contain the same event words.*

Where event words (EW) is defined as:

**Definition 2.** *Nonus/named entities with time or place labels.*

There are three types of labels for event words including time, place and both (time&place). We use five dimensions to represent a story, including nouns/ named entities, time featured nouns/named entities, place featired nouns/named entities, time&place featured nouns/named entities and the publication date, where all the featured nouns/named entities are event words. Cosine similarity, resemblance function and date similarity function are used to calculate similarity of each dimension. A combined story similarity function is used to calculate the similarity of two stories. Experimental results show that our approach gain a significant improvement over pure text similarity method and each dimension has its own contribution.

The following section contains the previous studies on story link detection. Section 3 shows our multidimensional model for representing stories and describes how to calculate similarities between two stories. Experimental results and discussions are described in Section 4. We give our conclusions in the last section.

## 2   Related Work

There were two kinds of methods for story link detection. One is based on vector space model and the other is based on language model.

Based on vector space model, Chen and Farahat et al. used incremental tf-idf instead of traditional static tf-idf in vector space model, and used several similarity method including Cosine, Hellinger, Tanimoto and Clarity to find out linked stories [4,5,7]. They also proposed a source-pair specific method for story link detection to avoid linking two stories from the same media because of the customary words. Shah et al. used named entities and traditional vector space model to represent a document [14]. They also used a graph based method to extend named entities for each document. Zhang et al. used an event model, which is actually a multi-vector model, to represent a story including time, number, person, location, organization, abstract and content description [16]. Brown et al. proposed a method to ignore common event to discriminate among similar events [3]. Chen et al. considered several important issues for monolingual and multilingual link detection [6]. They used nouns, verbs, adjectives and compound nouns to represent news stories, and used story expansion, topic segmentation and a translation model to help the detection process. Ferret et al. proposed a method to combines word repetition and the lexical cohesion stated by a collocation network to compensate for the respective weaknesses of segmentation and link detection [8].

For the second method, Nallapati et al. used a semantic language model for story link detection, they used named entities and part of speech tag to classify features into different categories, defined a semantic class for each document and used likelihood as features with perceptron learning algorithm to distinguish stories [13]. Yu et al. defined a semantic domain as a collection of semantic related terms for Chinese corpus [9]. With the semantic domain language model, two stories will be linked if they have similar semantic domains, the distance calculation is based on Kullback-Leibler divergency. Lavrenko et al. proposed a relevance model for story link detection and also used Kullback-Leibler divergency to calculate distance between two stories [11].

Our method is similar with traditional vector space model but different from it. We use event words to improve the performance of story link detection. Besides using Cosine similarity like previous studies, we use a resemblance function and define a date similarity function to help link detection.

## 3    Multidimensional Model for Story Representation

### 3.1    Event Words (EW)

Time and place information is important for story link detection. Our work is to maximize the using of the time and place information. In our study, time and place information is used to establish event words in order to distinguish same words in different documents, such as the "earthquake" at Sichuan and the "earthquake" at Yunnan. All the labels are divided into three types:

1. **time:** Nouns/named entities with only time label.
2. **place:** Nouns/named entities with only place label.
3. **time&place:** Nouns/named entities with both time and place labels.

For example, "earthquake@2008" is featured with time, "earthquake@Sichuan" is featured with place and "earthquake@Sichuan@2008" is featured with time-&place.

**Nouns/Named Entities Featured with Time ($NN_{time}$).** We use only date format as time information, because words like "today" or "yesterday" is hard to distinguish between different stories. Any time in documents can be represented as a triple in our model:

$$< year, month, day >$$

Regular expressions are used to extract time from documents. Only four types of combinations of $year$, $month$ and $day$ are used to label nouns/named entities, they are $year$, $year.month$, $year.month.day$ and $month.day$. For example, "earthquake@2008.05.12" is a time featured event words.

**Nouns/Named Entities Featured with Place ($NN_{place}$).** A place is a structure rather than a word in our model. Like time information, it can also be presented as a triple:

$$< city, region, country >$$

We have a place database which contains places information with triple format above. With a word presenting a place, our approach extends it to a triple. The *city* and *region* may be null if the word originally represent a *region* or *country*. After extension, a noun/named entity will be labeled with at most three places. For example, if a sentence contains "earthquake" and "Wenchuan", we will have three featured nouns, "earthquake@Wenchuan", "earthquake@Sichuan" and "earthquake@China", where Wenchuan is located in Sichuan, China.

**Nouns/Named Entities Featured with Time&Place (NN$_{time\&place}$).** We also use time&place labels in our method. "earthquake@Sichuan@2008" is more representable than "earthquake@Sichuan" and "earthquake@2008".

**How to Produce Event Words.** Two words are more related if they are close to each other in a document. So, in our research, nouns/named entities are labeled with time or places from the same sentence. For the label of time&place, all the possible combination of time and places within the same sentence will be used to label the nouns/named entities.

## 3.2 Modeling

Five dimensions are used to represent each document, which are shown in Table 1. Publication date is used in our method because it is an important feature to distinguish two news stories. For a publication date, we just care about *year*, *month* and *day*.

**Table 1.** Five Dimensions in Multidimensional Model for Story Link Detection

| Abbreviation | Description |
| --- | --- |
| NN | Nouns/Named Entities |
| NN$_{time}$(event words) | Nouns/Named Entities Featured with Time |
| NN$_{place}$(event words) | Nouns/Named Entities Featured with Place |
| NN$_{time\&place}$(event words) | Nouns/Named Entities Featured with Time&Place |
| PD | Publication date |

Different modeling method and similarity calculation approaches are used for different dimensions. Vector space model and Cosine similarity based on tf-idf is used for NN dimension. A resemblance function is used for event words and a date similarity function is used for publication date.

**tf-idf.** We use tf-idf in the NN dimension. The tf-idf is often used in information retrieval and text mining. This weight is a statistical measure used to evaluate how important a word is to a document in a collection or corpus. Equation 1 shows the calculation of tf-idf,

$$(tfidf)_{i,j} = \frac{n_{i,j}}{\sum_k n_{i,j}} \times \log \frac{|D|}{|\{d : t_i \in d\}|} \tag{1}$$

where $n_{i,j}$ is the number of occurrences of the considered term ($t_i$) in document $d_j$, the denominator is the sum of number of occurrences of all terms in document $d_j$, $|D|$ is total number of documents in the corpus, and $|\{d : t_i \in d\}|$ is number of documents where the term $t_i$ appears (that is $n_{i,j} \neq 0$).

### 3.3   Similarity Calculation

In our approach, we use similarity function to compare two news stories. We first calculate similarities for each dimension respectively. Since five dimensions are modeled differently, three similarity methods are used in our approach. Cosine similarity is used for the NN dimension, resemblance function is used for the event words dimensions and date similarity function is used for PD dimension.

**Cosine Similarity.** Cosine similarity is a measure of similarity between two vectors of $n$ dimensions by finding the cosine of the angle between them. Given two document vectors $d_1$ and $d_2$, the similarity can be represented as:

$$sim_{cosine}(d_1, d_2) = \frac{\sum_{i=1}^{n} w_{i,1} \times w_{i,2}}{\sqrt{\sum_{i=1}^{n} w_{i,1}^2} \sqrt{\sum_{i=1}^{n} w_{i,2}^2}} \tag{2}$$

where $w_{i,1}$ and $w_{i,2}$ are weights (tf-idf values here) of term $t_i$ in document $d_1$ and $d_2$, and $n$ is the total number of terms in the corpus. The similarity for the NN dimension between two news stories is:

$$sim_{NN}(i, j) = sim_{cosine}(d_{i_{NN}}, d_{j_{NN}}) \tag{3}$$

where $d_{i_{NN}}$ is the vector of the NN dimension of document $d_i$.

**Resemblance Function.** We choose the resemblance as our similarity metric for the dimensions of event words. The reason we use resemblance here is that featured nouns/named entities need more accurate comparison. The resemblance $r$ of two documents $d_1$ and $d_2$ is defined as follows:

$$r(d_1, d_2) = \frac{|d_1 \cap d_2|}{|d_1 \cup d_2|} \tag{4}$$

where $|d_1 \cap d_2|$ is the number of terms both occur in $d_1$ and $d_2$, and $|d_1 \cup d_2|$ is the number of all the distinct terms in $d_1$ and $d_2$. We use resemblance for the dimensions with featured nouns and named entities, so the similarities of these three dimensions can be represented as:

$$sim_{NN_{time}}(d_i, d_j) = r(d_{1_{NN_{time}}}, d_{2_{NN_{time}}}) \tag{5}$$

$$sim_{NN_{place}}(d_i, d_j) = r(d_{1_{NN_{place}}}, d_{2_{NN_{place}}}) \tag{6}$$

$$sim_{NN_{time\&place}}(d_i, d_j) = r(d_{1_{NN_{time\&place}}}, d_{2_{NN_{time\&place}}}) \tag{7}$$

**Date Similarity Function.** We first calculate time difference $time_{diff}$ between two date and represent in days. Then, the date similarity function $sim_{date}$ can be represented as:

$$sim_{date}(t_1, t_2) = \frac{1}{time_{diff}(t_1, t_2) + 1} \tag{8}$$

Where $t_1$ and $t_2$ is two different dates. We use date similarity for the PD dimension, so the similarity of the PD dimensions can be represented as:

$$sim_{PD}(d_1, d_2) = sim_{date}(d_{1_{PD}}, d_{2_{PD}}) \tag{9}$$

**Similarity Function for News Stories.** In order to calculate the similarity between two news stories, similarities for each dimension are combined together with the following equations.

$$sim_{i,j} = \sqrt[\lambda]{\frac{\alpha \cdot sim_{NN} + sim_{EW}}{\alpha + \beta + \gamma + \delta}} \times sim_{PD}^{\theta} \tag{10}$$

where

$$sim_{EW} = \beta \cdot sim_{NN_{time}} + \gamma \cdot sim_{NN_{place}} + \delta \cdot sim_{NN_{time\&place}} \tag{11}$$

We have radical sign here because we need to avoid the similarity falling into a too small interval. In our experiment, the parameters of the Equation (11) are respectively set to: $\alpha = 1$, $\beta = 2$, $\gamma = 4$, $\delta = 4$, $\theta = 2$ and $\lambda = 8$.

For some reason, some news stories may not have publication date. So we need a method to calculate similarities if there is no publication date in the corpus. The similarity function without $sim_{PD}$ is:

$$sim_{i,j_{withoutPD}} = \sqrt[\lambda]{\frac{\alpha \cdot sim_{NN} + sim_{EW}}{\alpha + \beta + \gamma + \delta}} \tag{12}$$

In our experiment, the parameters of the Equation (14) are respectively set to: $\alpha = 1$, $\beta = 2$, $\gamma = 4$, $\delta = 4$ and $\lambda = 4$.

We do some experiments over several group of parameters, and the above ones achieve the best result.

# 4   Experiment and Discussion

## 4.1   Data Set and Experimental Procedures

We use a Chinese corpus from SINA[1] which contains 1591 news stories on 148 topics. There are about 10 news stories for each topic. We assume that news stories from the same topic are linked with each other, because the topics are collected by people manually and each topic is refer to an event.

To get the result, we first do the word segmentation work and extract named entities from all the documents including recognizing time and places information. Then, we extract event words from all the processed news stories and establish multidimensional model for each story. We use similarity value to verify if the two stories are linked.

## 4.2   Evaluation Methods

**F-score.** The traditional F-measure or balanced F-score is the harmonic mean of precision and recall:

$$F = \frac{2pr}{p+r} \tag{13}$$

where $p$ is the number of correct results divided by the number of all returned results and $r$ is the number of correct results divided by the number of results that should have been returned.

**Detection Cost.** Detection cost is a evaluation method in TDT project. It can be represented as:

$$C_{det} = C_{miss} \cdot P_{miss} \cdot P_{target} + C_{fa} \cdot P_{fa} \cdot P_{non-target} \tag{14}$$

where $P_{miss} = \frac{number\ of\ missed\ detection}{number\ of\ targets}$, $P_{fa} = \frac{number\ of\ false\ alarms}{number\ of\ non\text{-}targets}$, $C_{miss}$ and $C_{fa}$ are the costs of a missed detection and a false alarm respectively, and are pre-specified, $P_{target}$ is the a priori probability of finding a target and $P_{non-target} = 1 - P_{target}$.

## 4.3   Experimental Results

Table 2 shows the results of our experiment. The first two rows are baseline systems using traditional vector space model with Cosine similarity while the first one is based on terms and the other is based on nouns/named entities. The last two methods are event words based methods (EWM) while the first one is the multidimensional model without PD dimension and the last one makes the final result.

Vector space model can get higher recall but lower precision, because it can not distinguish stories with similar contents but different events. With nouns/named

---

[1] http://www.sina.com.cn

**Table 2.** Experiment Results

|  | $p(\%)$ | $r(\%)$ | $F(\%)$ | $(C_{Det})_{Norm}$ |
|---|---|---|---|---|
| VSM | 50.66 | 67.35 | 58.54 | 0.5282 |
| VSM$_{NN}$ | 58.60 | 60.10 | 59.35 | 0.4202 |
| EMW$_{withoutPD}$ | 72.71 | 63.23 | 67.64 | 0.3795 |
| EMW$_{withPD}$ | **92.37** | **67.61** | **78.08** | **0.3266** |

entities, we can get higher precision but lower recall. With event words, we can achieve a higher precision and an acceptable recall, because event words describe stories more accurate and detailed. With publication date, a significant high precision can be achieved because it reduce the candidates to be linked.

### 4.4 Discussion

In our method, we choose five features each as a dimension to represent a document. Instead of these five features, there are lots of other information can be used to represent a document. We discuss here to show why we choose these five features instead of others, such as title, time words and place words. We use nouns/named entities as baseline here, and each time add one additional feature to represent documents. Table 3 shows the results. where double line arrow in

**Table 3.** Results with different dimensions

|  | $p(\%)$ | $r(\%)$ | $F(\%)$ | $(C_{Det})_{Norm}$ |
|---|---|---|---|---|
| NN | 58.60 | 60.10 | 59.35 | 0.4202 |
| NN + Title($\Downarrow$) | 57.22 | 50.93 | 53.89 | 0.5097 |
| NN + Time($\Downarrow$) | 43.36 | 57.48 | 49.43 | 0.4628 |
| NN + Place | 76.09($\uparrow$) | 47.56($\downarrow$) | 58.53 | 0.5318($\downarrow$) |
| NN + NN$_{time}$ | 65.02($\uparrow$) | 57.95 | 61.28($\uparrow$) | 0.4361 |
| NN + NN$_{place}$($\Uparrow$) | 67.60 | 64.37 | 65.94 | 0.3717 |
| NN + NN$_{time\&place}$ | 62.33($\uparrow$) | 59.38 | 60.82($\uparrow$) | 0.4241 |

the table means the feature has a significant impact on the result while single line arrow means it has a limited impact.

Generally, we may think that title is a good feature to distinguish news stories, since a title is an accurate summary of a story. But from the result above, we find out that all the performance decreases with titles taken into account. It is because different stories may have similar titles and titles are too short to distinguish when they have the same words.

We do not use time and places alone in our method. The time dimension make no improvements just implicate the performance. The place dimension gain a good performance in precision, but the recall and detection cost are unacceptable for story link detection. Actually, time and place alone may bring noises as well as title. Many events happen in the same place or at the same time.

From the result above, we can see that all the event words help to improve the performance especially for place labels. The reason why event words make such improvement is that they can describe a story more accurate and contains more information.

## 5    Conclusion

We propose a event words based method for story link detection in this paper. The main contribution of our work is:

1. Event words are used to distinguish stories with similar contents.
2. A multidimensional model based on event words is used in our approach.
3. A combined similarity method is used in our model.

Three similarity methods are used in our approach, Cosine similarity for nouns/ named entities, resemblance for event words and date similarity function for publication date. Our method gain a significant improvements over baseline systems and the results prove that:

1. Nouns/named entities are more helpful to story link detection than other content words.
2. Event words can improve the performance of story link detection.
3. Publication date is also a useful information for story link detection.

## Acknowledgement

The research is supported by the National Science Foundation of China under Grant No.60873134, Threads and topics detection for news events.

## References

1. Allan, J.: Topic detection and tracking: event-based information organization. Kluwer Academic Publishers, Norwell (2002)
2. Allan, J., Lavrenko, V., Swan, R.: Explorations within topic tracking and detection, pp. 197–224 (2002)
3. Brown, R.D.: Dynamic stopwording for story link detection. In: Proceedings of the Second International Conference on Human Language Technology Research, pp. 190–193. Morgan Kaufmann Publishers Inc., San Francisco (2002)
4. Chen, F., Farahat, A., Brants, T.: Story link detection and new event detection are asymmetric. In: NAACL 2003: Proceedings of the 2003 Conference of the North American Chapter of the Association for Computational Linguistics on Human Language Technology, pp. 13–15. Association for Computational Linguistics, Morristown (2003), doi:10.3115/1073483.1073488
5. Chen, F., Farahat, A., Brants, T.: Multiple similarity measures and source-pair information in story link detection. In: In HLT-NAACL 2004, pp. 2–7 (2004)

6. Chen, Y.-J., Chen, H.-H.: Nlp and ir approaches to monolingual and multilingual link detection. In: Proceedings of the 19th International Conference on Computational Linguistics, pp. 1–7. Association for Computational Linguistics, Morristown (2002)
7. Farahat, A., Chen, F., Brants, T.: Optimizing story link detection is not equivalent to optimizing new event detection. In: ACL 2003: Proceedings of the 41st Annual Meeting on Association for Computational Linguistics, pp. 232–239. Association for Computational Linguistics, Morristown (2003)
8. Ferret, O.: Using collocations for topic segmentation and link detection. In: Proceedings of the 19th International Conference on Computational Linguistics, pp. 1–7. Association for Computational Linguistics, Morristown (2002)
9. Hong, Y., Zhang, Y., Fan, J., Liu, T., Li, S.: Chinese topic link detection based on semantic domain language model. Journal of Software, 2265–2275 (2008)
10. Lavrenko, V., Allan, J., DeGuzman, E., LaFlamme, D., Pollard, V., Thomas, S.: Relevance models for topic detection and tracking. In: Proceedings of the Second International Conference on Human Language Technology Research, pp. 115–121. Morgan Kaufmann Publishers Inc., San Francisco (2002)
11. Luo, W., Liu, Q., Chen, X.: Development and analysis of technology of topic detection and tracking. In: Sun, M.S. (ed.) Proc. of the JSCL 2003, Beijing, China, pp. 560–566 (2003)
12. Nallapati, R.: Semantic language models for topic detection and tracking. In: NAACL 2003: Proceedings of the 2003 Conference of the North American Chapter of the Association for Computational Linguistics on Human Language Technology, Edmonton, Canada, pp. 1–6. Association for Computational Linguistics, Morristown (2003)
13. Schultz, J.M., Liberman, M.Y.: Towards a "universal dictionary" for multilanguage information retrieval applications, pp. 225–241 (2002)
14. Shah, C., Croft, W.B., Jensen, D.: Representing documents with named entities for story link detection (sld). In: CIKM 2006: Proceedings of the 15th ACM International Conference on Information and Knowledge Management, pp. 868–869. ACM, New York (2006)
15. Wayen, C.L.: Multilingual topic detection and tracking: Successful research enabled by corpora and evaluation. In: Proceedings of the Language Resources and Evaluation Conference (LREC), Athens, Greece, pp. 1487–1494 (2000)
16. Zhang, X., Wang, T., Chen, H.: Story link detection based on event model with uneven svm. In: Li, H., Liu, T., Ma, W.-Y., Sakai, T., Wong, K.-F., Zhou, G. (eds.) AIRS 2008. LNCS, vol. 4993, pp. 436–441. Springer, Heidelberg (2008)

# Ranking Multilingual Documents Using Minimal Language Dependent Resources

G.S.K. Santosh, N. Kiran Kumar, and Vasudeva Varma

International Institute of Information Technology, Hyderabad, India
{santosh.gsk,kirankumar.n}@research.iiit.ac.in, vv@iiit.ac.in

**Abstract.** This paper proposes an approach of extracting simple and effective features that enhances multilingual document ranking (MLDR). There is limited prior research on capturing the concept of multilingual document similarity in determining the ranking of documents. However, the literature available has worked heavily with language specific tools, making them hard to reimplement for other languages. Our approach extracts various multilingual and monolingual similarity features using a basic language resource (bilingual dictionary). No language-specific tools are used, hence making this approach extensible for other languages. We used the datasets provided by Forum for Information Retrieval Evaluation (FIRE)[1] for their 2010 Adhoc Cross-Lingual document retrieval task on Indian languages. Experiments have been performed with different ranking algorithms and their results are compared. The results obtained showcase the effectiveness of the features considered in enhancing multilingual document ranking.

**Keywords:** Multilingual Document Ranking, Feature Engineering, Wikipedia, Levenshtein Edit Distance.

## 1   Introduction

Multilingual Information Retrieval (MLIR) is desirable with the increase of information in different languages. With the rapid development of globalization and digital online information in Internet, a growing demand for MLIR has emerged. MLIR involves the subtask of Cross Lingual Information Retrieval (CLIR) separately for each desired language. The clear separation of the retrieved result lists between different languages makes it necessary to have a merging step in order to produce a single result list. However, merging is intertwined with ranking step that ranks the documents of multilingual result lists as per the relevancy to the information need.

The problem of CLIR has been well studied in the past decade especially with the help of CLEF, NTCIR, TREC and FIRE forums. In the realm of CLIR the problem of ranking multilingual result lists is a very challenging task. The task of identifying whether two different language documents talks about the same

---

[1] http://www.isical.ac.in/~clia/

A. Gelbukh (Ed.): CICLing 2011, Part II, LNCS 6609, pp. 212–220, 2011.

topic is itself very challenging. There are few early attempts on ranking multilingual documents (Round robin merging [1], raw-score merging [1]). These
merging processes have to make some simplifying assumptions. For example, one
may assume that the similarities calculated for different language result lists are
comparable; so the result lists can be merged according to their raw similarity
values [1]. One can also normalize the similarities first; but this approach implicitly assumes that the highly ranked documents in different languages are similar
to the query at a comparable level. These assumptions are not true. Until recent
past [2], [3], [4], [5], [6], there was little focus on merging multilingual result lists.
The recent work concentrated more on extracting semantic information such as
multilingual topics from documents. These methods are highly dependent upon
language specific tools like named-entity recognizer, part-of-speech tagger etc.,
hence they cannot be extended for languages with fewer resources, i.e., they do
not achieve high-multilinguality.

If there is a requirement for a ranking approach to be applied across various languages, language specific development pose major challenges. Also while
merging multilingual result lists, techniques that suit one language pair might
not be effective for another. For example, techniques for closely-related languages (Ex: Hindi, Marathi) might not be useful for a pair of languages in
widely different families (Ex: Spanish, Chinese). While some applications will
only be concerned with a small number of languages, others (e.g., foreign policy
or international patent law) will require systems that scale to tens of disparate
languages. Progress in developing language-independent approaches will greatly
benefit multilingual retrieval, and should therefore be encouraged within the
MLIR community.

In this paper, we extract simple and efficient features from multilingual documents and topics that enhance the performance of Multilingual Document Ranking (MLDR). We propose to exploit the similarities among candidate documents
to enhance MLDR. Because similar documents usually share similar ranks, cross-
lingual relevant documents can be exploited to enhance the relevance estimation
for documents of different languages. Given result lists of two different languages
along with their queries, various similarity measures are calculated among documents of same language and of different languages. Same set of similarity metrics are measured among monolingual documents and multilingual documents,
so that a document can be compared with any other document independent of
their languages.

It is known that while a given translation tool may produce acceptable translations for a given set of queries; it may perform poorly for other queries [7]. Also
the availability of language specific tools is very limited across languages. In this
vein, we have eliminated the usage of any language-specific tools while measuring document similarity metrics and other features. However, for calculating the
multilingual document similarity, only bilingual dictionaries are used. External
knowledge resource like wikipedia is also exploited in adding up to the efficiency
of similarity measurement. Provided the availability of the basic language resource (bilingual dictionary), this approach can be extended to other language

pairs. We carried out our experiments on FIRE 2010 corpus. Experiments are conducted by modelling several ranking algorithms on the extracted features. Their results are compared using the NDCG as the evaluation metric. The results obtained are verified against BM25 baseline ranking system and significant improvements are noticed in the ranking performance.

The rest of this paper is organized as follows. Section 2 reviews prior research on MLDR. Section 3 describes the features that generate similarity and relevance scores for all the query-document pairs. Section 4 describes the experimental results obtained using various ranking algorithms. Section 5 summarizes our findings and concludes with potential future work.

## 2   Related Work

The early attempts for merging multilingual documents were heuristic based approaches. Raw-score merging [1] tries to combine the result lists using the document scores that were previously assigned by the retrieval algorithms. As the scores were incomparable, efforts were made to normalize the scores [3] [4]. The documents with the normalized scores are then ranked. Round robin merging [1] tries to combine the result lists based on the ranks of the documents.

A 2-step merging strategy was proposed by Martinez-Santiago et al. [2] to rank multilingual result lists. In this approach, instead of directly merging the result lists, the multilingual result lists are first indexed and this indexed dataset is used to retrieve final multilingual result list. Recent work [6] [5] focused on implementing the learning approaches to the merge problem. They have extracted various features and trained the features using learning algorithms like FRank, Boltzmann. However, in the work presented by Tsai et al. [5], the constructed features identify person names, organization names and vocabulary terms. They are very much dependent upon the availability of language specific tools.

Although Gao et al. [6] laid emphasis on the importance of measuring multilingual document similarity, they have incorporated these similarities in implementing a clustering based approach to the problem of MLDR. However our approach focuses on measuring the direct influence of these similarities by incorporating them as features for a document.

The basic idea behind it is that a document can be similar to other documents in the result list of same language or in the result list of different language. Capturing the similarity between the documents gives us very useful information regarding the importance of the documents. If a document is found to be similar to many other documents and is also relevant enough to the query, then that document needs to be placed at a higher rank in the final result list. Hence, these similarity features are studied here in the context of enhancing MLDR performance.

## 3   System Overview

In a usual scenario of MLIR, given a query in a language, CLIR is performed on separate monolingual collections. Once monolingual result lists are obtained from each collection, all the lists are merged into a multilingual result list. Our work scenario considers a query in English and Hindi, along with their corresponding monolingual result lists as the starting point. In this context, the merging process does not have any prior information on how the original result lists are produced. However there are binary relevance judgements for all query-document pairs indicating whether a document is relevant or not. Features that effectively capture the document relevancy to the query and the similarity among the documents are extracted. Same set of features are considered for all the documents. Every document is represented in terms of a vector of these features. After the vectors are constructed for all the documents, these features are modelled using various ranking algorithms. The estimated relevance probabilities assigned to the documents by the ranking algorithms are used in ordering the documents.

## 3.1   Feature Engineering

In information retrieval and natural language processing (NLP), question answering (QA) is the task of automatically answering a question posed in natural language. To find the answer to a question, a QA computer program may use either a pre-structured database or a collection of natural language documents. Given a set of documents and a question, a QA system needs to address two interesting challenges [8].

1. *Estimating answer relevance.* Is a particular answer relevant to the question? How to identify relevant answer(s) among irrelevant ones? If irrelevant answers can be eliminated by using some knowledge base, then the remaining answers can be ranked better.
2. *Exploiting answer redundancy.* How do we exploit answer redundancy among answer candidates? If a particular relevant answer is found to repeat in a given list of answers, then that answer needs to be ranked higher.

Analyzing from this QA perspective, the problem of MLDR is indeed related to it. There is a need to identify relevant documents and also they need to be ranked higher when they are found similar to many other documents. Hence, we have emphasized on features that address these two challenges. Features are required to capture the relevancy of documents for a given query and the similarity among documents.

Such features are extracted from three levels 1) Query-Document similarity, 2) Monolingual Document similarity and 3) Multilingual document similarity. According to the extraction level, we describe these features in detail as follows.

### 3.1.1 Query-Document Similarity

There are many ways to calculate the relevancy among a document and a query. We used the tf and idf measures for measuring the relevancy of documents for a query. For every query term 'k' in a query Q, its tf-idf value is calculated with respect to a document 'j' ($tf - idf_{kj}$) in the document collection $D$ of the same language. For all the query terms, these scores are calculated and added up to get the tf-idf feature value for document 'j'. We normalized this tf-idf value within the document collection according to the following formula

$$TF - IDF = \frac{\sum_{\forall k \in Q} tf - idf_{kj}}{\sum_{\forall j \in D} \sum_{\forall k \in Q} tf - idf_{kj}} . \tag{1}$$

Each query topic has title, description and narration fields. The tf-idf measures are calculated for each of these fields and added as different features.

### 3.1.2 MonoLingual Document Similarity

Given two documents of the same language, the similarity can be measured in various ways as mentioned in the work of A. Huang [9]. Every document is represented in a vector notation. The terms of a document are assigned their tf-idf weights calculated within that document collection $D$. The wikipedia redirection terms corresponding to every term are also included in the vector. The concept of wikipedia redirections is explained below. The similarity measure for a document $d_i$ is calculated using the formula

$$sim_k(d_i) = \frac{\sum_{j=1(i \neq j)}^{D} sim'_k(d_i, d_j)}{\sum_{\forall i}^{D} \sum_{j=1(i \neq j)}^{D} sim'_k(d_i, d_j)} . \tag{2}$$

Each $sim'_k(d_i, d_j)$ is a similarity feature used to calculate document similarity between $d_i$ and $d_j$. Eucledian, Cosine, Jaccard, Pearson Correlation coefficient and Averaged Kullback-Leibler Divergence [9] are the different similarity features used in calculating the document similarity.

**Concept of Wikipedia Redirection:** Wikipedia is a free, web-based, collaborative, multilingual encyclopaedia. There are 262 language editions available as of now. So, the extensibility of our techniques across certain available languages is ensured even after using wikipedia knowledge base. Since Wikipedia is web-based and therefore worldwide, contributors of a same language edition may use different dialects or may come from different countries. These differences may lead to some conflicts over spelling differences, (e.g. color vs. colour) or points of view (e.g. sachin tendulkar, master blaster, little master, SRT).

All these representations conceptually refer to the same wikipedia page. When a user enters any one of these representations, the action gets redirected to that single wikipedia page which all these terms conceptually refer to. We collect all such redirections available in English and in Hindi document collections of wikipedia. Sometimes the redirections may contain phrases, in such cases the phrases are tokenized and are added to the list of redirections fetched for a given term. It was coined in [10], that these wikipedia redirections can be referred to as synonyms. These redirections are used in both monolingual and multilingual document similarity calculations.

### 3.1.3 MultiLingual Document Similarity

In order to compare two multilingual documents, we need to map them onto a common ground representation. Given a document in English and a document in Hindi, the English document terms are mapped into Hindi representation by using bilingual dictionary and wikipedia redirections. If a term is found in dictionary, it is replaced with its synonyms in Hindi. Each word may have more than one possible synonyms. For every term in the synonyms, its wikipedia redirections are also taken into account. Finally, the terms of a English document are represented in a vector of Hindi terms. Same set of similarity features used in measuring monolingual document similarities are used here. Similarities are calculated as per the Eq. (2).

Transliteration might be highly helpful in identifying the proper nouns, but it requires parallel transliterated (English-Hindi) word lists to build even a language-independent statistical transliteration technique [11]. Acquiring such word lists is a hard task when one of the language is a minority language. Including transliteration comes at the cost of reducing the extensibility of our approach. Priority is given to the latter.

**Modified Levenshtein Edit Distance Measure:** In all of the similarities calculated above, the terms are compared using the Modified Levenshtein edit distance as a string distance measure. In many languages, words appear in several inflected forms. For example, in English, the verb 'to walk' may appear as 'walk', 'walked', 'walks', 'walking'. The base form, 'walk', that one might look up in a dictionary, is called the lemma for the word. The terms are usually lemmatized to match the base form of that term. Lemmatizers are available for English and many other European languages. But the lemmatizers support is very limited in the context of Indian Languages. So, we have modified the levenshtein edit distance metric to replace the purpose of lemmatizers by adding certain language-independent rules. Henceforth, it can be applied for any language. This modified levenshtein edit distance would help us in matching a word in its inflected form with its base form or other inflected forms. The rules are very intuitive and are based on three aspects:

1. Minimum length of the two words
2. Actual Levenshtein distance between the words
3. Length of subset string match, starting from first letter.

## 4    Experiments and Evaluation

We have conducted experiments using the FIRE 2010 dataset available for the ad-hoc cross lingual document retrieval task. The data consists of news articles collected from 2004 to 2007 for each of the English, Hindi, Bengali and Marathi languages from regional news sources. There are 50 query topics represented in each of these languages. We have considered the English and Hindi articles for our experiments. There are only binary relevance judgements given for every topic-document pair. For every topic we have worked with a set of documents which contains all its relevant documents with a certain noise i.e., irrelevant documents. The noise considered is twice the number of relevant documents.

Different ranking algorithms like SVC (SVM Classification), RSVM (Ranking SVM), SVM Regression and Logistic Regression are used to learn ranking
functions by modelling the features extracted. The source codes of LibSVM[2],
SVM-Light[3], SVM-Rank[4] and Logistic Regression[5] are used to run SVC, SVM
regression, RSVM, and Logistic regression respectively. The probabilities predicted by these learning approaches are used in ranking the documents. In order
to evaluate the ranking order of the documents, a small set of documents are
annotated with ratings from 0(irrelevant) to 5(perfect) by human labellers. The
results of NDCG@5,10,15,20 (Normalized Discounted Cumulative Gain) are used
to compare the systems. MLIR ranking performance results of these learning approaches are compared in Table 1 with a BM25 baseline system.

**Table 1.** Comparison of MLDR Performances

| Method | NDCG@5 | NDCG@10 | NDCG@15 | NDCG@20 |
|---|---|---|---|---|
| SVC | 67.99 | 73.53 | 76.28 | 79.50 |
| SVM-Reg | 71.00 | 77.87 | 77.94 | **82.11** |
| RSVM | **75.04** | **78.84** | **78.03** | 79.83 |
| LogReg | 69.89 | 74.53 | 76.69 | 80.37 |
| BM25 | 64.19 | 69.38 | 68.82 | 72.21 |

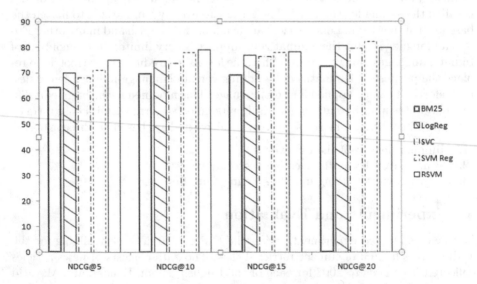

**Fig. 1.** A Graphical Comparison of the Performances of Ranking Algorithms

[2] http://www.csie.ntu.edu.tw/ cjlin/libsvm/
[3] http://svmlight.joachims.org/
[4] http://www.cs.cornell.edu/People/tj/svm_light/svm_rank.html
[5] http://komarix.org/ac/lr/lrtrirls

From the table 1 it is clear that all the learning algorithms has outperformed the BM25 baseline system. The numbers indicate that the ranking functions that modelled our features stand at par with the baseline system in terms of the performance. There is an overall enhancement in the performance of MLDR. Ranking SVM and SVM regression have achieved the best results. These accuracies provide a good indication that the features considered has proved to be effective in enhancing the MLDR performance.

## 5  Conclusion and Future Work

In this paper we have presented an approach to extract simple, effective features that enhances the MLDR performance. We have approached the problem of MLDR from a QA perspective. The features we considered extract the document relevancy to a query and various document similarity measures. These features are modelled using different ranking algorithms. From the results showcased in Table 1, it can be inferred that these features has considerably increased the MLDR performance. They dominated the BM25 baseline with a good margin.

As development of language specific tools pose major challenges across languages, we have come up with this approach without using any language tools. Compared to the existing approaches, this approach achieves high multilinguality; it is extensible to other languages and is easily reproducible provided the availability of minimum language resource (bilingual dictionary). The cost of reproducibility of this approach is the same for any other language pairs. This approach is not specific to any dataset, it can be applied to various other datasets. This approach takes into consideration the future growth of multilingual information need.

As there are dictionaries available for few other Indian Languages[6], we are currently working on extending our approach for other Indian languages using the FIRE 2010 datasets. As there are many other learning approaches available, we would like to explore their role in enhancing the performance of MLDR. We are planning to extend this approach to most-researched european languages and compare the MLDR accuracy of our approach with the existing state-of-art systems. We would like to work more on capturing query document relevancy by extracting key terms from the documents to get the topics of the documents and thereby assign more relevance to the documents that talks about the (almost)same topic as that of the query.

## References

1. Savoy, J., Calve, A.L., Vrajitoru, D.: Report on the TREC-5 experiment: Data fusion and Collection fusion. In: The Fifth Text Retrieval Conference (TREC-5), pp. 489–502 (1997)

---

[6] http://ltrc.iiit.ac.in/onlineServices/Dictionaries/Dict_Frame.html

2. Martinez-Santiago, F., Urena-Lopez, L., Martin-Valdiva, M.: A merging strategy proposal: The 2-step retrieval status value method. In: Information Retrieval, pp. 71–93 (2006)
3. Powell, A., French, J., Callan, J., Connell, M., Viles, C.: The impact of Database Selection on Distributed Searching. In: Proceedings of the 23rd Annual International ACM SIGIR Conference on Research and Development in Information Retrieval, pp. 232–239. ACM, New York (2000)
4. Lin, W., Chen, H.: Merging Mechanisms in Multilingual Information Retrieval. In: Peters, C., Braschler, M., Gonzalo, J., Kluck, M. (eds.) CLEF 2002. LNCS, vol. 2785, pp. 175–186. Springer, Heidelberg (2003)
5. Tsai, M., Wang, Y., Chen, H.: A Study of Learning a Merge Model for Multilingual Information Retrieval. In: Proceedings of SIGIR 2008, pp. 195–202. ACM, New York (2008)
6. Gao, W., Niu, C., Zhou, M., Wong, K.-F.: Joint ranking for multilingual web search. In: Boughanem, M., Berrut, C., Mothe, J., Soule-Dupuy, C. (eds.) ECIR 2009. LNCS, vol. 5478, pp. 114–125. Springer, Heidelberg (2009)
7. Savoy, J., Berger, P.-Y.: Selection and merging strategies for multilingual information retrieval. In: Peters, C., Clough, P., Gonzalo, J., Jones, G.J.F., Kluck, M., Magnini, B. (eds.) CLEF 2004. LNCS, vol. 3491, pp. 27–37. Springer, Heidelberg (2005)
8. Wo, J., Si, L., Nyberg, E., Mitamura, T.: Probabilistic Models for Answer-Ranking in Multilingual Question-Answering. ACM Transactions on Information Systems (2010)
9. Huang, A.: Similarity measures for Text Document Clustering. In: Proceedings of New Zealand Computer Science Research Student Conference, pp. 49–56 (2008)
10. Wu, F., Weld, D.: Autonomously semantifying Wikipedia. In: Proceedings of Sixteenth CIKM, CIKM 2007. ACM, New York (2007)
11. Ganesh, S., Harsha, S., Pingali, P., Varma, V.: Statistical Transliteration for Cross Language Information Retrieval using HMM alignment model and CRF. In: 2nd International Workshop on CLIA, 3rd International Joint Conference on Natural Language Processing (IJCNLP 2008) (2008)

# Measuring Chinese-English Cross-Lingual Word Similarity with *HowNet* and Parallel Corpus

Yunqing Xia[1], Taotao Zhao[1,2], Jianmin Yao[2], and Peng Jin[3]

[1] Department of Computer Science and Technology,
Tsinghua University, Beijing 100084, China
yqxia@tsinghua.edu.cn
[2] School of Computer Science and Technology,
Soochow University, Suzhou 215006, China
zhaott10@gmail.com, jyao@suda.edu.cn
[3] Lab of Intelligent Information Processing and Application,
Leshan Normal University, Leshan 614004, China
jandp@pku.edu.cn

**Abstract.** Cross-lingual word similarity (CLWS) is a basic component in cross lingual information access systems. Designing a CLWS measure faces three challenges: (i) Cross-lingual knowledge base is rare; (ii) Cross-lingual corpora are limited; and (iii) No benchmark cross-lingual dataset is available for CLWS evaluation. This paper presents some Chinese-English CLWS measures that adopt *HowNet* as cross-lingual knowledge base and sentence-level parallel corpus as development data. In order to evaluate these measures, a Chinese-English cross-lingual benchmark dataset is compiled based on the Miller-Charles' dataset. Two conclusions are drawn from the experimental results. Firstly, *HowNet* is a promising knowledge base for the CLWS measure. Secondly, parallel corpus is promising to fine-tune the word similarity measures using cross-lingual co-occurrence statistics.

**Keywords:** Cross-lingual word similarity, cross-lingual information access, *HowNet*, parallel corpus.

## 1 Introduction

Word similarity plays a vital role in natural language processing and information retrieval. In natural language processing, word similarity is widely used in word sense disambiguation [1], synonym extraction [2]. In information retrieval, word similarity is adopted in multimodal documents retrieval [5] and image retrieval [6]. Human-compiled linguistic knowledge base (e.g., *WordNet* [7] and *HowNet* [8]) were widely used to measure word similarity [2,9-12]. As thesauri are usually static and incomplete, corpora were then adopted to estimate word similarity based on co-occurrence statistics [13-15]. Li et al. (2003) proved that thesaurus and corpus can be integrated to yield a better performance [16].

Cross-lingual word similarity (CLWS) reflects semantic similarity between two words in different languages. Very recently, CLWS research started to attract

A. Gelbukh (Ed.): CICLing 2011, Part II, LNCS 6609, pp. 221–233, 2011.

attention when multi-lingual content is found surprisingly huge on the Internet. At least 150 languages are frequently used by people to communicate via the Internet, creating more than 313B web pages. Today, 68.4 percent web pages are written in English and around 16 percent in Japanese, German, Chinese and French[1]. This exhibits strong necessity to investigate on cross-lingual information access applications such as cross-lingual information retrieval and cross-lingual text classification/cluster, in which cross-lingual word similarity plays a vital role. Challenges in designing CLWS measure are summarized as follows.

(1) Cross-lingual knowledge base is rare. *WordNet* is widely adapted to measure similarity between English words [7]. However, *WordNet* is monolingual in nature thus is inapplicable to CLWS measuring. In contrast, *HowNet* is a Chinese-English bilingual knowledge. It defines concepts using Chinese words and their English counterparts. It is thus theoretically feasible to measure Chinese-English CLWS using *HowNet*. Nevertheless, coverage of *HowNet* should be expanded.

(2) Cross-lingual corpora are limited. Large-scale monolingual corpora (e.g., Gigaword from LDC[2]) can be easily obtained and have already been proven effective to measure monolingual word similarity [13-14]. But collecting large-scale cross-lingual corpora are difficult. The sentence-level parallel corpora used in machine translation research are applicable to measure cross-lingual word similarity [16,17]. However, size of the cross-lingual corpora is worrisome.

(3) No benchmark cross-lingual dataset is available. Miller-Charles dataset is a widely used benchmark dataset for evaluating English word similarity measures [18]. For Chinese, Liu and Li (2002) designed a different dataset [11]. But to the best of our knowledge, no benchmark dataset is available to evaluate CLWS measures.

This presents some Chinese-English CLWS measures using *HowNet* as linguistic knowledge base and parallel corpus as development data. Contributions of this work are summarized as follows.

(i)     The work is an earlier attempt to apply *HowNet* in CLWS measuring.
(ii)    Parallel corpus is first used in CLWS measuring.
(iii)   A benchmark dataset is created based on the *MC* dataset to evaluate CLWS measures.

The rest of this paper is organized as follows. Related work is summarized in Section 2. Then *HowNet*-based and corpus-based CLWS measures are presented in Section 3 and Section 4, respectively. Evaluation and discussion are given in Section 5. We draw conclusions in Section 6.

## 2 Related Work

The study on cross-lingual word similarity measuring is related to two research topics: monolingual word similarity measuring and cross-lingual information access.

---

[1] http://www.translate-to-success.com/online-language-web-site-content.html
[2] http://www.ldc.upenn.edu

## 2.1 Monolingual Word Similarity Measure

Monolingual word similarity measure started to attract research attention in early 1990's. In the past two decades, numerous word similarity measures have been designed for different languages. The measures can be grouped into three categories according to language resources the measures use: knowledge-based measures, corpus-based measures and the hybrid measures.

The knowledge-based measures rely on certain linguistic knowledge base to estimate conceptual distance between two words. For example, *WordNet* was used by Resnik (1995) and Lin (1998) to measure similarity between English words [9,10]. *HowNet* was used by Liu and Li (2002) and Dai et al. (2008) to measure similarity between Chinese words [11] according to concept definition of words. The corpus-based measures, on the other hand, make use of word co-occurrence statistics to calculate degree of relatedness between words. For example, Bollegala et al. (2007) proposed to measure word similarity using text snippets returned by Web search engines [13]. Cilibrasi and Vitanyi (2007) designed Google similarity distance with Google page counts [14]. Literatures show that corpus-based word similarity tends to be believable when large-scale corpora (e.g., Google corpus) are available. The hybrid measures were designed to further improve accuracy by incorporating strengths of thesauri and corpora. For example, Li et al. (2003) proposed to incorporate *WordNet* and text corpus [15], which is proved helpful in their experiments.

To the best of our knowledge, cross-lingual word similarity measure is not yet seen in literatures because most measures are not applicable because sound cross-lingual thesauri and large-scale parallel corpora are rare. Enlightened by the *HowNet*-based monolingual word similarity measures, we investigate in this work how *HowNet* and parallel corpus are combined to measure cross-lingual word similarity.

## 2.2 Cross-Lingual Information Access

Research work on cross-lingual information access is also related to this work. Two applications on cross-lingual information access are worth mentioning. Firstly, cross-lingual information retrieval (CLIR) seeks to retrieve documents in languages other than the query language. An early solution is to transfer CLIR task into monolingual information retrieval task by translating query or documents using machine translation (MT) systems [19]. However, Nie et al. (1999) proved that MT is inappropriate for CLIR. A probabilistic MT model was further developed using parallel texts that are automatically collected from the Web [16]. Few efforts were reported to measure cross-lingual query-document relevance based on cross-lingual word similarity. This ascribes mainly to shortage of cross-lingual language resources.

Secondly, cross-lingual text classification and clustering seek to find cross-lingual category information from mixed-language text collections. The key challenge is cross-lingual document similarity. Pouliquen et al. (2004) adopted multilingual thesaurus (i.e., Eurovoc) to convert documents into language-independent internal vectors in cross-lingual news topic tracking [19]. However, multilingual linguistic knowledge base is rare. We believe this problem can be appropriately dealt with when cross-lingual word similarity is achieved.

## 3  Measuring CLWS Using *HowNet*

### 3.1  Overview of *HowNet*

*HowNet* is deemed promising to Chinese-English word similarity measuring due to two reasons. Firstly, *HowNet* was proven effective in measuring similarity between two Chinese words [11] or two English words [12]. Secondly, *HowNet* defines concepts bilingually to explain both Chinese words and English words. For example, *HowNet* concept for Chinese word "海岸(*coast*)" is illustrated in Fig.1. Concept definition (i.e., *DEF* in Fig.1) of word "海岸(*coast*)" indicates that *coast* is a *land* and it *is near waters*. Meanwhile, Fig.1 shows that the *DEF* structure contains two parts: (1) *primary sememe*, e.g., "land|陆地", and (2) *modifier*, e.g., "{*BeNear*|靠近: *existent*={~}, *partner*={*waters*|水域}}}". The modifier comprises of several secondary sememes, which differs this words from others.

```
NO.=048973              // Concept ID
W_C=海岸                 // Chinese word
G_C=N [hai3 an4]        // Part of speech tag and Pinyin
W_E=coast               // Corresponding English word
G_E=N                   // Part of speech for English
DEF={land|陆地:{BeNear|靠近:existent={~},partner={waters|水域}}}
                        // Concept definition
```

**Fig. 1.** *HowNet* concept definition for word 海岸 (*coast*)

*HowNet* is deemed unique due largely to its bilingual nature. For example, slot W_C in Fig.1 specifies the Chinese word and W_E the English word. Sememe within *DEF* structure is also bilingual, e.g., "land|陆地", "{*BeNear*|靠近" and "*waters*|水域". Therefore, *HowNet* is applicable to measure Chinese-English CLWS.

### 3.2  The *HowNet*-Based CLWS Measure

A Chinese word similarity measure was designed by Liu and Li (2002) using *HowNet*. Dai et al. (2008) further adopted *HowNet* in an English word similarity measure. Enlightened by the previous work, we propose to use *HowNet* to measure Chinese-English cross-lingual word similarity based on concept definitions. *HowNet* provides an API to find concepts using Chinese or English words. For example, with the Chinese word海岸(*coast*), we can find the concept shown in Fig.1. Meanwhile, this concept can also be found using the English word *coast*. Given Chinese word $w^{CN}$ and English word $w^{EN}$, the following four steps are executed in the CLWS measure.

***Step* 1.** Search in *HowNet* with $w^{CN}$ to find a set of related concepts $C^{CN} = \{c_i^{CN}\}(i = 1...N^{CN})$, a set of related primary sememes $P^{CN} = \{p_i^{CN}\}$ and a set of secondary sememes $S^{CN} = \cup_{i=1}^{N^{CN}} S_i^{CN} = \{s_{i,k}^{CN}\}(k = 1..., M^{CN})$. We also obtain $C^{\mathrm{EN}} = \{c_j^{\mathrm{EN}}\}(j = 1...N^{\mathrm{EN}})$, $P^{\mathrm{EN}} = \{p_j^{\mathrm{EN}}\}$ and $S^{\mathrm{EN}} = \cup_{j=1}^{N^{\mathrm{EN}}} S_j^{\mathrm{EN}} = \{s_{j,l}^{\mathrm{EN}}\}$ $(l = 1..., M^{\mathrm{EN}})$.

**Step 2.** Calculate sememe similarity $Sim^{\mathrm{SEM}}$ between $s^{\mathrm{CN}} \in (P^{\mathrm{CN}} \cup S^{\mathrm{CN}})$ and $s^{\mathrm{EN}} \in (P^{\mathrm{EN}} \cap S^{\mathrm{EN}})$ based on distance and depth within *HowNet* as follows.

$$Sim^{\mathrm{SEM}}(s^{\mathrm{CN}}, s^{\mathrm{EN}}) =$$
$$\max\left\{0, 1 - \lambda_{dis} \cdot V^{dis}(s^{\mathrm{CN}}, s^{\mathrm{EN}})\right\} \cdot \min\left\{\lambda_{dep} \cdot V^{dep}(s^{\mathrm{CN}}, s^{\mathrm{EN}}), 1\right\} \tag{1}$$

where $V^{dis}(s_1, s_2)$ and $V^{dep}(s_1, s_2)$ are two functions returning distance and depth between the two sememes within *HowNet*; $\lambda_{dis}$ and $\lambda_{dep}$ are normalization factors for distance and depth, respectively. We assign $\lambda_{dis} = 40$ and $\lambda_{dep} = 4$ according to Dai et al. (2008)'s experiments.

**Step 3.** Calculate concept similarity $Sim^{\mathrm{CON}}$ between $c_i^{\mathrm{CN}}$ and $c_j^{\mathrm{EN}}$ based on sememe similarity $Sim^{\mathrm{SEM}}$ as follows.

$$Sim^{\mathrm{CON}}(c_i^{\mathrm{CN}}, c_j^{\mathrm{EN}}) = Sim^{\mathrm{SEM}}(p_i^{\mathrm{CN}}, p_j^{\mathrm{EN}}) \cdot Sim^{\mathrm{SEM}}(S_i^{\mathrm{CN}}, S_j^{\mathrm{EN}}) \tag{2}$$

$$Sim^{\mathrm{SEM}}(S_i^{\mathrm{CN}}, S_j^{\mathrm{EN}}) = \frac{\sum_{m=1}^{M^{\mathrm{CN}}} \sum_{n=1}^{M^{\mathrm{EN}}} Sim^{\mathrm{SEM}}(s_m^{\mathrm{CN}}, p_n^{\mathrm{EN}})}{M^{\mathrm{CN}} \times M^{\mathrm{EN}}} \tag{3}$$

**Step 4.** Calculate word similarity $Sim^{\mathrm{HN}}$ between $w^{\mathrm{CN}}$ and $w^{\mathrm{EN}}$ based on concept similarity as follows.

$$Sim^{\mathrm{HN}}(w^{\mathrm{CN}}, w^{\mathrm{EN}}) = \max_{i=1}^{N^{\mathrm{CN}}} \max_{j=1}^{N^{\mathrm{EN}}} Sim^{\mathrm{CON}}(c_i^{\mathrm{CN}}, c_j^{\mathrm{EN}}) \tag{4}$$

# 4 Incorporating Parallel Corpus to *HowNet*-Based CLWS Measure

## 4.1 The Measure

Parallel corpora are found helpful to cross-lingual information retrieval by Nie et al. (1999) [17]. Enlightened by this work, we propose to use parallel corpus to improve the *HowNet*-based CLWS measure. Four steps are followed.

**Step 1.** Locate relevant sentences within parallel corpus. Translation pairs in parallel corpora are organized at sentence level. That is, sentences in one language are aligned to their translations in another language on by one. By applying full text search, we can retrieve all aligned Chinese-English sentence pairs containing word $w^{\mathrm{CN}}$ or $w^{\mathrm{EN}}$ from the parallel corpus.

**Step 2.** Generate context word sets. Intuitively, word similarity can be reflected by their context. We thus propose to create context word set for each given word. Every word is assigned a weight, which is proportional to how many times the context word co-occurs with the given word. To reduce complexity, we select the top weighted 10 words to form a context set. Finally, context word set $E^{\text{CN}} = \{w_i^{\text{CN}}\}$ and $E^{\text{EN}} = \{w_j^{\text{EN}}\}(i, j = 1, 2, ..., N)$ are obtained for the Chinese and English words, respectively. Each context word is assigned a weight, which is automatically estimated from the corpus (see details in Section 4.3) based on the co-occurrences assumption. For example, the context word set for word *hill* is {(*land*, 175), (*top*, 159), (*area*, 153), (*mountain*, 123), (*high*, 119), (*country*,104),...}, in which the numbers represent times that the words co-occur with word *hill*.

**Step 3.** Calculate CLWS $Sim^{\text{DATA}}$ between $w^{\text{CN}}$ and $w^{\text{EN}}$ with the cross-lingual context word sets as follows.

$$
\begin{aligned}
Sim^{\text{DATA}}(w^{\text{CN}}, w^{\text{EN}}) = \\
\frac{1}{3}\Big(Sim^{\text{WS}}(w^{\text{CN}}, E^{\text{EN}}) + Sim^{\text{WS}}(w^{\text{EN}}, E^{\text{CN}}) + Sim^{\text{SET}}(E^{\text{CN}}, E^{\text{EN}})\Big)
\end{aligned}
\tag{5}
$$

in which $Sim^{\text{WS}}$ denotes cross-lingual word-set similarity and $Sim^{\text{SET}}$ cross-lingual set-set similarity (see Fig.2).

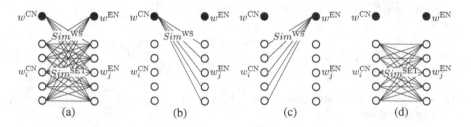

**Fig. 2.** Cross-lingual word-set similarity is measured with statistics from parallel corpus. (a) An overall measure; (b) Chinese word vs. English word set; (c) English word vs. Chinese word set; (d) Chinese word set vs. English word set.

**Step 4.** Calculate overall CLWS $Sim(w^{\text{CN}}, w^{\text{EN}})$ using a weighted sum equation as follows.

$$
Sim(w^{\text{CN}}, w^{\text{EN}}) = \alpha Sim^{\text{HN}}(w^{\text{CN}}, E^{\text{EN}}) + (1 - \alpha)Sim^{\text{DATA}}(w^{\text{CN}}, E^{\text{EN}})
\tag{6}
$$

in which $\alpha$ is an empirical combination weight.

### 4.2 Cross-Lingual Word-Set Similarity

Cross-lingual similarity between word and its counterpart context word set is illustrated in Fig.2 (b) and (c). Let $v_i^{\text{CN}}$ denote weight for the Chinese word $w_i^{\text{CN}}$ within the context, $v_j^{\text{EN}}$ weight for the English word $w_j^{\text{EN}}$, and $v_{ij}^{\text{CO}}$ weight for the word pair $(w_i^{\text{CN}}, w_j^{\text{EN}})$ that is contained in the parallel sentences. The following two approaches are proposed to calculate cross-lingual word-set similarity.

(1) Averaging (*AVG*)
In the *AVG* approach, cross-lingual word-set similarity is calculated by averaging cross-lingual word-word similarity as follows.

$$Sim^{\mathrm{WS}}(w^{\mathrm{CN}}, E^{EN}) = \frac{1}{N}\sum_{j=1}^{N} v_j^{\mathrm{EN}} Sim^{\mathrm{HN}}(w^{\mathrm{CN}}, w_j^{EN}) \tag{7}$$

$$Sim^{\mathrm{WS}}(w^{\mathrm{EN}}, E^{CN}) = \frac{1}{N}\sum_{j=1}^{N} v_i^{\mathrm{CN}} Sim^{\mathrm{HN}}(w^{EN}, w_i^{\mathrm{CN}}) \tag{8}$$

(2) Maximization (*MAX*)

In the *MAX* approach, cross-lingual word-set similarity is calculated by maximizing cross-lingual word-word similarity as follows.

$$Sim^{\mathrm{WS}}(w^{\mathrm{CN}}, E^{EN}) = \max_{j=1}^{N} \left\{ v_j^{\mathrm{EN}} Sim^{\mathrm{HN}}(w^{\mathrm{CN}}, w_j^{EN}) \right\} \tag{9}$$

$$Sim^{\mathrm{WS}}(w^{\mathrm{EN}}, E^{CN}) = \max_{i=1}^{N} \left\{ v_i^{\mathrm{CN}} Sim^{\mathrm{HN}}(w^{\mathrm{EN}}, w_i^{CN}) \right\} \tag{10}$$

### 4.3 Cross-Lingual Set-Set Similarity

Cross-lingual set-set similarity is illustrated in Fig.2 (d). Three approaches are given as follows.

(1) Enumerated Pairwise Averaging (*EPA*)

The *EPA* approach first enumerates all the cross-lingual word pairs. Illustrated in see Figure 2(a), $N \times N$ cross-lingual word pairs are created. The approach then calculates average of the $N \times N$ cross-lingual word similarity values. The *EPA* equation is given as follows.

$$Sim^{\mathrm{EPA}}(E^{\mathrm{CN}}, E^{\mathrm{EN}}) = \frac{1}{N \times N}\sum_{i=1}^{N}\sum_{j=1}^{N} v_i^{\mathrm{CN}} v_j^{\mathrm{EN}} v_{ij}^{\mathrm{CO}} Sim^{\mathrm{HN}}(w_i^{\mathrm{CN}}, w_j^{\mathrm{EN}}) \tag{11}$$

(2) Maximized Pairwise Averaging (*MPA*)

The *MPA* approach first produces the $N$ cross-lingual $(w_{i_k}^{\mathrm{CN}}, w_{j_k}^{\mathrm{EN}})$ pairs that satisfy two conditions: (1) The $N$ Chinese words and $N$ English words appear in the pairs only once; (2) Average of the $N$ cross-lingual word similarity values is maximal. The *MPA* equation is given as follows.

$$Sim^{\mathrm{MPA}}(E^{\mathrm{CN}}, E^{\mathrm{EN}}) = \frac{1}{N}\sum_{i=1}^{N} v_{i_k}^{\mathrm{CN}} v_{j_k}^{\mathrm{EN}} v_{i_k j_k}^{\mathrm{CO}} Sim^{\mathrm{HN}}(w_{i_k}^{\mathrm{CN}}, w_{j_k}^{\mathrm{EN}}) \tag{12}$$

(3) Pairwise Maximization (*PMX*)

The *PMX* approach seeks to find the cross-lingual $(w_{i_K}^{\mathrm{CN}}, w_{j_K}^{\mathrm{EN}})$ pairs that hold the globally biggest similarity value. The *PMX* equation is given as follows.

$$Sim^{\mathrm{PMX}}(E^{\mathrm{CN}}, E^{\mathrm{EN}}) = \max_{i=0}^{N} \max_{j=0}^{N} \left\{ v_i^{\mathrm{CN}} v_j^{\mathrm{EN}} v_{ij}^{\mathrm{CO}} Sim^{\mathrm{HN}}(w_i^{\mathrm{CN}}, w_j^{\mathrm{EN}}) \right\} \tag{13}$$

### 4.4 Parameter Estimation

The following three parameters are used in the corpus-based CLWS measure:

(1) $v_i^{\mathrm{CN}}$: weight for the Chinese word $w_i^{\mathrm{CN}}$ within the context;

(2) $v_j^{\text{EN}}$: weight for the English word $w_j^{\text{EN}}$ within the context;

(3) $v_{ij}^{\text{CO}}$: weight for the word pair $(w_i^{\text{CN}}, w_j^{\text{EN}})$ within sentence context.

Obviously, $v_i^{\text{CN}}$ and $v_j^{\text{EN}}$ are monolingual weights for a Chinese word and an English word, respectively. With any monolingual corpus, we are able to estimate the two weights using pointwise mutual information (PMI) [22], as follows.

$$v_i^{\text{CN}} = PMI(w_i^{\text{CN}}, w^{\text{CN}}) = \frac{f(w_i^{\text{CN}}, w^{\text{CN}})}{f(w_i^{\text{CN}}) \cdot f(w^{\text{CN}})} \tag{14}$$

$$v_j^{\text{EN}} = PMI(w_j^{\text{EN}}, w^{\text{EN}}) = \frac{f(w_j^{\text{EN}}, w^{\text{EN}})}{f(w_j^{\text{EN}}) \cdot f(w^{\text{EN}})} \tag{15}$$

in which $f(\cdot)$ returns frequency of given word(s) within corpus. $v_{ij}^{\text{CO}}$ is a weight that reflects how constantly a Chinese word co-occurs with an English word within parallel corpus, calculated as follows.

$$v_{ij}^{\text{CO}} = PMI(w_i^{\text{CN}}, w_j^{\text{EN}}) = \frac{f^*(w_i^{\text{CN}}, w_j^{\text{EN}})}{f(w_i^{\text{CN}}, w^{\text{CN}}) \cdot f(w^{\text{EN}}, w_j^{\text{EN}})} \tag{16}$$

Different from $f(\cdot)$ that requires the word(s) appearing in one monolingual sentence, $f^*(\cdot)$ counts times that the given Chinese-English word pair appear within the aligned cross-lingual sentence pairs. Traditionally, we apply logarithm on weight calculation. For those $f(\cdot) = 0$ or $f^*(\cdot) = 0$, we replace them with 0.01.

# 5 Evaluation

## 5.1 Setup

**Benchmark Dataset**

There is no benchmark dataset so far with the purpose of English-Chinese CLWS evaluation. In this work, we invited language experts to extend the *MC* dataset [18] so that the dataset can be applied to evaluate CLWS measures. For each word pair in the *MC* dataset, the experts first search within *HowNet* to locate the English word. Then the experts discuss and finally reach an agreement on its only one Chinese translation. The three rules are followed to select cross-lingual word pairs: (1) The English word must appear within *HowNet*; (2) The English word and its Chinese translation must both appear within parallel corpus for more than 10 times; (3) When multiples Chinese translations occur, the most frequently used translation (e.g., within the parallel corpus) is selected.

To avoid extra manpower, we continue using the human-judged similarity values that assigned in Miller and Charles' experiment [18]. Finally, the *MC* extended dataset contains 28 Chinese-English pairs are obtained (see Table 1).

**Table 1.** The extended *MC* dataset with Chinese-English cross-lingual word similarity values. CW represents Chinese word, EW English word and RAT rating.

| CW | EW | RAT | CW | EW | RAT | CW | EW | RAT | CW | EW | RAT |
|----|----|-----|----|----|-----|----|----|-----|----|----|-----|
| 微笑 | chord | 0.13 | 巫师 | lad | 0.42 | 飞鸟 | crane | 2.97 | 旅程 | Voyage | 3.84 |
| 公鸡 | voyage | 0.08 | 墓地 | forest | 0.84 | 飞鸟 | cock | 3.05 | 海岸 | Shore | 3.70 |
| 中午 | string | 0.08 | 公鸡 | food | 0.89 | 水果 | food | 3.08 | 工具 | implement | 2.95 |
| 玻璃 | magician | 0.11 | 海岸 | hill | 0.87 | 弟兄 | monk | 2.82 | 男孩 | Lad | 3.76 |
| 奴隶 | monk | 0.55 | 轿车 | journey | 1.16 | 疯人院 | asylum | 3.61 | 汽车 | Car | 3.92 |
| 海岸 | forest | 0.42 | 旅程 | car | 1.16 | 火炉 | furnace | 3.11 | 正午 | Noon | 3.42 |
| 先知 | monk | 1.10 | 吊车 | implement | 1.68 | 巫师 | magician | 3.50 | 宝石 | Jewel | 3.84 |

## Development data

The Chinese-English sentence-aligned bilingual corpus from Chinese LDC[3] was used in this work. Meanwhile, 2 million Chinese-English parallel sentences were collected from online bilingual documents, web pages, DVD subtitles and scientific literatures. Finally, a parallel corpus containing 3 million Chinese-English parallel sentences are used as development data.

## Evaluation metric:

In this experiment, we follow the traditional metric in word similarity evaluation, i.e., correlation coefficient [16].

### 5.2 Experiments

This experiment seeks to evaluate two CLWS measures that are implemented in this work: Measure $M1$ is *HowNet*-based and measure $M2$ incorporates the development data (i.e., a parallel corpus) to improve measure $M1$. Four parameters are involved in measure $M2$: (i) Word-set equations, see Equation (7)~(1); (ii) Set-set equations, see Equation (11)~(13); (iii) Combination weight, see Equation (6); and (iv) Size of the parallel corpus being used by measure $M2$ as development data . A set of experiments are conducted to compare measure $M1$ and $M2$ and to evaluate the parameters.

Note that no Chinese-English CLWS measures are reported so far. So we conduct experiments to compare parameters that might influence the proposed method.

### I. *HowNet*-based vs. Corpus-Incorporated

To compare the measure $M1$ and $M2$, we need to fix parameters in measure $M2$. We make use of on all development and adopt *MAX* in word-set similarity calculation and *MPA* in set-set similarity calculation because the two equations make measure $M2$ perform best (see Experiment II). The combination weight is set 0.5 also because an optimal performance is achieved. Experimental results are presented in Fig.3.

Two observations are made on Fig.3. Firstly, *HowNet* is able to yield a reasonable CLWS measure alone. However, the measure produces subnormal results on three pairs: {水果(fruit), food}, {轿车(car), journey} and {旅程(journey), voyage}. Looking into the corresponding concept definitions, we find these concept definitions can

---

[3] Chinese LDC: http://www.chinip.csdb.cn/

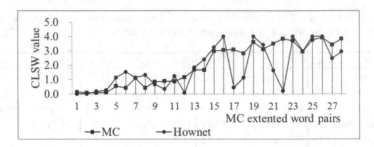

**Fig. 3.** CLWS value curves for the CLWS measures on the *MC* extended dataset. The *HowNet*-based measure (i.e., *M*1) yields 0.741 on correlation coefficient with the MC Extended dataset. When parallel corpus is incorporated (i.e., *M*2), the correlation coefficient is improved to 0.806.

be improved. Secondly, the *HowNet*-based measure is improved by 0.06 on correlation coefficient when parallel corpus is incorporated. It can thus be concluded that parallel corpus is promising to improve CLWS measures.

### II. *Word-Set Similarity Equations*
In this experiment, we concentrate on evaluation of measure *M*2 with two word-set similarity equations: (i) *AVG* with Equation (7) and (8); (ii) *MAX* with Equation (9) and (10). Experimental results are presented in Fig.4.

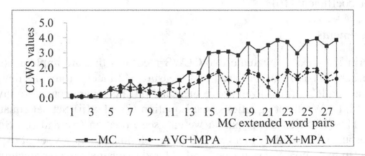

**Fig. 4.** CLWS value curves for the *M*2 measure on the *MC* extended dataset using two word-set similarity equations: *AVG* and *MAX*. Correlation value of the *M*2 measure is 0.747 with AVG and 0.806 with *MAX*.

Seen from Fig.4, curves for *M*2 measure with *AVG* and *MAX* are nearly consistent while *MAX* performs better. This proves that the maximization is a better equation to calculate cross-lingual word-set similarity.

### III. *Set-Set Similarity Equations*
This experiment seeks to compare three set-set similarity equations in measure *M*2: (i) *EPA* with Equation (11); (ii) *MPA* with Equation (12); (iii) *PMX* with Equation (13). Experimental results in Fig.5 show that measure *M*2 performs best with *MPA* equation. We can thus safely conclude that MPA is the best equation for cross-lingual set-set similarity calculation.

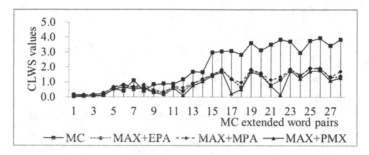

**Fig. 5.** CLWS value curves for the $M2$ measure on the $MC$ extended dataset using three set-set similarity equations: $EPA$, $MPA$ and $PMX$. Correlation value of the $M2$ measure is 0.805 with $EPA$, 0. 806 with $MPA$ and 0.754 with $PMX$.

## IV. Weight of Combination between *HowNet* and Parallel Corpus
This experiment seeks to evaluate how $\alpha$ in Equation (6) influences performance of the $M2$ measure. To achieve this goal, we range $\alpha$ from 0.1 to 0.9 in measure $M2$. Experimental results are given in Fig.6.

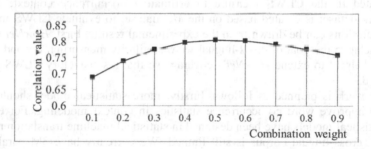

**Fig. 6.** Correlation values between the $MC$ extended dataset and $M2$ measure with various combination weights ($\alpha$) for *HowNet* and parallel corpus.

Seen from Fig.6, correlation value between $M2$ measure and the $MC$ extended dataset increase to 0.806 until $\alpha$ reaches 0.5 and drops when $\alpha$ is bigger than 0.5. It can thus be concluded that 0.5 is an appropriate weight that combines *HowNet* and parallel corpus in measure $M2$.

## V. Size of Parallel Corpus
This experiment intends to evaluate how size of parallel corpus influences the CLWS measures. To achieve this goal, we use parallel corpus with size ranging from 2M to 3M in measure $M2$. Experimental results are given in Fig.7.

It is shown in Fig.7 that performance of the measure remains between 0.75 and 0.76 when corpus size is smaller than 2.7M. The sharp increase discloses that 3M parallel sentences might be far from sufficient. In other words, performance of the measure can be further improved when more parallel sentences are available. Unfortunately, the 3M sentences are all we have. So we hope in the future work, more Chinese-English parallel sentences will be compiled to find the maximal performance of the measure.

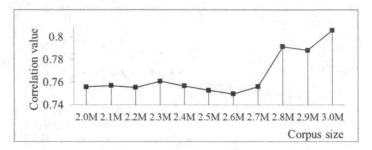

**Fig. 7.** Correlation values between the *MC* extended dataset and *M2* measure with different size of parallel corpus

## 6 Conclusion and Future Work

This paper presents a study on Chinese-English cross-lingual word similarity. Three contributions are made. First, this is an earlier attempt to apply *HowNet* in CLWS measuring. Second, in order to handle coverage problem of *HowNet*, parallel corpus is incorporated in the CLWS measure to estimate co-occurrence context. Third, a benchmark dataset is created based on the *MC* dataset to evaluate CLWS measures. Two conclusions can be drawn from the experimental results. First, *HowNet* is promising to achieve the goal of cross-lingual word similarity measuring. Second, parallel corpus is helpful to extend *HowNet*'s coverage so that a satisfactory CLWS measure is obtained.

Future work is planned as follows. Firstly, more statistical model should be explored to improve word co-occurrence statistics in context modeling. For example, many statistical models have been designed in statistical machine translation research. Secondly, cross-lingual corpus is still limited. We currently have 3M parallel Chinese-English sentence, which is proven insufficient to find an optimal performance. In the future work, more cross-lingual sentences will be collected so that some thorough evaluation can be conducted on our cross-lingual measure. At last, we are aware that efficiency is an important issue for CLWS. More experiments will be conducted on performance measuring.

## Acknowledgments

This work is partially supported by NSFC (60703051, 60970057, 61003152, 61003206) and MOST of China (2009DFA12970). We thank the reviewers for the valuable comments and advices. We thank the reviewers for the valuable comments and advices.

## References

1. Resnik, P.: Semantic similarity in a taxonomy: An information based measure and its application to problems of ambiguity in natural language. Journal of Artificial Intelligence Research 11, 95–130 (1999)

2. Lin, D.: Automatic retrieval and clustering of similar words. In: Proc. of COLING 1998, pp. 768–774 (1998a)
3. Srihari, R.K., Zhang, Z.F., Rao, A.B.: Intelligent Indexing and Semantic Retrieval of Multimodal Documents. Information Retrieval 2, 245–275 (2000)
4. Smeulders, A.W.M., Worring, M., Santini, S., Gupta, A., Jain, R.: Content-Based Image Retrieval at the End of the Early Years. IEEE Trans. Pattern Analysis and Machine Intelligence 22(12), 1349–1380 (2000)
5. Miller, G.A.: WordNet: A Lexical Database for English. Communication of ACM 38(11), 39–41 (1995)
6. Dong, Z., Dong, Q.: HowNet and the Computation of Meaning. World Scientific Publishing Co. Inc., River Edge (2006)
7. Resnik, P.: Using information content to evaluate semantic similarity in a taxonomy. In: Proc. of IJCAI 1995, pp. 448–453 (1995)
8. Lin, D.: An Information-Theoretic Definition of Similarity. In: Proc. of the 15th ICML 1998, pp. 296–304 (1998b)
9. Liu, Q., Li, S.: Word similarity computing based on *HowNet*. Computational Linguistics and Chinese Language Processing 17(2), 59–76 (2002)
10. Dai, L., Liu, B., Xia, Y., Wu, S.: Measuring Semantic Similarity between Words Using *HowNet*. In: Proc. of ICCSIT 2008, pp. 601–605 (2008)
11. Bollegala, D., Matsuo, Y., Ishizuka, M.: Measuring Semantic Similarity between Words using Web Search Engines. In: Proc. of WWW 2007, pp. 08–12 (2007)
12. Cilibrasi, R., Vitanyi, P.: The Google Similarity Distance. IEEE Transactions on Knowledge and Data Engineering 19(3), 370–383 (2007)
13. Li, Y., Bandar, Z.A., McLean, D.: An Approach for Measuring Semantic Similarity between Words Using Multiple Information Sources. IEEE Transactions on Knowledge and Data Engineering 15(4), 871–882 (2003)
14. Nie, J.-Y., Simard, M., Isabelle, P., Durand, R.: Cross-Language Information Retrieval based on Parallel Texts and Automatic Mining of Parallel Texts from the Web. In: Proc. of SIGIR 1999, pp. 74–81 (1999)
15. Rapp, R.: Automatic Identification of Word Translations from Unrelated English and German Corpora. In: Proc. of ACL 1999, pp. 519–526 (1999)
16. Miller, G.A., Charles, W.G.: Contextual Correlates of Semantic Similarity. Language and Cognitive Processes 6(1), 1–28 (1991)
17. Oard, D.: A Comparative Study of Query and Document Translation for Cross-Language Information Retrieval. In: Farwell, D., Gerber, L., Hovy, E. (eds.) AMTA 1998. LNCS (LNAI), vol. 1529, pp. 472–483. Springer, Heidelberg (1998)
18. Pouliquen, B., Steinberger, R., Ignat, C., Käsper, E., Temnikova, I.: Multilingual and Cross-lingual News Topic Tracking. In: Proc. of COLING 2004, vol. 2, pp. 959–965 (2004)
19. Terra, E., Clarke, C.L.A.: Frequency Estimates for Statistical Word Similarity Measures. In: Proc. of HLT-NAACL 2003, pp. 165–172 (2003)

# Comparing Manual Text Patterns and Machine Learning for Classification of E-Mails for Automatic Answering by a Government Agency

Hercules Dalianis[1], Jonas Sjöbergh[2], and Eriks Sneiders[1]

[1] Department of Computer and Systems Sciences (DSV),
Stockholm University, Forum 100, SE-164 40 Kista, Sweden
{eriks,hercules}@dsv.su.se
[2] KTH CSC,
SE-100 44 Stockholm, Sweden
jsh@kth.se

**Abstract.** E-mails to government institutions as well as to large companies may contain a large proportion of queries that can be answered in a uniform way. We analysed and manually annotated 4,404 e-mails from citizens to the Swedish Social Insurance Agency, and compared two methods for detecting answerable e-mails: manually-created text patterns (rule-based) and machine learning-based methods. We found that the text pattern-based method gave much higher precision at 89 percent than the machine learning-based method that gave only 63 percent precision. The recall was slightly higher (66 percent) for the machine learning-based methods than for the text patterns (47 percent). We also found that 23 percent of the total e-mail flow was processed by the automatic e-mail answering system.

**Keywords:** automatic e-mail answering, text pattern matching, machine learning, SVM, Naïve Bayes, E-government.

## 1 Introduction

Many governmental agencies and companies are today overwhelmed with e-mails with queries from citizens or customers that need an answer. Many of these e-mails are easy to reply to and do not need more advanced manual processing. The reply can even be made available on the web site of the government agency or the company. We studied the Swedish Social Insurance Agency (SSIA) (in Swedish "Försäkringskassan"[1]).

SSIA receives 350,000 e-mails per year, which are answered by 640 handling officers who also answer phone calls, use Internet chat, meet citizens and make decisions. The e-mail answering work in total corresponds to 25 full-time employees. If we could automatically answer even a fraction of these e-mails then much would be gained: citizens would obtain immediate answers and the workload of the handling officers would be reduced as they would not need to answer the most basic and monotonous e-mail queries and could focus on the more demanding ones and help citizens more effectively.

---

[1] http://www.forsakringskassan.se

A. Gelbukh (Ed.): CICLing 2011, Part II, LNCS 6609, pp. 234–243, 2011.

We have a joint research project with the SSIA within an E-government framework, where one of the goals is to help SSIA to answer some of these e-mails automatically. We have received 4,404 e-mails sent from citizens to the SSIA. These e-mails contain questions regarding parental benefit, housing allowance, pensions, sickness benefit, etc. Most questions are about the amount of money involved or when it will be paid to the individual, but there are also more general questions such as where one can find the correct application forms. We believe that around 20 to 30 percent of the e-mails can be answered automatically. A pattern-matching system called the *E-mail interceptor* can answer e-mails in categories similar to these automatically. We wanted to evaluate the precision and recall of the previously constructed *E-mail interceptor* system in a new domain, improve it, and compare it with standard machine learning methods.

## 2  Related Research

The research area of automatic e-mail answering is a rather novel research area, but work has been carried out for example by Busemann et al. [1], who constructed an automatic mail answering system for a German call centre. They used 4,777 e-mails that were manually divided into 47 categories with at least 30 e-mails in each. The average length of a document was 60 words. A number of natural language processing techniques were used to identify the core of the e-mails and to find the answer to be used for the automatic e-mail answering system. Techniques such as stemming on the e-mails using a lexicon of 100,000 stems were used to normalise the contents of the e-mails. Note that German is a highly inflected language. Shallow parsing techniques, negation detection, yes-no question detection, and wh-question detection were also used. A number of machine learning techniques were used to train the system. The best performance was given by SVM (Support Vector Machines); SVM-light obtained 56.2 percent accuracy and a top five accuracy of 78.2 percent [1]. The classification tool described in Busemann et al. [1] was included in an e-mail client where categorised messages were assigned a standard answer that could be further edited by a human.

Scheffer [2] constructed a system to reply to frequently answered questions for a German education provider (TELES European Internet Academy). Scheffer used only 528 e-mails in German for training and evaluation. 72 percent of the e-mails could be answered using the nine pre-defined standard answers. The classification of the e-mails was based on a combination of Naïve Bayes- and SVM-based classification.

Mercure is an automatic e-mail answering system developed as a research system for the customer service of a Canadian telecom company. The system is described by Lapalme and Kosseim in [3]. Lapalme and Kossiem used 1,000 e-mails in English for training and evaluation of the system. There was great variation in the complexity of the queries in these e-mails, ranging from basic factual queries to complex queries needing several sources and research before a reply could be given. Lapalme and Kosseim focused on a small topic area regarding investor relations. They tested e-mail classification with K-nearest neighbours, Naïve Bayes, and Ripper, with and without stop word removal, with and without stemming and truncation of words. The success rate was 90 percent for five categories, 80 percent for ten categories, and 67 percent for 22 categories. The Mercure system experienced difficulties with messages that covered several topics; performance measurements cover single subject messages only. It is not clear how the results of the classification were used. Apparently messages in some categories were forwarded to domain experts, and 'messages of the report category are answered by simply mailing the desired report'.

Sneiders [4] describes a text pattern-based e-mail answering system that is applied to two types of e-mail: e-mails from customers to a Latvian telecom operator, and e-mails from customers to a Swedish insurance company. The system for Latvian was semi-automatic, preparing answers for the support officers to send out, whereas the Swedish system was fully automated. Generally the e-mails to the two customer services were fairly uniform in style and thus suitable for automatic e-mail answering.

## 3 De-identification and Ethics

The e-mails sent to SSIA from citizens may contain sensitive information that should not be divulged outside the SSIA. Sensitive information is information that can reveal the sender of the e-mail.

Information that can identify the sender includes for example social security numbers, phone numbers, e-mail addresses, web addresses, street addresses and postal codes as well as personal names. Before SSIA handed over the e-mails to our research group the e-mails were de-identified by a de-identification program. The de-identification program was executed on 4,404 e-mails from the period from March to August 2009 and the de-identified e-mails were handed over to our research group by SSIA.

## 4 Data Collection: E-Mails from SSIA

Analysing 4,404 de-identified e-mails from SSIA we found that they were of varied length and complexity. Most of the e-mails were written in Swedish, which is a Germanic language with rich morphology and very productive compounding (creation of new long compound words).

Some e-mails generate very long threads with several tens of e-mails with queries and replies, sometimes up to 40 iterations. The majority, 96.2 percent, of the e-mails had only up to four threadings. The e-mail texts were no more complex than those processed in [4] but the topic diversity here was slightly larger. In [5] an experiment was carried out on clustering e-mails. E-mails were clustered both with and without the threads but also with the query and the first answer in the thread. The authors did not find any difference between using the whole e-mail with all threadings or just the query when using the K-Means algorithm.

We therefore decided to cluster the e-mails using only the query (without the threadings). A clustering process where the e-mails were clustered using the K-Means algorithm was carried out with the aim of identifying similar and relevant query groups. Eleven clusters of frequent queries were identified, hereafter called categories (see Fig. 1).

| | |
|---|---|
| • **When will you decide my housing allowance?** | **138** |
| • **I want an estimate of my future pension.** | **59** |
| • I want to change the taxation on my pension. (To avoid tax arrears.). | 39 |
| • **When do I get the money?** | **631** |
| • **How many days of parental benefits remain for my child?** | **100** |
| • Questions concerning child allowances. | 125 |
| • **Want a form (application form or otherwise).** | **170** |
| • Want a beneficiary certificate (used to get discounts). | 61 |
| • Want an EU card (entitles the holder to medical care in the EU). | 32 |
| • A question in any language other than Swedish. | 11 |
| • **Miscellaneous** | **3,205** |
| SUMMARY | 4,571[2] |

**Fig. 1.** Eleven answering categories (with the five selected for automatic answering in bold). The number represents the number of categorised e-mails.

Of these eleven categories, five categories were selected for automatic answering (Fig. 1 in bold); questions in these categories could be answered with a short answer that included a redirection to the SSIA website, which was convenient for demonstration purposes. In these five categories several similar e-mail queries could be answered using our text pattern matching system, the E-mail interceptor (see Section 6).

One observation is that 30 percent of the e-mails (see Fig 1.) fall into one of the nine top categories that can be answered automatically (excluding the categories 'language other than Swedish' and 'Miscellaneous') and 24 percent of the e-mails fall into the categories handled by the E-mail interceptor (see Fig. 1).

## 5 Annotation

To make it possible to evaluate the E-mail-interceptor and to the train the machine learning systems we needed annotated e-mails. We extracted the last message of the citizen from each e-mail, i.e., stripped the text from the previous conversations, and annotated the extracted texts. Four annotators started the annotation process with annotating the same small set of e-mails containing only 100 e-mails and then met for a discussion on how the annotation should be carried out and obtained a consensus. We finally annotated a total of 4,404 e-mails in eleven classes (categories).

The 4,404 e-mails (with only queries) encompass 296,855 tokens, i.e., an average of 65 tokens per e-mail. The e-mail tokens are on average 4.5 characters long.

We used part of the annotated e-mails as a training set (2,437 e-mails) and the remaining part (1,967 e-mails) as an evaluation set.

---

[2] The sum is 4,571 classifications of e-mails since the 4,404 e-mails can be in more than one category.

# 6  The Text Pattern-Based System

Our E-mail interceptor uses a set of FAQs (Frequently Asked Questions) specifying the questions that are to be answered automatically [4]. In this paper, the FAQ are the five categories detailed in Fig. 1, Section 4. For each question in the FAQ there is a set of hand-crafted text patterns that match wordings in query e-mails. The strengths of these patterns are the following:

- the text patterns capture relevant phrases, not just a set of keywords;
- each concept in a text pattern is described by a set of synonyms, generalisations, specialisations, etc., which can be single words or phrases;
- the synonyms are narrow context-dependent, rather than general, as in synonym dictionaries;
- since text patterns do not depend on each other, e-mails containing several questions can be assigned to several categories;
- the technique has been tested for three languages.

Before the E-mail interceptor can start operating, it is 'trained' to recognise e-mail texts that fit a given standard answer. We put 'trained' in quotes because this is not training as understood in machine learning. The training e-mails, in total 2,437, were analysed by a human and the text patterns linked to each text class were created manually. Currently, there is no method or tool for creating these text patterns automatically.

There were 1,967 e-mails in the evaluation set for the E-mail interceptor. 250 e-mails matched the patterns for at least one of the five relevant text classes, and three e-mails matched two, which makes 256 emails placed into an automated answer category, and 1970 total email placements.

Table 1 (and Table 2 for graphical form) shows the number of messages in each text class, the number of messages that the E-mail interceptor placed into each text class, and the precision and recall for each class. For the five relevant classes, precision ranges from 84 to 97 percent. Recall is just above 50 percent, except for one class with 41 percent.

The average precision and recall for the five classes, calculated by dividing all the correctly placed messages by the total number of relevant messages for these five classes, were 89 percent and 47 percent, respectively. The reason of such low recall is a lack of opportunity to perform iterative improvement of the text patterns. That is, we did not have an opportunity to observe what mistakes the system makes on a larger test corpus and correct these mistakes in the patterns. We believe our recall values would have been higher than the current 50 percent if several 'training' and testing iterations had been performed.

We also have the category of 'Miscellaneous' e-mails, with quite high precision and recall values. In the context of automated e-mail answering, these would be the messages sent on to manual processing, and therefore the precision and recall values of this text class are not particularly interesting.

**Table 1.** Results from the E-mail interceptor, the text pattern-based system

| No | Category | Placed in category | Placed in category, relevant | Total relevant | Precision | Recall | F-measure |
|---|---|---|---|---|---|---|---|
| 1 | When will you decide my housing allowance? | 34 | 33 | 62 | 0.97 | 0.53 | 0.69 |
| 2 | I want an estimate of my future pension. | 17 | 15 | 30 | 0.88 | 0.50 | 0.63 |
| 3 | When do I get the money? | 132 | 111 | 269 | 0.84 | 0.41 | 0.55 |
| 4 | Want a form (application form or otherwise). | 45 | 41 | 76 | 0.91 | 0.54 | 0.68 |
| 5 | How many days of parental benefits remain for my child? | 28 | 27 | 49 | 0.96 | 0.55 | 0.70 |
| | **Total** | 256 | 227 | 486 | | | |
| | **Average** | | | | **0.89** | **0.47** | 0.61 |
| 6 | Does not match above | 1714 | 1467 | 1490 | 0.86 | 0.99 | 0.91 |
| | **Total** | 1970 | 1694 | 1976 | | | |
| | **Average** | | | | 0.86 | 0.86 | 0.86 |

**Table 2.** Bar chart table presentation of Table 3, the results from the E-mail interceptor

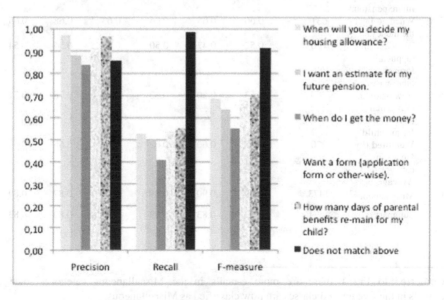

# 7 Applying Machine Learning Techniques

Apart from the handwritten pattern matching rules in the E-mail interceptor, we also applied machine learning methods for classifying the 4,404 e-mails. We used the WEKA framework [7].

We used Naïve Bayes and Support Vector Machines in WEKA in a standard text classification setting, i.e., the features for the machine learning were word vectors (TF/IDF), evaluated using tenfold cross-validation on the whole data set in Fig 1.

We classified the e-mails into six categories, the five used by the E-mail interceptor and one large category termed 'Miscellaneous'; see Table 3 (and Table 4 for graphical form) for the results. We used splitting of compound words into their components, lemmatisation of words, shallow parsing (chunking) of the text into phrases, and automatic spelling correction of misspelled words. Again, there was no real difference between the two machine learning methods SVM and Naïve Bayes or language technology preprocessing.

**Table 3.** The top five categories classified with both SVM and Naïve Bayes, using ten-fold cross-validation

| No | Categories | Manually Classified E-mails | SVM Precision | Recall | F-Score | Naïve Bayes Precision | Recall | F-Score |
|---|---|---|---|---|---|---|---|---|
| 1 | When will you decide my housing allowance? | 138 | 0.62 | 0.59 | 0.61 | 0.60 | 0.68 | 0.64 |
| 2 | I want an estimate of my future pension. | 59 | 0.55 | 0.41 | 0.47 | 0.39 | 0.59 | 0.47 |
| 3 | When do I get the money? | 631 | 0.66 | 0.65 | 0.66 | 0.63 | 0.68 | 0.65 |
| 4 | Want a form (application form or other-wise). | 170 | 0.52 | 0.47 | 0.50 | 0.54 | 0.54 | 0.54 |
| 5 | How many days of parental benefits remain for my child? | 100 | 0.63 | 0.60 | 0.62 | 0.65 | 0.78 | 0.71 |
| | **Weighted** (by #mails in cate-gory) **Average** | 220 | 0.63 | 0.60 | 0.62 | 0.60 | 0.66 | 0.63 |
| 6 | Miscellaneous[3] | 3473 | 0.89 | 0.90 | 0.89 | 0.90 | 0.87 | 0.89 |
| | Summary | 4571 | 0.82 | 0.83 | 0.83 | 0.83 | 0.82 | 0.83 |

---

[3] Compared to Fig 1 there are more e-mails in the Miscellaneous category, since all e-mails in the five unused classes are now classified as Miscellaneous.

**Table 4.** Graphical overview of the results of Table 3, showing the differences between SVM and Naïve Bayes using tenfold cross-validation

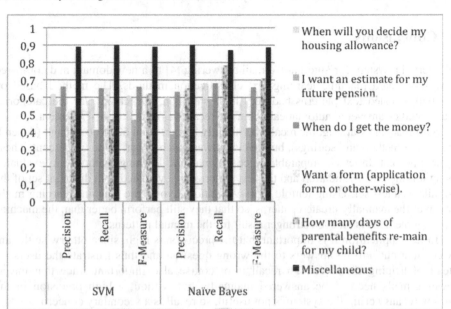

## 8  Error Analysis

One problem is that the 'miscellaneous category' is large compared with all other categories. This makes the machine learning methods focus on this (least interesting) category and perform poorly on the others. Other problems include questions from non-native speakers that contain very many writing mistakes, thus making it hard for the system to understand what the user means. Native speakers also make many mistakes, and there are very many 'creative' abbreviations that make simple word matching as well as other language processing methods difficult.

The precision and recall values do not seem to be directly connected to the number of messages in the category 'When do I get the money?' turned out to be a rather broad category in which the question was raised in many different contexts and different ways. Thus, the recall for the E-mail interceptor is low, at 41 percent, whereas precision is still high, at 84 percent (see Table 1).

The E-mail interceptor generally does better than the machine learning methods but there are of course also e-mails that the machine learning methods get right but where the E-mail interceptor fails. Most such examples stem from for example the SVM being much more aggressive in classifying e-mails into the smaller classes than the E-mail interceptor that is tuned for high precision on these. Another fairly typical example is an e-mail talking about how the person found the correct form online but does not have a printer so he would like to know if the form can be sent to him through normal post instead. The E-mail interceptor classified it as a request for information on finding forms online since the text is very similar to such e-mails while it is mentioning that this did not work. The SVM correctly classified it as 'Miscellaneous' based on there also being a lot of text not fitting any category in particular.

# 9  Conclusions

We ported a system for automatic e-mail answering [4] to a new domain and compared it with standard machine learning text classification methods. The method based on manually-created text patterns had very high precision, 89 percent, for the categories that would be answered automatically, but the recall was quite low at 47 percent.

With machine learning, the recall was slightly higher, from 60 to 66 percent (depending on the method and settings), but precision was much lower, 60 to 63 percent. These figures are not directly comparable since the machine learning methods are evaluated using tenfold cross-validation and the text patterns are evaluated on a different set of the e-mails as a test set. Using tenfold cross-validation favours the machine learning methods over the manually-created patterns, so that they still perform better than the machine learning methods is an even stronger result for the manual patterns.

In this application it is important that the precision is high, since otherwise the answers sent out will be answers to the wrong question and thus frustrate the users instead of helping them. A high recall is of course also important since that means fewer e-mails need to be answered manually, but without a high precision in the automatic answering the system is not useful, so recall is a secondary concern.

Annotation of the 4,404 e-mails used took about 40 hours. Training the machine learning systems took a few minutes, whereas manually constructing the patterns for the E-mail interceptor took around 40 hours. Although more work is required for the manually-constructed patterns, this method clearly outperforms the machine learning methods, both in terms of total classification accuracy, and, most importantly, its very high precision in the categories that are answered automatically.

Both methods would benefit from more annotated e-mails. The machine learning methods lack data for many categories, and the manually-created patterns would benefit from iterations of testing on new e-mails to discover and correct mistakes made by the current patterns.

One possibility of increasing the amount of training material without the need for manual annotation is to use active learning [7] In an active learning scenario the small amount of manual annotated material would hopefully boost performance both for the machine learning-based system and for the text pattern-based E-mail interceptor. To improve the performance of our approach we will also look into [8] if we can use action request classification features. A text pattern-based system is most advantageous in settings where the correctness of replies is crucial, where we want to maximise the end-user experience, and where a list of ten candidate answers is not an option, for example in fully automated e-mail answering without any human mediation. This approach is especially advantageous for e-mail flows with a high ratio of recurring inquiries.

The text patterns however have at least two limitations. First, the technique is designed for narrow and stable domains only. It should not be considered for text classification tasks in arbitrary text collections. Second, there is insufficient automation of the text pattern generation, which lessens the practical value of the technique until at least partial automation of this process is achieved.

One of the benefits of the machine learning methods is that much less manual work is needed. The e-mails do however contain very many misspellings of important words, non-standard abbreviations, grammatical mistakes of many kinds, etc. This makes their automatic processing difficult.

## Acknowledgments

We would like to thank Anne-Lie Karlsson at Försäkringskassan, SSIA, for her warm support of the IMAIL research group. We would also to give a special thanks to our annotators Viggo Kann, Ola Knutsson and Magnus Rosell for their devoted work.

Finally we would like to thank VINNOVA (The Swedish Governmental Agency for Innovation Systems) for the funding of the IMAIL-project.

## References

1. Busemann, S., Schmeier, S., Arens, R.G.: Message classification in the call center. In: Proceedings of the Sixth Conference on Applied Natural Language Processing, Seattle, Washington, pp. 158–165. ACL (2000)
2. Scheffer, T.: E-mail answering assistance by semi-supervised text classification. Intelligent Data Analysis 8(5), 481–493 (2004)
3. Lapalme, G., Kosseim, L.: Mercure: Towards an automatic e-mail follow-up system. IEEE Computational Intelligence Bulletin 2(1), 14–18 (2003)
4. Sneiders, E.: Automated E-mail Answering by Text Pattern Matching. In: Loftsson, H., Rögnvaldsson, E., Helgadóttir, S. (eds.) IceTAL 2010. LNCS, vol. 6233, pp. 381–392. Springer, Heidelberg (2010)
5. Dalianis, H., Rosell, M., Sneiders, E.: Clustering E-mails for the swedish social insurance agency – what part of the E-mail thread gives the best quality? In: Loftsson, H., Rögnvaldsson, E., Helgadóttir, S. (eds.) IceTAL 2010. LNCS, vol. 6233, pp. 115–120. Springer, Heidelberg (2010)
6. Hall, M., Frank, E., Holmes, G., Pfahringer, B., Reutemann, P., Witten, I.H.: The WEKA Data Mining Software: An Update. SIGKDD Explorations 11(1) (2005)
7. Cohn, D.A., Zoubin, G., Michael, I.J.: Active learning with statistical models. Journal of Artificial Intelligence Research 4, 129–145 (1996)
8. Lampert, A., Dale, R., Paris, C.: Detecting Emails Containing Requests for Action. In: The Proceeding of Human Language Technologies: The 2010 Annual Conference of the North American Chapter of the Association for Computational Linguistics, NAACL-HLT, Los Angeles, pp. 984–992 (2010)

# Using Thesaurus to Improve
# Multiclass Text Classification

Nooshin Maghsoodi and Mohammad Mehdi Homayounpour

Laboratory of Intelligent Signal and Speech Processing, Faculty of Computer Engineering,
Amirkabir University of Technology, Tehran, Iran
{n_maghsoodi,homayoun}@aut.ac.ir

**Abstract.** With the growing amount of textual information available on the Internet, the importance of automatic text classification has been increasing in the last decade. In this paper, a system was presented for the classification of multiclass Farsi documents which uses Support Vector Machine (SVM) classifier. The new idea proposed in the present paper, is based on extending the feature vector by adding some words extracted from a thesaurus. The goal is to assist classifier when training dataset is not comprehensive for some categories. For corpus preparation, Farsi Wikipedia website and articles of some archived newspapers and magazines are used. As the results indicate, classification efficiency improves by applying this approach. 0.89 micro F-measure were achieved for classification of 10 categories of Farsi texts.

**Keywords:** Text classification, Support vector machine, Thesaurus, Farsi.

## 1 Introduction

The task of assigning natural language documents to a set of predefined categories is known as text classification. Due to the wide availability of online information in the World Wide Web, it may be impossible to classify the documents manually; so automatic classification of text documents seems to be inevitable. The workflow in most of the text classification systems is to train the classification system using a training dataset including many text documents whose categories are known. In the test phase, the system assigns a category to a new document. Each document in training dataset consists of a great number of relevant and irrelevant words corresponding to its category. One way to decrease the complexity of a text classifier and to increase its speed is to discard the irrelevant words and to render more weight on relevant ones. This phase which is called feature selection, is considered in many classification systems and different approaches such as the selection of features based on information gain, $tf\_idf$ criterion and $x^2$ test have been applied [1, 2, 3]. Classifier component in such systems is often one of the statistical methods or machine learning techniques including multivariate regression model, nearest neighbor classifier [3, 4], probabilistic Bayesian models [5, 6], decision tree [5, 7] and support vector machine (SVM) [8, 9, 10]. According studies carried out by Yang [10], SVM outperforms other machine learning methods. Therefore, in the approach presented in this paper, SVM is applied

A. Gelbukh (Ed.): CICLing 2011, Part II, LNCS 6609, pp. 244–253, 2011.

as a classifier using the SVM$^{light}$ software. As told before, the selection of appropriate features can improve classifier performance. However, it should be noticed that extracting appropriate features, necessitates a comprehensive corpus covering a wide range of words which are relevant to each category and this important requirement is not always fully satisfied. In this paper, we aim to compensate the deficiency of training dataset by using a thesaurus for text classification in the Persian language. The thesaurus that we used consists of the common words in the Persian language as heading terms and a list of relevant words near in meaning to that heading term. In our approach, features related to each category are selected from the training dataset and a feature vector is constructed, then more words relevant to the categories are extracted from a thesaurus and are added to the constructed feature vector. Therefore, by extending the feature vector, its coverage on the words of each category augments. In the next phase, the final feature vector is used to train the classifier. Afterwards, the system can assign one of the predefined categories to any new document. According to the new approach proposed in this paper, it will be shown that the application of a thesaurus, improves classifier efficiency without the necessity for a comprehensive training dataset.

The rest of this paper is organized as follows. Related works is briefly reviewed in section 2. In section 3, the Persian thesaurus used in our experiments named Farhang-e Teyfi is introduced. Our approach for text classification is detailed in section 4. Section 5 describes some conducted experiments to demonstrate the suitability of the proposed approach and finally section 6 concludes this paper and presents some suggestions as future works.

## 2 Related Work

Improving accuracy of text classifiers has been an important issue and many studies have been conducted in this area. Much work has been conducted to find out effective approaches to represent document for text classification. Traditional "Bag of Words" (BOW) approach, which represents a document as a vector of weighted occurrence frequencies of individual terms, is limited because it only accounts for term frequency in the documents and can only use pieces of information that are explicitly mentioned in the training dataset. To overcome this limitation, some methods for extending the feature vector are developed. Most of these methods use an existing ontology or thesaurus. In [11] documents representation is done using WordNet, the MeSH (Medical Subject Headings) Tree Structure Ontology. They have shown that summarizing words in documents by synonyms in WordNet can improve the performance of TC. Hotho et al. [12] utilized a term ontology structured from WordNet to improve the BOW text representation. The authors adopted various strategies to enrich text document representation with synonyms and hyponyms from WordNet. Proposed method in [13], automatically construct a thesaurus of concepts from Wikipedia and then a unified framework is introduce to expand the BOW representation with semantic relations (synonymy, hyponymy, and associative relations). In [14] Feature generation is performed completely automatically, using machine readable hierarchical repositories of knowledge such as the Open Directory Project (ODP), Yahoo! Web Directory, and the Wikipedia encyclopedia. In [15], based on two thesaurus HowNet and Tongyici Cilin (hereinafter referred to Cilin), a semantic vector to describe a document is constructed. The method in [16] creates a Bayesian network to model the thesaurus and uses probabilistic inference to select the set of descriptors having high posterior probability of being relevant given the available document to be classified.

## 3   Farhang-e Teyfi Thesaurus

"Farhang-e Teyfi" or "Farhang-e Maqulei" Thesaurus [17] is the name of the first categorized lexicon for Farsi vocabulary. This thesaurus has been designed according to Roget thesaurus [18] and its headings are similar to Roget's. The first version of this thesaurus was published in 1998, titled "Farhang-e Teyfi". This version contains 6 classes, 39 sections, 990 heads and 5500 subheads that are semantically linked. Subheads are some nouns, verbs, adjectives or adverbs that are related to a head. These words are not exactly synonyms, but can be viewed as colors or connotations of a meaning or as a spectrum of a concept. Similar to the Roget thesaurus, this thesaurus provides rapid access to synonyms, facilitating the choice of appropriate words to express thoughts. As an example, a head word in the thesaurus means "equivalence" while it contains subheads such as "equality of value", "equality of importance", "sameness", "flat", "symmetry" and "compatibility" of which the last three terms are not synonyms of "equivalence". This structure, which is undesired in our project, will be explained more in the following sections.

## 4   Text Classification

Text classification is defined as assigning predefined categories to text documents. Commonly, document representation and classifier training are two main phases for construction of a text classifier. In the first phase, features that are more appropriate are selected and a numeric weight is assigned to each of them. Therefore, at the end of this phase a numeric vector to represent each document is achieved. In the second phase, these numeric vectors are used to train classifier.

### 4.1   Preliminary Feature Selection

Feature selection is an appropriate way to reduce dimensionality of the feature space, which reduces computational complexity while retaining essential information. In text classification, features should be able to discriminate between categories, and selected features for different categories should not overlap. In this paper, the mechanism of feature selection is based on category frequency, i.e. more repetition of a word in a category and less repetition of that word in other categories. In other words, our measure indicates the degree that a term corresponds to a specific category and is defined as term frequency times inverse category frequency (*tf_icf*) according to Eq. (1):

$$\text{tf\_icf}(t_{ij}) = t_{ij} \times \log \frac{\sum d_{ij}}{d_{ij}}. \tag{1}$$

where $d_{ij}$ is the number of documents in which term $i$ occurrs in category $j$ and $t_{ij}$ is the number of term $i$ which occurs in category $j$. Therefore, while a greater $t_{ij}$ indicates more repetition of a word in the documents. Multiplying $t_{ij}$ by a category frequency reduces its value if the word appears in most of the categories. The next problem is finding an appropriate threshold for $tf\_icf$. According to our study, $5/\log_{10}^{C}$ is a good value for this threshold and leads to good results. In the threshold equation, $C$ is the number of categories.

## 4.2 Document Representation

The selected features from the previous phase can be represented by the vector space model (VSM) [19]. The elements of this vector are the feature weights, which can be calculated using different weighting schemes. The most commonly used method is the so-called method $tf\_idf$ which stands for term frequency ($tf$) multiplied by inverse document frequency ($idf$), where $tf$ shows the relative frequency of a certain word appearing in a document and $df$ shows the proportion of documents presenting this word among all documents [2]. The product of $tf$ and inverse $df$ ($tf\_idf$) can be used to represent the importance of a keyword in the document, so higher frequency of a word in a document versus lower appearance of that word in other documents increases its weight. The $tf\_idf$ measure is defined as follows:

$$\text{tf\_idf}(t_i, d_j) = \text{tf}(t_i, d_j) \times \log(\frac{N}{N(t_i)}). \tag{2}$$

where $N$ is the number of all documents and $N(t_i)$ is the number of documents in the collection in which the term $t_i$ occurs at least once and $tf(t_i, d_j)$ is the frequency of the word $t_i$ in document $d_j$. Efficiency of this method on different corpuses has been proved [20]. This weighting method is also applied in this paper.

## 4.3 Using the Thesaurus to Expand Feature Vector

In this paper, the "Farhang-e Teyfi" thesaurus was used to extend the feature vector to increase text classification performance. According to the thesaurus structure, for each category, the thesaurus can be searched to find a set of related terms that lead to a better representation of that category. Word selection from the thesaurus is done based on features obtained from preliminary feature selection. In this mechanism, each selected feature is searched in the thesaurus and all the thesaurus heads with the feature as a subhead are selected, and subheads in those heads are candidate to be added to the feature vector. Finally, if the frequency of a candidate subhead in the documents that belong to a category in training dataset is larger than a predefined threshold, and that subhead has not already been one of the features in the feature vector, that subhead is added to the feature vector. This thresholding can partially prevent adding irrelevant subheads to the feature vector. Fig. 1 shows a pseudo code for this process. A point that should be considered in this method is appropriate definition of the threshold. Assigning a small value to this threshold adds more irrelevant terms to feature vector, while choosing a large threshold filters related terms.

$C_j \rightarrow$ the set of documents belonging to category j
$F_j \rightarrow$ the set of features assigned to category j
$T_k \rightarrow$ terms belonging to head k in the thesaurus
$V \rightarrow$ current feature vector
For each $F_j$
    for each feature$_i \in F_j$
        select all term$_h \in T_k$ where feature$_i \in T_k$ & TF(term$_h$,C$_j$)>THRESHOLD
           & ~(term$_h \in F_j$)
        Add  term$_h$ to V
    end for
end for

**Fig. 1.** The pseudo code for feature selection from thesaurus

Therefore, the overall procedure of the presented method in this paper is as follows. In the first step, preliminary feature selection selects appropriate features according to training data. Then the thesaurus is processed to expand the feature vector and the new features are added. The next step is feature weighting. If a feature is added to the feature vector from the preliminary feature selection, the weighting process is done according to Eq. (2) while, if the feature is selected from the thesaurus, Eq. (3) and Eq. (4) are used for feature weighting. If for a given feature, the value obtained from Eq. (3) is zero, Eq. (4) will be used.

$$\text{weight}(f_{ij}) = \frac{d_{ij}}{\sum d_{ij}}. \tag{3}$$

$$\text{constWeight} = \frac{1}{c}. \tag{4}$$

where $d_{ij}$ is the frequency of this feature in all the documents belonging to category j and C is the number of categories. After the weighting process, final feature vector is used to train the SVM classifier.

## 5   Implementation and Experimental Results

A variety of experiments are conducted to test the performance and behavior of the proposed algorithm. In the following sections, the training and test data used in the experiments are first explained, and the obtained results are then presented and analyzed. Each result is obtained by averaging the results of experiments conducted in a 5-fold cross-validation procedure.

### 5.1   Training Dataset

Due to the lack of Persian corpora for text classification, a text corpus was initially collected from Farsi Wikipedia, the Hamshahri newspaper archive and Soroush and

Roshd journals. Wikipedia is a web-based, free content encyclopedia written collaboratively by regular editing contributors. Each article in Wikipedia is related to a specific category. To access Soroush and Roshd articles we used the 100_million_ word corpus licensed by the Research Center of Intelligent Signal Processing [21]. To extract articles from Wikipedia, a crawler was written and after removing HTML tags, documents were labeled in 10 categories. The detailed specification of the corpus is shown in Table 1. These documents were encoded using the UTF-8 encoding system.

The following preprocessing was performed after corpus collection:

- Removing numbers, punctuations and foreign letters (English letters) from each document.
- Deletion of stop words.
- Stemming the words (Perstem stemmer was used [22]).
- Setting the threshold.

**Table 1.** Detailed description of the corpus

| | Number of training documents | Number of training documents | Size of total documents (KB) |
| --- | --- | --- | --- |
| Economy | 209 | 52 | 572 |
| Politic | 180 | 45 | 701 |
| Sport | 141 | 35 | 540 |
| Theology | 148 | 36 | 579 |
| Medicine | 110 | 23 | 452 |
| Art | 219 | 54 | 874 |
| Agriculture | 200 | 50 | 787 |
| Chemistry | 130 | 32 | 508 |
| Mathematics | 106 | 24 | 422 |
| Sociology | 214 | 52 | 877 |
| **Totality** | **1657** | **403** | **6319** |

## 5.2 Evaluation Measures

In the text classification, the most commonly used performance measures are precision, recall and F-measure. Precision on a category is the number of correct assignments to this category by the classifier divided by the total number of the classifier's assignments to this category and recall on a category signifies the rate of correct classified documents to this category among the total number of documents belonging to this category. There exists a trade-off between precision and recall of a system [20]. The F-measure is the harmonic mean of precision and recall and takes into account effects of both precision and recall measures. To evaluate the overall performance over the different categories, micro and macro averaging can be used. In macro averaging the average of precision or recall is computed over all categories. Macro averaging gives the same importance to all the categories [23]. On the other hand micro averaging considers the number of documents in each category and computes the average in proportion to these numbers. It gives the same importance to all the documents [23]. When the corpus has unbalanced distribution of documents into categories, by using macro averaging, classifier deficiency in classifying a category with fewer documents is emphasized. Since an imbalanced corpus is being dealt with, it seems more reasonable to use micro averaging.

## 5.3 Experimental Results

In this section, the experimental results are presented. The experiments consist of evaluating classifier performance when thesaurus are used as well as analyzing the text classification performance versus the number of categories, the average number of words in test files and the volume of information available in the training dataset.

In the first experiment, the number of categories is changed from 5, to 7 and 10, and classifier performance is evaluated with and without the thesaurus usage. Table 2 shows the results. It can be observed that, when a thesaurus is used, both precision and recall are increased, especially when the number of categories increases. The reason for this behavior refers to the preliminary feature selection deficiency in extraction of enough features corresponding to Eq. (1) when the number of classes increases. Using the thesaurus compensates the deficit of features and by the extension of the feature vector helps the system to cover test documents having a wide spectrum of words.

**Table 2.** Text classification performance, with and without using the thesaurus versus different number of categories

| Num. of categories | Without using the thesaurus | | | Using the thesaurus | | |
|---|---|---|---|---|---|---|
| | Micro_Pr | Micro_Re | Micro_F | Micro_Pr | Micro_Re | Micro_F |
| 5 | 0.89 | 0.92 | 0.90 | 0.94 | 0.95 | 0.945 |
| 7 | 0.88 | 0.87 | 0.875 | 0.93 | 0.92 | 0.925 |
| 10 | 0.79 | 0.8 | 0.795 | 0.88 | 0.90 | 0.89 |

In the next experiment, the same conditions as the previous experiment are obtained, except that only half of the training documents belonging to each category are used. The results, which are summarized in Table 3, illustrate noticeable decrease in performance measures. Of course, it should be noted that when the thesaurus is used, the reduction in classification efficiency is minor. Therefore, it can be concluded that when the volume of information included in the training dataset is not enough, the improving effect of the thesaurus will be more noticeable. Whenever the number of training documents is less, finding enough features for representation of categories will be more difficult, whereas the thesaurus can help represent categories more distinctly through feature vector expansion. In fact, this experiment proves the strength of our thesaurus based approach to handle the shortage in training documents.

**Table 3.** Comparison of using the thesaurus and not using the thesaurus, when the training documents are halved

| Num. of categories | Without using the thesaurus | | | Using the thesaurus | | |
|---|---|---|---|---|---|---|
| | Micro_Pr | Micro_Re | Micro_F | Micro_Pr | Micro_Re | Micro_F |
| 5 | 0.83 | 0.85 | 0.84 | 0.90 | 0.91 | 0.905 |
| 7 | 0.79 | 0.78 | 0.785 | 0.87 | 0.87 | 0.87 |
| 10 | 0.70 | 0.69 | 0.695 | 0.80 | 0.83 | 0.815 |

The third experiment, studies the effect of average number of words in test files on the text classification performance. The aim of this experiment is to find a balance between the average number of words in a test document and the system performance. Using test files including a reasonable number of words results in better performance. f it can be observed from Table 4, that when the number of words in test files is 700 words, the categorization performance is 0.89 using the thesaurus (0.795 without using the thesaurus). This performance decreases to 0.83 (0.7 without using the the-saurus) when the average number of words in test files is reduced to 350 words. Doubling the average number of words to 1400 words slightly improves the perfor-mance (0.91 and 0.82 using and without using the thesaurus respectively). It is worth mentioning that when the thesaurus is not used, the system needs longer test files for correct text classification, but when the thesaurus is used, the volume of information that is trained to the classifier will increase and therefore, the system can correctly classify short test files.

**Table 4.** Text classification performance with and without using the thesaurus versus average number of words in test files

| | Averaged number of words in test files | | | |
|---|---|---|---|---|
| | 350 | 700 | 1400 | 1800 |
| Using the thesaurus | 0.83 | 0.89 | 0.91 | 0.90 |
| Without using the thesaurus | 0.70 | 0.795 | 0.82 | 0.83 |

The two previous experiments show that our proposed algorithm, which uses a the-saurus, is more robust against the deficit of training and test data and presents more stable behavior with regard to variations in the size of test and training data.

# 6  Conclusion

Nowadays the Web and other Media contain an enormous amount of text information. Therefore, in mining and searching text information, the process of automatic text classification seems to be inevitable.

The proposed approach in this paper aims to enhance the classification accuracy by using the "Farhang-e Teyfi" thesaurus, which extends preliminary selected features. The results show that this approach improves the classification performance especially when accessible training datasets are not sufficiently comprehensive; i.e. the existing words are not able to distinguish a category form other categories. In this case, the use of a subsidiary knowledge resource such as a thesaurus may compensate the insufficiency in the existing information. In fact, some relevant features (words) obtained from the thesaurus is added to the existing feature vector which has been obtained by processing the training documents belonging to the desired category.

We believe that the more the relation between a given category and the words selected from the thesaurus, the more improvement in text classification performance will be achieved.

## Acknowledgment

The authors would like to thank Iranian Supreme Council of ICT (SCICT) for supporting this work.

## References

1. Huang, Y.L.: A theoretic and empirical research of cluster indexing for mandarin Chinese full text document. The Journal of Library and Information Science 24, 1023–2125 (1998)
2. Lee, C., Lee, G.: Information gain and divergence-based feature selection for machine learning-based text categorization. Information Processing and Management 42, 155–165 (2006)
3. Yang, Y., Pedersen, J.O.: A comparative study on feature selection in text categorization. In: Proceedings of 14th International Conference on Machine Learning, pp. 412–420. Morgan Kaufmann, Nashville (1997)
4. Dumais, S.: Inductive learning algorithms and representations for text categorization. In: Proceedings of the 7th International Conference on Information and Knowledge Management of Contents, pp. 148–155. ACM, Bethesda (1998)
5. Lewis, D.D., Ringuette, M.: A comparison of two learning algorithms for text categorization. In: Proceedings of the Third Annual Symposium on Document Analysis and Information Retrieval, pp. 81–93. University of Nevada, Las Vegas (1994)
6. McCallum, A., Nigam, K.: A comparison of event models for Naive Bayes text classification. In: Proceedings of the Workshop on Learning for Text Categorization, pp. 41–48 (1998)
7. Schutze, H., Hull, D., Pedersen, J.O.: A comparison of classifiers and document representations for the routing problem. In: Proceedings of the 18th International Conference on Research and Development in Information Retrieval, pp. 229–237. ACM, Seattle (1995)
8. Joachims, T.: Text categorization with support machines: learning with many features. In: Nédellec, C., Rouveirol, C. (eds.) ECML 1998. LNCS, vol. 1398, pp. 137–142. Springer, Heidelberg (1998)
9. Wang, T., Chiang, H.: Fuzzy support vector machine for multi-class text categorization. Information Processing and Management 43(4), 914–929 (2007)
10. Yang, Y.: An evaluation of statistical approaches to text categorization. Information Retrieval 1, 69–90 (1999)

11. Bloehdorn, S., Hotho, A.: Boosting for text classification with semantic features. In: Workshop on Text-based Information Retrieval (TIR 2004) at the 27th German Conference on Artificial Intelligence, pp. 149–166 (2004)
12. Hotho, A., Staab, S., Stumme, G.: Wordnet improves text document clustering. In: Proceedings of the Semantic Web Workshop at SIGIR 2003, pp. 61–69 (2003)
13. Wang, P., Hu, J., Zeng, H., Chen, Z.: Using Wikipedia knowledge to improve text classification. Knowledge Information System 19(3), 265–281 (2009)
14. Gabrilovich, E., Markovitch, S.: Feature Generation for Text Categorization Using World Knowledge. In: Proceedings of the 19th International Joint Conference on Artificial Intelligence, pp. 1048–1053 (2005)
15. Song, X., Huang, J., Zhou, J., Chen, X.: Research of Chinese Text Classification Methods Based on Semantic Vector and Semantic Similarity. In: International Forum on Computer Science-Technology and Applications, pp. 187–190 (2009)
16. Campos, L., Romero, A.: Bayesian network models for hierarchical text classification from a thesaurus. Approximate Reasoning 50, 932–944 (2009)
17. Fararuy, J.: Farhang-e maqulei (thesaurus) and electronic transmission of Farsi content. In: Proceeding of the First Workshop on Farsi Language and Computer (2004) (in persian)
18. Roget's Thesaurus, http://www.rain.org/~karpeles/rogfrm.html
19. Salton, G., Yang, C., Wang, A.: A vector space model for automatic indexing. Communications of the ACM 18(11), 613–620 (1975)
20. Sebastiani, F.: Machine learning automated text categorization. ACM Computing Surveys 34(1), 1–47 (2002)
21. Bijankhan, M.: 100 millions word Farsi Corpus. Technical Report, Research Center for Intelligent Signal Processing (2008)
22. Stemmer, P. (Version 0.9.7) [Computer Progtam], http://www.ling.ohio-state.edu/~jonsafari/persian_nlp.html
23. Diaz, I., Ranilla, J., Montanes, E., Fernandez, J., Combarro, E.F.: Improving performance of text categorization by combining filtering and support vector machines. Journal of the American Society for Information Science and Technology 55(7), 579–592 (2004)

# Adaptable Term Weighting Framework for Text Classification

Dat Huynh, Dat Tran, Wanli Ma, and Dharmendra Sharma

Faculty of Information Sciences and Engineering
University of Canberra
ACT 2601, Australia
{dat.huynh,dat.tran,wanli.ma,dharmendra.sharma}@canberra.edu.au

**Abstract.** In text classification, term frequency and term co-occurrence factors are dominantly used in weighting term features. Category relevance factors have recently been used to propose term weighting approaches. However, these approaches are mainly based on their own-designed text classifiers to adapt to category information, where the advantages of popular text classifiers have been ignored. This paper proposes a term weighting framework for text classification tasks. The framework firstly inherits the benefits of provided category information to estimate the weighting of features. Secondly, based on the feedback information, it is able to continuously adjust feature weightings to find the best representations for documents. Thirdly, the framework robustly makes it possible to work with different text classifiers on classifying the text representations, based on category information. On several corpora with SVM classifier, experiments show that given predicted information from TFxIDF method as initial status, the proposed approach leverages accuracy results and outperforms current text classification approaches.

**Keywords:** Text representation, feature weighting approach, term category dependency, classifier, and text classification.

## 1 Introduction

In text classification (TC), a single or multiple category labels are automatically assigned to a new text document based on category models, which are created after learning a set of labelled training text documents. TC methods normally convert a text document into a relational tuple using the popular vector-space model to obtain a list of terms with corresponding frequencies. A term-by-frequency matrix, interpreted as a relational table, is used to represent a collection of documents.

Due to the huge challenges and the difficulties of classifying document representations to a list of categories, a large number of classification algorithms have been developed to address the challenges in different degrees. Some of the popular algorithms have been currently used in TC such as multivariate regression

A. Gelbukh (Ed.): CICLing 2011, Part II, LNCS 6609, pp. 254–265, 2011.
© Springer-Verlag Berlin Heidelberg 2011

models, nearest neighbour classification [18], Bayes probabilistic approaches [9], decision trees [1], neural networks [10], boosting methods [13], and support vector machines (SVM)[6]. Among those approaches, SVM is considered as the state-of-the-art algorithm for text classification [14].

The effectiveness of TC method not only depends on the advantages of classifiers, but also, more importantly, how documents are represented. Previous studies of information retrieval [12,7] showed that appropriate term weighting approaches were crucial to the performance of information retrieval systems.

Term frequency (TF) has long been used to measure the importance levels of terms in a document [12]. TF is considered as a key component to evaluate term significance in a specific context [11]. The more a term is encountered in a certain context, the more it contributes to the meaning of the context. Whereas, Inverse Document Frequency (IDF) is regarded as the information value of terms in the collection. The combination of TF and IDF is typically used on weighting terms for building document representations [6,19,20].

Taking benefits of relationships among terms such as co-occurrence relations [17,5], Wikipedia-link relations [4,15,16], some term weighting approaches have been proposed to construct the representations of documents based on the relation information.

Recognising the benefits of category relevance factor, some term weighting approaches have been proposed to address the problem of TC [21,8,3,2]. However, because of the lack of category information from the new testing documents, some of the approaches consider the training category information only. The others focus on their own classifiers to predict the label set of the testing documents. Therefore, the advantages of popular text classifiers have been ignored from the category-based approaches, or the category information has not considered on weighting terms on both training and testing documents.

To address the deficiencies of those category-based term weighting approaches, this paper proposes a new term weighting framework for text classification, which firstly inherits the benefits of category information to estimate feature weights. Secondly, it provides abilities to continuously adjust feature weights based on the feedback information to find the best representations for documents. Thirdly, the framework robustly makes it possible to work with different text classifiers on classifying the text representations based on category information.

In the remaining of this paper, firstly the term weighting framework is introduced in Section 2. Then, how to apply the term weighting framework is presented in Section 3. In Section 4, experiments on several corpora are presented. The performance and discussion are detailed in Section 5. The conclusion is presented in Section 6.

## 2   Term Weighting Framework Based on Category Information

The idea of the framework is to consider the predicted category information of testing documents to justify their feature weights until a convergence is reached.

An aggregation label set will be an ideally predicted label to finally justify the feature weights for TC.

Similarity to machine learning-based methods, the term weighting framework includes two main phases (Fig. 1).

- **Training phase:** The purpose of the training phase is to build a training model based on training documents and their provided categories. The training model is then used to predict label set of testing documents.
- **Testing phase:** The purpose of this phase is to obtain the aggregated labels of the testing documents. There are two main processes in this phase: *"Initialising document representation"* and *"Round-robin justifying document representation"*. While the purpose of the former is to generate initial feature vectors for testing documents, the purpose of the later is to continuously predict label set from a given feature vector set, to check the convergence status, and to justify weighting features of testing documents if the label set is still divergent.

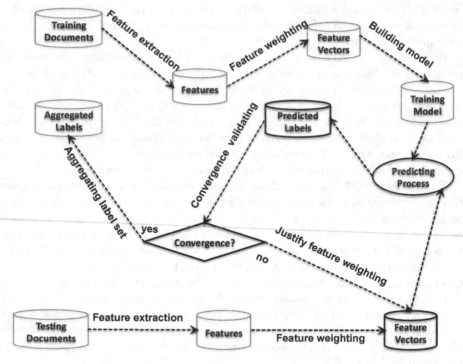

**Fig. 1.** The proposed term weighting framework

## 2.1   Training Phase

The training phase consists of the following steps:

1. A feature extraction process is applied to extract feature units from training documents.

2. A feature weighting method is applied to estimate the weight for each feature unit. The feature weights are calculated based on the importance levels of the features in documents and the category information profile of the containing category.
3. From the training data, a set of feature vectors which is constructed and is used to build a training model, represented for the training documents.

## 2.2   Testing Phase

The testing phase is divided into two processes: *"Initialising Document Representation Process"* and *"Round-robin Justifying Document Representation Process"*. The purpose of the former is to extract the representation of the testing documents, whereas the role of the later is to continuously justify the weights of features in respect to the predicted category information from the testing data.

### Initialising Document Representation Process

1. The feature extraction method is re-applied to extract feature units for testing documents.
2. Because there is no information about category of testing data, at this initial step, one of the popular term weighting methods will be applied (ex. TF×IDF) to initialise the weights of document features. Each document will be represented by a feature vector, and a set of feature vectors from testing set is ready for the predicting process.

### Round-robin Justifying Document Representation Process

1. With the initial set of feature vectors given from the initial process, using the training model, the predicting process employs a classifier (ex. SVM) to predict a set of labels.
2. If the predicted label set reaches a convergence status, the round-robin process will be terminated. The list of predicted label sets is aggregated into an aggregation label set, which is considered as the ideal predicted label. The accuracy can be calculated based on the aggregation label set.
3. Otherwise, based on the previous predicted label set, the round-robin process continues re-calculating the weights of features. Then, a new feature vector set is constructed and sent to the predicting process *(go back to step 1)* for continuing the round-robin process until reaching the convergence status.

## 2.3   How to Apply the Framework for Text Classification

In regard to the framework in text classification, the following details need to be implemented.

1. *Feature extraction:* Choosing a feature extraction method for training and testing documents.

2. *Initialising Term Weighting Method:* Choosing a normal term weighting approach for initially weighting testing documents.
3. *Term Weighing Approach based on Category Information:* Choosing a term weighting approach based on category information for training documents and for justifying weighting loops.
4. *Checking Convergence Status:* Designing an algorithm to identify convergence status.
5. *Aggregating Label Set:* Proposing a method to aggregate the final predicted label set for calculating accuracy.
6. *Classifiers:* Choosing text classifiers building training model, predicting and testing results.

# 3  Apply Term Weighting Framework for Text Classification Task

## 3.1  Term Weighting Method Based on Category Information

It can be seen in the training phase of the framework that feature weights need to be estimated based on a list of given categories (label set). Similarly, in the round-robin process, based on the predicted label set,the feature weights have to be recalculated for continuing the next loop of the round-robin process. Therefore, in this section, a term weighting method based on provided category information is proposed. Popularly, the weight of a term is determined by the importance of the term in a local context (document) and its information value from the collection. Thus, the weight of a term just only reflects relations between the term with the local context and the global context (collection). However, naturally dividing information into categories means to group the information with certain related topics. As a result, the term weight needs to be assigned under the consideration of its relations to the topics (categories). In other words, when weighting terms of documents, the dependencies between terms and categories should be taken into account.

Given a list of categories from a collection $C$ as $\{c_1, c_2, \ldots, c_{|C|}\}$ and a text document $d_j$ from the collection $C$, the list of terms represented for the document $d_j$ is defined as:

$$d_j = \left\{ (t_1, w_1), (t_2, w_2), \ldots, (t_{|d_j|}, w_{|d_j|}) \right\} \qquad (1)$$

where $t_i$ is a term $i$ in the document $d_j$ and $w_i$ is its weight value.

The weight of a term $t_i$ from the document $d_j$ under a category $c_k$ is defined as

$$tf.tcd(t_i, d_j, c_k) = tf(t_i, d_j) * tcd(t_i, c_k) \qquad (2)$$

where $tf(t_i, d_j)$ is normalised TF value of the term $t_i$ in the document $d_j$, and $tcd(t_i, c_k)$ is the term category dependency value of the term $t_i$ and the category $c_k$. The $tf(t_i, d_j)$ value is given as

$$tf(t_i, d_j) = \frac{freq(t_i, d_j)}{\max_{k=1\ldots|d_j|} freq(t_k, d_j)} \qquad (3)$$

where $freq(t_i, d_j)$ is the co-occurrence numbers of $t_i$ and $d_j$.

The idea of calculating term category dependency (TCD) is that terms representing a document are normally in a relation to its categories. If a term frequently occurs in many documents of one class and infrequently occurs in other classes, it should be considered as a representative for the class if its ranking value from the document is also comparable. Thus, a formula to measure the dependency degree of a term $t_i$ and a category $c_i$ is defined as

$$tcd(t_i, c_j) = \exp\left(\frac{tf(t_i, c_j) * df(t_i, c_j)}{\sqrt{\sum_{k=1}^{|C|}(t_i, c_k) * df(t_i, c_k)}}\right) \qquad (4)$$

## 3.2   Convergence Status Checking

In this section, an heuristic algorithm is proposed to check whether a convergence status exists. As described in the framework (Section ?), after a loop of round-robin, there is a new set of labels predicted. Let $L_{n-1} = \{l_1, l_2, \ldots, l_{n-1}\}$ is a predicted label sets after $n - 1$ loop. The current loop is $n^{th}$ and its predicted label set $l_n$ is under checking its convergence status. The convergence status at $n^{th}$ loop is defined by the formula $cvg$ as:

$$cvg(l_n) = min_{i=1}^{n-1}\left(diff(l_n, l_i)\right) \qquad (5)$$

where label set $l_n$ achieves the convergence status iff $cvg(l_n) <= \delta$, where $\delta$ is a given threshold; $diff(l_n, l_i)$ is the number of differences between two label sets $l_n$ and $l_i$ when line by line comparing in respect to a certain order *(the order of document IDs)*.

The intuition of the convergence status is that when choosing $\delta = 0$ and also the convergence status happening, the circle-circuit of generating label sets is created and label set at loop $n^{th}$ is the beginning of the circle-circuit. It can be seen that at the convergence status, the current predicted label set $l_n$ is the same as the label set $l_i$. The next coming loop will generate the set of label $l_{n+1}$, which will be completely as same as the label set $l_{i+1}$. This is happening because the label sets $l_{n+1}$ and $l_{i+1}$ are generated using information of the label set $l_n = l_i$. In other words, the feature vector set generated from the $(n + 1)^{th}$ loop will be exactly the same as the feature vector set generated from the $(i + 1)^{th}$ loop. Therefore, there is no chance to get a new set of labels from $n^{th}$ loop. That is why it is called the *"convergence status"*.

In the ideally situation, as choosing $\delta = 0$ and the convergence status exists, it is concluded that the further loops of round-robin process can be terminated. However, it takes time to achieve the convergence status. Thus, in practice, choosing $\delta$ as a trade off value to reduce time and computation consuming should be considered.

### 3.3   Label Set Aggregation

Supposing that there are $n$ loops completed, and a list of $n$ predicted label set $L$ is given as

$$L_n = \{l_1, l_2, \ldots, l_{n-1}, l_n\} \tag{6}$$

where $l_i$ is the label set produced at the loop $i^{th}$, and the structure of each label set $l_i$ is defined as

$$l_i = \{l_i^1, l_i^2, \ldots, l_i^m\} \tag{7}$$

where $l_i^j$ is predicted label of the document $j^{th}$ at the loop $i^{th}$, and $m$ is the number of documents in the collection.

Supposing that the label set $l_n$ is detected as reaching the convergence status. Therefore, the list of label set $L_{n-1} = \{l_1, \ldots, l_{n-1}\}$ is considered to estimate the aggregation label set $lg$ as follows

$$lg = \{lg^1, lg^2, \ldots, lg^m\} \tag{8}$$

where $lg^j$ ($j = 1 \ldots m$) is the aggregated label for the document $j^{th}$. The $lg^j$ is estimated as below

$$lg^j = \left\{ l_i^j : p(l_i^j | l_1^j \ldots l_n^j) \to max \right\} \tag{9}$$

If the document $d_j$ has a maximum prediction to be assigned the label $l_i^j$ with $n$ loops, that label is chosen to be the aggregation label for document $d_j$. The set of labels $lg$ is considered as the aggregation label set for the testing documents. It is used to estimate the accuracy of text classification.

## 4   Experiments

### 4.1   Corpora

**Reuters-21578.** In this experiment, the ApteMode version of Reuters-21578 is used. The collection was obtained by eliminating unlabelled documents and selecting categories which have at least one document in training set and one in testing set. The result contains *90* categories for both testing and training set. The number of documents in testing set is *3,019* and the number of documents in training set is *7,769*.

**Ohsumed Corpus.** The Ohsumed collection includes medical abstracts from the Medical Subject Heading (MeSH) categories. Based on the experiment of Joachims [6], the first *20,000* documents in year *1991* was used and divided into *10,000* documents for testing and *1,000* documents for training. The TC tasks is to classify those documents into the *23 cardiovascular diseases* categories. After selecting the set of categories, there were *6286* unique documents for training set and *7643* unique documents for testing set.

**NSFAwards Corpus.** This data set consists of *129,000* relatively short abstracts in English, describing awards granted for basic research by the US National Science Foundation (NSF) during the period *1990-2003*. For each abstract, there is a considerable amount of meta-data available, including the abbreviation code of the NSF division that processed and granted the award in question. We used this NSF division code as the class of each document. The title and the content of the abstract were used as the main content of the document for classification tasks. We used part of the corpus for the experiment by selecting *100* different documents for each single category, testing set and training set. After the pre-processing step, there were *19,018* documents from training set and *19,072* documents from testing set were used, which were contained in *199* categories.

In the experiment, all the corpora were pre-processed in the same way. The pure text of documents firstly was processed to eliminate their stop-words. Then, the remaining words were stemmed[1] to increase the statistic information for those being in the same word family. Lastly, rare words[2] were removed to avoid the large number of features. Table *1* shows the statistic information about the corpora after the pre-processing step.

**Table 1.** Numbers of documents and numbers of selected features (unique words with $DF \geq 3$) from the pre-processing step

| Corpora | # documents | | # selected features | |
|---|---|---|---|---|
| | Training Set | Testing Set | Training Set | Testing Set |
| Reuters-21578 | 7,769 | 3,019 | 7,177 | 3,817 |
| Ohsumed | 6,286 | 7,643 | 8,457 | 7,198 |
| NSFAwards | 19,018 | 19,072 | 16,407 | 13,126 |

## 4.2 Specified Term Weighting Framework for Text Classification Task

The term weighting framework for TC and the condition for applying are presented in Section *2*. In this section, the detailed discussion for how to apply the framework to classify text documents is presented.

1. *Feature extraction:* After preprocessing step as describing on Section *4.1*, single term as feature was selected.
2. *Initialising Term Weighting Method:* We chose TF×IDF as the term weighting method to assign weighting for terms. The initial feature was conducted in this step before sending to predicting process.
3. *Term Weighing Approach based on Category Information:* We chose term weighting method as describing in Formula *2* to estimate the weights of terms using information from the given training categories.

---

[1] Stemming tool http://tartarus.org/~martin/PorterStemmer
[2] Words whose document frequencies are less than *3*.

4. *Checking Convergence Status:* Applying the algorithm in Section *3.2* to check convergence status after generating predicted label set. To increase speed of processing, it is necessary to choose the threshold $\delta$, in this case $\delta = 20$ was selected for all the testing corpora.
5. *Aggregating Label Set:* The aggregation label set was generated by using the algorithm in Section *3.3*, after detecting the convergence status.
6. *Classifiers:* The SVM classifier was used to build training model and to predict data. The classifier operated under linear kernel, one of the best classifiers for text classification tasks [19]. Moreover, instead of scanning to find the best parameter for the SVM classifier, the default parameter value $c = 1$ was selected for every Linear SVM tasks.

## 5   Performance and Discussion

We have implemented the framework of weighting terms for TC with three different corpora. In the Reuters-21578 corpus, Table *2* showed that the proposed approach outperforms those reported on the same corpus by Yang and Liu [19]. Using SVM, the authors re-examined the experiment conducted by Joachims [6] and reported *.8599* micro-averaged $F_1$ score, which is slightly lower than *.8707* micro-averaged $F_1$ score of this experiment. When comparing to the SVM scores within *10* biggest categories of Reuters-21578, the proposed approach achieves the comparable results in comparison to the state-of-the-art text classification approach on Reuters-21578 corpus [6].

Similarly, in the Ohsumed corpus, the approach has achieved higher micro-averaged $F_1$ score in comparison to the reported results on this corpus. In fact, Yang and Pedersen [20] conducted their experiments and reported their classification results, which was under *60%* accuracy with KNN classifier. While the highest SVM score with the Ohsumed corpus was reported by Joachims just about *66.00%* on RBF kernel. Those results are considerably lower than

**Table 2.** Classification results from the Reuters-21578 corpus. The $\star$ indicates that the precision column was reported by Joachims [6].

| Category | Precision$\star$ | Precision | Recall |
|---|---|---|---|
| earn | .985 | **.9972** | .8999 |
| acq | .953 | **.9803** | .9666 |
| money-fx | .754 | **.8555** | .8268 |
| grain | **.919** | .6541 | .8121 |
| crude | **.890** | .8824 | .9524 |
| trade | .780 | **.8678** | .8974 |
| interest | .750 | **.9070** | .8931 |
| ship | **.865** | .7765 | .7416 |
| wheat | **.859** | .7089 | .7887 |
| corn | **.857** | .6875 | .7857 |
| **microavg-$F_1$** | .8650 | **0.8707** | |

**Table 3.** Classification results on all corpora from the round-robin process. The table shows results of first 5 loops of the process and the final convergence step. The number in brackets indicates the position that convergence status is detected. The aggregated results are the outcome results of the framework.

| Corpora | Initialise | Loop 1 | Loop 2 | Loop 3 | Loop 4 | Loop 5 | Convergence | Aggregation |
|---------|-----------|--------|--------|--------|--------|--------|-------------|-------------|
| Reuters | .7415 | .7887 | .8662 | .8665 | **.8718** | .8659 | .8707[8] | **.8707** |
| Ohsumed | .4299 | .7511 | .7814 | .7814 | **.7840** | .7824 | .7839[10] | **.7847** |
| NFSAward | .6608 | .7751 | .7787 | .7788 | .7782 | .7790 | **.7799[14]** | **.7795** |

micro-averaged $F_1$ score in this experiment, which is maintained about *77.47%* on the same collection.

Moreover, Fig. *2* shows clearly the distributions of classification results under category aspects. With the *"fuzzy"* and *"confusing"* Ohsumed corpus [20], reaching the very comparable classification results from each category and from the collection, the framework has demonstrated its strengths on classifying text documents.

In addition to Ohsumed corpus, the NSFAwards corpus is one of challenge corpora for text classification tasks. In this experiment, the framework has achieved considerable results. Within *199* categories and the equivalently number of documents per each category, the framework has worked well on the corpus to reach about *77.95%* micro-averaged $F_1$ score. More importantly, the meaningful number *78.44%* macro-averaged $F_1$ score has indicated that the approach achieves equivalently classification results from the category aspects.

With the purpose of evaluating the effectiveness of the framework and also considering the performance on every single loop of the framework, the result from the predicted label set of each loop is calculated, and the aggregation

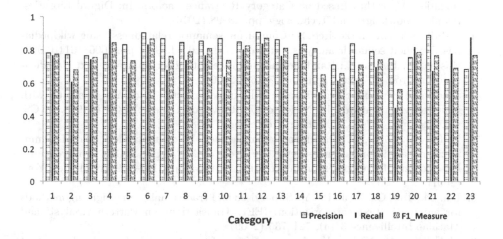

**Fig. 2.** Classification results from Ohsumed corpus within *23* categories

result is also estimated. Table *3* shows the micro-averaged $F_1$ score from each corpus. Starting with a "low accuracy" predicted label set, the $F_1$ score rapidly increases from the initialised step and reaching highest score before achieving convergence status. These experiment contributes to the conclusion that the convergence status has covered the expectable results of the classification tasks and the aggregation results also comparable with the highest results during the chains.

# 6    Conclusion

We have presented a framework for estimating weights of terms in text classification. The approach takes the predicted category information for justifying feature weightings to achieve high classification accuracies. Moreover, the framework is robust and can be applied to any text classifiers. With the predicted category information and repeatedly justifying weighting features, the approach has demonstrated its strength on classifying documents in three reliable corpora with the outstanding classification results in comparison with some other approaches on text classification.

# References

1. Apte, C., Damerau, F., Weiss, S.: Text mining with decision rules and decision trees. In: Proc. of Conference on Automated Learning and Discovery, Workshop 6: Learning from Text and the Web (1998)
2. Debole, F., Sebastiani, F.: Supervised term weighting for automated text categorization. In: Proc. of the 2003 ACM Symposium on Applied Computing, pp. 784–788. ACM, New York (2003)
3. Deng, Z., Tang, S., Yang, D., Zhang, M., Wu, X., Yang, M.: A Linear Text Classification Algorithm Based on Category Relevance Factors. In: Digital Libraries: People, Knowledge, and Technology, pp. 88–98 (2010)
4. Gabrilovich, E., Markovitch, S.: Computing semantic relatedness using wikipedia-based explicit semantic analysis. In: Proc. of the 20th IJCAI, pp. 1606–1611 (2007)
5. Hassan, S., Banea, C.: Random-walk term weighting for improved text classification. In: Proc. of TextGraphs, pp. 53–60 (2006)
6. Joachims, T.: Text categorisation with support vector machines: Learning with many relevant features. In: Nédellec, C., Rouveirol, C. (eds.) ECML 1998. LNCS, vol. 1398, pp. 137–142. Springer, Heidelberg (1998)
7. Lafferty, J., Zhai, C.: Document language models, query models, and risk minimization for information retrieval. In: Proc. of 24th ACM SIGIR Conference on Research and Development in Information Retrieval, pp. 111–119. ACM, New York (2001)
8. Lan, M., Tan, C., Su, J., Lu, Y.: Supervised and traditional term weighting methods for automatic text categorization. IEEE Transactions on Pattern Analysis and Machine Intelligence 31(4), 721–735 (2009)
9. McCallum, A., Nigam, K.: A comparaison of event models for naive bayes text classfication. In: AAA 1998 Workshop on Learning for Text Categorization (1998)

10. Ng, H., Goh, W., Low, K.: Feature selection, perceptron learning, and a usability case study for text categorization. In: Proc. $20^{th}$ ACM SIGIR Conference on Research and Development in Information Retrieval, pp. 67–73 (1997)
11. Robertson, S., Jones, K.S.: Simple, proven approaches to text retrieval. Tech. rep., University of Cambridge (1997)
12. Salton, G., Buckley, C.: Approaches term-weighting in automatic text retrieval. In: Proc. of Information Processing and Management, pp. 513–523 (1988)
13. Schapire, R., Singer, Y.: Boostester: A boosting-based system for text categorization. Machine Learning, 135–168 (2000)
14. Sebastiani, F.: Machine learning in automated text categorization. ACM computing surveys (CSUR) 34(1), 1–47 (2002)
15. Strube, M., Ponzetto, S.P.: Wikirelate! computing semantic relatedness using wikipedia. In: Proc. of the $21^{st}$ AAAI, pp. 1419–1424 (2006)
16. Wang, P., Hu, J., Zeng, H.J., Chen, L., Chen, Z.: Improving text classification by using encyclopaedia knowledge. In: The Seventh IEEE ICDM, pp. 332–341 (2007)
17. Wang, W., Do, D.B., Lin, X.: Term graph model for text classification. In: Li, X., Wang, S., Dong, Z.Y. (eds.) ADMA 2005. LNCS (LNAI), vol. 3584, pp. 19–30. Springer, Heidelberg (2005)
18. Yang, Y.: Effective and efficient learning for human decision in text categorization and retrieval. In: Proc. $17^{th}$ ACM SIGIR Conference on Research and Development in Inforation Retrieval, pp. 13–22 (1994)
19. Yang, Y., Liu, X.: A re-examination of text categorisation methods. In: Proc. of the $22^{nd}$ Annual International ACM SIGIR Conference on Research and Development in Information Retrieval, pp. 42–49 (1999)
20. Yang, Y., Pedersen, J.O.: A comparative study on feature selection in text categorisation. In: Proc. of the $14^{th}$ ICML, pp. 412–420 (1997)
21. Yu, S., Zhang, J.: A class core extraction method for text categorisation. In: Proc. of the $6^{th}$ FSKD, pp. 3–7 (2009)

# Automatic Specialized vs. Non-specialized Sentence Differentiation

Iria da Cunha[2,3,1], M. Teresa Cabré[1], Eric SanJuan[3],
Gerardo Sierra[2], Juan Manuel Torres-Moreno[2,3,4], and Jorge Vivaldi[1]

[1] Institut Universitari de Linguistique Applicada - UPF
Roc Boronat, 138 E-08018 Barcelona (España)
[2] Grupo de Ingeniería Lingüística - Instituto de Ingeniería UNAM
Torre de IngenieríÂa Basamento, Ciudad Universitaria Mexico, D.F. 04510 Mexico
[3] Laboratoire Informatique d'Avignon - UAPV
339 chemin des Meinajaries, BP91228 84911 Avignon Cedex 9, France
[4] École Polytechnique de Montréal - Département de génie informatique
CP 6079 Succ. Centre Ville H3C 3A7 Montréal (Québec), Canada
http://www.lia.univ-avignon.fr, http://www.iula.upf.edu,
http://www.iling.unam.mx

**Abstract.** Compilation of Languages for Specific Purposes (LSP) corpora is a task which is fraught with several difficulties (mainly time and human effort), because it is not easy to discern between specialized and non-specialized text. The aim of this work is to study automatic specialized vs. non-specialized sentence differentiation. The experiments are carried out on two corpora of sentences extracted from specialized and non-specialized texts. One in economics (academic publications and news from newspapers), another about sexuality (academic publications and texts from forums and blogs). First we show the feasibility of the task using a statistical n-gram classifier. Then we show that grammatical features can also be used to classify sentences from the first corpus. For such purpose we use association rule mining.

**Keywords:** Specialized Text, General Text, Corpus, Languages for Specific Purposes, Statistical Methods, Association Rules, Grammatical features.

## 1 Introduction

Compilation of Languages for Specific Purposes (LSP) corpora, that is, corpora including specialized texts, is necessary to carry out several tasks, such as: terminology extraction, compiling specialized dictionaries, lexicons or ontologies. This corpora compilation is human time effort consuming. Until now, professionals or specialists have to decide if the text is specialized or not.

But what is a specialized text? [1] mentions some features to be considered in order to answer this question: the text author, the potential reader, the structural organization and the lexical units' selection. There are two types of variability in specialized texts: horizontal determined by the subject and vertical determined

A. Gelbukh (Ed.): CICLing 2011, Part II, LNCS 6609, pp. 266–276, 2011.

by the specialization level. With regard to the second one, as shown in [2], three specialization levels can be considered: high (specialized writer and specialized receiver), medium (specialized writer and semi-specialized receiver, that is, for example, students) and low (specialized writer and non-specialized receiver, that is, general public).

For example, articles in newspapers should be considered as a low specialization level because they may deal with technical subjects, as, economics, medicine or law. However, they don't share the "conceptual and lexical control" of the domain.

There are several theoretical works about differences between general and specialized texts. Most of them consider that lexicon is the most discriminative factor (besides being the most visible) to carry out this differentiation. It is well-known that terms (units of the lexicon with a precise meaning in a particular domain [3]) show the specialized content of a subject; therefore, they appear in texts of their domain. But there are other features of specialized texts (as grammatical features, both morphological and syntactic) that can be considered as specific of these texts. Features as verbal flexion related to grammatical person, verbal tense or verbal mode have been underlined in some works [4]. Some authors, using small corpora, have established some grammatical phenomena that may differentiate specialized texts. In some cases, they have considered only a very limited number of features of a single category; in other cases, a scarce number of texts has been analysed manually. [5] analyses the frequency of names and verbs into a general corpus and a specialized corpus. Some authors have studied verbs into specialized French corpora [6,7,8,9]. The works of [10,11] are the first ones where this subject is studied using a bigger corpus (two millions of words). They conclude that certain grammatical features, besides lexicon, have a strong potential to differentiate specialized texts from non-specialized texts.

The aim of this work is to study automatic specialized vs. non-specialized sentence differentiation. The experiments are carried out on two corpora of sentences extracted from specialized and non-specialized texts. One in economics (academic publications and news from newspapers), another about sexuality (academic publications, and texts from forums and blogs).

This paper is organized as follows. In Section 2 we explain the methodology of our work. First, in Section 3, we show the feasibility of the task using a statistical n-gram classifier. Then, in Section 4, we show that grammatical features can also be used to classify sentences from the economics corpus. Finally, section 5 exposes the conclusions of the paper and the future work.

## 2   Methodology

We have compiled two corpora: a corpus including economic texts and a corpus including texts from the sexuality domain. Each one was divided into two subcorpora: specialized vs. non-specialized (or general).

The economic corpus was divided as follows:

1. A sub-corpus including texts from the specialized domain of economics, mainly scientific papers, books, theses, etc. (with 292,804 tokens included in 9,243 sentences).
2. A sub-corpus with non-specialized texts from the economics subsection of Spanish newspapers (with 1,232,512 tokens corresponding to 36,236 sentences).

These texts have been extracted from the Technical Corpus of the Institute for Applied Linguistics[1] (IULA-CT) of the Universitat Pompeu Fabra of Barcelona. It consists of documents in Catalan, Spanish, English, German and French, although the search through bwanaNet is at the moment restricted to the first three of these languages. It contains texts of several domains (economics, law, computing, medicine, genome and environment) as well as texts from newspapers. All the texts are POS tagged. This corpus is accessible on-line via http://bwananet.iula.upf.edu/. Further details on these resources are shown at [12].

The sexuality corpus was divided as follows:

1. A sub-corpus including texts from the specialized domain of sexuality, mainly scientific papers, books, theses, etc. (with 127,903 tokens included in 6,368 sentences).
2. A generic sub-corpus with texts from html pages, blogs and forums about sexuality (with 384,659 tokens corresponding to 31,475 sentences).

These texts have been extracted from the Sexuality Corpus of the *Grupo de Ingeniería Lingüistica* (GIL)[2] at the *Universidad Nacional Autonóma de México* (UNAM) [13]. In this corpus, texts are divided into five levels:

1. Level 1: Texts from Google Scholar.
2. Level 2: Texts from sexuality associations.
3. Level 3: PDF texts.
4. Level 4: Word and html texts.
5. Level 5: Blogs and forums texts.

For our experiments we have used texts from level 1 (specialized) and texts from level 4 and 5 (non-specialized). All the texts were tagged with POS tags.

Both corpora (economic and sexuality corpora) contain specialized and non-specialized texts. However there is an important difference between them. The first one includes academic or journalistic texts, so all the texts are well-written with a defined style, since text's authors are journalists or specialists from the domain. The second one, the sexuality corpus, includes (mainly) academical texts as well (into the specialized sub-corpus), but it contains texts from blogs and forums about sexuality, where the sentences are not always well-written and

---

[1] http://www.iula.upf.edu
[2] http://www.iling.unam.mx/

sometimes they are not complete sentences. This is a more "ambiguous" corpus, more difficult to characterize, which is interesting for our experiments as well.

Finally, we are interested in working at sentence level instead of entire documents. Indeed, documents can be classified using contextual information about their structure or statistical information about their specific vocabulary. At sentence level, none of these informations can be used. Clearly, we target an application that can look for technical/non-technical statements inside any document type. We first show that this is possible, at least using a statistical n-gram approach, then we study how grammatical information can be used to generate intuitive decision rules.

## 3  Sentence Classification Based on n-grams

We have developed an algorithm based on a ranking of $n$-grams. Two language models ($LM$) are constructed: one $LM_{\text{spe}}$ over the specialized corpus and another $LM_{\text{gen}}$ over the non-specialized corpus.

### 3.1  Algorithm

The $n$-grams distance algorithm is simple. It is inspired by the methods used in DEFT [15]. A language model is generated using a sliding-window of $n$ characters, with $n = 1, ..., 15$. This produces two language models $LM_{\text{spe}}$ and $LM_{\text{gen}}$. In the same way, we also consider the language model $LM_{\text{X}}$ generated by an unknown sentence $X$. To classify $X$ we compute the distance (absolute value of the ranking) $LM_{\text{X}} || (LM_{\text{spe}}; LM_{\text{gen}})$ and we choose the category closer to $X$.

### 3.2  Results

From the economics corpus we have randomly selected 9,000 sentences from each category (specialized and non-specialized). From the sexuality corpus we have randomly selected 544 sentences from the non-specialized category and 635 sentences from the specialized one. Therefore the experiment has been carried out on a set of 1,179 sentences corresponding. We have used the 90% of both corpora for training and the 10% for test, replicating this split 30 times at random.

Table 1 includes results obtained by the $n$-grams algorithm over the economics corpus. Performances using the first 20,000 and 30,000 $n$-grams are shown. Table 2 contains results obtained over the sexuality corpus. In this case, $n = 14$ and the number of $n$-grams is 500,000. These results show that the use of a higher $n$ and a high quantity of $n$-grams has a positive influence on the results. These results (average F-score of 0.8715 over the economics corpus and 0.8258 over sexuality corpus) are interesting, because they mean that a simple $n$-grams distance strategy is suitable to distinguish specialized and non-specialized texts correctly.

**Table 1.** Results of $n$-grams classifier over the economics corpus

|  | 20K 6-grams | | | 30K 6-grams | | |
|---|---|---|---|---|---|---|
|  | Precision | Recall | F-Score | Precision | Recall | F-Score |
| GEN | 0.6341 | 0.8312 | 0.7194 | 0.6744 | 0.8475 | 0.7511 |
| SPE | 0.9532 | 0.8776 | 0.9138 | 0.9583 | 0.8955 | 0.9259 |
| Average | 0.7937 | 0.8544 | 0.8166 | 0.8164 | 0.8715 | **0.8385** |

**Table 2.** Results of $n$-grams classifier over the sexuality corpus

|  | 400K 13-grams | | | 500K 15-grams | | |
|---|---|---|---|---|---|---|
|  | Precision | Recall | F-Score | Precision | Recall | F-Score |
| GEN | 0.7999 | 0.8121 | 0.8058 | 0.8102 | 0.8156 | 0.8128 |
| SPE | 0.8370 | 0.8257 | 0.8312 | 0.8412 | 0.8361 | 0.8385 |
| Average | 0.8184 | 0.8189 | 0.8185 | 0.8257 | 0.8258 | **0.8257** |

**Table 3.** Sample of 15-grams of specialized vs. non-specialized model of language

| Rank | $n$-gram (SPE) | ocurrencies | $n$-gram (GEN) | ocurrencies |
|---|---|---|---|---|
| 1 | e | 73472 | e | 57000 |
| ... | ... | ... | ... | ... |
| 254 | sexual | 1549 | a_de | 1140 |
| ... | ... | ... | ... | ... |
| 272 | e_s | 1444 | _sexual | 1062 |
| ... | ... | ... | ... | ... |
| 1890 | porno | 247 | de_es | 187 |
| ... | ... | ... | ... | ... |
| 2652 | s_pe | 182 | porno | 142 |
| ... | ... | ... | ... | ... |
| 4351 | a_a_l | 123 | _orgasmo | 92 |
| ... | ... | ... | ... | ... |
| 6767 | el_condón | 86 | iolencia | 54 |
| ... | ... | ... | ... | ... |
| 7757 | _orgasmo | 76 | nfecci | 55 |
| ... | ... | ... | ... | ... |
| 499999 | uinaria_porn | 2 | _de_una_put | 2 |

Table 3 shows a sample of $n$-grams of both language models, ordered by rank and with the number of occurrences. The smaller is the ranking, the less discriminant is the corresponding $n$-gram.

With regard to the sexuality corpus, the $n$-grams strategy maintains its performance, obtaining an average F-score of 0.8257, that is, a 0.0128 less than over the economics corpus.

# 4 Grammatical Features for Specialized vs. Non-specialized Sentences Differentiation

We have selected some linguistic features that may be characteristic of specialized texts and non-specialized texts.

## 4.1 Feature Description

We have used the features detected by [10] and [11]. Table 4 shows them. The full meaning of these POS tags can be seen on the following URL: `http://www.iula.upf.edu/corpus/etqfrmes.htm` Some POS tags are produced by subespecification of the full tag (ex. "A" is a subespecification of "AMS", "AMP", etc.). The machine learning approach that we have used is based on association rules, one of the most-known methods to detect relations among variables into large symbolic (i.e. non numerical) data [14].

**Table 4.** Linguistic features used in our work

| POS | Tag meaning |
|-----|-------------|
| A   | Determiner |
| C   | Conjunction |
| D   | Adverb |
| E   | Especifier |
| JQ  | Qualifier adjective |
| J   | Adjective |
| N4  | Proper noun |
| N5  | Common noun |
| P   | Preposition |
| R   | Pronoun |
| T   | Date |
| VC  | Verb (participle) |
| V1P | Verb (first person, plural) |
| V1S | Verb (first person, singular) |
| V2  | Verb (second person) |
| V   | Verb |
| X   | Number |

Table 5 shows an example of plain text and its corresponding generated test corpus text. In bold we have marked the category GEN, which indicates that this sentence is classified as part of a non-specialized text. Observe that "Plain text" section includes the sentence as found in the general corpus while the "Attributes generated from text" section includes just a list of the lemmas/tags found in such sentence.

**Table 5.** Example of economic plain text and attributes generated from text

| Plain text |
| --- |
| Tras el acuerdo con los pilotos, la dirección de Alitalia concluyó ayer de madrugada la negociación con los sindicatos del personal de tierra, que aceptaron 2.500 despidos (la propuesta inicial era de 3.500), la congelación de los salarios durante dos años y el bloqueo del fondo de previsión social durante el mismo periodo, para evitar la quiebra de la compañía. |

| Attributes generated from text |
| --- |
| **GEN** ser congelación despido previsión tierra dos dirección el tras para quiebra periodo negociación mismo piloto bloqueo = salario A Alitalia C D de N4 N5 personal compañía fondo P R que JQ V propuesta num X social con ayer aceptar madrugada sindicato concluir año inicial durante acuerdo y evitar |

## 4.2 Association Rules

We consider association rules of the form $X \Rightarrow D$, where X is a set of at most 5 lemmas and/or tags, D is the decision: SPE for specialized and GEN for general. For a rule to be valid, X has to be included in more than 0.5% of the sentences (this is called the support of the rule) and more than 90% of these sentences that include X have to be in category D (this is called the confidence of the rule). Since the right part of the rule is restricted to a few numbers of categories, we shall refer to these rules as decision rules. This kind of rules can be computed using "Apriori", a standard GPL packages by Christian Borgelt (http://www.borgelt.net/apriori.html). Our experiments over the economic corpus show that this strategy allows us to obtain 46,148 decision rules. It appears that:

- 60% of the rules induce category SPE, which means that there are more implicit decision rules among specialized texts than non specialized ones.
- 78% of the rules include at least one grammatical tag which shows that this information is significant to distinguish between these two categories.

Here is a sample set of 10 rules randomly extracted from the total list of decision rules for the economic corpus. Rules are given in Prolog format: the decision is on the left and the two figures on the right give respectively the support and the confidence of the rule.

```
SPE ← europea N4 JQ N5 (50, 100.0)
SPE ← millones X JQ P (70, 100.0)
GEN ← anunciar N4 P = (80, 98.3)
GEN ← ayer uno R N4 (10, 100.0)
SPE ← función C JQ D (12, 93.1)
GEN ← Gobierno haber VC V (60, 100.0)
GEN ← España que P = (100, 100.0)
```

SPE ← embargo sin de N5 (70, 100.0)
SPE ← internacional a R N5 (12, 90.8)
GEN ← presidente en R JQ (80, 93.0)

Therefore each rule indicates that if a given set of lemmas and tags is included in one sentence, there is a specific probability to classify the sentences as general (GEN) or specialized (SPE). As an example, the first rule may be read as follows: if the sentence under analysis includes the lemma "europea" and words with the POS tags "N4", "JQ" and "N5", then such sentence may be classified as specialized (SPE). The coverage of this rule is 50% with a 100% of precision.

## 4.3 Classifiers Based on Decision Rules

Once this set of rules is available, it is possible to build a classifier that, given a sentence, looks for the set of rules that match the sentence and chooses the rule that has the highest confidence. One important feature of this type of classifier is that it indicates when it cannot take a decision.

As a variant of this basic classifier (Classifier 1) we have developed a variant that only takes into accout those rules including at least one POS tag (Classifier 2). In this way it is possible to evaluate the actual impact of using POS tags as a classifier atribute.

## 4.4 Results

To evaluate the results of both algorithms we have used classical precision, recall and F-Score measures.

Results of this algorithm over the economics corpus are shown in Table 6.

**Table 6.** Results of Classifier 1 over the economics corpus

|         | Precision | Recall | F-Score |
|---------|-----------|--------|---------|
| GEN     | 0.7602    | 0.8671 | 0.8137  |
| SPE     | 0.8875    | 0.7239 | 0.8057  |
| Average | 0.8190    | 0.7890 | **0.8040** |

We have carried out another experiment over the economics corpus, using for the classifier (Classifier 2) only the association rules including at least one grammatical feature (POS tag). This is a subset of 36,217 rules (78%). Results obtained by Clasifier 2 over the economics corpus are shown in Table 7.

This evaluation shows that elimination of rules exclusively based on lemmas does not significantly degrade classifier performance. In fact, is seems that it lightly improves the average F-score (from 0.8040 to 0.8051).

**Table 7.** Results of Classifier 2 over the economics corpus

|  | Precision | Recall | F-Score |
|---|---|---|---|
| GEN | 0.7582 | 0.8959 | 0.8213 |
| SPE | 0.8749 | 0.7182 | 0.7889 |
| Average | 0.8166 | 0.8071 | **0.8051** |

**Table 8.** Results of Classifier 1 over the sexuality corpus

|  | 3 word rules | | | 1 word rules | | |
|---|---|---|---|---|---|---|
|  | Precision | Recall | F-Score | Precision | Recall | F-Score |
| GEN | 0.7573 | 0.6944 | 0.7245 | 0.7455 | 0.7371 | 0.7412 |
| SPE | 0.7258 | 0.7843 | 0.7539 | 0.7478 | 0.7559 | 0.7518 |
| Average | 0.7416 | 0.7393 | 0.7392 | 0.7466 | 0.7465 | **0.7465** |

Table 8 includes results obtained by Classifier 1 over the sexuality corpus, using very short association rules (with 3 tokens and 1 token). Results show that Classifier 1 performance is better over the economics corpus than over the sexuality corpus (with an average F-Score of 0.8040 and 0.7465, respectively). This fact would mean that grammatical (POS tags) and lexical features (tokens) included into the specialized texts in economics are quite different (that is, more discriminant) to the ones included into the non-specialized texts. However, although these features allow Classifier 1 to discriminate between specialized and non-specialized texts from the sexuality domain, they are less representative of each one of these corpora.

Our results show that both strategies (association rules and $n$-grams distances) work better over the economics corpus than over the sexuality corpus. This is due to the fact the economics corpus is a "real" specialized non-specialized corpus, including texts where all the sentences are well-written, they have a very well-defined style and the order of grammatical tags are correct. This is normal because the authors of these texts were specialists from the domain (in the case of academical texts) or journalists (in the case of news from newspapers), respectively.

Obtained results with both strategies are good over the economics corpus, although results with $n$-grams distances are a bit better than using association rules (0.8051 vs. 0.8385). Nevertheless, the association strategy has one advantage: the generated rules are humanly understandable and interpretable. The $n$-grams strategy offers only $n$-grams of characters, that is, unintelligible textual short passages (as the information included in Table 3 shows).

However, the association rules strategy over the sexuality corpus does not obtain so good results as over the economics corpus (average F-score of 0.7465 vs. 0.8051, respectively; that is a 0.0586 less). This is due to the fact that the specialized and non-specialized sexuality corpora contains texts extracted from very different sources (academic vs. forums and blogs) but the vocabulary they contain is very similar. This situation makes the differentiation task more difficult.

# 5   Conclusion and Future Work

The results we have obtained until now show that both strategies we have used in this work (n-grams distances and association rules based on lexical and grammatical features) are suitable to differentiate sentences from specialized and non-specialized texts. Results of the first experiment, employing a simple n-grams distance algorithm (generating language models for both corpora), show that performance using this strategy is high. Results of the second experiment, using lexical and grammatical features, show that grammatical features are discriminant enough for this task. We have shown that both approaches are useful to classify texts as specialized/non-specialized. The obtained F-scores for both methods are similar on the corpus from economics, but the classifier based on n-gram distances is clearly better when it is applied to the sexuality corpus. Such results seem to show that linguistic information is not as useful as foreseen. But specific characteristics of the texts included in the sexuality corpus may be the origin of this behaviour. These texts come from a source quite different from the texts in economics, since they come mainly from blogs, forums, associations, etc. that produce non-structured texts (incomplete or even non-grammatical sentences or wrong words). This requires additional experimentation in other domains as well as texts coming from equivalent sources.

We plan as well to develop an automatic tool able to detect sentences from specific domains (ex. medicine, economics, law, biology or physics), giving to the user the option to choose between specialized and non-specialized texts.

We consider that our results constitute an innovative perspective to research on domains related with terminology, specialized discourse and computational linguistics, like for example automatic compilation of LSP corpora or optimization of search engines.

# References

1. Cabré, M.T.: Textos especializados y unidades de conocimiento: metodología y tipologización. In: Garía Palacios, J., Fuentes, M.T. (eds.) Texto, terminología y traducción, pp. 15–36. Ediciones Almar, Salamanca (2002)
2. Pearson, J.: Terms in context. John Benjamin, Amsterdam (1998)
3. Cabré, M.T.: La terminología. Representación y comunicación. IULA-UPF, Barcelona (1999)
4. Kocourek, R.: La langue française de la technique et de la science. Vers une linguistique de la langue savante. Oscar Branstetter, Wiesbaden (1991)
5. Hoffmann, L.: Kommunikationsmittel Fachsprache - Eine Einführung. Sammlung Akademie Verlag, Berlin (1976)
6. Coulon, R.: French as it is written by French sociologists. Bulletin pédagogique des IUT (18), 11–25 (1972)
7. Cajolet-Laganière, H., Maillet, N.: Caractérisation des textes techniques québécois. Présence francophone (47), 113–147 (1995)
8. L'Homme, M.C.: Contribution á l'analyse grammaticale de la langue d'espécialité: le mode, le temps et la personne du verbe dans quelques textes, scientifiques é crits á vocation pédagogique. Université Laval, Québec (1993)

9. L'Homme, M.C.: Formes verbales de temps et texte scientifique. Le langage et l'homme 2-3(31), 107–123 (1995)
10. Cabré, M.T., Bach, C., da Cunha, I., Morales, A., Vivaldi, J.: Comparación de algunas características lingüísticas del discurso especializado frente al discurso general: el caso del discurso económico. In: XXVII Congreso Internacional de AESLA: Modos y formas de la comunicación humana (AESLA 2009), Universidad de Castilla-La Mancha, Ciudad Real (2010)
11. Cabré, M.T.: Constituir un corpus de textos de especialidad: condiciones y posibilidades. In: Ballard, M., Pineira-Tresmontant, C. (eds.), pp. 89–106. Artois Presses Université, Arras (2005)
12. Vivaldi, J.: Corpus and exploitation tool: IULACT and bwanaNet. In: Cantos Gómez, P., Sánchez Pérez, A. (eds.) I International Conference on Corpus Linguistics (CICL 2009), A survey on corpus-based research, Universidad de Murcia, pp. 224–239 (2009)
13. Medina, A., Sierra, G.: Criteria for the Construction of a Corpus for a Mexican Spanish Dictionary of Sexuality. In: 11th Euralex International Congress, vol. 2. Université de Bretagne-Sud. Lorient, Francia (2004)
14. Amir, A., Aumann, Y., Feldman, R., Fresko, M.: Maximal Association Rules: A Tool for Mining Associations in Text. Journal of Intelligent Information Systems 5(3), 333–345 (2005)
15. Stanislas, O., Mickael, R., Nathalie, C., Kessler, R., Lefèvre, F., Torres-Moreno, J.-M.: Système du LIA pour la campagne DEFT 2010: datation et localisation d'articles de presse francophones. In: DEFT 2010, Montréal (2010)
16. Kocourek, R.: La langue française de lá technique et de la science, 2nd edn. Oscar Branstetter, Wiesbaden (1991)

# Wikipedia Vandalism Detection: Combining Natural Language, Metadata, and Reputation Features

B. Thomas Adler[1], Luca de Alfaro[2], Santiago M. Mola-Velasco[3],
Paolo Rosso[3], and Andrew G. West[4,*]

[1] University of California, Santa Cruz, USA
thumper@soe.ucsc.edu
[2] Google and UC Santa Cruz, USA
luca@dealfaro.com
[3] NLE Lab. - ELiRF - DSIC. Universidad Politécnica de Valencia, Spain
{smola,prosso}@dsic.upv.es
[4] University of Pennsylvania, Philadelphia, USA
westand@cis.upenn.edu

**Abstract.** Wikipedia is an online encyclopedia which anyone can edit. While most edits are constructive, about 7% are acts of vandalism. Such behavior is characterized by modifications made in bad faith; introducing spam and other inappropriate content.

In this work, we present the results of an effort to integrate three of the leading approaches to Wikipedia vandalism detection: a spatio-temporal analysis of metadata (STiki), a reputation-based system (WikiTrust), and natural language processing features. The performance of the resulting joint system improves the state-of-the-art from all previous methods and establishes a new baseline for Wikipedia vandalism detection. We examine in detail the contribution of the three approaches, both for the task of discovering fresh vandalism, and for the task of locating vandalism in the complete set of Wikipedia revisions.

## 1 Introduction

Wikipedia [1] is an online encyclopedia that anyone can edit. In the 10 years since its creation, 272 language editions have been created, with 240 editions being actively maintained as of this writing [2]. Wikipedia's English edition has more than 3 million articles, making it the biggest encyclopedia ever created. The encyclopedia has been a collaborative effort involving over 13 million registered users and an indefinite number of anonymous editors [2]. This success has made Wikipedia one of the most used knowledge resources available online and a source of information for many third-party applications.

The open-access model that is key to Wikipedia's success, however, can also be a source of problems. While most edits are constructive, some are *vandalism*,

---

* Authors appear alphabetically. Order does not reflect contribution magnitude.

A. Gelbukh (Ed.): CICLing 2011, Part II, LNCS 6609, pp. 277–288, 2011.

the result of attacks by pranksters, lobbyists, and spammers. It is estimated that about 7% of the edits to Wikipedia are vandalism [3]. This vandalism is removed by a number of dedicated individuals who patrol Wikipedia articles looking for such damage. This is a daunting task: the English Wikipedia received 10 million edits between August 20 and October 10, 2010[1], permitting the estimation that some 700,000 revisions had to be reverted in this period.

Wikipedia vandalism also creates problems beyond the effort required to remove it. Vandalism lends an aura of unreliability to Wikipedia that exceeds the statistical extent of the damage. For instance, while Wikipedia has the potential to be a key resource in schools at all levels due to its breadth, overall quality, and free availability – the risk of exposing children to inappropriate material has been an obstacle to adoption [4,5]. Likewise, the presence of vandalism has made it difficult to produce static, high-quality snapshots of Wikipedia content, such as those that the Wikipedia 1.0 project plans to distribute in developing countries with poor Internet connectivity[2].

For these reasons, autonomous methods for locating Wikipedia vandalism have long been of interest. The earliest such attempts came directly from the user community, which produced several *bots*. Such bots examine newly-created revisions, apply hand-crafted rule sets, and detect vandalism where appropriate. Over time, these approaches grew more complex, using a vast assortment of methods from statistics and machine learning. Feature extraction and machine-learning, in particular, have proven particularly adept at the task – capturing the top spots at the recent PAN 2010 vandalism detection competition[3].

In this paper, we present a system for the automated detection of Wikipedia vandalism that constitutes, at the time of writing, the best-performing published approach. The set of features includes those of the two leading methodologies in PAN 2010: the Mola-Velasco system [6] (NLP) and the WikiTrust system [7] (reputation). Further, the features of the STiki system [8] (metadata) are included, which has academic origins, but also has a GUI frontend [9] enabling actual on-Wikipedia use (and has become a popular tool on English Wikipedia).

Since the systems are based largely on non-overlapping sets of features, we show that the combined set of features leads to a markedly superior performance. For example, 75% precision is possible at 80% recall. Moreover, fixing precision at 99% produces a classifier with 30% recall – perhaps enabling autonomous use.

Most importantly, we investigate the relative merit of different classes of features of different computational and data-gathering costs. Specifically, we consider (1) metadata, (2) text, (3) reputation, and (4) language features. *Metadata* features are derived from basic edit properties (*e.g.,* timestamp), and can be computed using straightforward database processing. *Text* features are also straightforward, but may require text processing algorithms of varying sophistication. *Reputation* features refer to values that analyze the behavior history of some entity involved in the edit (*e.g.,* an individual editor). Computing such

---

[1] http://en.wikipedia.org/wiki/User:Katalaveno/TBE
[2] http://en.wikipedia.org/wiki/Wikipedia:Wikimedia_School_Team
[3] Held in conjunction with CLEF 2010. See http://pan.webis.de

reputations comes with a high computational cost, as it is necessary to analyze large portions of Wikipedia history. Finally, *language* features are often easy to compute for specific languages, but require adaptation to be portable.

Moreover, we consider two classes of the vandalism detection problem: (1) the need to find *immediate* vandalism, (*i.e.*, occurring in the most recent revision of an article), and (2) *historical* vandalism, (*i.e.*, occurring in any revision including past ones). Immediate vandalism detection can be used to alert Wikipedia editors to revisions in need of examination. The STiki tool [9], whose features are included in this work, has been successfully used in this fashion to revert over 30,000 instances of vandalism on the English Wikipedia.

Historical vandalism detection can be used to select, for each article, a recent non-vandalized revision from the entire article history. The WikiTrust system (whose features are also included in this work) was recently used to select the revisions for the Wikipedia 0.8 project, a static snapshot of Wikipedia intended to be published in DVD form[4]. We consider historical detection to be an interesting variation of the standard Wikipedia vandalism detection problem, as it has the potential to use *future* information in edit analysis.

Combining the feature-vectors of the three systems, our meta-detector produces an area under the precision-recall curve (AUC-PR) of 81.83% for *immediate* vandalism detection. This is a significant improvement over the performance achieved from using any two of the systems in combination (performance ranges between 69% and 76%). Moreover, the meta-detector far exceeds the best known system in isolation (whose features are included), which won the PAN 2010 competition with 67% AUC-PR. Similar improvements were seen when performing the *historical* detection task. In a 99% precision setting, the meta-system could revert 30% of vandalism without human intervention.

The remainder of the work is structured as follows: Section 2 overviews related work. Section 3 describes our features and their categorization. Section 4 presents results. Finally, we conclude in Section 5.

## 2    A Brief History of Wikipedia Vandalism Detection

Given the damage that vandalism causes on Wikipedia, it is no surprise that attempts to locate vandalism automatically are almost as old as Wikipedia itself. The earliest tools consisted of *bots* that would labeled vandalism using handcrafted rule systems – encoding heuristic vandalism patterns. Examples of such bots include [10,11,12,13,14]. Typical rules were narrowly targeted, including: the amount of text inserted or deleted, the ratio of capital letters, the presence of vulgarisms detected via regular expressions, *etc.*.

Given the community's low tolerance for accidentally categorizing a legitimate edit as vandalism, such systems operated with high precision, but low recall. For instance, ClueBot was found to have 100% precision in one study, but fairly low

---

[4] http://blogs.potsdam.edu/wikipediaoffline/2010/10/26/wikipedia-version-0-8-is-coming/

recall: below 50% for any vandalism type, and below 5% for insertions [15]; a different study confirmed this low recall [16].

The idea that an edit's *textual content* is a likely source of indicative features has been investigated by several different research groups [15,16,17,18,19]. Casting the problem as a machine-learning binary classification problem, Potthast *et al.* [15] used manual inspection to inspire a feature set based on meta-data and content-level properties and built a classifier using logistic regression. Smets *et al.* [16] used Naïve Bayes applied to a bag-of-words model of the edit text. Chin *et al.* [19] delve deeper into the field of natural language processing by constructing statistical language models of an article from its revision history.

A different way of looking at edit content is the intuition that appropriate content somehow "belongs together." For example, *cohesion* can be measured via compression rates over consecutive editions of an article [16,18]. If inappropriate content is added to the article, then the compression level is lower than it would be for text which is similar to existing content. A drawback of this approach is that it tends to label as vandalism any large addition of material, regardless of its quality, while overlooking the small additions of insults, racial epithets, pranks, and spam that comprise a significant portion of vandalism.

The idea of using reputation systems to aid in vandalism detection was advanced in [20,21,22]. West *et al.* [8] apply the idea of reputations to editors and articles, as well as spatial groupings thereof — including geographical regions and topical categories.

Many previous works have some small dependence on metadata features [15,17,23], but only as far as it encoded some aspect of human intuition about vandalism. Drawing inspiration from email spam research, West *et al.* [8] demonstrated that the broader use of metadata can be very effective, suggesting that there are more indicators of vandalism than are apparent to the human eye.

The first systematic review and organization of features was performed by Potthast *et al.* [24] as part of the vandalism detection competition associated with PAN 2010. Potthast *et al.* conclude their analysis by building a meta-classifier based on all nine competition entries, and finds it significantly outperforms any single entry. As our own work will confirm, a diverse array of features is clearly beneficial when attacking the vandalism detection problem. Our work extends that of Potthast by concatenating entire feature vectors (not just the single variable output) and by analyzing the effectiveness of unique feature classes.

## 3   Vandalism Detection

On Wikipedia, every article is stored as a sequence of revisions in chronological order. Visitors to Wikipedia are shown the latest revision of an article by default; if they so choose, they can edit it, producing a new revision. Some of these revisions are *vandalism*. Vandalism has been broadly defined as any edit performed in bad faith, or with the intent to deface or damage Wikipedia. In this work, we do not concern ourselves with the definition of vandalism; rather, we use the PAN-WVC-10 corpus as our ground-truth. The corpus consists of over 32,000 edits (some 2,400 vandalism), each labeled by 3 or more annotators from Amazon Mechanical Turk. See [24] for additional details.

In order to detect vandalism, we follow a classical architecture: feature extraction, followed by data-trained classification. Features can be obtained from: (1) the revision itself, (2) from comparison of the revision against another revision (*i.e.*, a diff), or (3) from information derived from previous or subsequent revisions. For instance, the ratio of uppercase to lowercase characters inserted is one feature, as is the edit distance between a revision and the previous one on the same article. The feature vectors are then used to train and classify. As a classifier, we use the Random Forest[5] model [26]. We perform evaluation using 10-fold cross-validation over the entire PAN-WVC-10 corpus.

We consider two types of vandalism detection problem: immediate and historic. *Immediate* vandalism detection is the problem of detecting vandalism in the most recent revision of an article; *historic* detection is the problem of finding vandalism in any past revision. For immediate vandalism detection, one can only make use of the information available at the time a revision is committed. In particular, in immediate vandalism detection, information gathered from *subsequent* revisions cannot be used to decide whether a particular revision is vandalism or not. In contrast, historical vandalism detection permits the use of any feature. We propose one such possible feature: the implicit judgements made by later editors in deciding whether to keep some or all text previously added.

We divide our features into classes, according to the complexity required to compute them, and according to the difficulty of generalizing them across multiple languages. These classes are: Metadata, Text, Reputation, and Language, abbreviated as **M**, **T**, **R**, and **L**, respectively. Our work is based directly on the previous works of [6,7] and [8,9]. What follows is a discussion of representative features from each class. For a complete feature listing, see Table 1.

## 3.1   Metadata

*Metadata* (M) refers to properties of a revision that are immediately available, such as the identity of the editor, or the timestamp of the edit. This is an important class of features because it has minimal computational complexity. Beyond the properties of each revision found directly in the database (*e.g.* whether the editor is anonymous, used by nearly every previous work), there are some examples that we feel expose the unexpected similarities in vandal behavior:

- **Time since article last edited** [8]. Highly-edited articles are frequent targets of vandalism. Similarly, quick fluctuations in content may be indicative of edit wars or other controversy.
- **Local time-of-day** and **day-of-week** [8]. Using IP geolocation, it is possible to determine the *local* time when an edit was made. Evidence shows vandalism is most prominent during weekday "school/office hours."

---

[5] We used the Random Forest implementation available in the Weka Framework 3.7 [25], available at http://www.cs.waikato.ac.nz/ml/weka/

- **Revision comment length** [6,7,8]. Vandals decline to follow community convention by leaving either very short revision comments or very long ones.

## 3.2    Text

We label as *Text* (T) those language-independent features derived from analysis of the edit content. Therefore, very long articles may require a significant amount of processing. As the content of the edit is the true guide to its usefulness, there are several ideas for how to measure that property:

- **Uppercase ratio** and **digit ratio** [6,8]. Vandals sometimes will add text consisting primarily of capital letters to attract attention; others will change only numerical content. These ratios (and similar ones [6]) create features which capture behaviors observed in vandals.
- **Average** and **minimum edit quality** [7] (Historic only). Comparing the content of an edit against a future version of the article provides a way to measure the Wikipedia community's approval of the edit [17,22]. To address the issue of edit warring, the comparison is done against several future revisions. This feature uses edit distance (rather than the blunt detection of reverts) to produce an implicit quality judgement by later edits; see [22].

## 3.3    Language

Similar to text features, *Language* (L) features must inspect edit content. A distinction is made because these features require expert knowledge about the (natural) language. Thus, these features require effort to be re-implemented for each different language. Some of the features included in our analysis are:

- **Pronoun frequency** and **pronoun impact** [6]. The use of first and second-person pronouns, including slang spellings, is indicative of a biased style of writing discouraged on Wikipedia (non-neutral point-of-view). *Frequency* considers the ratio of first and second-person pronouns relative to the size of the edit. *Impact* is the percentage increase in first and second-person pronouns that the edit contributes to the overall article.
- **Biased** and **bad words** [6]. Certain words indicate a bias by the author (*e.g.* superlatives: "coolest", "huge"), which is captured by a list of regular expressions. Similarly, a list of bad words captures edits which appear inappropriate for an encyclopedia (*e.g.* "wanna", "gotcha") and typos (*e.g.* "seperate"). Both these lists have corresponding frequency and impact features that indicate how much they dominate the edit and increase the presence of biased or bad words in the overall article.

## 3.4    Reputation

We consider a feature in the *Reputation* (R) category if it necessitates extensive historical processing of Wikipedia to produce a feature value. The high cost of this computational complexity is sometimes mitigated by the ability to build on earlier computations, using incremental calculations.

- **User reputation** [7] (Historic only[6]) User reputation as computed by Wiki-Trust [22]. The intuition is that users who have a history of good contributions, and therefore high reputation, are unlikely to commit vandalism.
- **Country reputation** [8]. For anonymous/IP edits, it is useful to consider the geographic region from which an edit originates. This feature represents the likelihood that an editor from a particular country is a vandal, by aggregating behavior histories from that same region. Location is determined by geo-locating the IP address of the editor.
- **Previous** and **current text trust histogram** [7]. When high-reputation users revise an article and leave text intact, that text accrues reputation, called "trust" [7]. Features are, (1) the histogram of word trust in the edit, and (2) the difference between the histogram before, and after, the edit.

## 4    Experimental Results

In this section, we present results and discussion of our experiments using different combinations of meta-classifiers. Table 2 summarizes the performance of these subsets per the experimental setup described in Section 3. We present the results in terms of area under curve[7] (AUC) for two curves: the precision-recall curve (PR), and the receiver operating characteristics (ROC) curve. The results in terms of AUC-ROC are often presented for binary classification problem (which vandalism detection is), but AUC-PR better accounts for the fact that vandalism is a rare phenomenon [27], and offers a more discriminating look into the performance of the various feature combinations.

In Figure 1 we show precision-recall curves for each system, distinguishing between immediate and historic vandalism cases. Only [7] considers features explicitly for the historic cases. We find a significant increase in performance when transitioning from immediate to historical detection scenarios.

Analysis of our feature taxonomy, per Figure 2, leads to some additional observations in a comparison between immediate and historic vandalism tasks:

- Most obvious is the improvement in the performance of the Language (L) set, due entirely to the **next comment revert** feature. The feature evaluates whether the revision comment for the next edit contains the word "revert" or "rv," which is used to indicate that the prior edit was vandalism [7].

---

[6] In a live system, user reputation is available at the time a user makes an edit, and therefore, user reputation is suitable for immediate vandalism detection. However, since WikiTrust only stores the current reputation of users, *ex post facto* analysis was not possible for this study.

[7] http://mark.goadrich.com/programs/AUC/

[8] Note that performance numbers reported for [6] and [7] differ from those reported in [24] due to our use of 10-fold cross validation over the entire PAN2010 corpus and differences in ML models (*e.g.*, ADTree vs. Random Forest). We do not list the performance of the PAN 2010 Meta Detector because it was evaluated with an unknown subset of the PAN 2010 corpus, and is therefore not precisely comparable.

[9] Note that statistics for the "West *et al.*" system are strictly the metadata ones described in [8], and not the more general-purpose set used in the online tool [9].

**Table 1.** Comprehensive listing of features used, organized by class. Note that features in the "!Z" (not zero-delay) class are those that are only appropriate for historical vandalism detection.

| FEATURE | CLS | SRC | DESCRIPTION |
|---|---|---|---|
| IS_REGISTERED | M | [6,7,8] | Whether editor is anonymous/registered (boolean) |
| COMMENT_LENGTH | M | [6,7,8] | Length (in chars) of revision comment left |
| SIZE_CHANGE | M | [6,7,8] | Size difference between prev. and current versions |
| TIME_SINCE_PAGE | M | [7,8] | Time since article (of edit) last modified |
| TIME_OF_DAY | M | [7,8] | Time when edit made (UTC, or local w/geolocation) |
| DAY_OF_WEEK | M | [8] | Local day-of-week when edit made, per geolocation |
| TIME_SINCE_REG | M | [8] | Time since editor's first Wikipedia edit |
| TIME_SINCE_VAND | M | [8] | Time since editor last caught vandalizing |
| SIZE_RATIO | M | [6] | Size of new article version relative to new one |
| PREV_SAME_AUTH | M | [7] | Is author of current edit same as previous? (boolean) |
| REP_EDITOR | R | [8] | Reputation for editor via behavior history |
| REP_COUNTRY | R | [8] | Reputation for geographical region (editor groups) |
| REP_ARTICLE | R | [8] | Reputation for article (on which edit was made) |
| REP_CATEGORY | R | [8] | Reputation for topical category (article groups) |
| WT_HIST | R | [7] | Histogram of text trust distribution after edit |
| WT_PREV_HIST_N | R | [7] | Histogram of text trust distribution before edit |
| WT_DELT_HIST_N | R | [7] | Change in text trust histogram due to edit |
| DIGIT_RATIO | T | [6] | Ratio of numerical chars. to all chars. |
| ALPHANUM_RATIO | T | [6] | Ratio of alpha-numeric chars. to all chars. |
| UPPER_RATIO | T | [6] | Ratio of upper-case chars. to all chars. |
| UPPER_RATIO_OLD | T | [6] | Ratio of upper-case chars. to lower-case chars. |
| LONG_CHAR_SEQ | T | [6] | Length of longest consecutive sequence of single char. |
| LONG_WORD | T | [6] | Length of longest token |
| NEW_TERM_FREQ | T | [6] | Average relative frequency of inserted words |
| COMPRESS_LZW | T | [6] | Compression rate of inserted text, per LZW |
| CHAR_DIST | T | [6] | Kullback-Leibler divergence of char. distribution |
| PREV_LENGTH | T | [7] | Length of the previous version of the article |
| VULGARITY | L | [6] | Freq./impact of vulgar and offensive words |
| PRONOUNS | L | [6] | Freq./impact of first and second person pronouns |
| BIASED_WORDS | L | [6] | Freq./impact of colloquial words w/high bias |
| SEXUAL_WORDS | L | [6] | Freq./impact of non-vulgar sex-related words |
| MISC_BAD_WORDS | L | [6] | Freq./impact of miscellaneous typos/colloquialisms |
| ALL_BAD_WORDS | L | [6] | Freq./impact of previous five factors in combination |
| GOOD_WORDS | L | [6] | Freq./impact of "good words"; wiki-syntax elements |
| COMM_REVERT | L | [7] | Is rev. comment indicative of a revert? (boolean) |
| NEXT_ANON | !Z/M | [7] | Is the editor of the *next* edit registered? (boolean) |
| NEXT_SAME_AUTH | !Z/M | [7] | Is the editor of *next* edit same as current? (boolean) |
| NEXT_EDIT_TIME | !Z/M | [7] | Time between current edit and *next* on same page |
| JUDGES_NUM | !Z/M | [7] | Number of later edits useful for implicit feedback |
| NEXT_COMM_LGTH | !Z/M | [7] | Length of revision comment for *next* revision |
| NEXT_COMM_RV | !Z/L | [7] | Is *next* edit comment indicative of a revert? (boolean) |
| QUALITY_AVG | !Z/T | [7] | Average of implicit feedback from judges |
| QUALITY_MIN | !Z/T | [7] | Worst feedback from any judge |
| DISSENT_MAX | !Z/T | [7] | How close QUALITY_AVG is to QUALITY_MIN |
| REVERT_MAX | !Z/T | [7] | Max reverts possible given QUALITY_AVG |
| WT_REPUTATION | !Z/R | [7] | Editor rep. per WikiTrust (permitting future data) |
| JUDGES_WGHT | !Z/R | [7] | Measure of relevance of implicit feedback |

**Table 2.** Performance of all meta-classifier combinations

| Features | Immediate | | Historic | |
|---|---|---|---|---|
| | AUC-PR | AUC-ROC | AUC-PR | AUC-ROC |
| Adler et al.[8] | 0.61047 | 0.93647 | 0.73744 | 0.95802 |
| Mola-Velasco[12] | 0.73121 | 0.94567 | 0.73121 | 0.94567 |
| West et al.[9] | 0.52534 | 0.91520 | 0.52534 | 0.91520 |
| Language | 0.42386 | 0.74950 | 0.58167 | 0.86066 |
| Metadata | 0.43582 | 0.89835 | 0.66180 | 0.93718 |
| Reputation | 0.59977 | 0.92652 | 0.64033 | 0.94348 |
| Text | 0.51586 | 0.88259 | 0.73146 | 0.95313 |
| M+T | 0.68513 | 0.94819 | 0.81240 | 0.97121 |
| M+T+L | 0.76124 | 0.95840 | 0.85004 | 0.97590 |
| M+T+R | 0.76271 | 0.96315 | 0.81575 | 0.97140 |
| All | 0.81829 | 0.96902 | 0.85254 | 0.97620 |

- Both Metadata (M) and Text (T) show impressive gains in going from the *Immediate* task to the *Historic* task. For Metadata, our investigation points to NEXT_EDIT_TIME as being the primary contributor, as pages more frequently edited are more likely to be vandalized. For Text, the set of features added in the *historic* task all relate to the implicit feedback given by later editors, showing a correlation between negative feedback and vandalism.
- A surprise in comparing the feature sets is that the predictive power of [M+T] and [M+T+R] are nearly identical in the historic setting. That is, once one knows the future community reaction to a particular edit, there is much less need to care about the past performance of the editor. We surmise that bad actors quickly discard their accounts or are anonymous, so reputation would be useful in the *immediate* detection case, but is less useful in *historic* detection.

One of the primary motivations for this work was to establish the significance of Language (L) features as compared to other features, because language features are more difficult to generate and maintain for each language edition of Wikipedia. In the case of immediate vandalism detection, we see the interesting scenario of the AUC-PR for [M+T+L] being nearly identical to that of [M+T+R]. That is, the predictive power of Language (L) and Reputation (R) features is nearly the same when there are already Metadata (M) and Text (T) features present. The improvement when all features are taken together is indicative of the fact that Language (L) and Reputation (R) features capture different behavior patterns which only ocassionally overlap.

We chose to use the features of [6] as being representative of a solution focused on Language (L) features due to its top-place performance in the PAN 2010 competition [24]. Yet Figure 2 visualizes that the Language (L) class of features performs only marginally well. Inspection of Table 2 shows that Language (L)

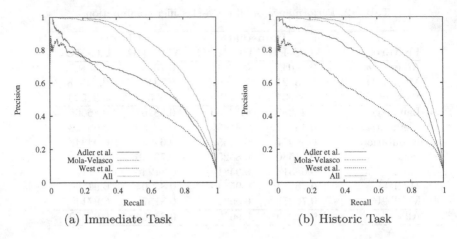

Fig. 1. Precision-Recall curves for the three systems and their combination

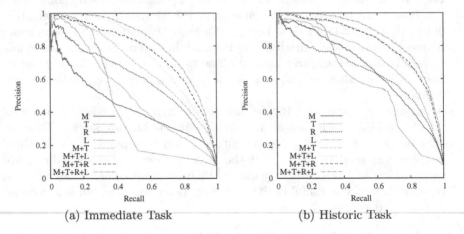

Fig. 2. Precision-Recall curves for feature categories

features have the worst PR-AUC, but the combined features of [6] have the highest performance. This suggests that the key to the performance beyond the that portion Language (L) features can detect lies in metadata and text features.

## 5    Conclusions

The success of a machine learning algorithm depends critically on the selection of features that are inputs to the algorithm. Although the previous works on the problem of Wikipedia vandalism detection utilize features from multiple categories, each work has individually focused predominantly on a single category.

We proposed that solving the vandalism detection problem requires a more thorough exploration of the available feature space. We combined the features of three previous works, each representing a unique dimension in feature selection. Each feature was categorized as either metadata, text, reputation, or language, according to the nature of how they are computed and roughly corresponding to their computational complexity.

We discovered that language features only provide an additional 6% of performance over the combined efforts of language-independent features. This has important ramifications for the development of vandalism detection tools across the other Wikipedia language editions. Moreover, our results outperform the winning system of the PAN 2010 competition, showing that the feature combination explored in this work considerably improves the state of the art (67% vs. 82% AUC). Finally, our meta-classifier could be suitable for the autonomous reversion of *some* bad edits – in a 99% precision setting, 30% recall was achieved.

**Acknowledgments.** The authors would like to thank Ian Pye of CloudFlare Inc. as well as Insup Lee and Sampath Kannan of the University of Pennsylvania. These contributors were integral in the development of the original/component systems. Additionally, Martin Potthast deserves acknowledgment for his development of the vandalism corpus and for generating interest in the vandalism detection problem. The authors from Universidad Politécnica de Valencia thank also the MICINN research project TEXT-ENTERPRISE 2.0 TIN2009-13391-C04-03 (Plan I+D+i). UPenn contributions were supported in part by ONR MURI N00014-07-1-0907. This research was partially supported by award 1R01GM089820-01A1 from the National Institute Of General Medical Sciences, and by ISSDM, a UCSC-LANL educational collaboration. The content is solely the responsibility of the authors and does not necessarily represent the official views of the National Institute Of General Medical Sciences or the National Institutes of Health.

# References

1. Wikimedia Foundation: Wikipedia (2010) [Online; accessed December 29, 2010]
2. Wikimedia Foundation: Wikistats (2010) [Online; accessed December 29, 2010]
3. Potthast, M.: Crowdsourcing a Wikipedia Vandalism Corpus. In: Proc. of the 33rd Intl. ACM SIGIR Conf. (SIGIR 2010). ACM Press, New York (July 2010)
4. Gralla, P.: U.S. senator: It's time to ban Wikipedia in schools, libraries, http://blogs.computerworld.com/4598/u_s_senator_its_time_to_ban_wikipedia_in_schools_libraries [Online; accessed November 15, 2010]
5. Olanoff, L.: School officials unite in banning Wikipedia. Seattle Times (November 2007)
6. Mola-Velasco, S.M.: Wikipedia Vandalism Detection Through Machine Learning: Feature Review and New Proposals. In: Braschler, M., Harman, D. (eds.) Notebook Papers of CLEF 2010 LABs and Workshops, Padua, Italy, September 22-23 (2010)
7. Adler, B., de Alfaro, L., Pye, I.: Detecting Wikipedia Vandalism using WikiTrust. In: Braschler, M., Harman, D. (eds.) Notebook Papers of CLEF 2010 LABs and Workshops, Padua, Italy, September 22-23 (2010)
8. West, A.G., Kannan, S., Lee, I.: Detecting Wikipedia Vandalism via Spatio-Temporal Analysis of Revision Metadata. In: EUROSEC 2010: Proceedings of the Third European Workshop on System Security, pp. 22–28 (2010)

9. West, A.G.: STiki: A Vandalism Detection Tool for Wikipedia (2010),
   `http://en.wikipedia.org/wiki/Wikipedia:STiki`
10. Wikipedia: User: AntiVandalBot – Wikipedia,
    `http://en.wikipedia.org/wiki/User:AntiVandalBot` (2010) [Online; accessed
    November 2, 2010]
11. Wikipedia: User:MartinBot – Wikipedia (2010),
    `http://en.wikipedia.org/wiki/User:MartinBot` [Online; accessed November 2,
    2010]
12. Wikipedia: User:ClueBot – Wikipedia (2010),
    `http://en.wikipedia.org/wiki/User:ClueBot` [Online; accessed November 2,
    2010]
13. Carter, J.: ClueBot and Vandalism on Wikipedia (2008),
    `http://www.acm.uiuc.edu/~carter11/ClueBot.pdf` [Online; accessed November
    2, 2010]
14. Rodríguez Posada, E.J.: AVBOT: detección y corrección de vandalismos en
    Wikipedia. NovATIca (203), 51–53 (2010)
15. Potthast, M., Stein, B., Gerling, R.: Automatic Vandalism Detection in Wikipedia.
    In: Macdonald, C., Ounis, I., Plachouras, V., Ruthven, I., White, R.W. (eds.) ECIR
    2008. LNCS, vol. 4956, pp. 663–668. Springer, Heidelberg (2008)
16. Smets, K., Goethals, B., Verdonk, B.: Automatic Vandalism Detection in
    Wikipedia: Towards a Machine Learning Approach. In: WikiAI 2008: Proceedings
    of the Workshop on Wikipedia and Artificial Intelligence: An Evolving Synergy,
    pp. 43–48. AAAI Press, Menlo Park (2008)
17. Druck, G., Miklau, G., McCallum, A.: Learning to Predict the Quality of Contribu-
    tions to Wikipedia. In: WikiAI 2008: Proceedings of the Workshop on Wikipedia
    and Artificial Intelligence: An Evolving Synergy, pp. 7–12. AAAI Press, Menlo
    Park (2008)
18. Itakura, K.Y., Clarke, C.L.: Using Dynamic Markov Compression to Detect Van-
    dalism in the Wikipedia. In: SIGIR 2009: Proc. of the 32nd Intl. ACM Conference
    on Research and Development in Information Retrieval, pp. 822–823 (2009)
19. Chin, S.C., Street, W.N., Srinivasan, P., Eichmann, D.: Detecting Wikipedia Van-
    dalism with Active Learning and Statistical Language Models. In: WICOW 2010:
    Proc. of the 4th Workshop on Information Credibility on the Web (April 2010)
20. Zeng, H., Alhoussaini, M., Ding, L., Fikes, R., McGuinness, D.: Computing Trust
    from Revision History. In: Intl. Conf. on Privacy, Security and Trust (2006)
21. McGuinness, D., Zeng, H., da Silva, P., Ding, L., Narayanan, D., Bhaowal, M.:
    Investigation into Trust for Collaborative Information Repositories: A Wikipedia
    Case Study. In: Proc. of the Workshop on Models of Trust for the Web (2006)
22. Adler, B., de Alfaro, L.: A Content-Driven Reputation System for the Wikipedia.
    In: WWW 2007: Proceedings of the 16th International World Wide Web Confer-
    ence. ACM Press, New York (2007)
23. Belani, A.: Vandalism Detection in Wikipedia: a Bag-of-Words Classifier Approach.
    Computing Research Repository (CoRR) abs/1001.0700 (2010)
24. Potthast, M., Stein, B., Holfeld, T.: Overview of the 1st International Competition
    on Wikipedia Vandalism Detection. In: Braschler, M., Harman, D. (eds.) Notebook
    Papers of CLEF 2010 LABs and Workshops, Padua, Italy, September 22-23 (2010)
25. Hall, M., Frank, E., Holmes, G., Pfahringer, B., Reutemann, P., Witten, I.: The
    WEKA Data Mining Software: An Update. SIGKDD Explorations 11(1) (2009)
26. Breiman, L.: Random Forests. Machine Learning 45(1), 5–32 (2001)
27. Davis, J., Goadrich, M.: The relationship between Precision-Recall and ROC
    curves. In: ICML 2006: Proc. of the 23rd Intl. Conf. on Machine Learning (2006)

# Costco: Robust Content and Structure Constrained Clustering of Networked Documents

Su Yan[1], Dongwon Lee[2], and Alex Hai Wang[3]

[1] IBM Almaden Research Center
San Jose, CA 95120, USA
[2] The Pennsylvania State University
University Park, PA 16802, USA
[3] The Pennsylvania State University Dumore, PA 18512, USA
syan@us.ibm.com,
{dongwon,hwang}@psu.edu

**Abstract.** Connectivity analysis of networked documents provides high quality link structure information, which is usually lost upon a content-based learning system. It is well known that combining links and content has the potential to improve text analysis. However, exploiting link structure is non-trivial because links are often noisy and sparse. Besides, it is difficult to balance the term-based content analysis and the link based structure analysis to reap the benefit of both. We introduce a novel networked document clustering technique that integrates the content and link information in a unified optimization framework. Under this framework, a novel dimensionality reduction method called COntent & STructure COnstrained (Costco) Feature Projection is developed. In order to extract robust link information from sparse and noisy link graphs, two link analysis methods are introduced. Experiments on benchmark data and diverse real-world text corpora validate the effectiveness of proposed methods.

**Keywords:** link analysis, dimensionality reduction, clustering.

## 1 Introduction

With the proliferation of the World Wide Web and Digital Libraries, analyzing "networked" documents has increasing challenge and opportunity. In addition to text content attributes, networked documents are correlated by links (e.g., hyperlinks between Web pages, citations between scientific publications etc.). These links are useful for text processing because they convey rich semantics that are usually independent of word statistics of documents [8].

Exploiting link information of networked documents to enhance text classification has been studied extensively in the research community [3,4,6,14]. It is found that, although both content attributes and links can independently form reasonable text classifiers, an algorithm that exploits both information sources has the potential to improve the classification [2,10]. Similar conclusion has been

A. Gelbukh (Ed.): CICLing 2011, Part II, LNCS 6609, pp. 289–300, 2011.

drawn for text clustering by a growing number of works [1,2,7,11,13,19]. However, the fundamental question/challenge still remains

*How to effectively couple the content and link information to get the most of both sources?*

Existing work either relies on heuristic combination of content and links, or assumes a link graph to be dense or noise-free, whereas link graphs of real-world data are usually sparse and noisy. To this end, we propose a novel clustering approach for networked documents based on the *COntent and STructure COnstrained (Costco) feature projection*, and cluster networked documents from a *dimension reduction* perspective. Compared to existing work, Costco has the following advantages

1. Couples content and link structure in a unified objective function, and hence avoids heuristic combination of the two information sources;
2. Alleviates the curse-of-dimensionality problem by constrained dimensionality reduction;
3. Does not rely on dense link structure and is robust to noisy links, which suits the method well for real-world networked data;
4. Is very simple to implement, so can be used for exploratory data analysis before any complicated in-depth analysis.

## 2    Related Work

The techniques for analyzing networked documents can be broadly categorized as content-based, link-based, and combined approaches. As more and more work confirm the effectiveness of using link structure to enhance text analysis, novel approaches to merge content and link information attract increasing interest in the text mining domain.

[6] proposes generative probabilistic models for document content and links. [4] uses factorized model to combine the content model and the link model. [14] tackles the problem by using the relaxation labeling technique. Besides the vast amount of work on link-enhanced text classification, there are increasing number of work focusing on link-enhanced clustering. [1] extends the relaxation labeling method to text clustering. The cluster assignment for each document is not only determined by content attributes, but is also influenced by the assignments of neighborhood documents on the link graph. [2] focuses on clustering scientific literature, and weights words based on link information. [11] extends the term-based feature space with in-link and out-link features. [7] treats networked document clustering as a spectral graph partitioning problem. [13] shares a similar idea of adopting graph-partitioning techniques, but merges content and links by weighting the link graph with a content similarity metric. Our technique is orthogonal to all the existing work by clustering networked documents from a dimension reduction perspective and is robust to sparse and noisy link graphs.

# 3   Main Proposal

## 3.1   Problem Statement

Text data, usually represented by the bag-of-words model, have extremely high-dimensional feature space (1000+). A feature projection approach can greatly reduce the feature space dimensionality while still preserve discriminative information. In the networked environment, seman-tically related documents tend to cite each other. If the link structure is noise-free and dense enough, then link-based clustering augmented by textual content [1,2], will generally yield well separated clusters . However, the link structure is often noisy and sparse. For instance, many links in Web pages are for navigational purpose and therefore not indicators of semantic relations [15]. We introduce an algorithm to bridget the disconnect between text and link structure from a feature projection perspective.

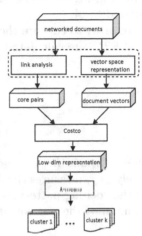

The overall clustering framework is outlined in Figure 1. Given networked documents, two preprocessing steps are performed. On the one hand, link analysis is performed to extract *core pairs*, which are pairs of documents strongly correlated with each other according to the link structure. On the other hand, the vector space model is employed to convert documents into high-dimensional vectors. Each dimension is a

**Fig. 1.**    Framework    of Costco-based    networked document clustering

word after preprocessing (stopping, stemming etc.). Core pairs and document vectors are then input into the feature projection module Costco. The generated low-dimensional data are partitioned by the traditional $k$-means clustering method into $k$ clusters, where $k$ is the desired number of clusters provided by users.

## 3.2   Local Link Analysis

The link graphs of real-world networked documents are usually sparse and noisy. Instead of naively assuming a pair of connected documents being similar in topic, we need schemes to extract more robust link information from the graph. A local link analysis scheme is introduced in this section.

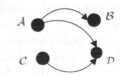

We model a link graph as *directed and unweighted*, denoted by $G(\mathbb{V}, \mathbb{E})$, where $\mathbb{V}$ is the set of the vertices/documents, and $\mathbb{E}$ is the set of edges/links between vertices. If document $d_i$ links to/cites document $d_j$, then there is an edge of unit weight starting from

**Fig. 2.**    Cociting    vs. Cocited

$d_i$ and pointing to $d_j$. Let matrix $L \in \mathbb{R}^{n \times n}$, where $n$ is the number of documents, be the corresponding *link matrix* defined as

$$L_{i,j} = \begin{cases} 1 & d_i \ cites \ d_j \\ 0 & otherwise. \end{cases} \qquad (1)$$

$L$ embodies two types of document concurrences: *cociting* and *cocited*, as illustrated in Figure 2. For example, both $\mathcal{A}$ and $\mathcal{C}$ cites $\mathcal{D}$, and $\mathcal{B}$ and $\mathcal{D}$ are being cocited by $\mathcal{A}$.

In order to capture the concurrences, two adjacency matrices $X \in \mathbb{R}^{n \times n}$ and $Y \in \mathbb{R}^{n \times n}$ are calculated

$$X_{i,j} = \frac{|L_{i*} \cap L_{j*}|}{|L_{i*} \cup L_{j*}|}, \quad 0 \le X_{i,j} \le 1 \qquad (2)$$

$$Y_{i,j} = \frac{|L_{*i} \cap L_{*j}|}{|L_{*i} \cup L_{*j}|}, \quad 0 \le Y_{i,j} \le 1 \qquad (3)$$

where $L_{i*}$ and $L_{*i}$ represent the $i$-th row vector and column vector of $L$ respectively. $X_{i,j}$ measures the Jaccard similarity of two documents $d_i$ and $d_j$ in terms of the cociting pattern, and $Y_{i,j}$ measures the similarity of the cocited pattern. Combining the two concurrences patterns, we have

$$Z = \alpha X + (1 - \alpha)Y \qquad (4)$$

where $\alpha \in [0, 1]$ is the parameter that controls the contribution of each individual link pattern to the overall structure-based similarity. Given $Z$, the set $\mathbb{C}$ of core pairs is then defined as

$$\mathbb{C} = \{(d_i, d_j) | Z_{i,j} > \theta\} \qquad (5)$$

where $\theta$ is a threshold that controls the reliability of link-based similarities.

### 3.3    Global Link Analysis

The link analysis scheme introduced in the previous section is a "local" method in the sense that for any query vertex/document in the graph, only the links between the query vertex and its direct neighbors are considered. Local analysis can miss some informative document pairs. For example in Figure 3, the relations among $\mathcal{A}$, $\mathcal{B}$, $\mathcal{D}$ and $\mathcal{E}$ are lost.

In the global scheme, we define a Markov random walk on the link graph. The link graph is modeled as *undirected and weighted*, denoted as $\tilde{G} = (\tilde{V}, \tilde{E})$. If

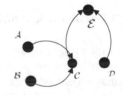

**Fig. 3.**    Local    method misses informative pairs

there is a link between two documents $d_i$ and $d_j$, we consider a relation (thus an edge) exits between them, no matter who starts the link. The edge is further weighted by the pairwise similarity $\mathfrak{D}(d_i, d_j)$ of the two documents. Let matrix $W \in \mathbb{R}^{n \times n}$, where $w_{i,j} = \mathfrak{D}(d_i, d_j)$, be the weight matrix. The one-step transition probabilities $p_{ik}$, which are the probabilities of jumping from any state (vertex) $i$ to one of its adjacent state $k$, are obtained directly from these weights $p_{ik} = W_{ik}/\sum_j W_{ij}$. We can organize the one step transition probabilities as a matrix $P$ whose $i, k$-th entry is $p_{ik}$.

Due to the sparseness of a link graph, two documents that are strongly correlated in topics may not be linked together. For example, a scientific article can not cite all the related work, and several Web pages with similar topics may scatter in the Web without any link among them. To remedy this problem, for each vertex whose degree is below the average, we add artificial links between the vertex and its $s$ nearest neighbors where $s$ is a small number.

For the augmented link graph, the transition matrix $P$ has the property that $Pe = e$, i.e., $P$ is stochastic, where $e$ is the vector with all 1 elements. We can now naturally define the Markov random walk on the undirected graph $\tilde{G}$ associated with $P$. The relation between two documents is evaluated by an important quantity in Markov chain theory, the *expected hitting time* $h(j|i)$, which is the expected number of steps for a random walk started at state $i$ to enter state $j$ for the first time. Formally, $h(j|i)$ is defined as

$$\begin{cases} h(i|i) = 0 \\ h(j|i) = 1 + \sum_{k=1}^{n} p_{ik} h(j|k) \ i \neq j \end{cases} \tag{6}$$

The choice of using expected hitting time to evaluate the correlation between two documents is justified by the desired property that the hitting time from state $i$ to state $j$ decreases when the number of paths from $i$ to $j$ increases and the lengths of the paths decrease. The core pairs can be naturally defined as

$$\mathbb{C} = \{(d_i, d_j) | (h(j|i) + h(i|j))/2 < \gamma\} \tag{7}$$

for some threshold $\gamma$.

### 3.4   Content and Structure Constrained Feature Projection (Costco)

Let matrix $D \in \mathbb{R}^{f \times n}$ be the document-term matrix where each column $d_i$ is a vector in the $f$-dimensional space. Let $\{(d_{j,1}, d_{j,2})\}_{j=1...m}$ be the set of $m$ document pairs that have been identified as core pairs at the link analysis step. Since these pairs of documents are strongly connected according to the link structure, there is a high probability that a core pair of documents are also semantically similar. We then desire a projection direction, such that any two documents of a core pair will be *more similar* to each other after being projected along the direction. To achieve this goal, we can minimize the variance between a pair of documents. Let us define the covariance matrix $V$ to encode the pooled variances for all the core pairs

$$V = \frac{1}{m} \sum_{\{(d_{j,1}, d_{j,2})\} \in \mathbb{C}} (d_{j,1} - d_{j,2})(d_{j,1} - d_{j,2})^T \tag{8}$$

Then the desired projection is

$$S^* = \arg \min_{S} Tr(S^T V S) \tag{9}$$

where $S \in \mathbb{R}^{f \times r}$ denotes the optimal transformation matrix, $r$ is the desired subspace dimensionality provided by users, and $Tr(\cdot)$ is the *trace* of a square matrix, defined as the summation of the diagonal elements.

Directly minimizing Eq. 9 leads to trivial solutions. For example, if the entire data set is projected to one point, then the covariance between core pair documents is minimized. To avoid trivial solution, we can put constrains on the variance of the entire data set to prevent all the data points huddle together. The covariance matrix of the entire data set is defined as

$$U = \frac{1}{n} \sum_{i=1}^{n} (d_i - \mu)(d_i - \mu)^T \tag{10}$$

where $\mu = \sum_{i=1}^{n} d_i$ is the global mean. Accordingly, we define the following objective

$$\begin{aligned} S^* &= \arg\max_{S} Tr \frac{S^T U S}{S^T V S} \\ &= \arg\max_{S} Tr((S^T V S)^{-1}(S^T U S)) \end{aligned} \tag{11}$$

The objective function defines a linear feature projection direction that both maximally preserves the variations of the entire data set and minimizes the total variances of core pairs. Simply put, after being projected along the optimal projection direction, the documents that are strongly connected (according to link structure) will be more similar to each other, while the rest documents are still well separated.

After the transformation matrix $S$ is solved, the high-dimensional ($f$-dim) data can be optimally represented in the $r$-dim subspace as $\widehat{D} = S^T D$, where $\widehat{D} \in \mathbb{R}^{r \times n}$, $r \ll f$. The optimization problem of Eq. 11 is a general eigenvector problem. Usually a regularization term is added to solve an ill-posed problem or to prevent overfitting [12]. We skip detailed discussion about it due to space limit. The overall clustering scheme is outlined in Algorithm 1.

---

**Algorithm 1.** Networked Document Clustering Based on Costco.

---

**Input**  : A set of $n$ networked documents
              Desired # clusters $k$
              Desired # dimensionality $r$
**Output**: a set of clusters
**begin** link analysis
  | Extract *core pairs* $\mathbb{C}$ by local link analysis (Eq. 5)
  |  or global link analysis (Eq. 7)
**end**
**begin** content analysis
  | Represent $n$ documents using vector space model to get $D \in \mathbb{R}^{f \times n}$;
**end**
Construct covariance matrix $U$ (Eq. 10);
Construct covariance matrix $V$ (Eq. 8);
Solve Eq. 11 to get low-dimensional data as $\widehat{D} = S^T D$;
Clustering low-dimensional data: $k$-means($\widehat{D}$, $k$);
**return** a set of clusters;

---

**Table 1.** UCI data sets

| Datasets | # classes | # instances | # features |
|---|---|---|---|
| balance | 3 | 625 | 4 |
| vehicle | 4 | 846 | 18 |
| breast-cancer | 2 | 569 | 30 |
| sonar | 2 | 208 | 60 |
| ionoshpere | 2 | 351 | 34 |
| soybean | 4 | 47 | 35 |

**Table 2.** 20-Newsgroups data sets

| Datasets | topics | # features |
|---|---|---|
| difficult | comp.windows.x, comp.os.ms-windows.mis, comp.graphics | 3,570 |
| mediocre | talk-politicis.misc, talk.politics. guns, talk.politics.mideast | 4,457 |
| easy | alt.atheism, sci.space, rec.sprot.baseball | 4,038 |

**Table 3.** Reuters data sets

| Datasets | # classes | # instances | # features |
|---|---|---|---|
| reu4 | 4 | 400 | 2,537 |
| reu5 | 5 | 500 | 2,257 |
| reu6 | 6 | 600 | 2,626 |

**Table 4.** WebKB and Cora Data sets

| Datasets | # classes | # instances | # features | # links |
|---|---|---|---|---|
| WebKB | 5 | 877 | 1,703 | 1,608 |
| Cora | 7 | 2,708 | 1,433 | 5,429 |

# 4  Performance Evaluations

## 4.1  Set-up

The proposed networked document clustering framework has been evaluated on 6 UCI benchmark data sets[1], 3 data sets generated from the 20-Newsgroups document corpus[2], 3 data sets generated from the Reuters document corpus[3], the WebKB data sets[4] of hypertext, and the Cora data set[4] of scientific publications. Statistics of the data sets are listed in Table 1 to Table 4.

For the 20-Newsgroups document corpus, 3 data sets are generated, each of which is a balanced combination of documents about 3 topics. Depending on the similarities in the topics, the 3 data sets show various levels of clustering difficulties. To generate the Reuters data sets, for a given number of topics $b$, firstly, $b$ topics are randomly sampled, and then about 100 documents of each topic are randomly sampled and mixed together. Table 3 shows the average statistics of 5 sets of independently generated data sets.

Spherical $k$-means [5] the Normalized Cut (NC) [18][5] are chosen as baseline clustering methods. Both techniques have shown success in clustering text data [9]. Costoco and nr-Costco are our proposals with and without regularization respectively. For competing dimensionality reduction techniques, we compare to two well-known unsupervised dimensionality reduction methods, the principal component analysis (PCA)[16] which is a linear method and the locally linear embedding (LLE)[17][6] which is a non-linear method. For competing techniques that couple content and link information, we implement *Augmented*[11] and

---

[1] http://archive.ics.uci.edu/ml/

[2] http://people.csail.mit.edu/jrennie/20Newsgroups/

[3] http://www.daviddlewis.com/resources/testcollections/reuters21578/

[4] http://www.cs.umd.edu/~sen/lbc-proj/LBC.html

[5] Original authors' implementation is used
http://www.cis.upenn.edu/~jshi/software/

[6] Original authors' implementation is used
http://www.cs.toronto.edu/~roweis/lle/

*L-Comb* [7,13]. *Augmented* augments the content-based vector space model with link features and applies $k$-means to the augmented document vectors. *L-Comb* linearly combines content similarity with link similarities and uses NC as the underlying clustering scheme. The method *Links* is a $k$-means clustering based on link similarity only.

To avoid biased accuracy results using a single metric, we used three widely-adopted clustering evaluation metrics: 1) *Normalized Mutual Information* (NMI), 2) *Rand Index* (RI), and 3) *F-measure*.

## 4.2   Controlled Experiments

In controlled experiment, given a data set, artificial links are generated and inserted between data points. In this way, we can control the density of a link graph as well as the error rate of links, and evaluate a method with various settings. Every method that uses link information will take use of all the available links instead of pruning out some links with preprocessing steps. With controlled experiments, clustering schemes can be evaluated in a fair setting without being influenced by preprocessing.

**Table 5.** Performance on UCI data sets measured by RI and F (noise-free)(best results are bold-faced)

| Datasets | # of links | FF(kmeans) | PCA | LLE | Augmented | FF(NC) | L-Comb(NC) | Costco | nr-Costco |
|---|---|---|---|---|---|---|---|---|---|
| balance | | 0.1806 | 0.6177 | 0.5730 | 0.5911 | 0.6706 | 0.6772 | **0.7151** | 0.7132 |
| vehicle | | 0.6462 | 0.6408 | 0.6507 | 0.6431 | 0.6709 | 0.6761 | **0.7404** | 0.7180 |
| breast-cancer | 400 | 0.7504 | 0.7504 | 0.6356 | 0.7504 | 0.7554 | 0.7541 | **0.8008** | 0.7486 |
| sonar | RI | 0.5032 | 0.5032 | 0.5031 | 0.5041 | 0.5043 | 0.5046 | **0.6700** | 0.5749 |
| ionosphere | | 0.5889 | 0.5889 | 0.5933 | 0.5889 | 0.5841 | 0.5841 | **0.6509** | 0.6196 |
| soybean | | 0.8283 | 0.8291 | 0.7761 | 0.9065 | 0.8372 | 0.8372 | **1.0000** | **1.0000** |
| balance | | 0.4629 | 0.5010 | 0.4506 | 0.4658 | 0.5686 | 0.5771 | **0.6290** | 0.6270 |
| vehicle | | 0.3616 | 0.3650 | 0.3597 | 0.3635 | 0.3594 | 0.3730 | **0.5365** | 0.4785 |
| breast-cancer | 400 | 0.7878 | 0.7878 | 0.6520 | 0.7878 | 0.7914 | 0.7905 | **0.8330** | 0.7866 |
| sonar | F | 0.5028 | 0.5028 | 0.6042 | 0.5064 | 0.5041 | 0.5048 | **0.6828** | 0.5945 |
| ionosphere | | 0.6049 | 0.6049 | 0.6580 | 0.6049 | 0.5997 | 0.5997 | **0.7346** | 0.7188 |
| soybean | | 0.6761 | 0.6805 | 0.5485 | 0.8282 | 0.6716 | 0.6716 | **1.0000** | **1.0000** |

To generate artificial links, we sample the cluster membership relation of pairs of documents and uniformly pick $x$ pairs to add links in. Given an error rate $e$ of links, we control the samples such that $\lceil x * e \rceil$ pairs of documents belong to different topic, which means these links are noise.

**Coupling Content and Links.** We first fix the error rate of links to be zero $e = 0$, and vary graph density by introducing $x = 100$ to 800 links between documents. This experiment measures the performance of a method in the noise-free setting with various levels of graph density. Figure 4, 5 and 6 show the clustering performance measured by NMI for the UCI, 20-Newsgroups, and Reuters data sets respectively. Table 5, 6 and 7 show the same result measured by RI and F score, with fixed 400 pairs of links. For all the data sets and different graph density levels, Costco consistently and significantly outperforms other competing

**Table 6.** Performance on 20-Newsgroup data sets measured by RI and F (noise-free) (best results are bold-faced)

| Datasets | # links | FF(kmeans) | PCA | Augmented | (FF)NC | L-Comb(NC) | Costco | nr-Costco |
|---|---|---|---|---|---|---|---|---|
| difficult | | 0.5231 | 0.3910 | 0.4111 | 0.4493 | 0.4506 | **0.7868** | 0.5543 |
| mediocre | 400 | 0.5865 | 0.4579 | 0.4674 | 0.7105 | 0.7499 | **0.9375** | 0.6488 |
| easy | RI | 0.6858 | 0.2350 | 0.1610 | 0.9251 | **0.9431** | 0.9256 | 0.5565 |
| difficult | | 0.4424 | 0.4792 | 0.4786 | 0.4681 | 0.4660 | **0.7157** | 0.5444 |
| mediocre | 400 | 0.5299 | 0.4926 | 0.5088 | 0.6686 | 0.7072 | **0.9064** | 0.5978 |
| easy | F | 0.8375 | 0.4725 | 0.4725 | 0.9781 | **0.9833** | 0.9746 | 0.6370 |

**Table 7.** Performance on Reuters data sets measured by RI and F (noise-free) (best results are bold-faced)

| Datasets | # links | FF(kmeans) | PCA | Augmented | (FF)NC | L-Comb(NC) | Costco | nr-Costco |
|---|---|---|---|---|---|---|---|---|
| Reu4 | | 0.6422 | 0.6694 | 0.6227 | 0.8141 | 0.8241 | **0.9891** | 0.8996 |
| Reu5 | 400 | 0.8172 | 0.7484 | 0.6626 | 0.8358 | 0.8405 | **0.9781** | 0.8973 |
| Reu6 | RI | 0.8563 | 0.6127 | 0.5433 | 0.9046 | 0.8791 | **0.9888** | 0.8809 |
| Reu4 | | 0.4932 | 0.5125 | 0.5297 | 0.6842 | 0.6977 | **0.9779** | 0.8323 |
| Reu5 | 400 | 0.6084 | 0.5285 | 0.4921 | 0.642 | 0.6403 | **0.0442** | 0.7701 |
| Reu6 | F | 0.6092 | 0.3966 | 0.3596 | 0.7240 | 0.6882 | **0.9657** | 0.6974 |

**Table 8.** Perormance on Cora and WebKB data sets (best results are bold-faced)

| Datasets | kmeans | PCA | Costco | nr-Costco | Links | Augmented | L-Comb(NC) |
|---|---|---|---|---|---|---|---|
| Cornell | 0.2163 | 0.3058 | **0.3809** | 0.2054 | 0.1365 | 0.2105 | 0.3544 |
| Texas | 0.2276 | 0.3291 | 0.3755 | 0.2163 | 0.1643 | 0.3149 | **0.4121** |
| Wisconsin | 0.3977 | 0.4067 | **0.4846** | 0.2609 | 0.0977 | 0.3982 | 0.4592 |
| Washington | 0.3469 | 0.3352 | **0.3885** | 0.1599 | 0.1991 | 0.3221 | 0.3404 |
| Cora | 0.1361 | 0.1592 | **0.3712** | 0.1631 | 0.0336 | 0.1496 | 0.1817 |

methods. Notice that, L-Comb and Augmented improve clustering accuracy for some data sets i.e., *vehicle*, *balance*, *easy*, but do not consistently perform well for all the data sets.

**Robustness to link errors.** Follow a similar setting of the previous experiment, we now fix the density of link graphs to have $x = 400$ pairs of links, but vary the error rate $e$ of links from 0 to 1. Figure 7 shows the behavior of Costco for 3 representative data sets (results on other data sets show similar patterns and thus omitted). As long as most of the links are informative (i.e., the percentage of noisy links is below 50%), without any link-pruning preprocessing steps, regularized Costco always improve clustering accuracy. These results indicate the robustness of Costco to noisy link graphs.

**Local vs. Global Link Analysis.** In this experiment, instead of using all the available links, Costco adopts the local and global link analyses to extract robust core pairs of documents and does dimensionality reduction accordingly. With fixed 400 links and an error rate of 0.5, Figure 8 shows the clustering

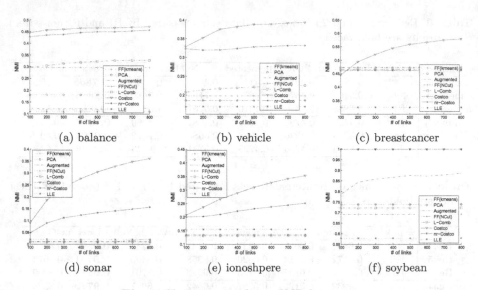

(a) balance     (b) vehicle     (c) breastcancer

(d) sonar     (e) ionoshpere     (f) soybean

**Fig. 4.** Clustering results on UCI data sets

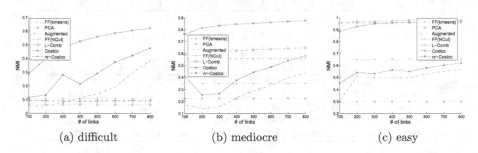

(a) difficult     (b) mediocre     (c) easy

**Fig. 5.** Clustering results on 20 Newsgroups data sets

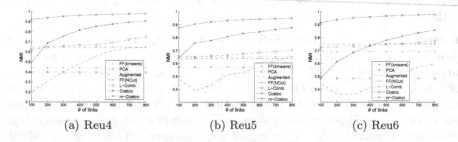

(a) Reu4     (b) Reu5     (c) Reu6

**Fig. 6.** Clustering results on Reuters data sets

results. In most cases, both link analysis methods can prune noise in links and improve clustering performance. Global link analysis usually outperforms local analysis as can be expected.

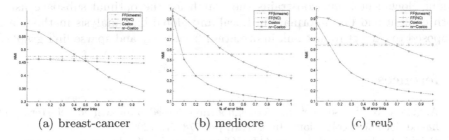

(a) breast-cancer          (b) mediocre          (c) reu5

**Fig. 7.** Clustering results on Reuters data sets

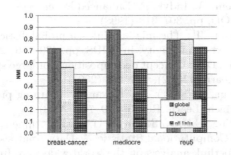

**Fig. 8.** Link analysis: global vs. local methods

## 4.3   Unrestrained Experiments

We evaluate all the methods with real-world networked documents. Experimental results are are shown in Table 8. Basically, similar patterns to controlled experiments are observed. For example, in most cases, Costco outperforms competing clustering methods and dimensionality reduction methods. The regularization improves the robustness in clustering performance, and dimensionality reduction in general alleviates the curse-of-dimensionality problem related to text data and generates more accurate data partitions. Note that, because all our data sets have very sparse and noisy link structures, the clustering method *Links*, which entirely relies on link structures, has the worst performance. But when combining link structure with content information, all the three content and link coupling techniques improve clustering performance. This observation confirms the usability of link structure (can be sparse and noisy) in text analysis.

## 5   Conclusion

A novel clustering model for networked documents is proposed. The Costco feature projection method is designed to represent high dimensional text data in an optimal low-dimensional subspace, and adopts the traditional $k$-means clustering method to partition the reduce-dimension data. Instead of using a stiff weighted combination of content-based and link-based similarities, Costco explores the correlation between the link structure and the semantic correlations among documents, and constrains the search for the optimal subspace using both content and link information. Local and global link analysis methods are proposed to extract robust link information from noisy and sparse link graphs.

# References

1. Angelova, R., Siersdorfer, S.: A neighborhood-based approach for clustering of linked document collections. In: CIKM, pp. 778–779 (2006)
2. Bolelli, L., Ertekin, S., Giles, C.L.: Clustering scientific literature using sparse citation graph analysis. In: PKDD, pp. 30–41 (2006)
3. Chakrabarti, S., Dom, B., Indyk, P.: Enhanced hypertext categorization using hyperlinks. In: SIGMOD, pp. 307–318 (1998)
4. Cohn, D.A., Hofmann, T.: The missing link - a probabilistic model of document content and hypertext connectivity. In: NIPS, pp. 430–436 (2000)
5. Dhillon, I.S., Modha, D.S.: Concept decompositions for large sparse text data using clustering. Mach. Learn. 42(1-2), 143–175 (2001)
6. Getoor, L., Friedman, N., Koller, D., Taskar, B.: Learning probabilistic models of link structure. J. Mach. Learn. Res. 3, 679–707 (2003)
7. He, X., Zha, H., Ding, C.H.Q., Simon, H.D.: Web document clustering using hyperlink structures. Computational Statistics & Data Analysis 41(1), 19–45 (2002)
8. Henzinger, M.: Hyperlink analysis on the world wide web. In: Hypertext, pp. 1–3 (2005)
9. Ji, X., Xu, W.: Document clustering with prior knowledge, pp. 405–412 (2006)
10. Menczer, F.: Lexical and semantic clustering by web links. JASIST 55(14), 1261–1269 (2004)
11. Modha, D.S., Spangler, W.S.: Clustering hypertext with applications to web searching. In: Hypertext, pp. 143–152 (2000)
12. Neumaier, A.: Solving ill-conditioned and singular linear systems: A tutorial on regularization. SIAM Review 40, 636–666 (1998)
13. Neville, J., Adler, M., Jensen, D.: Clustering relational data using attribute and link information. In: Proceedings of the IJCAI Text Mining and Link Analysis Workshop (2003)
14. Oh, H.-J., Myaeng, S.H., Lee, M.-H.: A practical hypertext catergorization method using links and incrementally available class information. In: SIGIR, pp. 264–271 (2000)
15. Park, H.W., Thelwall, M.: Hyperlink analyses of the world wide web: A review. J. Computer-Mediated Communication 8(4) (2003)
16. Pearson, K.: On lines and planes of closest fit to systems of points in space. Philosophical Magazine 2(6), 559–572 (1901)
17. Roweis, S.T., Saul, L.K.: Nonlinear dimensionality reduction by locally linear embedding. Science 290, 2323–2326 (2000)
18. Shi, J., Malik, J.: Normalized cuts and image segmentation (2000)
19. Wang, Y., Kitsuregawa, M.: Evaluating contents-link coupled web page clustering for web search results. In: CIKM, pp. 499–506 (2002)

# Learning Predicate Insertion Rules
# for Document Abstracting

Horacio Saggion

TALN
Department of Information and Communication Technologies
Universitat Pompeu Fabra
C/Tanger 122-140 Campus de la Comunicación Poble Nou - Barcelona
08018 - Spain
horacio.saggion@upf.edu

**Abstract.** The insertion of linguistic material into document sentences
to create new sentences is a common activity in document abstracting.
We investigate a transformation-based learning method to simulate this
type of operation relevant for text summarization. Our work is framed
on a theory of transformation-based abstracting where an initial text
summary is transformed into an abstract by the application of a number
of rules learnt from a corpus of examples. Our results are as good as
recent work on classification-based predicate insertion.

## 1 Introduction

The problem of generating summaries by automatic means started in the early
fifties [16] and continues nowadays to be a research topic receiving lot of attention [12,31,23,17,28,10]. The problem of generating "abstracts" – summaries
containing linguistic material not necessarily present in the document to be
summarized – has however received comparatively less attention. In this work
we aim at simulating the way abstracts are produced and try to capture from
textual data models of abstract production [26,11]. An example of the kind of
abstract we aim to produce is shown in Figure 1. It is an abstract from the ERIC
abstracting database which contains information extracted from the abstracted
document together with rhetorical predicates inserted during abstract writing.
These predicates inserted into the abstract by professional abstractors have specific communicative functions such as introducing the topic of the document,
elaborating information, discussing particular issues, concluding, etc. used sometimes to improve the abstract and make it more objective [20]. Here we focus on
this relatively new problem of combining document fragments with a limited set
of linguistic expressions to create quasi-extractive summaries. The inserted predicates "glue" together the extracted fragments, thus creating a quasi-extractive
summary. It is important to note that predicates can be prepended or appended
to the sentence fragments, in the later case using a passive construction (e.g.
"The state program in Rode Island is outlined"), note however, that we have

A. Gelbukh (Ed.): CICLing 2011, Part II, LNCS 6609, pp. 301–312, 2011.
© Springer-Verlag Berlin Heidelberg 2011

Describes a group of unconventional text retrieval systems which improve on conventional retrieval strategies by using innovative software and hardware to increase ... Says the software systems, AIDA, CLARIT, Metamorph, SIMPR, ... Reports that the hardware systems, CAFS-ISP, the Connection Machine... are all based on a parallel architecture ... Concludes that the major advantages of these software and hardware systems are that they either retrieve documents on the basis of language, ....

**Fig. 1.** Professional Abstracts with Inserted Predicates from ERIC Abstracting Service

concentrated on abstracts whose predicates have been prepended, and passive constructions will be dealt with in future work.

This abstracting problem was recently introduced and addressed as classification [25] where the predicates to be inserted were predicted based on sentence content and sentence context, reported prediction accuracy for the problem was 60%. Here, instead, we frame our work in a transformation-based learning methodology which is grounded in two observations from human abstracting studies:

1. Professional abstractors have *an internalised structure of abstracts* and rely on set of recurring lexical clues which reveal the presence and nature of components in the abstract 's structure [14];
2. The last stage in abstract writing consist on the application of *editing operations* (e.g., local revisions) tranforming an initial draft abstract into a consistent unit [4].

These two observations are used to ground a transformation-based learning methodology for the simulation of some of the processes observed in human abstracting. Observation (1) corresponds to the existence of a default or baseline abstract structure while observation (2) corresponds to the refinement/edition of the structure based on additional knowledge. These observations give rise to a method of generating predicates based on a Transformation-based learning (TBL) method where an initial baseline (e.g. a typical initial structure) is transformed into a final structure by the ordered application of a set of re-writing rules learned from experience. TBL has been applied before to a number of natural language processing problems such as parts-of-speech tagging [2], text chunking [24], clause boundary identification [6], and dialogue act tagging [29]. To the best of our knowledge, this is the first attempt to use TBL in a text summarization task. Additionally, we adopt the methodology proposed in [19] to overcome the template creation problem associated with TBL. As it will be shown in the experiments reported in this work, this technique achieves interesting results comparable to classification systems. However our main objective here is to demonstrate the applicability of the method to this new summarization task.

The rest of the paper is organized in the following way: In Section 2 we describe the corpus we use and the computational tools for corpus analysis and feature computation; then in Section 3 we describe how we apply Transformation-based learning to text abstracting. Section 4 presents a series of cross-validation

```
                        Decision tree structure

J48 pruned tree
------------------

noun_ford (-1) = ...
|    tok_1 (-1) = ...
|    |   tok_0 (0) = ...
....
|    tok_1 (-2)= ...
|    |   predicate (-1) = ...

         Generated feature sequences for template induction
noun_ford(-1) tok_1 (-1);
noun_ford(-1) tok_1 (-1)   tok_0 (0);
noun_ford(-1) tok_1 (-2);
noun_ford(-1) tok_1 (-2) predicate (-1);
...
```

**Fig. 2.** Decision Tree Generated from one Set of Abstracts

experiments and discusses the findings. In Section 5 we discuss related work and finally we close the paper in Section 6.

## 2 Data and Tools

We use a set of 219 abstracts in our experiments, all of them have a structure similar to the abstract in Figure 1; schematically this structure is $Pred_0\beta_0$ $Pred_1\beta_1... Pred_n\beta_n$; where $\beta_i$ are fragments from the document to be summarized and $Pred_i$ are rhetorical predicates prepended to the $\beta$ fragments. The set of predicates and their distributions are shown in Table 1.

The following tools are used for this research:

- The GATE system [18] is used to produce parts-of-speech and morphological information in each document;
- The WEKA machine learning [32] environment is used for decision-tree induction from corpora;
- The $\mu$-TBL environment [13] is used as infrastructure for transformation-based learning;
- Specialized programs implemented in GATE are used to compute features;
- Specialized programs are used to map GATE documents into the input required by $\mu$-TBL;
- Specialized programs based on [19] are used for template induction.

```
                          Induced Template (1)

tag:_>_ <- tag:_@[0] & size:_@[0] & tok_1:_@[0] & tag:_@[-1]
(or its equivalent for explanation purposes
tag:A>B <- tag:C@[0] & size:D@[0] & tok_1:E@[0] & tag:F@[-1]
)

                          Instantiated Rule (1)

tag:includes>contains <- tag:includes@[0] & size:5@[0] &
tok_1:screen@[0] & tag:concludes@[-1]

                          Induced Template (2)

tag:_>_ <- noun_ford:_@[-1] &  tok_1:_@[-2] & tag:_@[-1]

                          Instantiated Rule (2)

tag:says>adds <- noun_ford:'X'@[-1] & tok_1:overview@[-2] & tag:says@[-1]
```

**Fig. 3.** Induced Templates and Instantiated Rules

## 2.1 Features for Experimentation

We have computed a series of features we assume can help identify the predicates to be inserted at each position in the abstract because they have been used in different summarization tasks in the past. These features can be classified into content-based features, syntactic features, summarization features, semantic features, cohesion features, and discourse features.

- Content-based features: one feature for each of the three first lemmas of the sentence fragment $\beta_i$, so for example for the first sentence of the abstract in Figure 1 the following features/values will be produced: tok_0=a, tok_1=group, and tok_2=of;
- Syntactic features: the three first parts-of-speech tags of the sentence fragment;
- Summarization features: the relative position of the sentence from begin and end of the abstract, the size of sentence in number of words, the size of the abstract, the number of title words in the sentence (one feature for common nouns and another for proper nouns), the presence of title words in the sentence. There are various reasons for using these features: first, the position of information in an abstract is somehow correlated with the rhetorical status of the information and therefore at the begining of the abstract one may find introductory/presentation predicates while at the end of the abstract one may find predicates of conclusion/discussion; second, in relation to the size of the sentences, long sentences may well aggregate different pieces of information using predicates such as "include"; finally,

**Table 1.** Predicates and Distribution in the Corpus

| Predicate | Percent |
|-----------|---------|
| Points out | 2.17% |
| Contains | 2.80% |
| Indicates | 2.80% |
| Features | 2.89% |
| Describes | 2.98% |
| Notes | 3.07% |
| Reports | 3.07% |
| Mentions | 3.34% |
| Concludes | 3.52% |
| Explains | 4.79% |
| Adds | 5.51% |
| Discusses | 6.23% |
| Provides | 6.32% |
| Presents | 15.27% |
| Says | 16.71% |
| Includes | 18.52% |

the presence of title words in sentences may well predict the introduction of a key topic, and therefore the need for an introductory predicate such as "present", "describe" or "report".
- Semantic features: the presence of a definition pattern (one feature for definition proper and other for statement of usability). These features are recognised in sentence fragments with regular patterns such as "X is ...", "X is defined ...", "X consists of ...", etc. and "X is used ...", "use of X", etc. Presence of these patterns could assist the selection of predicates such as "explain" or "mention" or "note" to provide explanations about "X".
- Cohesion features: the presence of cohesive links between sentences (one feature for presence and one feature for number of cohesive links). These features indicate whether particular nouns co-occur in a sentence window. These features may be used to either predict the introduction of a topic (and therefore the need for a predicate such as "present"), or the continuation of a topic (and therefore the need for a predicate such as "add").
- Discourse features: the default predicate predicted by a baseline system(see below);

In our framework features from previous sentences (positions -1, -2, -3) are available for predicting the predicate in the current sentence (position 0). All features in the current sentence are also available for prediction.

## 3   Methodology

Given the abstracts annotated with predicates and summarization features, the following methodology is adopted:

1. Initial or baseline abstract structure:

   – Given a training set of summaries, a baseline system is induced; this baseline represents a default summary structure which is applied to each abstract in the corpus.

2. Template induction:

   – Training documents annotated with the baseline predictions are fed into a decision-tree classification algorithm [22] to obtain a decision-tree. We rely on the learning environment WEKA and in particular on its J48 implementation of decision trees for this purpose. A schematic representation of a tree obtained from a set of abstracts in our corpus can be seen in Figure 2 (note that we are only interested in the tree structure as shown in the figure). The number in parenthesis right after the feature names represents the position of the feature with respect to the current sentence (0 is the current sentence, -1 is the previous sentence, etc.). Features shown in the figure are: noun_ford to represent a cohesive link between nouns (e.g. noun repetition in two different sentences), tok_i to represent a token at position $i$, and predicate to represent a predicted predicate in the sentence,
   – The decision tree is used to induce a series of templates which are generated as sequences of $k$ features (for $k = 1, ...n$) obtained by traversing the decision tree from top (root) to bottom – depth first traversal. Figure 2 shows some of the generated sequences and Figure 3 shows templates represented in the $\mu - TBL$ learning framework we have adopted. As for the maximum number of features in a template, we have experimented with different template sizes from two to six and obtained better overall accuracy with a maximum of 5 features per template. The formalism used for template representation is explained in full in [13] where *tag* indicates in our case the predicate to correct, the _ symbol indicates a free variable to be instantiated and @[n] indicates the position of the feature in the text with respect to the current position. As an illustration, template (1) in Figure 3 represents a case of replacement in the current sentence of the abstract of predicate A by predicate B in the context of a predicted predicate C (e.g. predicted by the baseline system), an abstract of size D, the occurence of token E as second word in the sentence, and where the predicate predicted for the previous sentence is F.

3. Learning correction or re-writing rules:

   – Given the set of templates and the annotated training corpus, discourse correction rules are learnt using the TBL methodology. In TBL these correction rules are learnt for positions where the baseline system made an incorrect prediction. The rules have the general form: change predicate P1 by P2 in context C. The contexts are instantiations of the induced templates at the target positions in the corpus. Figure 3 shows rules instantiated from the templates in the $\mu$-TBL learning framework. Rule (2) for example indicates to change the predicate "says" into predicate "adds" in the current sentence (0) if in the previous sentence (-1) there is a noun which is also mentioned in the current sentence (0), the word "overview" is present two sentences before (-2) the current sentence, and the predicate "says" was used for the previous sentence (-1). This rule makes use of cohesion, lexical information, and discourse context.

– The correction accuracy of each rule (number of corrections minus numbers of mistakes made) is computed and the best rule selected for the final algorithm. Learning of rules ends when no improvements are observed or when all rules in the current iteration have accuracy below a certain threshold experimentally set.

The method produces a baseline system plus an ordered set of correction rules. During testing, given a set of ordered sentence fragments $\beta_0\ \beta_1...\ \beta_n$, the baseline is applied to obtain an initial abstract $Pred_0\beta_0\ Pred_1\beta_1...\ Pred_n\beta_n$, after this, the rules are applied in learnt order to predicates $Pred_i$ (in order of occurrence) to correct the baseline structure and generate the final structure $\overline{Pred_0}\beta_0$ $\overline{Pred_1}\beta_1... \overline{Pred_n}\beta_n$

## 4   Experiments and Results

The experiments reported here are cross-validation experiments over the corpus. Each experiment consists of the generation of a TBL system from a corpus annotated with summarization features and a baseline predicate predictor as indicated in the previous section. After this the TBL system is applied to a test set and the system evaluated. We follow the evaluation methodology adopted in [25]: for each abstract in the corpus we compute *accuracy* as the proportion of correctly predicted predicates and we compute statistics on the number of errors made by the algorithm: this is illustrated in Table 2 where the algorithm predicts 3 out of 5 predicates, thus obtaining accuracy of 60% and making 2 errors. Accuracy is averaged over all test cases to obtain an overall algorithmic accuracy; also computed are statistics on overall number of errors made by the algorithm (e.g., proportion of abstracts with $\leq n$ errors where $n = 0, 1, 2$). We have experimented with various baselines including baselines based on parts-of-speech information, word-level information, etc. Here, we present results for the use of three baselines based on positional information since it has been shown relevant for both content selection [15] and rhetorical classification [31] for summarization. The three baselines we use are:

– Begin: predicts predicate at position $b$ as the predicate that in the training corpus occurs most frequently at that position ($b$ is the position of the predicate from the beginning of the abstract);
– End: predicts predicate at position $e$ as the predicate that in the training corpus is observed most frequently at that position ($e$ is the position from the end of the abstract);

**Table 2.** Exemplification of Gold and Predicted Structures

| Gold Standard Abstract | Predicted Abstract |
|---|---|
| Presents... *Illustrates* ... **Consid-ers** .... **Concludes** ... *Includes* ... | Presents... *Exemplifies* ... **Consid-ers** .... **Concludes** ... *Contains* ... |

**Table 3.** Baseline and TBL Prediction Accuracy for each Predicate

| Predicate | Beg-base | Beg-TBL | End-base | End-TBL | B/E-base | B/E-TBL |
|---|---|---|---|---|---|---|
| Adds | 0% | 8% | 0% | 2% | 23% | 25% |
| Concludes | 0% | 26% | 0% | 5% | 38% | 38% |
| Contains | 0% | 13% | 0% | 0% | 0% | 0% |
| Describes | 0% | 0% | 6% | 6% | 6% | 6% |
| Discusses | 0% | 55% | 2% | 1% | 2% | 39% |
| Explains | 0% | 30% | 0% | 21% | 28% | 32% |
| Features | 0% | 37% | 0% | 3% | 0% | 53% |
| Includes | 97% | 91% | 85% | 86% | 90% | 89% |
| Indicates | 0% | 10% | 0% | 0% | 0% | 19% |
| Mentions | 0% | 11% | 0% | 0% | 0% | 5% |
| Notes | 0% | 6% | 0% | 3% | 0% | 0% |
| Points out | 0% | 87% | 0% | 17% | 0% | 71% |
| Presents | 95% | 86% | 57% | 56% | 90% | 88% |
| Provides | 0% | 46% | 6% | 29% | 0% | 34% |
| Reports | 0% | 0% | 0% | 0% | 0% | 0% |
| Says | 70% | 70% | 48% | 71% | 75% | 73% |

**Table 4.** Prediction Accuracy at Predicate Level and at Discourse Level (proportion of abstracts with $\leq n$ errors)

| Method | Accuracy | 0 errs | <= 1 errs | <= 2 errs |
|---|---|---|---|---|
| Begin-baseline | 51% | 11% | 39% | 61% |
| Begin-TBL | 61% | 21% | 46% | 71% |
| End-baseline | 43% | 7% | 31% | 43% |
| End-TBL | 49% | 11% | 33% | 51% |
| Beg/End-baseline | 53% | 11% | 39% | 61% |
| Begin/End-TBL | 61% | 21% | 43% | 72% |

– Begin/End: predicts predicate at position $b, e$ as the predicate that in the training corpus occurs most frequently at that position ($b, e$ are the combined positions of the predicate from beginning and end of the abstract).

For positions not observed in the training corpus, the most frequent predicate in the corpus is assigned. This method of prediction is similar to baselines used in TBL-based parts-of-speech tagging [2].

Table 4 presents accuracy results for baseline and TBL methods. Results are reported as total accuracy in predicting predicates and number of errors made when looking at the whole set of predicates. This is a hard evaluation metric since it does not take into account possible synonymy relations among predicates (e.g., predicate "Includes" and "Contains" can be considered synonyms in our context, we consider them different for evaluation purposes, however.). The best overall classification result is obtained by the TBL methods which correct a baseline based on position from the beginning of the abstract (Beg-TBL and Beg/End-TBL). All TBL methods outperform statistically their respective baselines. We have measured differences with a statistical $t$-test. There are also statistical differences between both Beg-TBL and End-TBL and between Beg/End-TBL and End-TBL. Where accuracy at the whole structure is concerned, 72% of all structures are correctly predicted by Beg-TBL and Beg/End-TBL. It is worth mentioning here that the Beg/End-TBL method is able to predict all structures in abstracts with one component, 80% correct structures with 2 components, 74% correct structures with 3 components, and 60% of all structured with 4 components.

Table 3 presents predicate classification accuracy for baselines and TBL methods. It can be observed that baseline systems can only predict 2 or 3 of the predicates. TBL methods are more uniform in their prediction across the set of predicates. The overall results, the results at individual predicate prediction, and the results at full structure prediction are similar to results using classification systems [25].

# 5   Related Work

Text abstracting operations have been studied in the context of professional summarization [4,21,5,20] for educational purposes or to capture professional expertise. In Computational Linguistics there have been attempts to simulate some of the operation in [9,27] and more recently in [25]. There have been a series of approaches to the generation of abstracts or pseudo abstracts in the literature: statistical methods (word based or syntactic) have been used to generate short title-like summaries in [1,30]. [33] concentrate on the generation of quasi-abstractive summaries by learning sets of sentences that could be used to generate a single sentence in a summary. For sentence realization they use $n$-gram probabilities computed over the sentence set outputting sentence fragments based on a bigram model. Where computer-assisted abstracting is concern, [7] studies a series of operations applied to "text extract" for the creation of abstracts. The work is based on the CAST summarization corpus [8] and the operations include insertion of new material into the abstracts. Predicates similar to those studied here have been used in the TEXT computer-assisted system for abstract writing [3].

# 6   Conclusions

The work presented here is the first to investigate the application of transformation-based learning to the problem of generating abstracts by simulating one insertion operation observed in human abstracting. We have adopted a methodology that does not suffer from the template acquisition problem, generating the templates automatically from data. The work is grounded on observations from text abstracting studies indicating that a draft abstract is usually edited to obtain a final text – a typical situation in human writing. We simulate the creation of both the draft and the final structure of the abstract in a transformation-based learning approach. We have carried out a series of experiments obtaining performance comparable (in terms of accuracy) to classification approaches to the same problem. However, our approach produces a series of rules which can be useful to understand why and when particular transformations occur. In our future work we intend to apply this methodology to the simulation of additional text abstracting operations.

## Acknowledgements

We thank two anonymous reviewers for feedback, corrections, and positive comments. Horacio Saggion is grateful to Programa Ramón y Cajal from Ministerio de Ciencia e Innovación, Spain and to a COMENÇA grant (#10.004) from Universitat Pompeu Fabra.

## References

1. Banko, M., Mittal, V.O., Witbrock, M.J.: Headline generation based on statistical translation. In: ACL 2000: Proceedings of the 38th Annual Meeting on Association for Computational Linguistics, Morristown, NJ, USA, pp. 318–325. Association for Computational Linguistics (2000)
2. Brill, E.: Some Advances in Transformation-Based Part of Speech Tagging. In: Proceedings of the Twelfth National Conference on AI (AAAI 1994), Seattle, Washington (1994)
3. Craven, T.C.: Human creation of abstracts with selected computer assistance too. Information Research 3(4) (April 1998)
4. Cremmins, E.T.: The Art of Abstracting. ISI PRESS (1982)
5. Endres-Niggemeyer, B.: SimSum: an empirically founded simulation of summarizing. Information Processing & Management 36, 659–682 (2000)
6. Fernandes, E.R., Pires, B.A., dos Santos, C.N., Milidiú, R.L.: and. Clause identification using entropy guided transformation learning. In: 2009 Seventh Brazilian Symposium in Information and Human Language Technology (STIL), pp. 117–124 (2009)
7. Hasler, L.: From extracts to abstracts: human summary production operations for computer-aided summarisation. In: The RANLP 2007 Workshop on Computer-Aided Language Processing (CALP), Borovets, Bulgaria, pp. 11–18 (2007)

8. Hasler, L., Orăsan, C., Mitkov, R.: Building better corpora for summarisation. In: Proceedings of Corpus Linguistics 2003, Lancaster, UK, March 28-31, pp. 309–319 (2003)
9. Jing, H., McKeown, K.: Cut and Paste Based Text Summarization. In: Proceedings of the 1st Meeting of the North American Chapter of the Association for Computational Linguistics, Seattle, Washington, USA, April 29-May 4, pp. 178–185 (2000)
10. Jones, K.S.: Automatic summarising: The state of the art. Inf. Process. Manage. 43(6), 1449–1481 (2007)
11. Kan, M.-Y., McKeown, K.R.: Corpus-trained text generation for summarization. In: Proceedings of the Second International Natural Language Generation Conference (INLG 2002), Harriman, New York, USA, pp. 1–8 (2002)
12. Kupiec, J., Pedersen, J., Chen, F.: A Trainable Document Summarizer. In: Proc. of the 18th ACM-SIGIR Conference, Seattle, Washington, United States, pp. 68–73 (1995)
13. Lager, T.: μ-tbl lite: a small, extendible transformation-based learner. In: Proceedings of the Ninth Conference on European Chapter of the Association for Computational Linguistics, Morristown, NJ, USA, pp. 279–280. Association for Computational Linguistics (1999)
14. Liddy, E.D.: The Discourse-Level Structure of Empirical Abstracts: An Exploratory Study. Information Processing & Management 27(1), 55–81 (1991)
15. Lin, C., Hovy, E.: Identifying Topics by Position. In: Fifth Conference on Applied Natural Language Processing, March 31-April 3, pp. 283–290. Association for Computational Linguistics (1997)
16. Luhn, H.P.: The Automatic Creation of Literature Abstracts. IBM Journal of Research Development 2(2), 159–165 (1958)
17. Mani, I.: Automatic Text Summarization. John Benjamins Publishing Company, Amsterdam (2001)
18. Maynard, D., Tablan, V., Cunningham, H., Ursu, C., Saggion, H., Bontcheva, K., Wilks, Y.: Architectural Elements of Language Engineering Robustness. Journal of Natural Language Engineering – Special Issue on Robust Methods in Analysis of Natural Language Data 8(2/3), 257–274 (2002)
19. Milidiú, R.L., Santos, C.N., Duarte, J.C.: Phrase chunking using entropy guided transformation. In: in Proc. of ACL 2008: HLT, pp. 647–655 (2008)
20. Montesi, M., Owen, J.M.: Revision of author abstracts: how it is carried out by LISA editors. Aslib Proceedings 59(1), 26–45 (2007)
21. Pinto Molina, M.: Documentary Abstracting: Towards a Methodological Model. Journal of the American Society for Information Science 46(3), 225–234 (1995)
22. Quinlan, J.R.: Learning decision tree classifiers. ACM Compututer Surveys 28(1), 71–72 (1996)
23. Radev, D.R., Jing, H., Budzikowska, M.: Centroid-based summarization of multiple documents: sentence extraction, utility-based evaluation, and user studies. In: ANLP/NAACL Workshop on Summarization, Seattle, WA (April 2000)
24. Ramshaw, L., Marcus, M.: Text Chunking Using Transformation-Based Learning. In: Proceedings of the Third ACL Workshop on Very Large Corpora (1995)
25. Saggion, H.: A classification algorithm for predicting the structure of summaries. In: UCNLG+Sum 2009: Proceedings of the 2009 Workshop on Language Generation and Summarisation, Morristown, NJ, USA, Association for Computational Linguistics (2009)

26. Saggion, H., Lapalme, G.: Concept Identification and Presentation in the Context of Technical Text Summarization. In: Proceedings of the Workshop on Automatic Summarization, ANLP-NAACL 2000, Seattle, WA, USA, April 30, Association for Computational Linguistics (2000)
27. Saggion, H., Lapalme, G.: Generating Indicative-Informative Summaries with SumUM. In: Computational Linguistics (2002)
28. Saggion, H., Gaizauskas, R.: Multi-document summarization by cluster/profile relevance and redundancy removal. In: Proceedings of the Document Understanding Conference 2004, Boston, USA, May 6-7. NIST (2004)
29. Samuel, K., Carberry, S., Vijay-Shanker, K.: An investigation of transformation-based learning in discourse. In: Proceedings of the Fifteenth International Conference on Machine Learning, ICML 1998, pp. 497–505. Morgan Kaufmann Publishers Inc., San Francisco (1998)
30. Soricut, R., Marcu, D.: Abstractive headline generation using WIDL-expressions. Inf. Process. Manage. 43(6), 1536–1548 (2007)
31. Teufel, S., Moens, M.: Argumentative classification of extracted sentences as a first step towards flexible abstracting. In: Mani, I., Maybury, M.T. (eds.) Advances in Automatic Text Summarization, pp. 155–171. The MIT Press, Cambridge (1999)
32. Witten, I.H., Frank, E.: Data Mining: Practical Machine Learning Tools and Techniques with Java Implementations. Morgan Kaufmann, San Francisco (October 1999)
33. Xie, Z., Di Eugenio, B., Nelson, P.C.: From extracting to abstracting: Generating quasi-abstractive summaries. In: Proceedings of LREC (2008)

# Multi-topical Discussion Summarization Using Structured Lexical Chains and Cue Words

Jun Hatori[1], Akiko Murakami[2,3], and Jun'ichi Tsujii[1,3,4,5]

[1] Graduate School of Information Science and Technology, University of Tokyo
7-3-1 Hongo, Bunkyo-ku, Tokyo, Japan
{hatori,tsujii}@is.s.u-tokyo.ac.jp
[2] IBM Research – Tokyo
1623-14 Shimotsuruma, Yamato, Kanagawa, Japan
akikom@jp.ibm.com
[3] Graduate School of Interdisciplinary Information Studies, University of Tokyo
7-3-1 Hongo, Bunkyo-ku, Tokyo, Japan
[4] School of Computer Science, University of Manchester
131 Princess Street, Manchester, M1 7DN, UK
[5] National Centre for Text Mining (NaCTeM), UK
131 Princess Street, Manchester, M1 7DN, UK

**Abstract.** We propose a method to summarize threaded, multi-topical texts automatically, particularly online discussions and e-mail conversations. These corpora have a so-called reply-to structure among the posts, where multiple topics are discussed simultaneously with a certain level of continuity, although each post is typically short. We specifically focus on the multi-topical aspect of the corpora, and propose the use of two linguistically motivated features: lexical chains and cue words, which capture the topics and topic structure. Particularly, we introduce the *structured lexical chain*, which is a combination of traditional lexical chains with the thread structure. In experiments, we show the effectiveness of these features on the Innovation Jam 2008 Corpus and the BC3 Mailing List Corpus based on two task settings: key-sentence and keyword extraction. We also present detailed analysis of the result with some intuitive examples.

## 1 Introduction

Online discussion has become a popular tool for collaboration among people as they discuss various topics online. However, with its increasing popularity, problems have arisen with information overload, which makes it difficult for people to catch up with up-to-date topics and central points of the discussion. Particularly, if organizers intend to draw out useful findings from the whole discussion, they often encounter a problem with obtaining the big picture of the content that is distributed among a large number of posts. Therefore, great demand exists for systems that provide users with an overview of the discussion.

Posts in an online discussion are typically organized in either a sequential or a tree-structured thread. Although the former has simpler structure, the latter allows division of many topics into smaller branches. For this reason, the tree-structured thread has been

A. Gelbukh (Ed.): CICLing 2011, Part II, LNCS 6609, pp. 313–327, 2011.

adopted in many large discussion fora (e.g. Slashdot[1]) as well as in internal discussions in enterprises. Fig. 1 is an excerpt of an online discussion thread in the IBM Corporation, where the main topic of the thread, "leave pool," is branched into two subtopics, "leave accumulation" and "maternity leave," which are clearly identified in the two distinct branches in the thread tree. In larger threads, it is even common that multiple topics are discussed alternately in the same sequential branch. This multi-topicality of texts is a challenge for both parcitipants and systems to comprehend the whole content of the discussion. Therefore, our approach to the overview of the discussion is twofold: we first try to recognize the topics discussed (*topic extraction*), and then incorporate the information of the topics into the task of *key-sentence extraction* (extractive summarization).

To address the problem of *multi-topicality* of texts, some researches have introduced *lexical chains* for the task of summarization (e.g. [1]). Lexical chains are chains of semantically related words; each is considered to render a topic in the document. Recently, the lexical chains have also been successfully applied to the task of multi-document summarization [2,3]. However, to the extent of our knowledge, they have never been applied to threaded texts such as online discussions and e-mail conversations.

To apply the lexical chains to summarization of online discussions, we focus on the use of the thread structure, by which we can infer the flow of the arguments and topics. In Fig. 1, we can observe that the chains of semantically related words, such as "leave (pool)," "accumulate(d)," and "maternity, paternity" characterize the topics in the thread, capturing the cohesive property of topics in the thread structure. This motivates the use of lexical chains with the thread structure: we introduce the *structured lexical chains*, by which we can combine the traditional lexical chains with a newly proposed scoring scheme that evaluates the importance of each sentence in the context of the thread structure.

Another characteristic of discussion corpora is that the writers tend to use typical expressions to clarify their statements in short posts. In Fig. 1, many underlined key sentences include (italicized) characteristic expressions that typically appear in sentences stating the writer's main opinion or proposal. For example, auxiliary verbs such as "should" and "could," and verbs such as "suggest" and "think" are examples of these expressions. These are considered to be examples of *cue words*, which have been discussed in the linguistic literature [4]. We propose to model these expressions explicitly with scores reflecting how strongly they contribute for a sentence to be a key sentence. In experiments, we explore a set of cue words that are effective for this task in both manual and automatic ways, and evaluate them using the proposed summarization model.

Because numerous online discussions exist with different domains and characteristics, it is not practical to construct a supervised system. For that reason, we construct an unsupervised model based on the graph-based multi-document summarization model presented by [5]. We then further extend this model to incorporate the structured lexical chains and cue words. The proposed model works with minimal supervision; we show that the almost-unsupervised, graph-based model with a few manually selected cue words works comparably with the supervised counterpart.

---

[1] http://slashdot.org

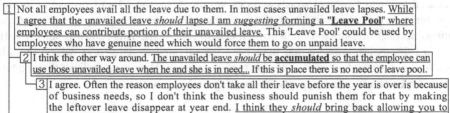

1. Not all employees avail all the leave due to them. In most cases unavailed leave lapses. While I agree that the unavailed leave *should* lapse I am *suggesting* forming a **"Leave Pool"** where employees can contribute portion of their unavailed leave. This 'Leave Pool' could be used by employees who have genuine need which would force them to go on unpaid leave.

2. I think the other way around. The unavailed leave *should* be **accumulated** so that the employee can use those unavailed leave when he and she is in need... If this is place there is no need of leave pool.

3. I agree. Often the reason employees don't take all their leave before the year is over is because of business needs, so I don't think the business should punish them for that by making the leftover leave disappear at year end. I think they *should* bring back allowing you to **accumulate** leave as necessary... [...]

4. I would have linked to have more paid maternity leave & I don't expect that IBM should necessarily give more than is currently provided. I *suggest* that we could have a policy that you could 'save leave' for **maternity** and **paternity**. I would have grabbed that early in my IBM career. Unsure if this could be implemented, or even if other staff would be interested? What do other IBMers think?

5. An IBM branch office allows (or did, the last time I checked) limited self-funded annual leave (expires annually). Maybe a similar scheme can be implemented for **maternity** leave. The big issue I see with this is the increased cost to the business, so maybe cap it to two years, then refund the money if it's still unused by then.

**Fig. 1.** A thread example from the Innovation Jam 2008 Corpus

As datasets, we mainly address the IBM's internal discussion, the "Innovation Jam 2008 Corpus" (hereinafter called "I-Jam 2008 Corpus"). We manually annotated key sentences and topics information on this corpus, and then used them to evaluate our model. To validate and compare the results, we also perform experiments on the BC3 Corpus, which is a collection of mailing list threads and is expected to share the multi-topical nature and conciseness of the expression with the I-Jam 2008 Corpus.

Here are several key terms that will be used throughout this paper.

**forum**: Discussion board with a specific theme for discussion.

**thread**: Series of posts which are mutually connected by reply-to relations.

**post**: Message written by a participant.

In what follows, in Section 2.2, we first introduce related works. We describe our model in detail in Section 3. We present our experimental settings and results in Section 4, and our conclusions in Section 5.

## 2 Background

### 2.1 Corpora

**Innovation Jam (I-Jam) 2008 Corpus.** The Innovation Jam (I-Jam) 2008 Corpus is a collection of online discussion called "Innovation Jam 2008," which was held by the IBM Corporation in 2008. Up to now, the company has held several sessions of a short-term, intensive discussion called "Jam." The Innovation Jam is one of those sessions, and is intended for not only IBM employees, but also for the customers and families of the employees and customers. In the Innovation Jam 2008, the participants discussed various topics related to the company's future plan; the session attracted 29,498 posts by 8,937 participants within five days. Such a relatively concentrated nature of the discussion naturally encouraged people to use simple and concise expression, which are even clarified using topical words and cue words. Also, the I-Jam Corpus is a *brain-storming*-type discussion in which the participants discuss various topics from various viewpoints in an attempt to obtain novel and inspiring ideas. This contrasts starkly with standard discussion corpora that have been investigated to date [6,7], which include *question–answer* and *problem–solution* type discussion. Hence, we believe that the targeted corpus of our research is also interesting to the community.

**BC3 Corpus.** As another corpus used for experiment, we used the BC3 Corpus [8], which is already annotated with extractive summaries. This is a collection of e-mail posts in the W3C Corpus. The annotation is done by three annotators, with a kappa agreement of 0.50 for the extractive summary sentences. The BC3 Corpus comprises 41 threads, which include 200 documents.

## 2.2  Related Work

Our method for extracting key sentences and topics is closely related to extractive summarization and keyword extraction research, particularly that for web texts, such as blogs, mailing lists, and discussion fora. The primary characteristics of these corpora are that the threads are updated dynamically as the discussion proceeds; also, they consist of documents linked by reply-to relations.

To reduce the number of documents that must be read to comprehend the ongoing discussion, some researchers (e.g. [9]) have emphasized evaluation of the importance of each document. Other researchers directly examined the summarization of threads: to date, research efforts have investigated blogs [10,11], e-mails [12,7,13], and discussion fora [11]. However, these studies have not explicitly emphasized the multi-topical aspects of the corpora.

Some models exploit corpus-specific reply-to structures. [14] exploits the thread structure to summarize mailing lists. In this method, the ancestral messages of a post are regarded as its context and are used in the summarization process. For summarization of a discussion thread, [15] used the thread structure indirectly to find successive appearances of the same *clue word*s. In this context, our method is more advanced in that we use the structural information to recognize *subchain*s of a lexical chain, with novel ideas of *subchain*s and *locality*, which we describe in Section 3.3 in detail.

Although the clue words are merely repetitions of the same word, a *lexical chain* considers semantically related words as well. Several researchers [16,17] have used this approach for the summarization of single documents. More recently, the lexical chain has also been applied to the multi-document summarization [2,3]. For keyword extraction, [18] reported success in applying lexical chains to topic extraction from a single document. They considered strong lexical chains to be prominent topics of a document. However, lexical chains were used without consideration of the structural information. Consequently, they have never been applied to the summarization of e-mail conversations nor online discussions. The *structured lexical chain*, which we propose in this paper, is the first method to combine lexical chains with a thread structure.

# 3   Model Description

In this section, we describe our model for the key-sentence and topic extraction task. We first describe a graph-based summarization model by [19] in Section 3.1; then introduce two features we propose: cue words and lexical chains, in Section 3.2 and Section 3.3, respectively. Finally, we briefly describe a supervised model that we use for comparison with the proposed (almost-)unsupervised model.

## 3.1   Graph-Based Summarization Model

First of all, let us briefly describe the graph-based models proposed by [19] and its extension by [5]. In these models, we first construct a graph, where each node represents a sentence and each vertex represents a word shared by two sentences. By calculating the PageRank [20] for the vertices in the graph, one can find which sentence is most likely to be a key sentence, based on the assumption that a sentence that includes more information shared by other *important* sentences is important. Despite the simple framework, their model achieved scores comparable to those of state-of-the-art models.

The extension by [5] is to incorporate the importance of documents and sentence–document correlations as modifications to the edge weights. Because the incorporation of the importance of the documents did not improve the performance in our preliminary experiment, we only used the sentence–document correlation in our model. The resulting PageRank value is given as

$$R(s) = (1 - d) + d \sum_{s' \in \mathcal{S}} \frac{f(s, s')R(s')}{\sum_{s'' \in \mathcal{S}} f(s, s'')} \qquad (1)$$

$$f(s, s') = \mathrm{Sim}(s, s') \cdot \frac{1}{2}(\mathrm{Imp}(s) + \mathrm{Imp}(s')) \qquad (2)$$

$$\mathrm{Imp}(s) = \mathrm{Sim}(s, doc(s)) \,, \qquad (3)$$

where we set $d = 0.5$ based on our preliminary experiment on the development set.

## 3.2   Cue Word

A cue word [4] is a characteristic expression that affects the extract-worthiness of a sentence. It is either a *bonus* word or a *stigma* word, which is respectively the indicator of an important or an unimportant sentence. Words and phrases such as 'important,' 'should,' and 'I propose' are examples of bonus words (phrases), whereas those such as 'for instance' and 'example' are considered to be stigma words. In the graph-based model, we incorporated information from cue words as a modification to the edge weights as

$$\mathrm{Imp}(s) = \mathrm{Sim}(s, doc(s)) \cdot \prod_{c \in \mathcal{W}(s)} \mathrm{CueScore}(c) \,, \qquad (4)$$

where $\mathcal{W}(s)$ denotes the set of cue words in sentence $s$.

## 3.3    Structured Lexical Chain

A *lexical chain* [1] is a sequence of semantically related words in a text. As described by [18], we assume that each lexical chain characterizes a topic of the thread. Because it captures a considerable part of the lexical cohesiveness in natural language texts and is easily incorporated, it has been widely used for various tasks including text summarization [16] and key phrase extraction [18].

We extended this by incorporating the information of thread structure, thereby introducing the idea of *structured lexical chains*. In constructing a structured chain, we first segment each chain into local substructures called *subchains*, and score each subchain with respect to the strength of the local structure. We describe this newly proposed method for constructing and scoring the subchains in Section 3.3 and Section 3.3.

Considering the contribution of the lexical chains, the score of an edge connecting sentences $s$ and $s'$ is modified as

$$f(s, s') = \mathrm{Sim}(s, s')\mathrm{Rel}(s, s') \cdot \frac{1}{2}(\mathrm{Imp}(s) + \mathrm{Imp}(s')) + \lambda \sum_{c \in \mathcal{LC}(s,s')} \mathrm{Score}(c) \ , (5)$$

where $\mathcal{LC}(s, s')$ is the set of lexical chains that includes words in both sentences $s$ and $s'$. Based on results of a preliminary experiment, we set $\lambda = 2.5$.

**Chain construction.** First, we describe a general method for constructing lexical chains that is also applicable for constructing the structured lexical chains. [21] proposed an efficient linear-time algorithm for recognizing lexical chains, which performs simple word sense disambiguation simultaneously. Their method comprises two steps: the first calculates the scores of all possible chains with no sense disambiguation; the second removes each word instance from any chain in which it does not maximally contribute in terms of the relation scores (i.e. simple word sense disambiguation). As semantic relations, we used synonym, hypernym, hyponym, and sibling relations in WordNet [22] following their approach; we additionally exploited holonym, antonym, and nominalization links. The weights are modified slightly from their original work: 0.95 for nominalizations, 0.9 for antonyms, 0.5 for siblings, 0.3 for hypernyms and hyponyms, and 0.2 for holonyms, which are set on the development set.

**Subchain.** We introduced the concept of *subchains*, which are maximal local structures of a lexical chain. Consequently, one lexical chain consists of one or more subchains. A subchain for a lexical chain $c$ is a local subgraph of documents, all of which include any element in the chain $c$, as shown in Fig. 2. It is constructed as follows. We connect, with a *direct edge*, each pair of directly connected documents that both include one or more words in the chain $c$. To increase the coverage, we also connect with an *indirect edge* each pair of documents that is connected via one intervening document node in the thread structure. A subchain is merged with other subchains until no more subchains can be merged via a direct or indirect edge. Eventually, the example in Fig. 2 consists of two subchains.

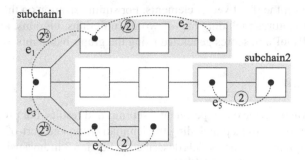

**Fig. 2.** Illustration of lexical chain scoring, where each box with a dot denotes a post that includes a word in the target chain

**Scoring.** After constructing the subchains for a lexical chain $c$, the chain score $\text{Score}(c)$ is calculated as

$$\text{Score}(c) = \text{Strength}(c) + \text{Locality}(c) \tag{6}$$

$$\text{Locality}(c) = \ln \sum_{c' \in \mathcal{SC}(c)} \prod_{e \in \mathcal{E}(c')} \text{EdgeScore}(e) \ , \tag{7}$$

where $\mathcal{SC}(c)$ stands for the set of subchains in the lexical chain $c$, $\mathcal{E}(c')$ signifies the set of (direct and indirect) edges in the subchain $c'$, $n(e)$ denotes the number of children of the document that includes the first word ($e_{\text{start}}$) of the edge $e$, and $\text{EdgeScore}(e)$ represents $2^{\frac{1}{n(e)}}$ if $e$ is a direct edge and $2^{\frac{1}{2n(e)}}$ if $e$ is an indirect edge. Here, we introduced *locality*, which measures the strength of the locally connected structure of the lexical chain. Because an actively discussed topic is more likely to have a more locally-concentrated structure, this metric helps differentiate a strong, topical lexical chain from unimportant chains (or chains with frequent but general words). The chain strength $\text{Strength}(c)$ is calculated similarly as [21]. Fig. 2 portrays locality calculation. For this thread structure, the chain locality is calculated as $\ln \left[ (2^{\frac{1}{3}} \cdot \sqrt{2}) \cdot (2^{\frac{1}{3}} \cdot 2) + 2 \right]$.

### 3.4 Regression-Based Summarization Model

We also construct a supervised regression-based summarization model based on the approach by [23]. In the experiment on the I-Jam 2008 Corpus, this is used to see the degree to which supervised training with lexical features can further improve the model over that with manually chosen cue words. On the other hand, on the BC3 Corpus, this framework is used as a main framework for the experiment because we need to evaluate our model on the extractive summaries in the corpus. Because the task of summarization requires the generation of fixed-length summaries, [23] mentioned that the regression-based approach is more suitable for this task than other frameworks. We use the support vector regression (SVR) classifier and Bagging with the RSTTree classifier, by which they reported superior results among several machine learning techniques.

For summarization on the BC3 Corpus, we used the same feature set as [23], including the position of the sentence and the post, number of words and recipients, and the average and sum of the tf-idf vector elements. For summarization on the I-Jam Corpus, we used lexical features including bag-of-words (unigrams, bigrams, and trigrams) in addition to the PageRank scores generated using the graph-based model.

# 4   Experiment

In this section, we describe our experimental settings and results. For proprocessing, we first performed a standard step including the lemmatization and part-of-speech tagging of words. We implemented the models described in Section 3 in Java using two machine learning libraries, Amis [24] and Weka[2].

## 4.1   Task Setting

Our models are evaluated on two task settings: key-sentence extraction (extractive summarization) and keyword extraction.

A *key sentence* is defined as a sentence that describes or which is most closely related to the main argument of a post. Intuitively, an important proposal or a new idea, which is most likely to be included in the summary of the whole thread, shall be a key sentence. We did not create human-annotated summaries because the scarcity of annotators (only two) complicates the creation of summaries with reasonable agreement.

A *topic* is a subject or theme that is discussed in a thread. Because each thread is allotted a theme for discussion, these topics are considered as subtopics related to the main theme of the thread. We define each topic using a set of key words or phrases, as exemplified in Table 1.

**Table 1.** Examples of annotated topics and their definitions with key words and phrases

| Topic | Definition |
|---|---|
| Desalination of sea water | desalination, desalinate |
| Water leakage from supply piping | dispersion, leak, leakage |
| Semantic web | semantic web, semiotic web |

## 4.2   Annotation

Because no human-annotated data for the I-Jam 2008 Corpus were available, we first created an annotated corpus from the corpus.

First, we annotated key sentence(s) of each post. We annotated at least one key sentence to each post. Although summarization and key-sentence extraction are fundamentally different tasks, a key-sentence extraction system can be evaluated by considering that the collection of extracted sentences comprises the (extractive) summary. We also noticed some cases in which multiple key sentences should be annotated. In this case, the annotators are allowed to annotate multiple key sentences when they think it really is necessary (e.g., cases in which multiple major arguments exist).

---

[2] http://www.cs.waikato.ac.nz/ml/weka/

Second, we annotated the major topics of each thread. The annotators were told to choose all the topics that they thought were discussed actively in the thread. As the guide for active topics, and to prevent the proliferation of minor topics, any topic they think is described in fewer than three posts was ignored. The maximum number of keywords for each topic is five, including variations of inflected forms.

## 4.3  Datasets

**I-Jam 2008 Corpus.** As the dataset used for the experiment on the I-Jam 2008 Corpus, we selected 10 threads from 10 fora in the corpus. From each forum, a thread was randomly selected from those with 15–80 posts. This is because smaller threads might have unclear, noisy thread structure, while larger threads are expensive to annotate. The average number of posts in these threads is 36.3, and each post consists of 6.7 sentences on average. The average number of key sentences per post was 1.54; the average number of topics per thread was 4.10. We performed a simple test of inter-annotator agreement between two annotators. The result was roughly 70%[3] for the key sentence extraction and 60% for the topic extraction.

We used two different experimental settings: HALF–HALF split and five-fold cross validation. For the HALF–HALF split setting, we divided the dataset into two halves, one for training (5 threads, 170 posts) and the other for evaluation (5 threads, 193 posts). In the five-fold cross validation setting, we divided the dataset into five parts: three for training, one for development, and one for evaluation. The reason that we use two different settings is that because the hand-coded cue words were taken from the training portion of HALF–HALF setting, it cannot be evaluated in the five-fold cross valida- tion setting, although the results with the cross validation is more reliable. Because the graph-based models require no supervised training, the training sets are used only in the supervised model. The development set that we used in the preliminary experiment consists of two threads other than any of the 10 threads described above.

**BC3 Corpus.** Among information of various kinds annotated in the BC3 Corpus, we only use information of the extractive summaries to evaluate the performance of our summarization model. We used the same normalization as [25], such as converting "I" and "us" into "[person]." We did not use the locality measure for calculating the lexical chain score because this corpus has no explicit tree structure.

The dataset is split into five balanced portions (A, B, C, D, and E). Each part is used either as training, development, or a test set by turns in a five-fold cross validation scheme. In each trial, three sets are used as the training set, one set as the development set, and the other set as the testing set. Each set includes eight threads with roughly 600–700 sentences. Using the development set, the regularization coefficient $\sigma$ for the regression-based model is set.

---

[3] In the measurement of the inter-annotator agreement, two annotators were requested to select only one sentence from a post if they annotated more than one sentence. This requirement is stricter than the annotation scheme, and is therefore lowering the agreement rate.

## 4.4    Baseline and Evaluation

For key sentence extraction, the first baseline we used is a simple but powerful classifier that extracts the first sentence from each post. We also reimplemented the graph-based method by [5], and used this as the second baseline. Each model outputs the sentence with the highest score as the key sentence for each post; the evaluation is based on whether or not this sentence is included in the gold standard sentences. In the BC3 Corpus, because annotations by three annotators exist, we used the average weighted recall, known as the pyramid precision [26], to calculate the final score[4]. The weighted recall is given as

$$\text{WeightedRecall} = \frac{\sum_{i \in \text{Sent}_{\text{Summary}}} \text{score}_i}{\sum_{i \in \text{Sent}_{\text{Gold}}} \text{score}_i} \; , \tag{8}$$

where $\text{score}_i$ is the number of annotators who selected the sentence in the extractive summary, normalized by the sentence length.

For the topic extraction task, we used two baselines: the TF-IDF and the *edge scores*. TF-IDF is a widely used metric for keyword extraction; it is calculated by the term frequency multiplied by the logarithm of the inverse document (post) frequency. Another baseline we propose to use is the *edge score* of a keyword $w$, which is calculated with the edge scores in the graph. After the PageRank is calculated for each vertex, the *edge score* of a word $w$ is calculated as

$$S(w) = \ln \frac{\#doc}{\text{DF}(w)} \sum_{e \in \mathcal{E}(w)} R(s_{\text{start}}) R(s_{\text{end}}) \; , \tag{9}$$

where $\mathcal{E}(w)$ represents the set of edges associated with word $w$, and $\text{DF}(w)$ denotes the document (post) frequency of the word $w$. In the topic-extraction task, the recall is used as the evaluation measure because the number of the topics in a thread is given to the model (i.e., The model always outputs the same number of topics as the gold standard.). Recall is calculated based on how many of the output topics are actually included in the gold summaries.

## 4.5    Cue Words

From the development set in the HALF–HALF setting, we chose 31 cue words and heuristically set weights for these words, as listed in Table 2. Most of these seems to be general expressions used in a braimstorming-type discussion. For example, conjunctions, such as 'therefore' and 'for this reason,' are obviously good indicators of concluding sentences, and phrases, such as "point/problem is" and "one thought is," are used to draw reader's attention.

---

[4] As [27] mentioned, the ROUGE score, which has been widely used in the summarization of newswire texts, reportedly does not correlate well with human evaluations in the meeting domain [28]. Therefore, we used the standard measure in the domain of the e-mail summarization, following [27].

## 4.6 Results and Discussion

**Key-sentence extraction on I-Jam 2008 Corpus.** Table 3 shows experimental results for key-sentence extraction on the I-Jam 2008 Corpus. (a)–(e) are the models without supervised training, while (f) is a supervised model. † denotes statistically significant improvement[5] over "(c) Graph (MDS)." Our model with cue words outperforms the baseline model by a substantial margin of 4.97%, even though the cue words we used were hand-coded and limited in size. The use of lexical chains further improved the performance by 3.32%. These results underscore the effectiveness of the proposed use of structured lexical chains and cue words. The SVR-based supervised model with lexical features showed a slight improvement over the graph-based models. However, this improvement is marginal compared to the improvement that is provided by use of the hand-coded cue words. This difference suggests that the manual annotation of a small number of cue words is effective, and that the unsupervised model with minimum human effort works sufficiently well.

Table 4 shows a list of the highest-weighted features for the I-Jam 2008 Corpus when we use a maximum-entropy classifier[6] with exactly the same feature set as in the

**Table 2.** Cue words and the associated weights used in the experiment on the I-Jam 2008 Corpus

| | |
|---|---|
| Bonus words | should (1.3), would (1.1), could (1.1), important (1.2), significant (1.2), real (1.2), now (1.2), proposal (1.1), idea (1.1), challenge (1.1), conclusion (1.3), suggest (1.2), propose (1.1), believe (1.1), need (1.1), thus (1.2), therefore (1.3), so (1.1), for this reason (1.3), I/my (1.1), so there (1.1), problem is (1.2), point is (1.2), is/are to (1.1), one thought is (1.3), it 's (1.1), I think (1.1), need to (1.2) |
| Stigma words | example (0.6), for example (0.7), for instance (0.4), agree (0.3) |

**Table 3.** Experimental results for key-sentence extraction on the I-Jam 2008 Corpus. The † denotes statistically significantly improvement over "(c) Graph (MDS)".

| | HALF–HALF | five-fold |
|---|---|---|
| (a) Baseline | 40.88% | 43.11% |
| (b) Graph (SDS) | 45.86% | 46.11% |
| (c) Graph (MDS) | 60.22% | 60.18% |
| (d) +Cue word | 65.19%† | - |
| (e) +Lex. chain | 68.51%† | 61.98% |
| (f) SVR + (e) | 69.61%† | 62.87% |

**Table 4.** Highest-weighted features for the I-Jam 2008 Corpus

| Expression | Label | $\alpha$ Value |
|---|---|---|
| the idea | F | 17.40 |
| translate | T | 9.19 |
| question | T | 8.59 |
| suggest | T | 8.20 |
| I would | F | 6.30 |
| idea | T | 5.73 |
| could you | T | 5.51 |
| you can | F | 5.44 |
| as I | F | 5.36 |
| such | F | 4.91 |

---

[5] All significance tests are based on McNemar's test.
[6] Note that an SVR model outputs no weight information.

SVR-based model. The first column shows the word forms of extracted expressions. The second column shows whether the feature is associated with *true* (i.e. included in the summary) or *false* (i.e. excluded from the summary). The third column shows the $\alpha$ weights in the maximum-entropy classifier. The result seems quite reasonable. The phrase "the idea" is shown to be stigmatic because it is typically used to mention the content of the last message in a precursive expression before stating the author's own idea. In contrast, expressions such as "question" and "suggest" are bonus words which are used to state the author's own question and suggestion. Thus, it seems apparent that the lexical features captured the importance of cue words, and contributed to the result.

**Topic extraction on I-Jam 2008 Corpus.** Table 5 presents the results for topic extraction. Our model with the structured lexical chains outperforms the two baselines by a large margin, and shows that the structured chains captured the topical information of the thread. However unfortunately, because the annotated data are too few, we were unable to infer the statistical significance of the improvements.

Table 6 presents an example of generated lexical chains. This example is taken from a thread on the I-Jam 2008 Corpus. Chains consisting only of the same word occurrences are excluded; the structure of chains is also omitted. It is apprent from this example that most chains seem to express a topic or theme of a discussion thread, and lexical chains are appropriately capturing semantically related words, such as near-synonyms "abuse–use" and antonyms "pessimists–optimists." In this example, only the bottom one seems wrong because the word "rules" is misclassified as having an incorrect sense "formulae."

**Table 5.** Experimental results for topic extraction on the I-Jam 2008 Corpus

|  | Micro Avg. | Macro Avg. |
|---|---|---|
| TF-IDF | 19.51% ( 8/41) | 23.08% |
| Edge score | 24.39% (10/41) | 25.92% |
| Proposed | 36.59% (15/41) | 34.17% |

**Table 6.** An example of generated lexical chains on the I-Jam 2008 Corpus

| Score | Chain |
|---|---|
| 1.39 | salary wage salary pay wage |
| 0.74 | abuse abuse misuse |
| 0.69 | alumni graduate |
| 0.64 | pessimists optimists |
| 0.10 | rules formulae formulae |

**Extractive Summarization on BC3 Corpus.** Table 7 presents the experimental results for key-sentence extraction on the BC3 Corpus. "ME" and "BAG" respectively corresponds to the maximum-entropy and bagging classifiers. "no-lex" models do not use lexical features, while "lex" models do. "lex-lc" models do use both lexical and lexical-chain features. Both in the maximum-entropy and bagging models, the use of lexical feature improved the performance by around 1.2%. The use of lexical chains further improved the model by 2.0%. The performance of "BAG (lex-lc)" was better than "BAG (no-lex)" with the statistical significance level of $p < 0.05$. In Table 7, "GOLD Avg." is the average of weighted recalls for three gold annotations, which is considered to be an upper limit of the score. Considering this fact, the highest recall of 67.54% is a fairly good result. The score seems to be lower than that of [27], who

**Table 7.** Experimental results for key-sentence extraction on the five-fold cross validation on the BC3 Corpus

| | |
|---|---|
| (a) Baseline | 43.24% |
| (b) Graph (MDS) | 57.42% |
| (c) ME (no-lex) | 61.40% |
| (d) ME (lex) | 62.61% |
| (e) BAG (no-lex) | 65.33% |
| (f) BAG (lex) | 65.50% |
| (g) BAG (lex-lc) | 67.54%† |
| GOLD Avg. | 74.60% |

reported approximately 80% weighted recall. However, considering that we used almost identical feature sets as those, and considering that 80% is higher than the performance of "GOLD Avg.," this difference is attributed to the difference in the evaluation criteria, probably the calculation of the weighted average recall. Therefore, we can conclude that the performance of our model is comparable to or better than the performance of [27]. Even if this were not the case, we at least demonstrated that the use of cue words and lexical chains is effective in both discussion and mailing list corpora.

## 5 Conclusion

In this paper, we have presented a key-sentence and topic extraction model for multi-topical, threaded corpora, using structured lexical chains and cue words. Particularly, we proposed to use the structured lexical chains, which can incorporate the locality and continuity of the topics with a thread structure. Evaluation of the model was performed on the two datasets: The Innovation Jam 2008 Corpus and the BC3 E-mail Conversation Corpus. On the I-Jam 2008 Corpus, the use of cue words greatly improved the extractive summarizer. The use of structured lexical chains further improved the performance. The experiment on the keyword extraction task also revealed the effectiveness of the structured lexical chains, which is also confirmed by manual analysis. It is remarkable that even a few cue words improved the model significantly, although the further improvement by a supervised machine learning technique was marginal. This represents a hopeful finding for constructing a model with minimal supervision. We also conducted an experiment on the BC3 Mailing List Corpus, again demonstrating that the use of lexical features and lexical chains improved the model. As a whole, we conclude that the summarization of structured discussion corpora can be accomplished using an unsupervised model with structured lexical chains and cue words, and manual selection of a handful of cue words is effective, saving time used for creating training data for supervised learning. In future works, we are planning to conduct the experiment on larger and more diverse corpora, to validate the current result and to analyze the domain dependence of the model further. Also, we think that a more probabilistic formalization is necessary to achieve better performance.

## Acknowledgement

We are grateful to the anonymous reviewers for their valuable comments. This work was partially supported by Grant-in-Aid for Specially Promoted Research (MEXT, Japan), and JSPS (Japan Society for the Promotion of Science) Research Fellowship.

## References

1. Morris, J., Hirst, G.: Lexical cohesion computed by thesaural relations as an indicator of the structure of text. Computational Linguistics 17(1), 21–43 (1991)
2. Chen, Y.M., Wang, X.L., Liu, B.Q.: Multi-document summarization based on lexical chains. In: International Conference on Machine Learning and Cybernetics (2005)
3. Li, J., Sun, L.: A lexical chain approach for update-style query-focused multi-document summarization. In: Li, H., Liu, T., Ma, W.-Y., Sakai, T., Wong, K.-F., Zhou, G. (eds.) AIRS 2008. LNCS, vol. 4993, pp. 310–320. Springer, Heidelberg (2008)
4. Edmundson, H.: New methods in automatic extracting. Journal of the ACM 16(2), 264–285 (1969)
5. Wan, X.: An exploration of document impact on graph-based multi-document summarizaion. In: EMNLP 2008: Proceedings of the Conference on Empirical Methods in Natural Language Processing, Morristown, NJ, USA, pp. 755–762. Association for Computational Linguistics (2008)
6. Farrell, R., Fairweather, P.G., Snyder, K.: Summarization of discussion groups. In: CIKM 2001: Proceedings of the Tenth International Conference on Information and Knowledge Management, pp. 532–534. ACM, New York (2001)
7. Mckeown, K., Shrestha, L., Rambow, O.: Using question-answer pairs in extractive summarization of email conversations. In: Gelbukh, A. (ed.) CICLing 2007. LNCS, vol. 4394, pp. 542–550. Springer, Heidelberg (2007)
8. Ulrich, J., Murray, G., Carenini, G.: A publicly available annotated corpus for supervised email summarization. In: AAAI 2008 EMAIL Workshop, Chicago, USA. AAAI, Menlo Park (2008)
9. Klaas, M.: Toward indicative discussion fora summarization. In: UBC CS TR-2005-04 (2005)
10. Hu, M., Sun, A., Lim, E.P.: Comments-oriented blog summarization by sentence extraction. In: CIKM 2007: Proceedings of the Sixteenth ACM Conference on Conference on Information and Knowledge Management, pp. 901–904 (2007)
11. Zhou, L., Hovy, E.: On the summarization of dynamically introduced information: Online discussions and blogs. In: AAAI Symposium on Computational Approaches to Analysing Weblogs (AAAI-CAAW), pp. 237–242 (2006)
12. Zajic, D.M., Dorr, B.J., Lin, J.: Single-document and multi-document summarization techniques for email threads using sentence compression. Inf. Process. Manage. 44(4), 1600–1610 (2008)
13. Rambow, O., Shrestha, L., Chen, J., Lauridsen, C.: Summarizing email threads. In: HLT-NAACL 2004: Proceedings of HLT-NAACL 2004: Short Papers on XX, Morristown, NJ, USA, pp. 105–108. Association for Computational Linguistics (2004)
14. Lam, D., Rohall, S.L., Schmandt, C., Stern, M.K.: Exploiting e-mail structure to improve summarization. In: ACM 2002 Conference on Computer Supported Cooperative Work (CSCW 2002) (2002)
15. Carenini, G., Ng, R.T., Zhou, X.: Summarizing email conversations with clue words. In: WWW 2007: Proceedings of the 16th International Conference on World Wide Web, pp. 91–100 (2007)

16. Barzilay, R., Elhadad, M.: Using lexical chains for text summarization. In: Proceedings of the ACL Workshop on Intelligent Scalable Text Summarization, pp. 10–17 (1997)
17. Brunn, M., Chali, Y., Pinchak, C.: Text summarization using lexical chains. In: Document Understanding Conference (DUC), pp. 135–140 (2001)
18. Ercan, G., Cicekli, I.: Using lexical chains for keyword extraction. Inf. Process. Manage. 43(6), 1705–1714 (2007)
19. Mihalcea, R.: Unsupervised large-vocabulary word sense disambiguation with graph-based algorithms for sequence data labeling. In: Proceedings of the Conference on Human Language Technology and Empirical Methods in Natural Language Processing (2005)
20. Page, L., Brin, S., Motwani, R., Winograd, T.: The pagerank citation ranking: Bringing order to the web (1999)
21. Silber, H.G., McCoy, K.F.: Efficiently computed lexical chains as an intermediate representation for automatic text summarization. Computational Linguistics 28(4), 487–496 (2002)
22. Fellbaum, C.: WordNet: An Electronic Lexical Database. MIT Press, Cambridge (1998)
23. Ulrich, J., Carenini, G., Murray, G., Ng, R.: Regression-based summarization of email conversations. In: 3rd Int'l AAAI Conference on Weblogs and Social Media (ICWSM 2009), San Jose, CA. AAAI, Menlo Park (2009)
24. Miyao, Y., Tsujii, J.: Maximum entropy estimation for feature forests. In: Proc. of Human Language Technology Conf. (HLT) (2002)
25. Carvalho, V.R., Cohen, W.W.: Improving "email speech acts" analysis via n-gram selection. In: ACTS 2009: Proceedings of the HLT-NAACL 2006 Workshop on Analyzing Conversations in Text and Speech, Morristown, NJ, USA, pp. 35–41. Association for Computational Linguistics (2006)
26. Carenini, G., Ng, R.T., Zhou, X.: Summarizing emails with conversational cohesion and subjectivity. In: Proceedings of ACL 2008: HLT, Columbus, Ohio, pp. 353–361. Association for Computational Linguistics (2008)
27. Ulrich, J.: Supervised machine learning for email thread summarization. Master's thesis, University of British Columbia (2008)
28. Liu, F., Liu, Y.: Correlation between rouge and human evaluation of extractive meeting summaries. In: HLT 2008: Proceedings of the 46th Annual Meeting of the Association for Computational Linguistics on Human Language Technologies, Morristown, NJ, USA, pp. 201–204. Association for Computational Linguistics (2008)

# Multi-document Summarization Using Link Analysis Based on Rhetorical Relations between Sentences

Nik Adilah Hanin Binti Zahri and Fumiyo Fukumoto

Interdisciplinary Graduate School of Medicine and Engineering
University of Yamanashi, Japan
{g09dh103,fukumoto}@yamanashi.ac.jp

**Abstract.** With the accelerating rate of data growth on the Internet, automatic multi-document summarization has become an important task. In this paper, we propose a link analysis incorporated with rhetorical relations between sentences to perform extractive summarization for multiple-documents. We make use of the documents headlines to extract sentences with salient terms from the documents set using statistical model. Then we assign rhetorical relations learned by SVMs to determine the connectivity between the sentences which include the salient terms. Finally, we rank these sentences by measuring their relative importance within the document set based on link analysis method, PageRank. The rhetorical relations are used to evaluate the complementarity and redundancy of the ranked sentences. Our evaluation results show that the combination of PageRank along with rhetorical relations among sentences does help to improve the quality of extractive summarization.

**Keywords:** Probability model, n-gram, link-based analysis, Support Vector Machine, extractive summarization, rhetorical relations.

## 1 Introduction

Due to rapid growth of information on the Internet recently, finding specific data from huge amount of document is crucial since it requires a lot of time and efforts for users to read each document. As a result, automatic summarization has become an important technique nowadays. Text summarization helps to simplify information search and cut the search time by pointing the most relevant information which allows users to quickly comprehend the information contained in a large document.

The general approach of automatic text summarization is extractive or abstractive summarization. Extractive summarization focuses in finding the most salient sentences from the original document, while abstractive summarization focuses on generating summary by selecting only important terms from documents and might not contain original phrase or word. Our work focuses on extractive summarization. Previous works in this area have proposed various techniques such as, centroid-based summarization method [1], automated document indexing based on statistical latent model [2] and most recent technique, text summarization based on Cross-document Structure Theory (CST) relationship between sentences[3][4][5] .

A. Gelbukh (Ed.): CICLing 2011, Part II, LNCS 6609, pp. 328–338, 2011.

Multiple documents describing the same topic present some tough challenges for text summarization. For instance, a multi-document summary must consist of coherent information which represents the entire document in the set. Another challenging issue is dealing with multi-document phenomena such as complementarity, redundancy and overlapping during summarization [5]. These phenomena are caused by some multi-document properties such as paraphrases, partial overlap and elaboration, which provides similar information from different documents [4]. This becomes more difficult when the document describes an event that evolves over time that might give several repeated and also contradicted information.

To generate a generic summary for multi-document, the first step is to extract the most salient sentences from an individual document, referred as local context. Since that the document headline of news articles usually gives an overview of the overall events, our methodology make use of the document headlines to extract the local context from each document [6]. Next, by creating connectivity of each sentence and justifying their connection/relation to each other, in this research, we hope to identify the global context, referring to most important and relevant sentences from the document set while dealing with multi-document phenomena.

In this paper, we propose a method, which consists of extraction of relevant sentences using statistical model and summary generation using link analysis incorporated with rhetorical relations between sentences. In the next section, we discuss the related works on multi-document summarization. In Section 3, we present the overview of our system. The experiment setup and its evaluation result are presented in Section 4.

## 2 Related Works

One of the previous works on text summarization is proposed by [6]. This work describes an approach for abstractive summarization which capable of generating shorter summaries compared to the original sentences. This method applies statistical models for content selection and term ordering process to produce short summaries. The system build a model of the relationship between the features appear in the document headline and document content in order to select local context from individual document. This method, however, is not applicable to extract global context that represents the whole document set.

Link based analysis algorithms also have been successfully used in text summarization. The most common algorithms are HITS (Hyperlinked Induced Topic Search) [7] and PageRank [8], which are designed for ranking Web Pages. HITS determine two values for a page: its authority, which estimates the value of the content of the page, and its hub value, which estimates the value of its links to other pages. Meanwhile, PageRank consider the impact of both coming and outgoing links into one single model. It has shown in [9] that both algorithms provide the best performance during automatic unsupervised sentence extraction in the context of text summarization task. Hence, we were inspired to use link based analysis algorithm to determine the global context by measuring the ranking score of local context using PageRank. However, the final summary extraction according to PageRank might cause some redundancy problems since that the similar sentences will be ranked close to other. Given this issue, we made some improvement to PageRank to eliminate this problem.

Besides link analysis, the most recent approach in text summarization is based on CST relationship. One of the earliest works on this area is the incorporation of CST relations with MEAD summarizer proposed by [4]. This method proposes the enhancement of text summarization by replacing low-salience sentences with sentences that maximize the total number of CST relationship in the final summary. The most recent work is a deep knowledge approach system, CSTSumm (CST-based SUMMarizer) [5]. This system produces multi-document summaries from CST-analyzed document, which ranks input sentences according to the number of CST relation presents. CSTSumm shows a great capability of producing informative summaries since the system deals better with multi-document phenomena mentioned above. However, these methods are fully relied on manually annotated corpus and require deep linguistic knowledge.

It has been shown in the previous works that the information obtained from CST can improve multi-document summarization. We follow this idea, but we only utilize the rhetorical relationship between sentences presented by CST instead of CST itself to overcome the limitation mentioned above. We apply the rhetorical relationship to PageRank to improve the performance of the summarizer by including complementarity and redundancy analysis during summary extraction. Our aim is to utilize only the surface features of the sentences and minimize the using of annotated sentences so that our method is applicable to any other domains or languages.

## 3   System Overview

Our summarization system structure is illustrated in Figure 1. Our method focuses on 2 main tasks, which are extraction of relevant sentences, and summary generation using link analysis, PageRank. The summary generation task is consist of i) identification of rhetorical relations between sentences, and ii) sentence ranking by PageRank based on the identified rhetorical relations.

### 3.1   Extraction of Relevant Sentences

Document headline describing the entire events written in a document can benefits the extraction of local context from an individual documents. We use statistical model proposed by [6] to learn the relationship between the features which occurred both in headlines and documents. The probability of the terms for summary candidate can be computed as the product of the probability of the terms in the candidate sentences, assuming that the likelihood of a word in the summary is independent with each other. Hence, the overall probability is computed as the product of the likelihood of (i) the selected term from document, (ii) the term length and (iii) the most likely sequencing of the terms in the document sets, shown as follows:

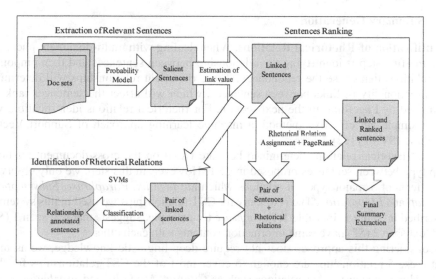

**Fig. 1.** System overview

$$P(w_i,..,w_n) = \prod_{i=1}^{n} P(w_i \in H \mid w_i \in D) \cdot P(len(H)=n) \cdot \prod_{i=2}^{n} P(w_i \mid w_1,...,w_{i-1}) \qquad (1)$$

where, $P(w_i \in H \mid w_i \in D)$ is derived from conditional probability of word occurred in the headline and documents, estimated as follows;

$$P(w_i \in H \mid w_i \in D) = \frac{P(w_i \in D \mid w_i \in H) \cdot P(w_i \in H)}{P(w_i \in D)} \qquad (2)$$

$H$ and $D$ represent the list of words contained in headline and document content. The model is computed for each word listed in the document sets and used to compute score for appropriate terms of candidate summary.

The term length $n$ is set from 3 to 12 words, and the normalization is then performed against the overall probability value. The probability of any word ordering for terms (iii) is computed by the word sequence probability model. Here, we used the simplest language model, the bigram model. The probability of a word sequence in a measured term is estimated by the product of the probabilities of seeing each word appear at the immediate left context. Meanwhile, the probability of the unseen word sequence in the training data are estimated by using back-off weight proposed by [11].

In order to extract sentences that relevant to the entire documents, we performed the overall probability measurement against all document in the data set. Finally, the sentences include terms with high probability score are extracted as local context for each document.

## 3.2  Summary Generation

**Identification of Rhetorical Relations.** When dealing with multi-document phenomena, the first step is identifying the relationship between sentences and then pinpoints the relations that cause the redundancy. These rhetorical relationships will determine the directionality of links between sentences, which will affect the sentences ranking estimated by PageRank in the next section. For rhetorical relations identification, we use a simple yet effective method, a machine learning approach of Support Vector Machine (SVMs) [12].

The rhetorical relations determined here are based on the cross-document relationship type between sentences proposed in CST. However, in this step, we only observed the effects of 5 major types of relations which are *Identity, Paraphrase, Subsumption, Overlap* and *Elaboration*. The taxonomy for CST relationships we used in this system is described in Table 1. In Table 1, *Paraphrase* for example, suggests that text span 1 (*S1*) and text span 2 (*S2*) have same information contents with each other.

Considering this approach does not require deep linguistic knowledge, we assume that these 5 relationships are enough to cover most of the CST relationships for this task. This is because other relations such as *Citation, Modality* and *Attribution* share similar characteristic of information content with *Identity* and *Paraphrase*, except for different version of event description. The rest of the relationships proposed by CST are covered by *Overlap* and *Elaboration* due to lack of significant characteristic presents unless with manual annotation.

We use CST-annotated sentences pair available at [13] as training data for the SVMs. We provide the following features of sentences pair to SVMs for learning purpose.

i)      Cosine similarity value between the sentences pair

$$\cos(S_1, S_2) = \frac{\sum s_{1,i} * s_{2,i}}{\sqrt{\sum (s_{1,i})^2} * \sqrt{\sum (s_{2,i})^2}} \tag{3}$$

ii)     Word overlap between the sentences pair

$$wol(S_1, S_2) = \frac{\#commonwords(S_1, S_2)}{\#words(S_1) + words(S_2)} \tag{4}$$

iii)    Lengths of both sentences;
        We set the feature vector for longer sentences as 1, and 0 for shorter
iv)     The overlap ratio of words from the first sentences in the second sentences, and vice versa

The input for the system will be the features derived from sentences pair in the test set. The output will be the identification of rhetorical relations between the sentences pair.

**Sentences Ranking.** The extracted local contexts are ranked according to their relative importance within the document set using PageRank to identify the global contexts. In search engines context, PageRank is a method to determine how important a

**Table 1.** Taxonomy of the rhetorical relationship used in the system

| Relationship | Description | Text span 1 (S1) | Text span 2 (S2) |
|---|---|---|---|
| Identity | The same text appears in more than one location | The Richter scale is a measure of ground motion as recorded on seismographs. | The Richter scale is a measure of ground motion as recorded on seismographs. |
| Paraphrase | Two text spans have the same information content | Wayne Tresemer, the county's Disaster Services director, said water was standing up to 5 feet deep in some streets in the town of Newark, and the fire department and other agencies used boats to evacuate some residents. | He said water was standing up to 5 feet deep in some streets on the city's east side, and the Newark Fire Department and other agencies used boats to evacuate some residents. |
| Subsumption | S1 contains all information in S2, plus additional information not in S2 | Thunderstorms swept through central and eastern Ohio, causing flooding that killed at least ten people, left dozens missing and forced hundreds of others from their homes, officials said today. | About 200 people were reported evacuated in central Ohio. |
| Elaboration | S1 elaborates or provides details of some information given more generally in S2 | As ferries increase in size, so the numbers on board at risk from mechanical failure or a crew error go up. | Ferries are among the safest vessels afloat. |
| Overlap (partial equivalence) | S1 provides facts X and Y while S2 provides facts X and Z; X, Y, and Z should all be non-trivial. | The high grain prices resulting from a lower supply would hit poor, food-importing countries the hardest, the U.N. agency warned. | Lyng said total consumption of 1988 crops will probably be somewhat less than previously estimated because prices are much higher. |

page can be on the Web according to the incoming hyperlinks counts from other pages. In this model, we assume that one sentence is linked to another sentences if there is a similarity value exists between them. Here, a sentence connectivity matrix is constructed based on cosine similarity (Eq.(3)) value between two sentences. We considered the link between sentences as vote of support. Therefore, the more links connected to the sentence, the more important the sentence become.

The directionality of this link is determined based on the rhetorical relationship between the two sentences estimated by SVMs. We considered the sentence pair belongs to *Overlap* type is having 2-way direction because of the partial equivalence information in both ways. However, for *Subsumption* and *Elaboration* type, the directionality is 1-way. For *Subsumption* type, the information contained in second sentences has been described in the first sentences along with other additional information; which makes the directionality is from second sentences to the first sentences. As for *Elaboration* type, the first sentences provides more details of information given more generally in the second sentences, which make the directionality same as *Subsumption*. However, in some cases, SVMs failed to identify the relationship between the sentences from data set. Here, we set the directionality in both ways to preserve the unidentified relations.

In addition, we make some modification of the directionality assignment for *Identity* and *Paraphrase*. Since these types of relations provide the same amount and quality of information within, we combined the incoming and outgoing links for both sentences, and estimate the PageRank score of a group of similar sentences. However, the values of incoming and outbound links do not change. We made this improvement to deal with the redundancy issue. Figure 2 demonstrates the assignment of relationship type with directionality and link modification against *Identity* and *Paraphrase* type.

Let the similarity value of both sentences be the value of each link. For a given sentence $S_i$, let $In(L_i)$ be the number of sentences that linked towards $S_i$, and let $Out(L_i)$ be the number of links from $S_i$ . The PageRank score for sentence, $S_i$ is defined as follows:

$$PR(S_i) = \frac{1-d}{N} + d \sum_{S_j \in In(S_i)} \frac{PR(S_j)}{|Out(S_j)|}$$ (5)

where $d$ is the optimum damping factor, set as 0.85 according to [8] and $N$ is the number of sentences in the document set.

Although PageRank score is computed only for the extracted local contexts, the score is determined by all incoming and outbound links from the entire sentences in document set. Finally, we applied sorting algorithm to rank the global contexts determined by the PageRank score in decreasing order. Here, we considered the sentences with high value of PageRank score contained high amount of information within them and the value indicates the relevance and importance level in the entire documents. These sentences are extracted in decreasing order of PageRank score as final summary according to the length of summary set in the system.

# 4    Evaluation

## 4.1  Data

We used 1 year of Reuters'96 corpus from August 20th, 1996 to August 19th, 1997 to train and build statistical model for local context extraction. The corpus is preprocessed using Brill's tagger [15] to POS-tag the sentences, extract content words and lemmas of the words. Our system is evaluated using 95 news articles from 11 document sets of test data obtained from Document Understanding Conference (DUC) 2002.

## 4.2  Summary Generation

For evaluation, we used similarity metrics to estimate the system performance. We experimented the cosine similarity measurement and compute the correlation between summaries extracted by the system and the summaries that manually produced by human. We used two types of manually-produced summaries, which are abstractive and extractive summaries adopted from DUC'2002. We referred the evaluation based on abstractive and extractive summaries as evaluation Task 1 and Task 2, respectively. Table 2 and Table 3 show the system performance for each task. According to

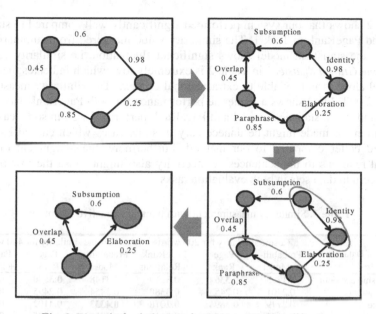

**Fig. 2.** Rhetorical relations assignment and link modification

the data availability of DUC'2002, the performance measurement based on abstractive summary is computed for 100 and 200 words summaries, and as for extractive summaries, the lengths are 200 and 400 words. The experimental results also include the similarity measurement of summaries generated by statistical model proposed by [6] and PageRank for comparison. The column *"Statistical Model"* and *"PageRank"* show the evaluation result by each method. The *"PageRank+Relations"* column shows evaluation result for proposed method. The bold font indicates the highest similarity measurement for each data set.

**Table 2.** Similarity measurement with manual abstractive summary

| Topic of Data Set | Cosine similarity for 100 words | | | Cosine similarity for 200 words | | |
|---|---|---|---|---|---|---|
| | Probability Model | Page Rank | PageRank + Relations | Probability Model | Page Rank | PageRank + Relations |
| Boat and ship accident | 0.1114 | 0.1091 | **0.1800** | 0.1739 | 0.1830 | **0.1949** |
| Drought | 0.1543 | 0.1588 | **0.2138** | 0.2355 | 0.2395 | **0.2633** |
| Earthquake | 0.2264 | 0.2210 | **0.2708** | 0.2513 | 0.3564 | **0.3671** |
| Flood | 0.1964 | **0.2590** | 0.1195 | **0.3145** | 0.2874 | 0.2103 |
| Hurricane Andrew disaster | 0.1837 | 0.2649 | **0.3091** | 0.2552 | 0.2434 | **0.2942** |
| Hurricane Gilbert disaster | 0.0637 | 0.2419 | **0.3043** | 0.1636 | 0.2784 | **0.3526** |
| Hurricane Hugo disaster | 0.1091 | 0.1922 | **0.2215** | 0.1764 | **0.3395** | 0.2461 |
| Earthquake in Iran | **0.2134** | 0.1627 | 0.1307 | 0.2024 | 0.2302 | **0.2428** |
| Earthquake in Nepal | 0.2327 | 0.3114 | **0.3237** | 0.1953 | **0.3259** | 0.2376 |
| Storm Twister | **0.2100** | 0.2058 | 0.1670 | **0.2120** | 0.1992 | 0.2078 |
| Volcano eruption | 0.1246 | **0.1892** | 0.1404 | 0.2369 | **0.2969** | 0.2364 |

Table 2 shows that our system performed significantly well compared to statistical model and PageRank in Task 1. The similarity value measured from summaries generated by the statistical model shows significant fluctuation, i.e similarity value for *"Hurricane Gilbert disaster"* in Table 2 is extremely low, which indicates the statistical model alone is not reliable to extract global context. The similarity measurement for Task 2 in Table 3 shows a moderate performance for both PageRank and our system. In both tasks, statistical model and PageRank performed well in some cases. The modification we made might be unnecessary in some cases which caused PageRank performed better compared to our method. Furthermore, the assignment of wrong rhetorical relations to the sentences connectivity also might cause the low performance for certain data sets in both evaluation tasks.

**Table 3.** Similarity measurement with manual extractive summary

| Topic of Data Set | Cosine similarity for 200 words | | | Cosine similarity for 400 words | | |
|---|---|---|---|---|---|---|
| | Probability Model | Page Rank | PageRank + Relations | Probability Model | Page Rank | PageRank + Relations |
| Boat and ship accident | 0.1793 | **0.3364** | 0.3279 | 0.3148 | **0.3530** | 0.3395 |
| Drought | 0.2401 | 0.2453 | **0.3380** | 0.3454 | 0.2495 | **0.3547** |
| Earthquake | 0.2779 | **0.3898** | 0.2078 | **0.4333** | 0.4102 | 0.3945 |
| Flood | 0.3677 | **0.3550** | 0.2111 | **0.3755** | 0.3575 | 0.3658 |
| Hurricane Andrew disaster | 0.2627 | 0.3504 | **0.3843** | 0.3154 | **0.3615** | **0.3615** |
| Hurricane Gilbert disaster | 0.1461 | 0.2524 | **0.2858** | 0.2813 | 0.3955 | **0.4359** |
| Hurricane Hugo disaster | 0.2301 | **0.3647** | 0.2606 | 0.2443 | 0.3194 | **0.3281** |
| Earthquake in Iran | 0.1443 | 0.3048 | **0.3464** | 0.3765 | **0.4069** | 0.3753 |
| Earthquake in Nepal | 0.2051 | **0.2697** | 0.2602 | 0.2837 | **0.3842** | 0.2819 |
| Storm Twister | **0.2523** | 0.2125 | 0.1906 | 0.3131 | **0.3533** | 0.2628 |
| Volcano eruption | **0.4076** | 0.3941 | 0.3183 | 0.3768 | 0.4636 | **0.4989** |

For reference, we include the similarity measurement between manually produced abstractive and extractive summaries from DUC`2002 in Table 4. The similarity value shows a moderate performance of summary generation considering that the extractive summary is manually produced by human. This indicates that our method shows a quite promising result despite of a few errors described in the above.

In addition, the application of rhetorical relationship during sentences ranking according to PageRank has improved the quality of extractive summarization. The indication and modification of the directionality of links between sentences not only helps to ranks the most salient sentences, but also helps to deal with multi-document phenomena faced by multi-document summarization. For example, in most redundancy cases, the sentences with similar content will be likely ranked closely to each other, as shown is Figure 3 (a), which is taken from the evaluation. As a result, the final summary content will be overlapped, and this does not serve the purpose of generating summary. The bold font in Figure 3 (b) indicates that our system has successfully eliminated this problem, and enables to extract more relevant sentences for final summary.

**Table 4.** Similarity measurement between manually-produced abstractive and extractive summaries (length : 200 words)

| Topic of Data Set | Cosine Similarity |
|---|---|
| Boat and Ship Accident | 0.3193 |
| Drought | 0.2408 |
| Earthquake | 0.3623 |
| Flood | 0.4190 |
| Hurricane Andrew disaster | 0.3334 |
| Hurricane Gilbert disaster | 0.3538 |
| Hurricane Hugo disaster | 0.3169 |
| Earthquake in Iran | 0.3566 |
| Earthquake in Nepal | 0.4206 |
| Storm Twister | 0.4567 |
| Volcano eruption | 0.3712 |

| | |
|---|---|
| 1) **The hurricane center said Gilbert was the most intense storm on record in terms of barometric pressure.**<br>2) **The center said Hurricane Gilbert was the most intense storm on record in terms of barometric pressure.**<br>3) The storm was headed west northwest, said National Hurricane Center director Bob Sheets.<br>4) Winds were still weakening as the storm moved west-northwest at 12 mph, he said.<br>5) Camille's storm surge was 25 feet high, but the hurricane center was forecasting a surge of only 8-12 feet for Gilbert, Zimmer said.<br>6) Flash floods were likely, it said.<br>7) ``Moisture and heat are what drives the hurricane, " Zimmer said. | 1) The hurricane center said Gilbert at one point was the most intense storm on record in terms of barometric pressure, which was measured Tuesday at 26.13 inches, breaking the 26.35 inches recorded for the 1935 hurricane that devastated the Florida Keys.<br>2) There were no reports of casualties.<br>3) Such storms have maximum sustained winds greater than 155 mph and can cause catastrophic damage.<br>4) Hundreds of homes were destroyed, he said.<br>5) Tropical storm force winds, at least 39 mph, extended out ward up to 250 miles to the north and 200 miles to the south of the center. |
| (a) Summay by PageRank | (b) Summary by the system |

**Fig. 3.** Redundancy elimination by the system

In future, the usage of lexical database such as *WordNet* would help to identify the synonyms or similar words in sentences in order to improve the automatic classification of rhetorical relationship by SVMs. Also, similarity metrics other than cosine similarity such as, the Blue metric might help to evaluate the performance of our system. Overall, our system is capable to extract generic summaries using PageRank based on rhetorical relation assignment by SVMs and requires no deep linguistic knowledge in the process.

# 5 Conclusion

This paper presented a novel method to automatic multi-document summarization using link analysis, PageRank based on rhetorical relations among sentences. According to the type of relations exists between the sentences, we assign directionality to each links that enhanced the overall PageRank score of each sentences. Our system also deals better with redundancy issue by modifying the connectivity of the sentences which successfully eliminates the redundancy problem. The most important feature is our system does not rely on fully annotated corpus and does not require deep linguistic knowledge during rhetorical relationship assignment to the sentences. The evaluation results show a quite promising potential of our summarization system. Future works will include (i) the improvement of rhetorical relations identification process, and (ii) expending the scope of summary generation.

# References

1. Radev, D.R., Jing, H., Sty, M., Tam, D.: Centroid-based summarization of multiple documents. Inf. Process. Manage. (40), 919–938 (2004)
2. Bhandari, H., Shimbo, M., Ito, T., Matsumoto, Y.: Generic Text Summarization Using Probabilistic Latent Semantic Indexing. In: The Third International Joint Conference on Natural Language Processing, Hyderabad, India (January 7-12, 2008)
3. Radev, D.R.: A common theory of information fusion from multiple text sources, step one: Cross-document structure. In: Proceedings of 1st ACL SIGDIAL Workshop on Discourse and Dialogue, Hong Kong (October 2000)
4. Zhang, Z., Blair-Goldensohn, S., Radev, D.R.: Towards CST-enhanced summarization. In: AAAI 2002 (August 2002)
5. Jorge, M.L.C., Pardo, T.S.: Experiments with CST-based Multidocument Summarization Workshop on Graph-based Methods for Natural Language Processing, ACL 2010, Uppsala, Sweden, pp. 74–82 (July 2010)
6. Banko, M., Mittal, V.O., Witbrock, M.J.: Headline Generation Based on Statistical Translation. In: ACL 2000, Proceedings of the 38th Annual Meeting of the Association for Computational Linguistics, Hong Kong (October 3-6, 2000)
7. Kleinberg, J.M.: Authoritative sources in a hyperlinked environment. Journal of the ACM 46(5), 604–632 (1999)
8. Brin, S., Page, L.: The anatomy of a large-scale hypertextual Web search engine. Computer Networks and ISDN Systems 30(1-7) (1998)
9. Mihalcea, R.: Graph-based Ranking Algorithms for Sentence Extraction, Applied to Text Summarization. In: Proceedings of the 42nd Annual Meeting of the Association for Computational Linguistics, companion volume (ACL 2004), Barcelona, Spain (July 2004)
10. Erkanand, G., Radev, D.: LexPageRank: Prestige in muli-document text summarization. In: Proceedings of EMNLP (2004)
11. Katz, S.: Estimation of probabilities from sparse data for the language model component of a speech recognizer. IEEE Trans. on Acoustics, Speech and Signal Processing (1987)
12. Vapnik, V.: The Nature of Statistical Learning Theory. Springer, New York (1995)
13. Radev, D.R., Otterbacher, J.: CSTBank PhaseI,
    http://tangra.si.umich.edu/clair/CSTBank/phase1.htm
14. Brill, E.: A Simple Rule-based Part-of-Speech Tagger. In: Proceedings of 3rd Conference on Applied Natural Language Processing, pp. 152–155 (1992)

# Co-clustering Sentences and Terms for Multi-document Summarization

Yunqing Xia[1], Yonggang Zhang[1,2], and Jianmin Yao[2]

[1] Department of Computer Science and Technology,
Tsinghua University, Beijing 100084, China
yqxia@tsinghua.edu.cn
[2] School of Computer Science and Technology,
Soochow University, Suzhou 215006, China
yonggang118@gmail.com, jyao@suda.edu.cn

**Abstract.** Two issues are crucial to multi-document summarization: diversity and redundancy. Content within some topically-related articles are usually redundant while the topic is delivered from diverse perspectives. This paper presents a co-clustering based multi-document summarization method that makes full use of the diverse and redundant content. A multi-document summary is generated in three steps. First, the sentence term co-occurrence matrix is designed to reflect diversity and redundancy. Second, the co-clustering algorithm is performed on the matrix to find globally optimal clusters for sentences and terms in an iterative manner. Third, a more accurate summary is generated by selecting representative sentences from the optimal clusters. Experiments on DUC2004 dataset show that the co-clustering based multi-document summarization method is promising.

**Keywords:** Co-clustering, multi-document summarization, term extraction.

## 1 Introduction

Handling a large set of topically-related articles manually is usually laborious and time-consuming. Aiming at generating a summary that covers the major themes in an article collection, multi-document summarization provides a promising solution to the information overload problem. An ideal multi-document summary should cover not only the key topic of the multi-documents but also the diverse views of the multi-documents. Two distinct characteristics make multi-document summarization rather different from single-document summarization: diversity and redundancy [1-5]. Content within some topically-related articles are usually redundant while the topic is delivered from diverse perspectives. This is because the writers usually show common interests on popular target but they tend to report the target from different perspectives. As a result, diversity among the articles tends to be significant. However, some background information is usually necessary for the readers to follow the story. Therefore, redundant sentences are constantly found within the articles.

A. Gelbukh (Ed.): CICLing 2011, Part II, LNCS 6609, pp. 339–352, 2011.

A variety of methods have been developed to address the above two issues. The common agreement is that diversity and redundancy should be appropriately addressed to find representative sentence. For example, maximal marginal relevance (MMR) was adopted by Lin and Hovy (2002) to penalize the sentences being highly redundant with the representative sentences [1]. A conceptual model was designed by Harabagiu and Lacatusu (2005) to condense diverse topic information [2]. Affinity graph was used by Wan and Yang (2008) to measure diversity penalty [3]. A sentence-based topic model was designed by Wang et al. (2009) [4] to manage diversity. In general, three common assumptions were made in multi-document summarization research: (1) Diversity and redundancy are independent of each other; (2) Diversity should be strengthened; and (3) Redundancy should be weakened.

We argue that diversity and redundancy actually work with each other perfectly to indicate representative sentences and terms. Two observations are enlightening. Firstly, diversity is reflected by terms, thus sentences can be grouped into different clusters. Meanwhile, redundancy is also reflected by terms, thus sentences about the same theme can be grouped into one cluster. Secondly, the sentence clusters facilitate term clustering. As a result, the diverse or redundant terms can be detected more precisely, which in turn helps to find sentence clusters more precisely. The mutual enhancement makes it possible to find optimal clusters for sentences and terms.

In this work, we propose to design a sentence-term co-occurrence matrix to represent the diversity and redundancy. The matrix is similar to the contingency table in the co-clustering theory [5]. Naturally, we choose the co-clustering algorithm to manage sentences and terms so as to identify important sentences and terms. One key issue for clustering algorithm is feature weighting. So we extend the sentence-term co-co-occurrence matrix to incorporate weights for sentences and terms. Thereafter, the co-clustering algorithm is applied on the matrix to find the optimal clusters and weights of sentences and terms. Finally, summary can be generated by selecting sentences according to the optimal sentence clusters and sentence weights. As a byproduct, key terms for the topic are also produced. Selecting representative sentences using clustering algorithm is not a new attempt in multi-document research. However, this is the first attempt to apply co-clustering algorithm to further improve clustering accuracy. Experiments on DUC2004 dataset show that the co-clustering algorithm produces satisfactory results.

The rest of this paper is organized as follows. Section 2 reviews related work. An example is presented in Section 3 to motivate the co-clustering solution. Section 4 illustrates the technical details of the co-clustering based multi-document summarization method. Experiments as well as discussions are presented in Section 5 and Section 6 concludes the paper.

## 2   Related Work

We review related work on multi-document summarization and the co-clustering theory, upon which our method is built.

## 2.1 Multi-document Summarization

Generally, there are two strategies for multi-documents summarization: the statistical-based analysis and the graph-based analysis.

The statistical-based analysis seeks to rank sentences using sentence statistics. In NeATS system, Lin and Hovy (2002) used sentence position, term frequency, topic signature and term clustering to rank sentences and applied MMR equation to remove redundancy [1]. In MEAD system, Radev et al. (2004) proposed the centroid-based method to score sentences based on sentence-level and inter-sentence features, including cluster centroids, position, TF-IDF (term frequency inverse document frequency) [6]. Harabagiu and Lacatusu (2005) investigated five popular topic representations and explored a conceptual representation of topics [2].

The graph-based analysis attempts to rank sentences based on votes or recommendations between each other. A graph-connectivity model was designed to find salient sentences in WebSumm system [7]. Mihalcea (2004) evaluated the performance of the most common graph-based ranking algorithms, such as Hyperlink-Induced Topic Search, Positional Power Function and PageRank [8]. In LexPageRank system, Erkan and Radev (2004) proposed to calculate sentence importance based on the concept of eigenvector centrality [9]. Wan and Yang (2008) also investigated on cluster-based link analysis to leverage the cluster-level information [3].

In this work, graph-based link analysis techniques are also adopted in sentence weighting. However, two major differences are notable. Firstly, the clusters are re-constructed in our method with refined weights of sentences and terms in an iterative way while previous works used the clusters invariably. Secondly, both diversity and redundancy are used to enhance weights of sentences and terms while most previous works tried to weaken the redundancy.

## 2.2 Co-clustering

Proposed by Dhillon et al. (2003), the co-clustering algorithm seeks to simultaneously cluster two discrete random variables that subject to an empirical joint probability distribution [5]. The task is accomplished by maximizing the mutual information between the clustered random variables. It was also proved by Dhillon et al. (2003) that the algorithm is effective in document clustering using word-document occurrence data. The co-clustering algorithm was also used to classify out-of-domain data [10] and to classify cross-domain knowledge extracted from Wikipedia [11].

In our work, co-clustering is employed to find optimal clusters of sentences and terms. One significant extension is that weights of sentences and terms are incorporated in the co-clustering iterations. That is, in each co-clustering iteration, besides the clustering of sentences and terms, weights for sentences and terms are refined, which are in turn used in the next iteration by the clustering algorithm to generate finer clusters.

## 3  Co-clustering Sentences and Terms

Diversity and redundancy can be represented in a sentence-term matrix, which shows the co-occurrence relation that sentences contain certain terms and terms appear within certain sentences. This exhibits the basic intuition that we adopt the co-clustering theory to find optimal clusters of sentences and terms simultaneously.

In this section, a small example, which includes only six sentences, is presented to illustrate the application of co-clustering on multi-document summarization. The six sentences, which are listed in Table 1, are selected from articles on the topic of Olympic Corruption in City Bidding for 2002 Winter Games. To save space, only eight terms, which are listed in Table 2, are selected as representatives to illustrate the idea. It is agreed by human judges that sentence $s_1$ and $s_6$ are more important to the topic while sentence $s_3$ is less important.

**Table 1.** Example sentences with the topic of Olympic Corruption in City Bidding for 2002 Winter Games

| ID | Sentence |
|---|---|
| $s_1$ | Moving quickly to tackle an escalating corruption scandal, IOC president Juan Antonio Samaranch questioned Salt Lake City officials Friday in the first ever investigation into Hodler's allegations of vote-buying in Olympic city selection. |
| $s_2$ | Samaranch said the IOC would possibly consider a new procedure to eliminate the temptations for corruption in the selection of host cities. |
| $s_3$ | Pound said there had been concern in the IOC for some time about agents. |
| $s_4$ | Samaranch expressed surprise at allegations of corruption made by the IOC executive board member Marc Hodler of Switzerland in selection of Salt Lake City. |
| $s_5$ | A top IOC official on Saturday made explosive allegations of widespread Olympic corruption, saying agents demand up to \$1 million to deliver votes in the selection of host cities. |
| $s_6$ | Tasuku Tsukada was responding to allegations by Marc Hodler, the Swiss member of the IOC executive board, of systematic corruption in the Olympic bidding. |

**Table 2.** Eight terms selected from the example sentences

| ID | Term | ID | Term |
|---|---|---|---|
| $t_1$ | selection | $t_5$ | allegations |
| $t_2$ | corruption | $t_6$ | Salt Lake City |
| $t_3$ | Samaranch | $t_7$ | Olympic city |
| $t_4$ | Marc Hodler | $t_8$ | IOC |

## 3.1 Co-occurrence Matrix

**Definition 1**: *Co-Occurrence Matrix*
Given that the articles contains $N$ sentences, denoted by $s_i (i = 1, ..., N)$, and $K$ terms, denoted by $t_j (j = 1, ..., K)$, the *co-occurrence matrix* is defined as an $N \times K$ matrix $V = (v_{ij})$, in which $v_{ij}$ represents the occurrence number of term $t_j$ in sentence $s_i$.

For the six sentences in Table 1 and eight terms in Table 2, a $6 \times 8$ co-occurrence matrix is built, which is shown in Table 3. The diversity and redundancy between sentences can be effectively discovered according to co-occurrence matrix. For example, in Table 3, sentence $s_1$ and $s_4$ are highly redundant as they overlap on seven terms and sentence $s_2$ and $s_6$ are more diverse as they overlap on only two terms.

**Table 3.** Sample co-occurrence matrix

| – | – | $t_1$ | $t_2$ | $t_3$ | $t_4$ | $t_5$ | $t_6$ | $t_7$ | $t_8$ |
|---|---|---|---|---|---|---|---|---|---|
| – | 0 | 0.509 | 0.940 | 0.521 | 0.288 | 0.521 | 0.290 | 0.288 | 0.940 |
| $s_1$ | 0.817 | 1 | 1 | 1 | 1 | 1 | 1 | 1 | 1 |
| $s_2$ | 0.791 | 1 | 1 | 1 | 0 | 0 | 0 | 0 | 1 |
| $s_3$ | 0.306 | 0 | 0 | 0 | 0 | 0 | 0 | 0 | 1 |
| $s_4$ | 0.817 | 0 | 1 | 1 | 1 | 1 | 1 | 1 | 1 |
| $s_5$ | 0.791 | 1 | 1 | 0 | 0 | 1 | 0 | 0 | 1 |
| $s_6$ | 0.817 | 0 | 1 | 0 | 1 | 1 | 1 | 0 | 1 |

## 3.2 Sentence-Term Matrix

**Definition 2**: *Sentence-Term Matrix*
Given a co-occurrence matrix $V$, sentence weights $W_s = \{w_i\}(i - 1, ..., N)$ and term weights $W_t = \{w_j\}(j - 1, ..., K)$, the *sentence-term matrix* $M$ is defined as:

$$M = \begin{pmatrix} 0 & W_t \\ W_s^T & V \end{pmatrix}$$

in which weights for sentences and terms refers to their importance to the topic that the articles address. In this work, the weights are calculated using certain weighting algorithm (see Section 4.2) starting from the co-occurrence matrix. For the example, the initial sentence-term matrix is given in Table 4.

**Table 4.** Sample initial sentence-term matrix

| – | – | $t_1$ | $t_2$ | $t_3$ | $t_4$ | $t_5$ | $t_6$ | $t_7$ | $t_8$ |
|---|---|---|---|---|---|---|---|---|---|
| – | 0 | 0.513 | 0.700 | 0.540 | 0.552 | 0.631 | 0.550 | 0.443 | 0.510 |
| $s_1$ | 0.793 | 1 | 1 | 1 | 1 | 1 | 1 | 1 | 1 |
| $s_2$ | 0.571 | 1 | 1 | 1 | 0 | 0 | 0 | 0 | 1 |
| $s_3$ | 0.250 | 0 | 0 | 0 | 0 | 0 | 0 | 0 | 1 |
| $s_4$ | 0.750 | 0 | 1 | 1 | 1 | 1 | 1 | 1 | 1 |
| $s_5$ | 0.582 | 1 | 1 | 0 | 0 | 1 | 0 | 0 | 1 |
| $s_6$ | 0.653 | 0 | 1 | 0 | 1 | 1 | 1 | 0 | 1 |

The iterative sentence/term co-clustering starts from the sentence-term matrix shown in Table 4. On the one hand, the co-occurrence values can be used to find both sentence clusters and term clusters according to their redundancy and diversity. On the other hand, once sentences or terms are clustered, sentence and term weights are re-calculated. The weights can be in turn used by the clustering algorithm to rebuild sentence and term clusters. Finally, the optimal clusters and weights are identified.

### 3.3 Co-clustering Algorithm

The co-clustering algorithm implements an iterative clustering model that makes use of row-column constraint to conduct mutual enhancement. In our case, we view the sentence-term matrix as the row-column constraint and design the co-clustering procedure is designed to find optimal sentence/term clusters.

**Fig. 1.** Sample clusters of sentences

For the aforementioned example, sentences are first grouped into four clusters according to the pair-wise similarities (see Fig.1) and term weights. With the sentences clustered, the sentence-term matrix is updated in Table 5. The original sentence $s_1$ and $s_4$ are grouped into a cluster $s_{1,4}$, and the original sentences $s_2$ and $s_5$ into another cluster $s_{2,5}$.

**Table 5.** Sample sentence-term matrix with sentences clustered

| – | – | $t_1$ | $t_2$ | $t_3$ | $t_4$ | $t_5$ | $t_6$ | $t_7$ | $t_8$ |
|---|---|---|---|---|---|---|---|---|---|
| – | 0 | 0.414 | 0.670 | 0.443 | 0.540 | 0.607 | 0.540 | 0.353 | 0.710 |
| $s_{1,4}$ | 0.858 | 0.5 | 1 | 1 | 1 | 1 | 1 | 1 | 1 |
| $s_{2,5}$ | 0.634 | 1 | 1 | 0.5 | 0 | 0.5 | 0 | 0 | 1 |
| $s_3$ | 0.284 | 0 | 0 | 0 | 0 | 0 | 0 | 0 | 1 |
| $s_6$ | 0.653 | 0 | 1 | 0 | 1 | 1 | 1 | 0 | 1 |

The co-occurrence matrix changes accordingly by calculating new sentence cluster centroids. For example, the centroid for sentence cluster $s_{1,4}$ is (0.5,1,1,1,1,1,1,1). The refined weights for sentence clusters and terms are also updated with the weighting algorithm.

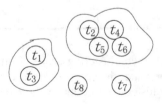

**Fig. 2.** Sample clusters of terms

Thereafter, with the updated sentence weights, the terms will be re-clustered. For the example, terms are grouped into four clusters (see Fig.2) according to the sentences weights. With the terms clustered, the sentence-term matrix is updated in Table 6. The original terms $t_1$ and $t_3$ are grouped into cluster $t_{1,3}$, and the original terms $t_2$, $t_4$, $t_5$ and $t_6$ into cluster $t_{2,4,5,6}$. Again, new weights for sentences and terms are calculated with the weighting algorithm.

**Table 6.** Sample sentence-term matrix with both sentence and terms clustered

| – | – | $t_{1,3}$ | $t_{2,4,5,6}$ | $t_7$ | $t_8$ |
|---|---|---|---|---|---|
| – | 0 | 0.503 | 0.680 | 0.412 | 0.858 |
| $s_{1,4}$ | 0.829 | 0.75 | 1 | 1 | 1 |
| $s_{2,5}$ | 0.626 | 0.75 | 0.375 | 0 | 1 |
| $s_3$ | 0.421 | 0 | 0 | 0 | 1 |
| $s_6$ | 0.642 | 0 | 1 | 0 | 1 |

Assuming weights for the clusters can be inherited by all their members, a brand refined sentence-term matrix, which is shown in Table 7, is obtained.

**Table 7.** Sample refined sentence-term matrix

| – | – | $t_1$ | $t_2$ | $t_3$ | $t_4$ | $t_5$ | $t_6$ | $t_7$ | $t_8$ |
|---|---|---|---|---|---|---|---|---|---|
| – | 0 | 0.503 | 0.679 | 0.503 | 0.680 | 0.679 | 0.679 | 0.412 | 0.860 |
| $s_1$ | 0.829 | 1 | 1 | 1 | 1 | 1 | 1 | 1 | 1 |
| $s_2$ | 0.626 | 1 | 1 | 1 | 0 | 0 | 0 | 0 | 1 |
| $s_3$ | 0.421 | 0 | 0 | 0 | 0 | 0 | 0 | 0 | 1 |
| $s_4$ | 0.829 | 0 | 1 | 1 | 1 | 1 | 1 | 1 | 1 |
| $s_5$ | 0.626 | 1 | 1 | 0 | 0 | 1 | 0 | 0 | 1 |
| $s_6$ | 0.642 | 0 | 1 | 0 | 1 | 1 | 1 | 0 | 1 |

By now, a co-clustering iteration has finished. We find sentence $s_1$ and $s_4$ as well as term $t_8$ are assigned larger weights, which accords with our observation well that sentence $s_1$ and $s_4$ are more important in addressing the topic.

The co-clustering procedure repeats the above iteration until all clusters remain stable. For the example, the optimal sentence clusters are $s_{1,4,6}$, $s_{2,5}$ and $s_3$. The optimal weights for sentences and terms are given in Table 8.

**Table 8.** Sample optimal sentence-term matrix

| – | – | $t_1$ | $t_2$ | $t_3$ | $t_4$ | $t_5$ | $t_6$ | $t_7$ | $t_8$ |
|---|---|---|---|---|---|---|---|---|---|
| – | 0 | 0.509 | 0.940 | 0.521 | 0.288 | 0.521 | 0.290 | 0.288 | 0.940 |
| $s_1$ | 0.817 | 1 | 1 | 1 | 1 | 1 | 1 | 1 | 1 |
| $s_2$ | 0.791 | 1 | 1 | 1 | 0 | 0 | 0 | 0 | 1 |
| $s_3$ | 0.306 | 0 | 0 | 0 | 0 | 0 | 0 | 0 | 1 |
| $s_4$ | 0.817 | 0 | 1 | 1 | 1 | 1 | 1 | 1 | 1 |
| $s_5$ | 0.791 | 1 | 1 | 0 | 0 | 1 | 0 | 0 | 1 |
| $s_6$ | 0.817 | 0 | 1 | 0 | 1 | 1 | 1 | 0 | 1 |

With the sentence clusters and weights, sentences are selected to generate the summary (see Section 4.3). For the example, the summary is generated by selecting sentences from clusters with largest weight (i.e., $s_1$ from cluster $s_{1,4,6}$ and $s_2$ from cluster $s_{2,5}$), given in Fig.3.

> ($s_1$)Moving quickly to tackle an escalating corruption scandal, IOC president Juan Antonio Samaranch questioned Salt Lake City officials Friday in the first ever investigation into Hodler's allegations of vote-buying in Olympic city selection. ($s_2$)Samaranch said the IOC would possibly consider a new procedure to eliminate the temptations for corruption in the selection of host cities.

**Fig. 3.** A sample multi-document summary

## 4 Co-Clustering Based Multi-document Summarization

The workflow for co-clustering (CoC) based multi-document summarization method is shown in Fig.4.

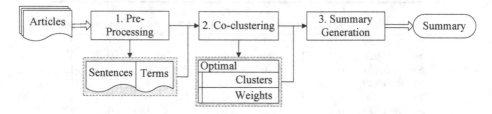

**Fig. 4.** Workflow for the CoC-based multi-document summarization method

The CoC-based method is composed of three modules: pre-processing, co-clustering and summary generation. The pre-processing module accepts multiple articles and outputs sentences and terms. The co-clustering module produces optimal clusters and weights for both sentences and terms. The summary generation module makes use of the optimal clusters and weights to generate a summary.

### 4.1 Pre-processing

The following three steps are performed to extract the sentences and terms out of the input articles:

- *Sentence segmentation.* Punctuation marks for period, question and exclamation are applied to detect sentences.
- *Stop word recognition.* As stop words are used in term extraction, a stop word list is applied to recognize stop words within text.

- *Term extraction.* Word strings that contain no stop word are considered as term candidate. Then c-value[1] is calculated for each candidate. A threshold is used to filter terms with smaller c-value. In this work, the threshold for c-value is set 2 empirically.

Finally, a sentence set and a term set are obtained for each topic.

## 4.2 Co-Clustering

I. *Algorithm*
The co-clustering module assembles the weighting and clustering functions using the co-clustering algorithm in Figure 5.

---

**Algorithm**: Co-Clustering

---

Input: Sentence set and term set.
Output: Optimal sentence weights and term weights.
**Initialize**:
  1. Build sentence-term co-occurrence matrix
  2. Compute sentence weights considering each sentence as a cluster.
  3. Compute term weights considering each term as a cluster
  4. Create initial sentence-term matrix.
  5. $\theta=1$.
**Repeat**:
  6. Find sentence clusters using terms as features.
  7. Re-compute sentence weights.
  8. Find term cluster using sentences as features.
  9.  Re-compute term weights.
  10. Update sentence-term matrix.
  11. $\theta$++.
**Until** clusters of sentences and terms do not change.

---

**Fig. 5.** Co-clustering algorithm for sentences and terms

In initialization, the sentence-term co-occurrence matrix is created by counting term occurrences in sentence (line #1). Then each sentence and each term is assigned a weight (line #2~3). The co-occurrence matrix and the weights are combined into a sentence-term matrix (line #4). Thereafter, the algorithm repeats the iteration for sentence/term clustering (line #6 and #8) and re-weighting (line #7 and #9) until all clusters remain stable. Once new clusters are obtained, terms and sentences are re-weighted. Weighting algorithms and clustering algorithms are given below.

---

[1] C-value is a domain-independent method for multi-word ATR which aims to improve the extraction of nested terms. The method takes as input an SL corpus and produces a list of candidate multi-word terms [12].

II. *Weighting*

Given that sentences and terms are clustered, certain weighting algorithm can be applied to calculate weights for the sentences and terms. Enlightened by (Mihalcea, 2004) [8], some graph-based algorithms are integrated in this work to enhance diversity in weighting. In our implementation, the graph is defined as $G = (S, T, E_0, E_{SC}, E_{TC})$, in which $S$ and $T$ are sentences nodes and term nodes respectively; $R_0$ refers to the edge between sentence and term; $R_{SC}$ and $R_{TC}$ denote edges between two sentences and two terms, respectively. Hence, a two-layer directed graph is built. We also use $w(s)$ to denote weight of sentence $s$ and $w(t)$ to denote weight of term $t$. The following three algorithms are explored to weight sentences or terms.

- *Hyperlink-Induced Topic Search* (HITS): Developed by Kleinberg (1999), HITS is an iterative algorithm designed to rank Web pages according to their *authority* degree [13].
- *Positional Power Function* (PPF): Introduced by Herings et al. (2001), PPF is a ranking algorithm that determines the score of a vertex as a function that combines both the number of its successors and the score of its successors [14].
- *PageRank* (PR): Developed by Brin and Page (1998), PageRank is a method for Web link analysis considering integrates the impact of both incoming and outgoing links [15].

III. *Clustering*

Sentences are clustered using terms as features and terms are clustered using sentences features. The co-clustering algorithm is thus adopted to manage the bootstrapping process until global optimal sentence clusters and term clusters are found. Three classical clustering algorithms are compared in the experiments: agglomerative clustering, divisive clustering and k-means [3].

In practice, it is hard to predict number of natural clusters. Therefore, an empirical similarity threshold $TH_S$ is set to guide the clustering process. Given a set of objects, we alter the clustering parameters so that every cluster satisfies the following termination condition.

$$TH_S \approx \frac{\sum_{i \in V} \sum_{j \in V, j \neq i} Sim(c_i, c_j)}{|V| \times (|V| - 1)}, (|V| > 1), \tag{1}$$

in which $V$ represents object set, $|V|$ total number of objects, and $c_i$ and $c_j$ centroids for two different object clusters.

Clusters are represented by their weighted centroids. So, the pair-wise cluster similarity is measured by calculating cosine similarity between centroid vectors for the two clusters. Obviously, weight plays an important role in similarity measuring. Intuitively, similarity can be measured more precisely with more accurate weights. Recall concept of co-clustering, we believe that weights of sentences and terms can be refined with the co-clustering theory in an iterative manner.

### 4.3 Summary Generation

Sentences are sequentially picked out of sentence clusters to generate a summary in the following steps:

*Step* 1: Clusters are ranked according to cluster weight, which is average sentence weight within the cluster.

*Step* 2: Sentences within each cluster are ranked according to the re-calculated sentence weight using the same weighting algorithm as in the co-clustering module. That is, term weights are used to calculate weights for sentences.

*Step* 3: One after another, the top-weighted sentence in every cluster is picked out to form the summary, until a user-preferred summary length is met. If the present summary is already longer than the length, the sentences from the clusters with smaller weights are excluded from the summary. Otherwise, a new round of sentence selection is performed until the summary length is met.

## 5   Evaluation

### 5.1   Setup

*Dataset*
The dataset in DUC2004 task #2 is used as test data. It contains 500 articles for 50 topics, namely, 10 articles for each topic. To build the gold standard, eight human judges were employed and each was asked to create summaries for 25 topics and every topic has four manually generated summaries.

*Evaluation Metrics*
Following DUC2004, we adopt the ROUGE measures to evaluate our method. Introduced by Lin (2004), the ROUGE measures count number of overlapping units between the computer-generated summary and the gold-standard summaries created by human judges. Then the counts are used to calculate precision, recall and F measure. Finally, the micro average is calculated to evaluate the performance on all topics. According to (Lin, 2004) [16], ROUGE-N and ROUGE-SU work well when stop words are excluded from matching.

### 5.2   Experiment I: Human Judges vs. Systems

This experiment aims to compare our system against human judges and the state-of-the-art systems.

Thirty-five systems participated in DUC2004 multi-document summarization task, in which S65 and S67 achieve the highest ROUGE scores. Our system (i.e., CoC) adopts HITS in weighting, agglomerative algorithm in clustering, and defines the clustering similarity threshold 0.8. Two known systems are compared: (1) S-MMR: a statistical-based system incorporating MMR equation [1]; (2) G-HITS: a graph-based system adopting HITS algorithm [3]. Experimental results are presented in Table 9.

According to Table 9, we find that the CoC system outperforms any other systems on DUC2004 dataset. Moreover, our system outperforms three human judges on

ROUGE-2. We attribute the outperformance to the involvement of term extraction because more than 30% terms in DUC2004 dataset are multi-word expressions, e.g., *Olympic city* and *federal antitrust law*. This results in a relatively higher performance under ROUGE-2 measure.

**Table 9.** Experimental results of systems and human judges (A~H)

|              |       | ROUGE-1   | ROUGE-2    | ROUGE-SU  |
|--------------|-------|-----------|------------|-----------|
| Systems      | CoC   | **0.38459** | **0.09382** | **0.13231** |
|              | S65   | 0.37938   | 0.09148    | 0.12975   |
|              | S67   | 0.37542   | 0.09215    | 0.13036   |
|              | S-MMR | 0.38268   | 0.08992    | 0.13191   |
|              | G-HITS | 0.37967  | 0.08888    | 0.12861   |
| Human judges | A     | 0.39596   | 0.08976    | 0.14266   |
|              | B     | 0.39971   | 0.09410    | 0.15162   |
|              | C     | 0.39804   | 0.09888    | 0.14677   |
|              | E     | 0.40908   | 0.09800    | 0.15212   |
|              | D     | 0.40796   | *0.10790*  | 0.15215   |
|              | F     | 0.40976   | 0.08904    | 0.15316   |
|              | G     | 0.39000   | 0.08570    | 0.13837   |
|              | H     | *0.41808* | 0.10475    | *0.15563* |

## 5.3  Experiment II: Weighting Algorithms

This experiment shows how weighting algorithm influences performance of our system. The three weighting algorithms described in Section 4.2 are investigated. For other components, we adopt the agglomerative algorithm in sentence/term clustering and the similarity threshold is set 0.8. Experimental results are presented in Table 10.

**Table 10.** Experimental results of CoC system with different weighting algorithms

| Algorithms | ROUGE-1   | ROUGE-2   | ROUGE-SU  |
|------------|-----------|-----------|-----------|
| HITS       | **0.38459** | **0.09382** | **0.13231** |
| PPF        | 0.36683   | 0.07771   | 0.11801   |
| PageRank   | 0.38412   | 0.09100   | 0.13177   |

Seen from Table 10, our method yields best performance with HITS algorithm on all ROUGE measures. In fact, the *PageRank* algorithm yields similar performance. The PPF algorithm has the worst performance. This proves that graph-based ranking algorithms for Web link analysis can be also effective for sentence extraction.

## 5.4  Experiment III: Clustering Algorithms

This experiment shows how clustering algorithm influences performance of our system. The three clustering algorithms described in Section 4.2 are investigated. For other components, we use HITS algorithm in weighting and set the clustering similarity threshold 0.8. Experimental results are presented in Table 11.

**Table 11.** Experimental results of CoC system with different clustering algorithms

| Algorithms | ROUGE-1 | ROUGE-2 | ROUGE-SU |
|---|---|---|---|
| K-means | 0.36912 | 0.08396 | 0.12401 |
| Divisive | 0.36017 | 0.08592 | 0.11987 |
| Agglomerative | **0.38459** | **0.09382** | **0.13231** |

Shown in Table 11, the agglomerative algorithm yields best performance on all ROUGE measures. When we looked into the clustering results produced by the three clustering algorithm, we found the k-means algorithm and divisive algorithm produce much more clusters than the agglomerative algorithm when the termination condition is applied. For the k-means algorithm, selection of the initial $K$ objects is tricky, resulting in much more errors. The divisive algorithm suffers the same problem as the k-means algorithm is applied in every dividing step.

### 5.5 Experiment VI: Similarity Thresholds in Clustering

This experiment intends to evaluate how similarity threshold influences system performance. Seven thresholds, ranging from 0.4 to 1.0, are investigated. For other components, *HITS* algorithm is adopted in weighting and agglomerative algorithm in clustering. Experimental results are presented in Fig.6.

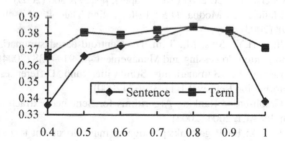

**Fig. 6.** Experimental results of CoC system with different similarity thresholds in clustering

It is shown in Figure 6 that system performance increases until similarity threshold meets 0.8. We address the observation as follows. In agglomerative clustering algorithm, increasing threshold results in more clusters and each cluster contains objects that are closer. But when the threshold is close to 1, say 0.9, too many clusters are created and redundancy in these clusters becomes very little, making further analysis rather difficult.

## 6  Conclusion

Diversity and redundancy are two major challenging issues for generic multi-document summarization. In this paper, sentence-term matrix is designed to represent diversity and redundancy within multiple articles. The matrix covers both sentence-term co-occurrences and their weights, in which the latter can be further refined to represent diversity and redundancy more accurately. Finding that diversity and redundancy enhance each other to form natural clusters, a co-clustering based method is thus proposed to apply the sentence-term matrix to yield the globally optimal clusters for sentences and terms in a few iterations. Finally, summary can be generated based on the optimal clusters and weights. Experimental results on DUC2004 dataset show that the co-clustering based method is promising in performing the task of generic multi-document summarization.

**Acknowledgment.** This work is partially supported by NSFC (60703051, 60970057, 61003152) and MOST of China (2009DFA12970). We thank the reviewers for the valuable comments and advices.

# References

1. Lin, C.-Y., Hovy, E.H.: From Single to Multi-document Summarization: A Prototype System and its Evaluation. In: ACL 2002, pp. 457–464 (2002)
2. Harabagiu, S., Lacatusu, F.: Topic themes for multi-document summarization. In: ACM SIGIR 2005, pp. 202–209 (2005)
3. Wan, X., Yang, J.: Multi-Document Summarization Using Cluster-Based Link Analysis. In: ACM SIGIR 2008, pp. 299–306 (2008)
4. Wang, D., Zhu, S., Li, T., Gong, Y.: Multi-Document Summarization using Sentence-based Topic Models. In: ACL 2009 (Short Paper), pp. 297–300 (2009)
5. Dhillon, I.S., Mallela, S., Modha, D.S.: In-formation-Theoretic Co-clustering. In: KDD 2003, pp. 89–98 (2003)
6. Radev, D.R., Jing, H.Y., Stys, M., Tam, D.: Centroid-based summarization of multiple documents. Information Processing and Management 40, 919–938 (2004)
7. Mani, I., Bloedorn, E.: Summarizing Similarities and Differences among Related Documents. Information Retrieval 1(1), 35–67 (2000)
8. Mihalcea, R.: Graph-based Ranking Algorithms for Sentence Extraction Applied to Text Summarization. In: ACL 2004 (2004)
9. Erkan, G., Radev, D.: LexPageRank: prestige in multi-document text summarization. In: EMNLP 2004, pp. 365–371 (2004)
10. Dai, W., Xue, G.-R., Yang, Q., Yu, Y.: Co-clustering based classification for out-of-domain documents. In: KDD 2007, pp. 210–219 (2007)
11. Wang, P., Domeniconi, C., Hu, J.: Using Wikipedia for Co-clustering Based Cross-domain Text Classification. In: ICDM 2008, pp. 1085–1090 (2008)
12. Frantzi, K.: Ananiadou S, A Hybrid Approach to Term Recognition. In: NLP+IA 1996(A), pp. 93–98 (1996)
13. Kleinberg, J.M.: Authoritative sources in a hyperlinked environment. Journal of the ACM 46(5), 604–632 (1999)
14. Herings, P.J., van der Laan, G., Talman, D.: Measuring the Power of Nodes in Digraphs. Technical report, Tinbergen Institute, 01-096/1 (2001)
15. Brin, S., Page, L.: The Anatomy of a Large-Scale Hypertextual Web Search Engine. Computer Networks and ISDN Systems 30(1-7) (1998)
16. Lin, C.-Y.: ROUGE: A Package for Automatic Evaluation of Summaries. In: ACL 2004, Workshop on Text Summarization Branches Out, pp. 74–81 (2004)

# Answer Validation Using Textual Entailment

Partha Pakray[1], Alexander Gelbukh[2], and Sivaji Bandyopadhyay[1]

[1] Computer Science and Engineering Department,
Jadavpur University, Kolkata, India
[2] Center for Computing Research, National Polytechnic Institute,
Mexico City, Mexico
parthapakray@gmail.com, www.gelbukh.com,
sbandyopadhyay@cse.jdvu.ac.in

**Abstract.** We present an Answer Validation System (AV) based on Textual Entailment and Question Answering. The important features used to develop the AV system are Lexical Textual Entailment, Named Entity Recognition, Question-Answer type analysis, chunk boundary module and syntactic similarity module. The proposed AV system is rule based. We first combine the question and the answer into Hypothesis (H) and the Supporting Text as Text (T) to identify the entailment relation as either "VALIDATED" or "REJECTED". The important features used for the lexical Textual Entailment module in the present system are: WordNet based unigram match, bigram match and skip-gram. In the syntactic similarity module, the important features used are: subject-subject comparison, subject-verb comparison, object-verb comparison and cross subject-verb comparison. The results obtained from the answer validation modules are integrated using a voting technique. For training purpose, we used the AVE 2008 development set. Evaluation scores obtained on the AVE 2008 test set show 66% precision and 65% F-Score for "VALIDATED" decision.

**Keywords:** Answer Validation Exercise (AVE), Textual Entailment (TE), Named Entity (NE), Chunk Boundary, Syntactic Similarity, Question Type.

# 1 Introduction

Answer Validation Exercise (AVE) is a task introduced in the QA@CLEF competition. AVE task is aimed at developing systems that decide whether the answer of a Question Answering system is correct or not. There were three AVE competitions AVE 2006 [1], AVE 2007 [2] and AVE 2008 [3]. AVE systems receive a set of triplets (Question, Answer and Supporting Text) and return a judgment of "SELECTED", "VALIDATED" or "REJECTED" for each triplet. The evaluation methodology was improved over the years and oriented to identify the useful factors for QA systems improvement. Thus, in 2007 the AVE systems were to select only one VALID answer for every question from a set of possible answers, whereas in 2006, several VALID answers were possible to be selected. In 2008[1], the organizers

---

[1] http://nlp.uned.es/clef-qa/ave/

A. Gelbukh (Ed.): CICLing 2011, Part II, LNCS 6609, pp. 353–364, 2011.
© Springer-Verlag Berlin Heidelberg 2011

increased the complexity of the data set by setting that all the answers to a question may be incorrect. The task of the participating systems was to ensure that all the answers to such questions are marked as "REJECTED".

There were three Recognizing Textual Entailment (RTE) competitions RTE-1 in 2005 [4], RTE-2 in 2006 [5] and RTE-3 in 2007 [6] which were organized by PASCAL (Pattern Analysis, Statistical Modeling and Computational Learning) - the European Commission's IST-funded Network of Excellence for Multimodal Interfaces. In 2008, the fourth edition (RTE-4) [7] of the challenge was organized by NIST (National Institute of Standards and Technology) in Text Analysis Conference (TAC). In every new competition several new features of RTE were introduced. The TAC RTE-5 [8] challenge in 2009 includes a separate search pilot along with the main task. The TAC RTE-6 challenge[2], in 2010, includes the Main Task and Novelty Detection Task along with RTE-6 KBP Validation Pilot Task. The RTE-6 does not include the traditional RTE Main Task which was carried out in the first five RTE challenges, i.e. there will be no task to make entailment judgments over isolated T-H pairs drawn from multiple applications. In 2010, Parser Training and Evaluation using Textual Entailment [9] was organized in SemEval-2. We have developed our own RTE system and have participated in RTE-2009, in Parser Training and Evaluation using Textual Entailment as part of SemEval-2 and also in TAC RTE-2010.

Related works are described in Section 2. Section 3 describes corpus statistics. Section 4 describes the Answer Validation system. The experiments carried out on the development and test data sets are described in Section 5 along with the results. The discussions on the experimental results are discussed in Section 6. The conclusions are drawn in Section 7.

## 2   Related Works

In the various AVE Challenges, several methods are applied on the AVE task. Most of these systems use some sort of lexical matching, e.g., simple word overlap, n-gram match and longest Common subsequence. A number of systems represent the text as parse trees (e.g., syntactic, dependency) before the actual task. Some of the systems use semantic relation (e.g., logical inference, Semantic Parsing) for solving the AVE problem.

Use of Textual Entailment recognition techniques [1][2] to do answer validation has shown a great success [10]. The system [11] utilizes a Recognizing Textual Entailment (RTE) system as a component to validate answers. The rules followed in building the patterns for question transformation, the generation of the corresponding hypothesis and final answer ranking are described in [12]. The AVE task was cast into a Recognizing Textual Entailment (RTE) problem in [13] and an existing RTE system was used to validate answers. Additional information from named-entity (NE) recognizer, question analysis component and other sources are also considered to make the final decision. Their approach is closest to the method used in the present work. But, a different scoring mechanism and a different set of features have been used in the present work. The scoring technique used in the present work is based on

---

[2] http://www.nist.gov/tac/2010/RTE/index.html

applying voting principle on the outputs generated from the NER system, TE system, Chunk Boundary, Syntactic similarity and the question type information.

## 3  Corpus Statistics

The corpus for English mono-lingual was made available by the AVE 2008 organizers. The corpus was organized as a set of triplets (Question, Answer, and Supporting Text) and the participating systems had to specify the answer correctness, i.e., whether "SELECTED", "VALIDATED" or "REJECTED". The AVE 2008 Development Set Data Format is shown in Figure 1.

```
<q id="83" lang="EN">
<q_str>Where is the Hermitage Museum?</q_str>
<a id="83_2" value="REJECTED">
<a_str>Birseck</a_str>
<t_str doc="en/p03/334819.xml">The Mesolithic period has some
examples of portable art, like painted pebbles (Azilien) from
Birseck, Eremitage in Switzerland, and in some areas, like the
Spanish Levant, stylized rock art.</t_str>
</a>
...
</q>
```

**Fig. 1.** AVE 2008 Test Gold Data Format

In the Figure 1, the data format "q_str" tag contains the question, "a" tags correspond to every possible answer, "a_str" tag contains the answer itself and justification text is in the "t_str" tag.

```
q_id  a_id [SELECTED | VALIDATED | REJECTED] confidence
```

**Fig. 2.** AVE 2008 Data Output Format

The AVE 2008 Data Output Format is shown in Figure 2. The output for a question – answer combination can be either VALIDATED, SELECTED or REJECTED which are described below.

   i. VALIDATED indicates that the answer is correct with respect to the supporting text. There is no restriction on the number of VALIDATED answers to a question.
   ii. SELECTED indicates that the answer is VALIDATED and it is the one chosen as the output of a hypothetical QA system. The SELECTED answers were evaluated against the QA systems of the Main Track. No more than one answer per question can

be marked as SELECTED. At least one of the VALIDATED answers must be marked as SELECTED.

iii. REJECTED indicates that the answer is incorrect or there is no enough evidence of its correctness. There is no restriction in the number of REJECTED answers.

For AVE 2008, separate sub-tasks for the following 11 languages were described: Basque, Bulgarian, German, English, Spanish, French, Italian, Dutch, Portuguese, Romanian, Greek.

## 4   System Description

In this section, we describe our Answer Validation (AV) system. The architecture of the proposed system is described in Figure 3. The various components of the AV system are Pattern Generation Module, Hypothesis Generation Module, Question Type Analysis Module, Named Entity Recognition (NER) Module, Textual Entailment (TE) Module, Chunk Boundary and syntactic similarity module. Each of these modules is now being described in subsequent subsections.

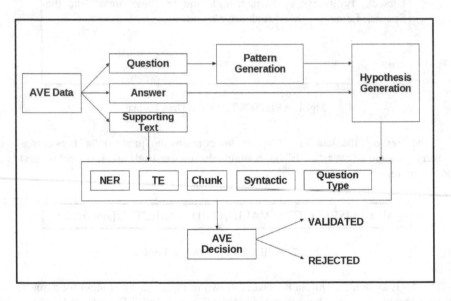

**Fig. 3.** Answer Validation System

### 4.1   Pattern Generation Module

At first we convert each question into an affirmative sentence that denotes the answer pattern and place the </answer> template in place of the appropriate answer. The pattern generation module is rule based.

For Example, question id 0061 (AVE-2008 Test set)

Question::  Where was Joseph Fourier born?
Template::  Joseph Fourier was born in </answer>.

## 4.2  Hypothesis Generation Module

After Pattern generation the </answer> template is replaced by the answer string forming the generated Hypothesis. Now, we have the Text (T), the Supporting Text and Hypothesis (H), the generated Hypothesis. For example, for question id 0061 (AVE-2008 Test set), we generate the following hypotheses for each of the alternative answers:

H0061_1:  Joseph Fourier was born in Paris.
H0061_2:  Joseph Fourier was born in France.

## 4.3  NER Module

It is based on the detection and matching of Named Entities (NEs) in the Supporting Text (T) - generated Hypothesis (H) pair. Once the NEs of the hypothesis and the text have been detected, the next step is to determine the number of NEs in the hypothesis that match in the corresponding text. The measure NE_Match is defined as NE_Match=number of common NEs between T and H/Number of NEs in Hypothesis.

If the value of NE_Match is 1, i.e., 100% of the NEs in the hypothesis match in the text, then the text-hypothesis pair is considered as an entailment. The T-H pair is assigned the value "VALIDATED", otherwise, the pair is assigned the value "REJECTED".

## 4.4  Textual Entailment Module (TE)

This TE module is based on three types of matching, i.e., WordNet based Unigram Match and bigram matching and Skip-bigram Matching.

**a.  WordNet based Unigram Match.** In this method, the various unigrams in the hypothesis for each Supporting Text (T) - generated Hypothesis (H) pair are checked for their presence in the text. WordNet synsets are identified for each of the unmatched unigrams in the hypothesis. If any synset for the H unigram match with any synset of a word in the T then the hypothesis unigram is considered as a successful WordNet based unigram match. If the value of Wordnet_Unigram_Match is 0.75 or more, i.e., 75% or more unigrams in the H match either directly or through WordNet synonyms, then the T-H pair is considered as an entailment. The text-hypothesis pair is then assigned the value  "VALIDATED", otherwise, the pair is assigned the value "REJECTED".

**b.    Bigram Match.** Each bigram in the hypothesis is searched for a match in the corresponding text part. The measure Bigram_Match is calculated as the fraction of the hypothesis bigrams that match in the corresponding text, i.e., Bigram_Match=(Total number of matched bigrams in a T-H pair /Number of hypothesis bigrams). If the value of Bigram_Match is 0.5 or more, i.e., 50% or more bigrams in the H match in the corresponding T, then the T-H pair is considered as an entailment. The text-hypothesis pair is then assigned the value "VALIDATED", otherwise, the pair is assigned the value "REJECTED".

**c. Skip-grams.** A skip-gram is any combination of n words in the order as they appear in a sentence, allowing arbitrary gaps. In the present work, only 1-skip-bigrams are considered where 1-skip-bigrams are bigrams with one word gap between two words in a sentence. The measure 1-skip_bigram_Match is defined as

1_skip_bigram_Match = skip_gram(T,H) / n,

where skip_gram(T,H) refers to the number of common 1-skip-bigrams (pair of words in order with one word gap) found in T and H and n is the number of 1-skip-bigrams in the hypothesis H. If the value of 1_skip_bigram_Match is 0.5 or more, then the T-H pair is considered as an entailment. The text-hypothesis pair is then assigned the value "VALIDATED", otherwise, the pair is assigned the value "REJECTED".

If all the three matches assign the "VALIDATED" value to the text-hypothesis pair then the entailment value for this pair is "VALIDATED", otherwise, the pair is assigned the value "REJECTED".

### 4.5 Question-Answer Type Analysis Module

The original questions are pre-processed using Stanford Dependency parser [14]. The question type and its expected answer type are generally identified by looking at the question keyword. Table 1 lists the question and the expected answer types. For example,  if the question type is "When", the expected answer type is a "DATE/TIME". The answer string "<a_str>" is parsed by the RASP Parser [15]. If the RASP parser generates the tag "<timex type=date>" then the answer string is "VALIDATED", otherwise it is "REJECTED". For "What" type questions we look for the keyword (e.g., Company) that is related to "What" through a dependency relation. If the keyword is "Company" the expected answer type is "Organization". If the corresponding answer string is tagged by the RASP parser as "Organization", the answer string is marked as "VALIDATED", otherwise it is "REJECTED".  If the question type is "How" and the answer string is tagged as "CD" by the RASP parser, the answer string is marked as "VALIDATED", otherwise it is "REJECTED".

**Table 1.** Question Keyword and Expected Answer

| Question Type | Expected Answer |
|---------------|-----------------|
| Who | PERSON |
| When | DATE / TIME |
| Where | LOCATION |
| What | OBJECT |
| How | MEASURE |

### 4.6 Chunk Module

The question sentences are pre-processed using Stanford dependency parser. The words along with their part of speech (POS) information are passed through a

Conditional Random Field (CRF) based chunker [16] to extract phrase level chunks of the questions. A rule-based module is developed to identify the chunk boundaries. Key chunks are identified for each question. The chunks that are related by each *prep* relation constitute the key chunks corresponding to that *prep* relation. Each verb chunk present in the question sentence is also a key chunk. If there are no *prep* relations present in the question sentence then all chunks present except the Wh chunk are considered as key chunks. These key chunks are searched in the supporting text associated with the question sentence. An example question with its chunk boundary information, dependency relations and set of extracted key chunks are shown below.

Question:: What was the nationality of Jacques Offenbach?
Chunk Boundary:: (What/WP/B-NP) (was/VBD/B-VP) (the/DT/B-NP nationality/ NN/I-NP) (of/IN/B-PP) (Jacques/NNP/B-NP Offenbach/NNP/I-NP )
Dependency::
[ attr was What
  det nationality the
  nsubj was nationality
  nn Offenbach Jacques
  prep_of nationality Offenbach ]
Extracted Key chunks:: (was/VBD/B-VP) (the/DT/B-NP nationality/NN/I-NP) (Jacques/NNP/B-NP Offenbach/NNP/I-NP )

The supporting text is parsed using the Stanford Dependency parser. The output of the parser is passed through a Conditional Random Field (CRF) based chunker to extract phrase level chunks for each sentence in the supporting text. The identified key chunks from the question are now matched in the supporting text associated with the question. If a complete key chunk matches, the weight heuristically assigned to the matching is defined as (chunk length + K) / (text length) where K is the chunk length to give more weight to the complete chunk match. If there is a partial chunk match, the weight is defined as (matching partial chunk length) / (text length). The weight of a question – supporting text pair is identified as the total weights corresponding to the key chunk weights. The question-supporting text pairs that achieve the maximum weight are identified and the corresponding answers are tagged as "VALIDATED". The question-supporting text pair that receives a Zero weight is tagged as "REJECTED".

## 4.7  Syntactic Similarity Module

This module is based on the Stanford dependency parser [14], which normalizes data from the corpus of text and hypothesis pairs, accomplishes the dependency analysis and creates appropriate structures. Our Entailment system uses the following features.

**a. Subject.** The dependency parser generates nsubj (nominal subject) and nsubjpass (passive nominal subject) tags for the subject feature.
**b. Object.** The dependency parser generates dobj (direct object) as object tags.
**c. Verb.** Verbs are wrapped with either the subject or the object.
**d. Noun.** The dependency parser generates nn (noun compound modifier) as noun tags.

**e. Preposition.** Different type of prepositional tags are prep_in, prep_to, prep_with etc. For example, in the sentence "A plane crashes in Italy.", the prepositional tag identified is prep_in(in, Italy).

**f. Determiner.** Determiner denotes a relation with a noun phase. The dependency parser generates *det* as determiner tag. For example, the parsing of the sentence "A journalist reports on his own murders." generates the determiner relation as det(journalist,A).

**g. Number.** The numeric modifier of a noun phrase is any number phrase. The dependency parser generates num (numeric modifier). For example, the parsing of the sentence "Nigeria seizes 80 tonnes of drugs." generates the relation num (tonnes, 80).

For the sentence, "Nigeria seizes 80 tonnes of drugs", the Stanford Dependency Parser generates the following set of dependency relations:

[
nsubj(seizes-2, Nigeria-1),
num(tonnes-4, 80-3),
dobj(seizes-2, tonnes-4),
prep_of(tonnes-4, drugs-6)
]

### 4.7.1 Matching Module

After dependency relations are identified for both the text and the hypothesis in each pair, the hypothesis relations are compared with the text relations. The different features that are compared are noted below. In all the comparisons, a matching score of 1 is considered when the complete dependency relations along with all of its arguments match in both the text and the hypothesis. In case of a partial match for a dependency relation, a matching score of 0.5 is assumed.

**a. Subject-Verb Comparison.** The system compares hypothesis subject and verb with text subject and verb that are identified through the *nsubj* and *nsubjpass* dependency relations. A matching score of 1 is assigned in case of a complete match. Otherwise, the system considers the following matching process.

**b. WordNet Based Subject-Verb Comparison.** If the corresponding hypothesis and text subjects do match in the subject-verb comparison, but the verbs do not match, then the WordNet distance between the hypothesis and the text is compared. If the value of the WordNet distance is less than 0.5, indicating a closeness of the corresponding verbs, then a match is considered and a matching score of 0.5 is assigned. Otherwise, the subject-subject comparison process is applied.

**c. Subject-Subject Comparison.** The system compares hypothesis subject with text subject. If a match is found, a score of 0.5 is assigned to the match.

**d. Object-Verb Comparison.** The system compares hypothesis object and verb with text object and verb that are identified through dobj dependency relation. In case of a match, a matching score of 0.5 is assigned.

**e. WordNet Based Object-Verb Comparison.** The system compares hypothesis object  with text object. If a match is found then the verb corresponding to the hypothesis object with text object's verb is compared. If the two verbs do not match then the WordNet distance between the two verbs is calculated. If the value of WordNet distance is below 0.5 then a matching score of 0.5 is assigned.

**f. Cross Subject-Object Comparison.** The system compares hypothesis subject and verb with text object and verb or hypothesis object and verb with text subject and verb. In case of a match, a matching score of 0.5 is assigned.

**g. Number Comparison.** The system compares numbers along with units in the hypothesis with similar numbers along with units in the text. Units are first compared and if they match then the corresponding numbers are compared. In case of a match, a matching score of 1 is assigned.

**h. Noun Comparison.** The system compares hypothesis noun words with text noun words that are identified through *nn* dependency relation. In case of a match, a matching score of 1 is assigned.

**i. Prepositional Phrase Comparison.** The system compares the prepositional dependency relations in the hypothesis with the corresponding relations in the text and then checks for the noun words that are arguments of the relation. In case of a match, a matching score of 1 is assigned.

**j. Determiner Comparison.** The system compares the determiner in the hypothesis and in the text that are identified through *det* relation. In case of a match, a matching score of 1 is assigned.

**k. Other relation Comparison.** Besides the above relations that are compared, all other remaining relations are compared verbatim in the hypothesis and in the text. In case of a match, a matching score of 1 is assigned.

API for WordNet Searching RiWordnet[3] provides Java applications with the ability to retrieve data from the WordNet database.

Each of the matches through the above comparisons is assigned some weight learned from the development corpus. A threshold of 0.3 has been set on the fraction of matching hypothesis relations based on the development set results that gives optimal precision and recall values for both "VALIDATED" and "REJECTED". The threshold score has been applied on the AVE test set using the same methods of dependency parsing followed by comparisons.

### 4.8 Answer Validation Decision Module

In this module, we use the voting technique as described in this section. At first we check if any named entity (NE) is present in the generated Hypothesis (H). The following conditions are checked:

i. If in the generated hypothesis (H) NE is present then we check the result of the NER module. If NER module generates "VALIDATED" tag to the answer, then the results of the Textual Entailment module, Question-Answer Type Analysis module, Chunk and Syntactic similarity module are checked. If all these modules generate the

---

[3] http://www.rednoise.org/rita/wordnet/documentation/index.htm

"VALIDATED" result, the answer is tagged as "VALIDATED". Otherwise, the answer is tagged as "REJECTED".

ii. If in the generated hypothesis (H), NE is not present then the results of the Textual Entailment module, Question-Answer Type Analysis module and Chunk and Syntactic similarity module are checked. If all these modules generate the "VALIDATED" result, the answer is tagged as "VALIDATED". Otherwise, the answer is tagged as "REJECTED".

# 5   Experiment and Result

The Answer Validation system has been tested on the AVE 2008 Development Set for English. The AVE 2008 development set consists of 195 pairs of which only 21 are positives (10.77% of the total number of pairs). The recall, precision and f-measure values on the Development data obtained over correct ("VALIDATED") answers are shown in Table 2.

**Table 2.** AVE 2008 Development Set Precision, Recall and F-Score

| AVE Development Set | Result |
|---|---|
| "VALIDATED" in the Development Set | 21 |
| "VALIDATED" in the proposed AV system | 34 |
| "VALIDATED" match | 15 |
| Precision | 0.44 |
| Recall | 0.71 |
| F-score | 0.54 |

The AVE 2008 English annotated test set consists of 1055 pairs and the number of correct "VALIDATED" answer is 79 (7.5% of the total). The recall, precision and f-measure values on the test data obtained over correct answers are shown in Table 3.

**Table 3.** AVE 2008 Test Set Precision, Recall and F-Score

| AVE Test Set | Result |
|---|---|
| "VALIDATED" in the Test Set | 79 |
| "VALIDATED" in the proposed AV system | 78 |
| "VALIDATED" match | 52 |
| Precision | 0.66 |
| Recall | 0.65 |
| F-score | 0.65 |

# 6   Discussion

In this section we compare our results with other systems that participated in the respective CLEF 2008@AVE tracks. Participating system results are shown in Table 4.

Table 4. Compare our Result with AVE 2008 Participating System

| Group Name | F-Score | Precision | Recall |
|---|---|---|---|
| DFKI | 0.64 | 0.54 | 0.78 |
| UA | 0.49 | 0.35 | 0.86 |
| UNC | 0.21 | 0.13 | 0.56 |
| IASI | 0.19 | 0.11 | 0.85 |
| **Our AVE Result** | **0.65** | **0.66** | **0.65** |

The results obtained by our AVE system on the respective AVE tracks are shown in bold. It is observed that the results obtained by our system have outperformed the participating systems based on lexical and syntactic approaches in the AVE-2008 on the basis of F-Score value.

# 7 Conclusion

In this work we have used the Lexical Entailment, Chunking, Syntactic Similarity and Named Entity recognition modules. The next step is to carry out detailed error analysis of the present system and identify ways to overcome the errors. Experiments have been started for a semantic based AVE system. Use of lexical information along with syntactic features and semantic features in the AVE system would be another set of interesting experiments to handle the correct decision making task.

**Acknowledgements.** We acknowledge the support of the CONACYT Mexico – DST India project "Answer Validation through Textual Entailment". The second author acknowledges is a Visiting Scholar at Waseda University, Japan, and acknowledges support of SIP-20100773 grant, CONACYT 50206-H grant, and CONACYT scholarship for Sabbatical stay 2010.

# References

1. Peñas, A., Rodrigo, Á., Sama, V., Verdejo, F.: Overview of the answer validation exercise 2006. In: Peters, C., Clough, P., Gey, F.C., Karlgren, J., Magnini, B., Oard, D.W., de Rijke, M., Stempfhuber, M. (eds.) CLEF 2006. LNCS, vol. 4730, pp. 257–264. Springer, Heidelberg (2007)
2. Peñas, A., Rodrigo, Á., Verdejo, F.: Overview of the Answer Validation Exercise 2007. In: Peters, C., Jijkoun, V., Mandl, T., Müller, H., Oard, D.W., Peñas, A., Petras, V., Santos, D. (eds.) CLEF 2007. LNCS, vol. 5152, pp. 237–248. Springer, Heidelberg (2008)
3. Rodrigo, Á., Peñas, A., Verdejo, F.: Overview of the answer validation exercise 2008. In: Peters, C., Deselaers, T., Ferro, N., Gonzalo, J., Jones, G.J.F., Kurimo, M., Mandl, T., Peñas, A., Petras, V. (eds.) CLEF 2008. LNCS, vol. 5706, pp. 296–313. Springer, Heidelberg (2009)
4. Dagan, I., Glickman, O., Magnini, B.: The PASCAL Recognising Textual Entailment Challenge. In: Proceedings of the First PASCAL Recognizing Textual Entailment Workshop, 2005 (2005)

5. Bar-Haim, R., Dagan, I., Dolan, B., Ferro, L., Giampiccolo, D., Magnini, B., Szpektor, I.: The Second PASCAL Recognizing Textual Entailment Challenge. In: Proceedings of the Second PASCAL Challenges Workshop on Recognising Textual Entailment, Venice, Italy, 2006 (2006)

6. Giampiccolo, D., Magnini, B., Dagan, I., Dolan, B.: The Third PASCAL Recognizing Textual Entailment Challenge. In: Proceedings of the ACL-PASCAL Workshop on Textual Entailment and Paraphrasing, Prague, Czech Republic (2007)

7. Giampiccolo, D., Dang, H.T., Magnini, B., Dagan, I., Cabrio, E.: The Fourth PASCAL Recognizing Textual Entailment Challenge. In: Proceedings of TAC 2008 (2008)

8. Bentivogli, L., Dagan, I., Dang, H.T., Giampiccolo, D., Magnini, B.: The Fifth PASCAL Recognizing Textual Entailment Challenge. In: TAC 2009 Workshop, National Institute of Standards and Technology Gaithersburg, Maryland, USA (2009)

9. Yuret, D., Han, A., Turgut, Z.: SemEval-2010 Task 12: Parser Evaluation using Textual Entailments. In: Proceedings of the SemEval 2010 Evaluation Exercises on Semantic Evaluation, 2010 (2010)

10. Rodrigo, Á., Peñas, A., Verdejo, F.: UNED at Answer Validation Exercise 2007. In: Peters, C., Jijkoun, V., Mandl, T., Müller, H., Oard, D.W., Peñas, A., Petras, V., Santos, D. (eds.) CLEF 2007. LNCS, vol. 5152, pp. 404–409. Springer, Heidelberg (2008)

11. Wang, R., Neumann, G.: DFKI-LT at AVE 2007: Using Recognizing Textual Entailment for Answer Validation", Working Notes of CLEF AVE 2007 (2007)

12. Iftene, A., Balahur, A.: Answer Validation on English and Romanian Languages. In: Peters, C., Deselaers, T., Ferro, N., Gonzalo, J., Jones, G.J.F., Kurimo, M., Mandl, T., Peñas, A., Petras, V. (eds.) CLEF 2008. LNCS, vol. 5706, pp. 448–451. Springer, Heidelberg (2009)

13. Wang, R., Neumann, G.: Information Synthesis for Answer Validation. In: Peters, C., Deselaers, T., Ferro, N., Gonzalo, J., Jones, G.J.F., Kurimo, M., Mandl, T., Peñas, A., Petras, V. (eds.) CLEF 2008. LNCS, vol. 5706, pp. 472–475. Springer, Heidelberg (2009)

14. de Marneffe, M.-C., MacCartney, B., Manning, C.D.: Generating Typed Dependency Parses from Phrase Structure Parses. In: 5th International Conference on Language Resources and Evaluation (LREC) (2006)

15. Briscoe, E., Carroll, J., Watson, R.: The Second Release of the RASP System. In: Proceedings of the COLING/ACL 2006 Interactive Presentation Sessions (2006)

16. Phan, X.-H.: CRFChunker: CRF English Phrase Chunker. In: PACLIC 2006 (2006)

# SPIDER: A System for Paraphrasing in Document Editing and Revision — Applicability in Machine Translation Pre-editing

Anabela Barreiro

Centro de Linguística da Universidade do Porto, Portugal
barreiro_anabela@hotmail.com

**Abstract.** This paper presents SPIDER, a system for paraphrasing in document editing and revision with applicability in machine translation pre-editing. SPIDER applies its linguistic knowledge (dictionaries and grammars) to create paraphrases of distinct linguistic phenomena. The first version of this tool was initially developed for Portuguese (ReEscreve v01), but it is extensible to different languages and can also operate across languages. SPIDER has a totally new interface, new resources which contemplate a wider coverage of linguistic phenomena, and applicability to legal terminology, which is described here.

**Keywords:** paraphrase, language composition tool, authoring aid, text processing application, pre-editing, revision, linguistic quality assurance.

## 1 Introduction

The relevance of paraphrases for natural language processing has been clearly defined, and paraphrases are being used in different types of applications for a variety of purposes. Paraphrasal knowledge plays a very important role in interpretation and generation of natural language. In *natural language interpretation*, dynamic semantics and identical parses resulting from paraphrases are important to successful applications. In *natural language generation*, the generation of paraphrases allows more varied and fluent text to be produced [Iordanskaja et al. 1991]. In *multi-document summarization*, the identification of paraphrases allows information across documents to be condensed [McKeown et al. 2002] and helps improve the quality of the generated summaries [Hirao et al. 2004]. In *question answering*, discovering paraphrased answers may provide additional evidence that an answer is correct [Ibrahim et al. 2003]. Paraphrases can also be useful in *text mining*, preventing a passage being discarded due to the inability to match a question phrase deemed as very important [Paşca and Dienes 2005]. In *information extraction*, paraphrases help text categorization tasks or mapping to texts with similar characteristics, lessening the disparity in the trigger word or the applicable extraction pattern [Shinyama and Sekine 2005]. Paraphrasing also helps *machine translation* performance [Callison-Burch 2007], in particular the translation of multi-word units [Barreiro 2011].

A. Gelbukh (Ed.): CICLing 2011, Part II, LNCS 6609, pp. 365–376, 2011.

The application of automated paraphrasing to authoring aids and text pre-editing has been on the wish-list of researchers and commercial enterprises for a long time. This paper represents a beginning in bringing paraphrasing into the hands of the text author. Section 2 describes different methods to achieve paraphrases. Section 3 presents the motivation for the existence of the SPIDER paraphrasing tool and, points out its novelty aspects compared to existing authoring aids, such as the Microsoft Word text editor. Section 4 describes the base linguistic resources used to build SPIDER and the methodology used to achieve the resulting paraphrasal resources. It describes, in particular, the morphosyntactic and semantic relations established at the lexicon level and the paraphrasing grammars created to apply those relations in texts. Section 5 presents SPIDER's technology, graphical user interface and modus operandi. Section 6 presents some preliminary evaluation results when SPIDER is applied to text pre-editing and then translated automatically, and discusses how translatability can improve by using paraphrases. Section 7 presents future work.

## 2   Methods and Applicability of Paraphrasing

Phrasal and sentence transformation can be achieved by means of different types of linguistic resources and techniques. Lexical resources, general language dictionaries and ontologies, such as WordNet [Fellbaum 1998] are relevant sources of knowledge for synonym and paraphrasing. There are three generally accepted methods for paraphrase acquisition: corpora-based, statistical-based and dictionary or rule-based. Statistical and corpora-based methods are intertwined.

Statistical methods for acquiring paraphrases are the most popular and lead to large collections of phrase or sentence alternatives. However, statistical methods do not deliver the precision that dictionary and rule-based methods provide. They use sophisticated algorithms and apply these algorithms to corpora, therefore depending either on large volumes of corpora or on training corpora. Corpora may be unavailable for some languages or of poor quality. Even though statistical methods do not produce clean data and require considerable human validation, they help finding empirical data to be validated by linguists. Empirical data is used as a base to create comprehensive linguistic rules that will enhance system precision. Statistical methods can also help speed up the process of linguistic annotation. They can recognize frozen expressions, but in general, they do not recognize multi-word units with a more flexible structure.

Dictionary and rule-driven paraphrasing is less popular because it is time-consuming and requires linguistic knowledge. However, dictionary and rule-based methods have proven highly effective for generating lexically related paraphrases. For example, [Barreiro 2011] demonstrates that (i) the strategy of paraphrasing support verb constructions with semantically equivalent single verbs (e.g. *to give a hug to/to hug*) produces a significant impact in machine translation; (ii) while addressing and providing a solution for a specific linguistic problem, the study is reproducible and extendable to distinct linguistic phenomena, and (iii) successfully applied to different purpose natural language processing applications, such as authoring aids, where paraphrases can be efficiently employed to help clarify texts, presenting obvious benefits to linguistic quality assurance in text processing. Based on this research,

SPIDER was created with consistency, precision and linguistic quality assurance in mind, to help the user change text at the word and at the phrase level (multi-word units), through the assimilation of linguistic knowledge, employing semi-automatically developed resources, such as electronic dictionaries and transformational grammars, designed and verified by linguists. It follows a systematic methodology and applies it to document editing and revision and machine translation pre-editing. However, SPIDER can be integrated in a wide variety of applications with multiple functionalities. The first version of SPIDER was developed for Portuguese (ReEscreve v01), but it is extensible to different languages and can also operate across languages. ReEscreve is being used as an authoring aid online public service, described in [Barreiro and Cabral 2009] and [Barreiro 2011]. SPIDER has evolved significantly with regards to ReEscreve: it has a completely new interface, enlarged and enhanced resources for a wider coverage of linguistic phenomena and includes terminology. Section 5 will illustrate its applicability to legal texts.

## 3  SPIDER: Motivation and Novelty

There were several reasons to develop SPIDER. First, there is lack of authoring aid tools to help users with stylistic issues and paraphrasing. While controlled language tools [Mitamura and Nyberg 2002] and stylistic checkers may offer some useful solutions for text editing, they use mostly manually crafted rules and they lack the characteristics and functionality that a paraphraser offers, namely the variety of alternatives for each expression and the possibility to choose among them, according to personal preferences, style, idiomacity, among others. Another important reason for SPIDER is directly related to translatability and the need to prepare the source text in terms of quality assurance so that it can cause a positive impact in translation. A poorly written text is difficult to understand, but its understandability becomes even more difficult after translation. We believe that it is more advantageous to "clean up the mess" before than after the translation process, and that pre-editing tools can minimize the time and effort spent on post-editing. Since machine translation became an indispensable tool for common users these days, there is a strong and urgent need to invent more sophisticated linguistic tools. A tool that serves the purpose of helping texts become clearer, and more understandable, will have applicability in most natural language processing areas. By using linguistically based automated paraphrasing and text-editing mechanisms, SPIDER can help users with their writing needs by providing suggestions for customized text authoring, being suitable for any word processing application, for summarization purposes, for stylistics, for machine translation, and so on.

SPIDER presents some novelty in relation to the state of the art paraphrasing and text editors, not only in terms of functionality, but also in terms of its linguistic knowledge management capability. Most text editors offer synonyms and simple linguistic constructions in a canonical or dictionary form as suggestions for the users to edit their text and the user still needs to adapt the suggestion according to gender, number, tense and other inflectional features. SPIDER permits linguistically more complex transformations, such as discontinuous multi-word units while maintaining the source text grammaticality and semantic coherence and preserves the source text inflectional traits in the suggestions provided. When used interactively, SPIDER retrieves contextualized suggestions for words or expressions in a user's text. Users can select their preferences by clicking on the words or expressions that better suit the purposes of their text, on the spot and without any further human intervention. SPIDER also allows users to suggest new expressions that can be immediately applied to their text, making the text editing process easier, more flexible, and upgradable. This method of text rewriting has proven effective in improving text quality, especially when employed as a machine translation pre-editor. Source and target text quality improvement is the most important goal in linguistic quality assurance. SPIDER can also help general users with writing difficulties or learners of English as a non-native language, being suitable as a pedagogical tool. When integrating terminologies, SPIDER can help writing in vertical or technical domains. Section 5 exemplifies the novel linguistic and functional aspects presented by SPIDER. But first, section 4 describes SPIDER's linguistic resources and methodology adopted to generate paraphrases.

## 4 Linguistic Resources and Methodology

SPIDER linguistic knowledge database, Eng4NooJ (version 1.0), was built with the NooJ freeware multilingual development environment [Silberztein 2004], which uses finite-state transducer technology. NooJ's tools support the development, testing, debugging, maintenance and gathering of other different types of linguistic resources, namely general or domain specific dictionaries or local grammars, and they assist linguistic research and the development of natural language processing applications, such as SPIDER. NooJ tools are also used to parse corpora, build sophisticated concordances, and apply large coverage linguistic resources to texts for distinct purposes, including the identification and analysis of local linguistic phenomena.

Eng4NooJ initial linguistic resources came from the OpenLogos machine translation system [Scott 1989] [Scott 2003], were enhanced with new properties, including derivational and morphosyntactic and semantic relations, using a similar methodology to that used in the development of Port4NooJ, described in [Barreiro 2008]. In sum, the adapted and enhanced linguistic knowledge in Eng4NooJ consists of several dictionaries and local grammars. The resources contain semantico-syntactic and ontological relations from the OpenLogos machine translation system. New inflectional and derivational rules were created from scratch, and applied into several new semantico-syntactic grammar types developed within the NooJ linguistic environment. The OpenLogos dictionary alpha-numeric data was converted into mnemonics, a way of representation closer to that used in NooJ. Even though SPIDER was designed as a Web application, in principle anyone with NooJ could also, independently, use the grammars and lexicons in their own environment.

Eng4NooJ dictionary entries include new properties that establish new linguistic relations and transformations. The dictionary contains morphosyntactic and semantic relations between verbs and nominal predicates (predicate nouns and predicate adjectives), between adjectives and adverbs, and between nouns and adverbs. It also includes syntactic relations between predicate nominals and support verbs that occur with them, and stylistic variations of those support verbs. This linguistic knowledge allows extensive transformation and re-writing, including recognition and rewriting of words or multi-word units into their synonyms or paraphrases, in the appropriate contexts (to *clear up (weather) > to become better/brighter*); support verb constructions into single verbs (*to make a decision > to decide*; *to give support to N(AN) > to support N(AN)*; *to go V-ing > to continue V-ing*; *to get into contact with > to contact*; *to turn on N(light) > to extinguish N*; *to become acid > to acidify*); support verb constructions into their stylistic variants (*to make an audit > to perform an audit*; *to make an impression > to cause an impression*); aspectual constructions into verbs, (*to launch an attack > to attack*); multiword adverbs and adverbial phrases into single adverbs (*in a constructive way > constructively*; *on purpose > purposely / deliberately*); relatives into possessives (*the position that the Church defends > the position of the Church*; *the role that the politicians play > the role of the politicians*); relatives into participial adjectives (*the president that was elected > the president elected*); relatives into compound nouns (a *container for the milk > a milk container*); phrases with *"made of"* (*a bottle made of plastic > a plastic bottle*); and certain passives into actives (*the young man is released by the police officer > the police officer releases the young man*).

Eng4NooJ inherited an interesting feature of the OpenLogos system that is uncharacteristic of other linguistic resources. These resources integrate what is called the Semantico-syntactic Abstraction Language (SAL) component in the dictionary. SAL is the representational language that permits easy mapping from natural language to symbolic language [Scott 2003] [Scott and Barreiro 2009] [Barreiro et al. 2011], setting both meaning (semantics), and structure (syntax) in a continuum. This representation language allows words to be represented at a higher level of semantic abstraction (a second order). For example, the noun (N) *table* can have different SAL properties (and therefore, different Portuguese (PT) transfers) in the Eng4NooJ dictionary entries represented in (1) and (2).

(1) table,N+FLX=BOOK+SAL=COsurf+PT=mesa

(2) table,N+FLX=BOOK+SAL=INdata+PT=tabela

In (1), *table* is classified with the SAL code [COsurf], which contains the properties "concrete" and "surface". In (2), table is classified with the SAL code [INdata], which stands for "information", "recorded data". SAL codes are embedded in the dictionary in the form of mnemonics and can be called by SAL rules, which we will also describe in this paper. All SAL categories (more than 1,000) can be consulted at the OpenLogos System Archives and be downloaded with the OpenLogos system. Table 1 shows a small sample of the main dictionary, with representation of all part-of-speech categories, variable and invariable entries and the SAL properties.

**Table 1.** General dictionary sample

| | |
|---|---|
| alligator,N+FLX=BOOK+SAL=AN+rept+PT=crocodilo | which,RELINT+INFORM+which+PT=o qual |
| enter,V+FLX=WALK+INMO+IntoType+PT=entrar | or,CONJ+JOIN+or+PT=ou |
| approximate,A+FLX=NATURAL+AV+quan+PT=aproximado | alongside,PREP+LOC+at+PT=ao lado de |
| yesterday,ADV+TEMP+punc+past+PT=ontem | many,DET+PL+many+PT=muitos |
| several,PRO+IMPERS+INDEF+PL+several+PT=vários | one third,ARITHM+NUMfrac+PT=um terço |

For example, the word *alligator* in Table 1 is classified as a noun (N) that inflects like the word *book* (FLX=BOOK), where *book* represents the morphological paradigm for regular nouns that take an *–s* in the plural. The SAL mnemonic [AN+rept] stands for animate, reptiles. It designates cold-blooded, egg-laying vertebrates. The corresponding Portuguese transfer is *crocodilo*. The word *enter* is a verb (V) which inflects like the verb *walk* (FLX=WALK), where *walk* represents the morphological paradigm for regular verbs like *walk, walked, walking*. The SAL mnemonic [INMO+IntoType] represents motional intransitive verbs [INMO], which comprise all verbs of motion, such as *depart, go, fly, run, walk*, among others. This SAL group of verbs take kinetic-type prepositions, e.g., *into, onto, up to*, etc., denoting directed motion. The corresponding Portuguese transfer is *entrar*.

In Eng4NooJ, the dictionary is associated with a textual ".nof" file that contains a description of the inflectional and derivational system. These textual descriptions are handcrafted rules. Inflectional and derivational rules apply to entries in the dictionary of lemmas. As represented in the dictionary entries illustrated in Table 1, all words that can inflect, i.e., all variable entries, specify their inflectional paradigm. The same kind of representation applies to multi-word units. For example, the noun "*table*" represented in (1) and (2) has assigned the paradigm BOOK, which is represented in the rule [BOOK = <E>/s + s/p]. This rule permits the word "*table*" to be recognized or annotated as the singular form (*s*) and the word "*tables*" as the plural form (*p*). Characters before the slash sign (/) represent the word endings. Therefore, *s/* means "add an *–s* to the lemma" to recognize or annotate the plural form of the word. The singular form has an empty string <E>, signifying that nothing should be added to the lemma because the singular form remains identical to the lemma.

Structurally similar to inflectional rules, derivational rules formalize nominalizations, adjectivalizations and adverbializations. Many derivative nouns and adjectives can be turned into action verbs or resultative verbs and many adjectives can turn into adverbs, etc. For example, *quotation* is recognized, or annotated as a nominalization derived from the verb *quote*; *applicable* is recognized, or annotated as derived from the verb *apply*; and *quickly* is recognized, or annotated as an adverbialization of the adjective *quick*. (Adverbial) nouns can also be semantically connected with adverbs, such as *with enthusiasm* is a synonym of *enthusiastically*, or *with optimism* is a synonym of *optimistically*. The link between noun and adverb in these examples are established by the derivational rules illustrated in (3)-(6).

(3) NDRV04 = <B>ion/Npred+Nom

(4) ADRV02 = <B>icable

(5) AVDRV01 = <E>ly/ADV

(6) AVDRV04 = <B>tically/ADV

In these rules, NDRV04, ADRV02, AVDRV01 and AVDRV04 are names of derivational paradigms; the command <B> means "remove the last character to the lemma" and add the string specified immediately after (*-ion*; *-icable*; *-tically*); the <E> command has been explained above. These operations transform the verb into a noun, in (3); the verb into an adjective, in (4); and the noun into an adverb, in (5) and (6). These rules permit to increase rapidly the dictionary with derivation words, whenever they share the same semantic values as the words they derived from.

The different types of derivation relations are established at the dictionary level. For example, derivational paradigms for predicate nouns (mostly nominalizations) and adverbializations (adjectival and nominal) are identified in the verb entries, as examples (7)-(9) show.

(7) impress,V+FLX=POLISH+SAL=PVPCpleasetype+PT=impressionar+**DRV=NDRV01:B OOK+VSUP=make+VSUP=cause+NPREP=on**

(8) aesthetic,A+FLX=NATURAL+SAL=AVstate+PT=estético+**DRV=AVDRV03**

(9) skepticism,N+FLX=BOOK+SAL=ABcause+PT=cepticismo+**DRV=NAVDRV02**

In (7), DRV=NDRV01 defines the rule for the paradigm *impression* (*impress > impression*). :BOOK defines the inflectional paradigm for the derived noun *impression* (it inflects like the noun *book*). VSUP defines the support verb that co-occurs with the derived predicate noun. Both the support verb *make* and *cause* can co-occur with the predicate noun *impression*. NPREP defines the adnominal preposition, *on*, for this particular support verb construction. The grounds for paraphrasing are therefore established. The support verb constructions *make an impression on* and *cause an impression on* can be paraphrases of the single verb *impress*. Semantically unrelated words (even if morphosyntactically related), such as in the expressions *to be in a hurry ≠ to hurry up; to be surprised ≠ to surprise;* or *to give affection ≠ to affect,* were obviously not mapped in the dictionary. In (8), DRV=AVDRV03 defines the rule for the paradigm *aesthetically* (*aesthetic > aesthetically*). The equivalence between the adverbials *in an aesthetic manner* and *aesthetically* is then established by means of a local grammar that uses the information in the dictionary. In (9), DRV=NADRV02 defines the rule for the paradigm *skeptically* (*skepticism > skeptically*). Similarly to the previous example, the equivalence between the adverbials *with skepticism* and *skeptically* is established through 'one' simple local finite-state grammar, represented in Figure 1, that applies the new dictionary properties that linked a derived adverb to an adjective, permitting the transformations required in paraphrasing.

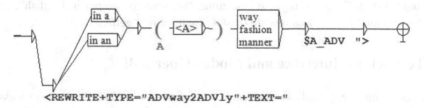

`<REWRITE+TYPE="ADVway2ADVly"+TEXT="`

**Fig. 1.** Grammar to recognize adverbial compounds and transform them into equivalent single adverbs

The grammar in Figure 1 recognizes adverbial compounds such as *in a modest way*, where the adjective *modest* is stored in a variable "A", represented in the grammar between parentheses. The rule then searches in the dictionary for an adverb that is linked to the adjective *modest* ($A_ADV). If this adverb is specified, the rule orders a rewriting of the input expression, the compound adverb, by the single adverb found in the dictionary, *modestly*. The rule illustrated in Figure 2 combines with the previous rule permitting rewriting to take place in a text.

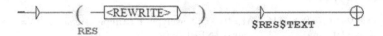

**Fig. 2.** Grammar to rewrite text

SPIDER was inspired by ParaMT, a prototype of a multilingual paraphraser (or translation system). ParaMT uses a similar methodology, except for that it provides an equivalent in a language different from the one of the source text (paraphrases across language). ParaMT can be used directly in machine translation. At the current stage of development, ParaMT aims at translating multi-word units efficiently, handling considerably well the translation of Portuguese support verb constructions into English verbs, as illustrated in Figure 3.

| a fazer um estágio para | dar aulas de / tutor | Religião |
| a fazer um estágio para | dar aulas de / lecture | Religião |
| a fazer um estágio para | dar aulas de / teach | Religião |
| começa a | dar exemplos / exemplify | : |
| sentia-se capaz de | dar um murro em / punch | quem quisesse detê-lo |
| gostávamos de lhe | dar uma palavrinha / speak | . |

**Fig. 3.** Portuguese support verb constructions translated as English verbs

Because Eng4NooJ contains Portuguese transfers for each entry, any grammar used to obtain monolingual transformations can be reused to obtain bilingual (or multilingual) transformations. The recycling of grammars is minimal, since the only parameter that needs to be added is the specification of the output language, as $EN for English or $PT for Portuguese (meaning, "retrieve the output in English", etc.). For monolingual transformations, no output language is specified.

## 5   Technology, Interface and Modus Operandi

SPIDER's web interface allows the users to insert text or upload files to be processed, by selecting either the interactive or the automated mode. The SPIDER allows the user to upload a document file in several formats (Microsoft Word versions up to Word 2007, PDF, PS, HTML and .txt text files). This capability uses a methodology similar to the one used in [Maia et al. 2005], which enables the extraction of text from several documents and process of these texts. Figures 4 and 5 show suggestions provided by SPIDER's paraphrasing system to user inserted texts, for general language linguistic phenomena (Figure 4) or for domain specific language (Figure 5).

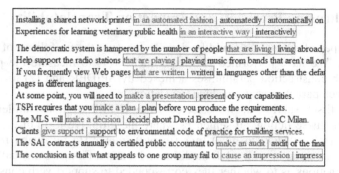

**Fig. 4.** SPIDER suggestions for general language linguistic phenomena

In Figure 4, next to the source text expressions, SPIDER retrieves single word suggestions. Users can select their preferences by just clicking in the original expressions or the suggestions that better suit the objectives/purposes of their texts. Users can also insert their own suggestions. As a result of this process, the user's text will be rewritten accordingly. The system gathers the user's feedback on preferences and can organize/order those preferences statistically. More than one option might be suitable, but only one can be selected. SPIDER permits the user to be shown where the text can be edited, helping maintain a coherent style, reduce wordiness, and clarify vague or imprecise terms or expressions. Terminological and domain specific dictionaries can be applied by SPIDER to facilitate technical writing, as illustrated in Figure 5. SPIDER presents several terminological possibilities and the user can choose among those by clicking on the word or expression that is the most appropriate to the context or just leave the source term untouched.

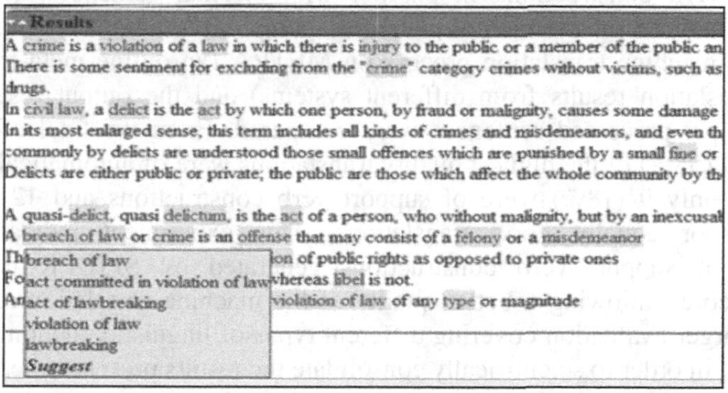

**Fig. 5.** Identification of legal terms and suggestions for the term "breach of law"

If no suitable suggestion is provided by SPIDER, or users themselves find more suitable ones, they can add them to their text. To insert new options, users click on the word "Suggest" in bold, and they can insert their options in the interactive popup box and use those options immediately in their text. However, user inserted options need to be approved and validated by a linguist and inserted in the database for the system to retrieve them in future uses. Terminological options need to be validated by domain experts.

# 6  Paraphrasing Precision and Impact on Translatability

Different types of evaluation could be done to measure the usefulness of a tool such as SPIDER. We have performed two experiments: (1) to measure the precision of support verb construction paraphrases achieved by SPIDER and (2) to measure the impact of such paraphrases in machine translation.

For our first experiment, we selected from COMPARA, a parallel corpus of English-Portuguese fiction, all sentences where the infinitive form of the Portuguese verbs *fazer* (*to do*), *dar* (*to give*), *pôr* (*to put*), *tomar* (*to take*) and *ter* (*to have*) occurred with a noun or with a left modifier and a noun. First, we manually classified these combinations as to whether they corresponded to support verb constructions or not. We confirmed that globally in 64.2% of their occurrence, these verbs are used as support verbs. Subsequently we selected randomly a sub-corpus with 500 sentences (100 for each selected verb), containing instances of only support verb constructions. We classified them manually and compared these results with the results obtained automatically by using SPIDER. We tried to have constraining recognition rules so that paraphrasing would be more precise. Currently, SPIDER can recognize 62.6% of the support verb constructions (SVC recognition recall) with high scores in precision, 98.4% (SVC recognition precision). Furthermore, it does not only recognize the support verb constructions, as it also paraphrases them with high degree of success, 93.4% (SVC paraphrasing precision).

In our second experiment, we measured the impact of the support verb constructions paraphrases in machine translation. From the same corpus (COMPARA), we selected 50 sentences, randomly. The procedures included the implementation of the following routines: (i) the automated pre-processing of support verb constructions with SPIDER and conversion into (strong) verb expressions, i.e., SPIDER suggested paraphrases for the support verb constructions in parallel with original sentences; (ii) both pre-processed sentences (automatically generated paraphrases) and the original text are submitted to machine translation process on METRA (an online meta-translator that retrieves translation results from different systems) and the output results for both original and pre-processed sentences were compared. From the total of 50 pairs of sentences, 29 (58%) of the highest quality translations were of automatically generated paraphrases, only 9 (18%) were of support verb constructions and 12 (24%) were equally bad or equally good translations. This experiment indicates that the paraphrases of support verb constructions generated by SPIDER help improve translation scores, allowing a better quality of the machine translation results in that context. A larger evaluation covering different types of linguistic phenomena needs to be performed in order to scientifically consolidate the results presented here.

# 7  Future Work

In this paper we presented SPIDER, a paraphrasing tool that suggests different ways of expressing content and ideas throughout the text, but also a learning tool for the user to compose text or help with the use of equivalent expressions in context. Future enhancements of SPIDER will enable automated phrase selection to choose the best candidate in a particular context, based on expert historical usage data (specialized crowdsourcing), where the most popular selections by experts will be offered as solutions. This would create a controlled resource based on user preferences. As these resources evolve, they will produce text that after editing is shorter and more comprehensible than the source text. While the meaning is preserved and often clarified by such editing, there is an additional advantage: the reduction of the number of words used in the text will diminish noise and improve the understandability of the text. Such improvement in text understandability will cause a positive impact in translation, and will help natural language processing applications in general.

In future developments, the capability of the local grammars in relation to transformational aspects needs to be expanded and enhanced in order to correct existing problems, improve precision of the paraphrases offered and extend paraphrasing to broader linguistic phenomena. It is also being considered the integration of *revision memories* (RM for short) in SPIDER. Inspired by the concept of translation memories, revision memories are existing aligned text edits, which facilitate the rewriting of sentences or fragments of sentences that have been previously edited, so that they do not have to be re-edited (source sentence > aligned reviewed sentence = revision memory). We finally hope that SPIDER helps bringing paraphrasing to user applications.

# References

Barreiro, A., Scott, B., Kasper, W., Kiefer, B.: OpenLogos Rule-Based Machine Translation: Philosophy, Model, Resources and Customization. In: Forcada, M.L., Sánchez-Martínez, F. (eds.) Machine Translation, Special Issue on Free/Open Source Machine Translation (2011)

Barreiro, A.: Make it simple with paraphrases: Automated paraphrasing for authoring aids and machine translation. Lambert Academic Publishing (2011) ISBN 978-3-8383-8565-5

Barreiro, A., Cabral, L.M.: ReEscreve: a translator-friendly multi-purpose paraphrasing software tool. In: Goulet, M.-J., Melançon, C., Désilets, A., Macklovitch, E. (eds.) Proceedings of the Workshop Beyond Translation Memories: New Tools for Translators, The Twelfth Machine Translation Summit, Château Laurier, Ottawa, Ontario, Canada, August 29, pp. 1–8 (2009)

Barreiro, A.: Port4NooJ: Portuguese Linguistic Module and Bilingual Resources for Machine Translation. In: Blanco, X., Silberztein, M. (eds.) Proceedings of the 2007 International NooJ Conference, Barcelona, Spain, June 7-9, 2007, pp. 19–47. Cambridge Scholars Publishing, Newcastle (2008)

Callison-Burch, C.: Paraphrasing and Translation. PhD Thesis. University of Edinburgh (2007)

Fellbaum, C.: WordNet: An Electronic Lexical Database (Language, Speech, and Communication). The MIT Press, Cambridge (1998)

Hirao, T., Fukusima, T., Okumura, M., Nobata, C., Nanba, H.: Corpus and Evaluation Measures for Multiple Document Summarization with Multiple Sources. In: Proceedings of the COLING 2004, pp. 535–541 (2004)

Ibrahim, A., Katz, B., Lin, J.: Extracting structural paraphrases from aligned monolingual corpora. In: Proceedings of the Second International Workshop on Paraphrasing (ACL 2003) (2003)

Iordanskaja, L., Kittredge, R., Polguère, A.: Lexical Selection and Paraphrase in a Meaning-Text Generation Model. In: Paris, C.L., Swartout, W.R., Mann, W.C. (eds.) Natural Language Generation in Artificial Intelligence and Computational Linguistics, pp. 293–312. Kluwer Academic Publishers, Dordrecht (1991)

Maia, B., Sarmento, L., Santos, D., Cabral, L., Pinto, A.S. (2005). The Corpógrafo - a Web-based environment for corpus research. In: Corpus Linguistics Conference, Birmingham, UK (July 14-17, 2005) ISSN: 1747-9398

McKeown, K., Barzilay, R., Evans, D., Hatzivassiloglou, V., Klavans, J., Nenkova, A., Sable, C., Schiffman, B., Sigelman, S.: Tracking and Summarizing News on a Daily Basis with Columbia's Newsblaster. In: Proceedings of the Human Language Technology Conference, San Diego, CA, USA (March 2002)

Mitamura, T., Nyberg, E.: Automatic rewriting for controlled language translation. In: NLPRS 2002, Workshop on Automatic Paraphrasing: Theories and Applications (2002)

Paşca, M., Dienes, P.: Aligning Needles in a Haystack: Paraphrase Acquisition Across the Web. In: Dale, R., Wong, K.-F., Su, J., Kwong, O.Y. (eds.) IJCNLP 2005. LNCS (LNAI), vol. 3651, pp. 119–130. Springer, Heidelberg (2005)

Scott, B., Barreiro, A.: OpenLogos MT and the SAL representation language. In: Pérez-Ortiz, J.A., Sánchez-Martínez, F., Tyers, F.M. (eds.) Proceedings of the First International Workshop on Free/Open-Source Rule-Based Machine Translation, November 2-3, pp. 19–26. Universidad de Alicante. Departamento de Lenguajes y Sistemas Informáticos, Alicante, Spain (2009)

Scott, B.: The Logos Model: An Historical Perspective. Machine Translation 18, 1–72 (2003)

Scott, B.: The Logos System. In: MT Summit II, Munich, Germany (1989)

Shinyama, Y., Sekine, S.: Using Repeated Patterns across Comparable Articles for Paraphrase Acquisition. New York University. Proteus Technical Report (2005)

Silberztein, M.: NooJ: A Cooperative, Object-Oriented Architecture for NLP. In: INTEX pour la Linguistique et le traitement automatique des langues. Presses Universitaires de Franche-Comté. Cahiers de la MSH Ledoux, Besançon, France (2004)

# Providing Cross-Lingual Editing Assistance to Wikipedia Editors

Ching-man Au Yeung, Kevin Duh, and Masaaki Nagata

NTT Communication Science Laboratories
2-4 Hikaridai, Seika-cho, Soraku-gun
Kyoto, 619-0237, Japan
{auyeung,kevinduh}@cslab.kecl.ntt.co.jp, nagata.masaaki@lab.ntt.co.jp

**Abstract.** We propose a framework to assist Wikipedia editors to transfer information among different languages. Firstly, with the help of some machine translation tools, we analyse the texts in two different language editions of an article and identify information that is only available in one edition. Next, we propose an algorithm to look for the most probable position in the other edition where the new information can be inserted. We show that our method can accurately suggest positions for new information. Our proposal is beneficial to both readers and editors of Wikipedia, and can be easily generalised and applied to other multi-lingual corpora.

## 1 Introduction

There are currently over 250 different language editions in Wikipedia. However, significant differences exist between different editions in terms of size and quality [6]. Several projects on Wikipedia have been initiated to bridge this information gap with the help of both human and machine translation [12,13,14]. Google also provides a translator toolkit that assists users to translate Wikipedia articles[1].

While existing efforts focused on translating whole articles, we believe maintaining existing articles across different languages is also a major challenge. Wikipedia is by no means a static encyclopedia. Articles are constantly being revised by editors. As different language editions are being developed separately, it is likely that different language editions will contain different information, depending on the focuses of the editors or interests of the respective community.

Although Wikipedia is not intended to be an encyclopedia in which different language editions are exact translations of one another [14], it is desirable to keep any article up-to-date and comprehensive. However, the effort required to identify what should be translated can be prohibitively expensive, especially when the target document already has substantial content. This requires editors to continuously monitor articles in different languages, which is clearly unscalable.

We propose a framework that assists Wikipedia editors or translators to transfer information from one language into another. We term this task **cross-lingual document enrichment**. Our proposed framework is completely automatic and

---

[1] Google Translator Toolkit: http://translate.google.com/toolkit

A. Gelbukh (Ed.): CICLing 2011, Part II, LNCS 6609, pp. 377–389, 2011.

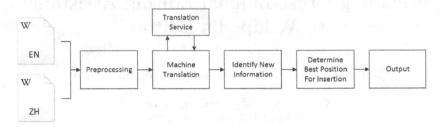

**Fig. 1.** System design of our proposed cross-lingual document enrichment framework

only requires the availability of a machine translation service. While we focus on Wikipedia in this paper, our techniques can be applied to any multi-lingual corpus where disparity of information in different languages is a problem. We believe this research has a positive impact on the creation and maintenance of huge multi-lingual corpora which have become more common nowadays.

## 2    Cross-Lingual Document Enrichment

While our proposal is independent of the languages involved, for concreteness of presentation, we assume that our source document is in English and the target document is in Chinese. We choose to treat *sentences* as the basic units that carry information. The two major processes in our proposed framework are:

1. **New information identification**: Given two sets of sentences (in Chinese and English), identify a subset (of English sentences) that contains information not found in Chinese. (Section 2.2)
2. **Cross-lingual sentence insertion**: Determine the best position where a translation of the new sentence should be inserted into the Chinese document, respecting the document's existing discourse structure. (Section 2.3)

Figure 1 depicts the overall system design of our framework. For each article, English and Chinese editions are preprocessed to remove formatting information. Sentences are extracted and labelled by section and paragraph IDs. To compare sentences in different languages, we make use of a machine translation tool[2]. We translate all Chinese sentences into English, so that the information content could be compared. However, in practice any process that maps the two editions to the same symbol set is possible. For example, we can translate the English to Chinese, translate both editions to French/Italian/Spanish, or any combination of the above methods[3].

---

[2] We use Google Translate http://translate.google.com/ in this work but in theory any broad-coverage translation service is possible.

[3] In fact, we can also translate the two editions to a latent mapping that is not reminiscent of any human language, using machine learning techniques like [3].

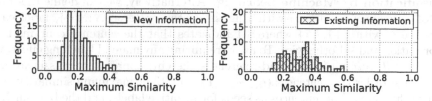

**Fig. 2.** Distribution of maximum similarity values of sentences with new or existing information for the article 'Angkor Wat'

In this paper, we use 'article' to refer to a particular topic in Wikipedia, such as 'Angkor Wat' or 'India'. An article has one or more language editions. We refer to each edition as a document. Let $E$ be the document in English and $C$ be the document in Chinese. We define a document as a sequence of sentences. Hence $E = (e_1, e_2, ..., e_M)$ and $C = (c_1, c_2, ..., c_N)$, where $M$ and $N$ are the numbers of sentences in English and Chinese respectively.

## 2.1 Measuring Sentence Similarity

To measure sentence similarity, we first submit the Chinese edition to a machine translation service and obtain an English translation. Then, for any document $D$, we extract a vocabulary $V_D$ after stop-word removal and stemming. Each sentence $s$ is represented by a term vector $s = (w_1, w_2, ..., w_{|V_D|})$, where $w_i$ is the weight of the word $v_i \in V_D$ in $s$, determined by TF-IDF. The similarity of two sentences is calculated by the cosine similarity $\cos(s_i, s_j)$ between the two vectors. In our implementation, we add terms appearing in section titles to the term vectors of the sentences in the corresponding sections. We find that section titles are indicative of the topics of the sentences, and are helpful in improving the similarity metric.

## 2.2 Identifying New Information

To determine whether an English sentence contains new information with respect to the Chinese edition, we consider the following two methods.

**Heuristic Method.** Intuitively, English sentences with existing information should have high similarity to at least one sentence in the Chinese edition, while those with new information should have low similarity to all Chinese sentences. A heuristic method is to rank the English sentences by their maximum similarity to any Chinese sentence ($\max_j \cos(e_i, c_j)$), and consider sentences with maximum similarity lower than a certain threshold as containing new information.

Figure 2 shows the maximum similarity values for the article 'Angkor Wat'. The extreme values are relatively well separated across positive and negative samples, and the heuristic method will provide correct answer to a certain extent. However, there are regions where both positive and negative samples can be found. We suspect that this is due to limitations in the machine translation process. Thus, we also consider the following more sophisticated method.

**Classification by Machine Learning.** Alternatively, we can define this task as follows: given an article with English sentences $(e_1, e_2, ..., e_M)$, label each sentence $e_i$ with $\{+1, -1\}$ where $+1$ indicates that the sentence contains new information and $-1$ otherwise. To avoid requiring any manual labelling effort, we adopt a *self-training* (see [1] and references therein) approach to classification.

We first order the sentences $(e_1, e_2, ..., e_M)$ by their maximum similarity, and choose the top $N\%$ of sentences as seeds for negative labels, and the bottom $N\%$ as seeds for positive labels. These labels are used to train a support vector machine (SVM) classifier. In this way, no manual annotations are required. The key assumption is that the extreme values of the maximum similarity are relatively reliable indicators of the true label. Based on our observations in Figure 2, we believe that the self-training assumption is reasonable on this kind of dataset.

We introduce several varieties of features for the SVM classifier. A feature vector is defined for each sentence $e_i$. The main types of features are:

- **Similarity**: Maximum cosine similarity of $e_i$. This is the feature used in the heuristic baseline (Section 2.2).
- **Neighbour**: Maximum cosine similarity of the neighbours, $e_{i+1}$ and $e_{i-1}$. The idea is that if the neighbours have low similarity, then more likely $e_i$ will contain new information, and the opposite is also likely to be true.
- **Entropy**: Entropy of similarity values of $e_i$, where similarity distribution is converted into probability distribution by $p(e_i|c_j) = \frac{\cos(e_i, c_j)}{\sum_{j'} \cos(e_i, c_{j'})}$. This feature counteracts situations where particular words lead to high cosine values for all sentences. Intuitively, an English sentence (if it contains existing information) should only be matched to a small number of Chinese sentences, and would achieve low entropy.

In practice, we have a total of 18 features, where each feature is a variant of one of the three main types listed above. For example, one feature is the entropy, while another (related) feature is the difference with the entropy averaged over the document.

## 2.3   Cross-Lingual Sentence Insertion

Our next task is to identify the positions in the Chinese edition where the sentences should be inserted. In some cases there may not be a single correct position for a new sentence as it can simply be inserted into a particular paragraph or section where the content matches that of the sentence. However, in other cases a sentence may elaborate an existing sentence and should be placed after that sentence. To accommodate this stricter requirement, we formulate our problem as finding a sentence $c_j$ in the Chinese edition after which (translation of) the new sentence $e_i$ should be inserted. To solve this problem, we consider several different methods as described below.

**Insertion by Manual Alignment.** As a first step, we consider that some sentences in English have been aligned to those in Chinese manually. Intuitively, the sentence should be inserted in a way that maintains the order of description

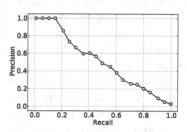

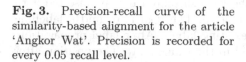

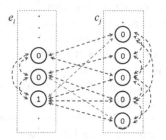

**Fig. 3.** Precision-recall curve of the similarity-based alignment for the article 'Angkor Wat'. Precision is recorded for every 0.05 recall level.

**Fig. 4.** A graph constructed based on the document structure and similarity values between sentences in two documents

or the flow of the article. Thus, a reasonable scheme is as follows. We look for an English sentence before $e_i$, say $e_{i-1}$ that is manually aligned to a Chinese sentence $c_j$. Since $e_{i-1}$ corresponds to $c_j$, it becomes natural that $e_i$ when translated into Chinese should follow $c_j$ as well. If $e_{i-1}$ has no corresponding sentence in the Chinese edition, we can repeat the process and check $e_{i-2}$ and so on. In practice, however, there will probably be no manually aligned sentences available for us to carry out this scheme. Hence, this method will mainly be used for comparison in evaluating our other proposed automatic methods.

**Insertion by Similarity-based Alignment.** When sentences are not manually aligned, we can generate some alignments by selecting pairs of sentences that achieve high values of similarity. Depending on the quality of translation, these pairs are very likely to be correct alignments. For example, after ranking the sentence pairs by their similarity values, we can heuristically choose the top 100 pairs as correct alignments. Figure 3 shows a typical precision-recall curve of one of the articles we manually aligned. In this figure, choosing the top 100 pairs corresponds to 47% precision and 55% recall, which is a well-balanced operating point. With these alignments, we can then apply the same method described in the above section. The limitation of this relatively simple method, of course, is that the sentence alignments may not be correct, leading to erroneous sentence insertions. In addition, highly similar sentences might be concentrated in a particular part of the article (e.g. the introductory sections).

**Label Propagation.** In view of the limitations of the above methods, we propose a method that is based on the technique of label propagation in classification and takes advantage of all similarity values among the sentences. Label propagation [17] uses a graph to incorporate similarity information for all pairwise examples in the data. If a label is known for an example, it is placed on the example and 'propagated' or 'diffused' to other examples that have no known labels. This can be seen as running a Markov chain over the graph.

We construct a graph $G = (V, E)$ where the set of vertices $V$ are English and Chinese sentences $(e_1, ..., e_M)$ and $(c_1, ..., c_N)$. There are then $M \times N$ graph edges between the Chinese and English sides, where the edge weights $w_{ij}$ represent the cosine similarity $\cos(e_i, c_j)$. Edges among sentences in the same language are also created to represent the document structure. We set $w_{ij} = 1/dist(c_i, c_j)$ if $c_i$ and $c_j$ are from the same paragraphs, where $dist$ is the distance (number of intervening sentences) between $c_i$ and $c_j$; if they are in different paragraphs, we set $w_{ij} = 0$. The graph allows us to represent global information about all similarity links and document structure. Figure 4 gives a pictorial example.

We initialise the graph by labelling the English sentence to be inserted with label +1, and all other sentences with label 0. The goal is to find a labelling over $(c_1, c_2, ..., c_N)$ by propagating the existing labels. After label propagation, each Chinese sentence will receive a label in the range $[0, 1]$. The position after the Chinese sentence with the maximum value is then chosen to be the place of insertion. Intuitively, Chinese sentences that have a high probability link to the English sentence with +1 label will more likely be the insertion position.

Label propagation can be performed by an iterative Markov chain computation, or by direct eigenvector computation [17]. In the latter case, the following objective can be used:

$$\min_{\mathbf{f}} \sum_{(i,j) \in E} w_{ij}(f_i - f_j)^2 \qquad (1)$$

where $f_i$ is the labelling on vertex $i$, which is capped at +1 or 0 for English sentences and left undetermined for Chinese sentences. $\mathbf{f}$ is an $(N + M)$-dimensional vector of labels. The objective accomplishes label propagation by forcing a pair of vertices $(i, j)$ to have similar labels $f_i$ and $f_j$ if the edge weight $w_{ij}$ is large. $\mathbf{f}$ is computed by taking the eigenvectors of the graph Laplacian [17].

## 3    Experiments

### 3.1    Data Set and Preprocessing

We collect a set of articles from Wikipedia to evaluate our proposed framework. We first found a set of 2,792 articles that are featured articles in English (as of 17 February 2010)[4]. Featured articles are well-developed and mature articles and they represent good source of new information for other language editions.

From within this set, we performed extensive manual annotation on nine articles on a broad range of topics. These articles contain a total of about 2,000 English sentences and 1,600 Chinese sentences. Two bilingual-speaking annotators work to identify which English sentences contain new information. If an English sentence does *not* provide new information, the annotators label which Chinese sentence it aligns to. Alignments of multiple Chinese sentences to one English sentence (and vice versa) are allowed. Further, when a Chinese sentence only contains partial information, it is also considered as aligned to the English.

---

[4] http://en.wikipedia.org/wiki/Wikipedia:Featured_articles

**Table 1.** Articles selected for manual inspection and alignment. The table shows the number of sentences in the English and Chinese editions. The 'aligned' column shows the number of sentences in English that are aligned to some sentences in Chinese.

| Article | Sent. (EN) | Sent. (ZH) | Aligned |
|---|---|---|---|
| Acetic acid | 194 | 169 | 155 |
| Angkor Wat | 149 | 222 | 71 |
| Australia | 258 | 229 | 72 |
| Ayumi Hamasaki | 227 | 306 | 114 |
| Battle of Cannae | 221 | 149 | 100 |
| Boeing 747 | 350 | 185 | 298 |
| H II region | 116 | 81 | 103 |
| India | 245 | 156 | 67 |
| Knights Templar | 156 | 119 | 39 |

The manual annotation is a laborious process since on average the featured articles selected have 210 sentences in one English document and substantial amounts in Chinese. The manual annotation took 2-3 hours on average per article. The inter-annotator agreement was high, with $\kappa = 0.826$, determined on 3 articles (732 sentences) of overlapping annotation.

## 3.2 Experimental Setup and Results

**Identifying New Information.** We first present experiments on identifying sentences that contain new information. Our test set contains the nine articles manually annotated. A sentence in the English edition is considered to be containing new information if it is not aligned to any Chinese sentence. We compare four different methods in this classification task:

1. **Heuristic**: Heuristic method that uses only cosine similarity information (Section 2.2)
2. **Self-train**: Linear SVM trained on top/bottom (N=30%) group of sentences (Section 2.2)[5].
3. **Cheat**: Similar to the SVM above, but the true labels are used in training. This is a diagnostic to see to what extent the assumption of self-training holds true.
4. **Random**: Randomly classifying a sentence as containing either new or existing information.

To avoid having to decide on a particular similarity threshold, we evaluate using the area under the precision-recall curve (AUC) for each annotated document. A higher value of AUC in general means that precision is higher for a given recall level. The results are shown in Table 2.

Both heuristic and the self-train SVM achieved high AUC values of 70-95%, for all but two articles, showing that most new information can be captured automatically. Performance on two articles gave surprisingly low AUC scores ('Boeing 747' and 'H II Region'). We discovered that they are the only articles in which the number of aligned pairs is larger than the number of Chinese sentences

---

[5] We use SVM-rank, a publicly-available SVM tool: http://svmlight.joachims.org

**Table 2.** Results of new information identification. Numbers refer to the area under the precision-recall curve in percentage

| Article | Heuristic | Self-train | Cheat | Random |
|---|---|---|---|---|
| Acetic Acid | 70.86 | **72.24** | 79.69 | 24.36 |
| Angkor Wat | 81.36 | **81.77** | 86.45 | 49.85 |
| Australia | 92.97 | **93.04** | 93.16 | 74.72 |
| Ayumi Hamasaki | **72.59** | 71.26 | 72.32 | 50.14 |
| Battle of Cannae | 84.60 | **85.14** | 83.14 | 54.69 |
| Boeing 747 | **54.18** | 52.95 | 54.16 | 19.22 |
| H II Region | 54.94 | **55.65** | 71.30 | 46.83 |
| India | 95.40 | **95.63** | 95.75 | 71.24 |
| Knights Templar | **89.38** | 88.59 | 93.63 | 79.17 |

**Table 3.** Percentage of weights assigned to features for different methods

| | Heuristic | Self-train | Cheat |
|---|---|---|---|
| Similarity features | 100 | 52 ± 3 | 35 ± 12 |
| Neighbour features | 0 | 16 ± 2 | 39 ± 15 |
| Entropy features | 0 | 32 ± 2 | 27 ± 12 |

(see Table 2). This implies that the Chinese sentences are longer, such that multiple English sentences were aligned to one Chinese sentence. As a result similarity between sentences in these two articles tend to be much lower than expected. Overall, we note that for articles in which lengths of sentences are not drastically different across languages, our methods gave reasonable results.

Further, we analysed the SVM models to see what kinds of features are important. We summed up (linearly) the SVM's weights corresponding to each of the three main types of features. Table 3 shows that for the self-trained SVM, similarity features are deemed most useful and account for 52% of the weights, and that entropy features are second. This is expected because all training samples have extreme similarity values. However, for the cheating SVM neighbour features are much more important, accounting for 39% of the weights. This shows that whether neighbouring sentences contain new information is a useful hint on how a sentence should be classified.

In summary, both heuristic and self-trained SVM give satisfactory performances by achieving over 70% AUC in most cases, with the former performing slightly better in some cases. The cheating SVM suggests that we can significantly improve classification results even if only some of the sentences are labelled manually.

**Cross-lingual Sentence Insertion.** Our second task is sentence insertion. For each article, we randomly select a number of sentences in English that has been manually aligned to some Chinese sentences. We then cover up these alignments, simulating the situation that we have a set of new sentences whose correct positions in the Chinese edition are known. We test the performance of the following four methods:

1. **Manual alignments**: A method based on the manually created alignments of other sentences (Section 2.3).

**Table 4.** Sentence insertion results, with 30% new information

| Method | Average Distance | Section Accuracy | Paragraph Accuracy |
|---|---|---|---|
| Manual alignment | 11.5 | 72.8% | 43.1% |
| Similarity alignment | 19.3 | 57.5% | 35.5% |
| Label prop (para) | **10.5** | **83.9%** | **72.7%** |
| Label prop (sec) | 13.2 | 81.7% | 71.5% |

**Table 5.** Sentence insertion results, with 50% new information

| Method | Average Distance | Section Accuracy | Paragraph Accuracy |
|---|---|---|---|
| Manual alignment | 14.9 | 70.7% | 39.7% |
| Similarity alignment | 17.6 | 59.3% | 34.3% |
| Label prop (para) | **11.3** | **82.9%** | **76.8%** |
| Label prop (sec) | 14.0 | 81.9% | 76.4% |

2. **Similarity alignments**: A method based on the alignments automatically created by sentence cosine similarity. We choose the top 100 pairs of sentences as correct alignments (Section 2.3).

3. **Paragraph-based label propagation**: The method described in Section 2.3.

4. **Section-based label propagation**: Similar to the above method, but we also create links between sentences appearing in the same section.

To measure performance, we use three different evaluation metrics, averaged over the nine test articles: (1) **Average Distance** (distance in number of sentences between the predicted position and the true insertion position), (2) **Section Accuracy** (whether sentence is inserted into the correct section), and (3) **Paragraph Accuracy** (whether sentence is inserted into the correct paragraph). We decide not to measure accuracy at the sentence level. This is because very often there is no single 'best position' where a sentence should be inserted. Instead, suggesting a paragraph to which a sentence should be inserted is already of great assistance to an editor.

We conduct experiments with different amounts of test data (30% and 50%). The results are in Table 4 and 5. We observe that label propagation outperforms both manual and similarity-based alignments in all three metrics. This is a nice result considering that manual alignment uses true alignments while label propagation does not. The implication is that true alignments are actually not necessary for global graph-based methods. The similarity-based alignment does not use manual information but it performs much worse (e.g. up to 26% decrease in section accuracy). In both tables, we see that different variants of the graph lead to slightly different accuracies for label propagation. It is well-known in the semi-supervised learning literature that optimising the graph structure may lead to better results [16], and we leave that as future work. Here we do not optimise the graph because we want to stay within an unsupervised learning setting where minimal human annotation is required for our methods.

---

**Etymology** [edit]

Before the Portuguese settlement in the early 16th century, Macau was known as *Haojing* (Oyster Mirror) or *Jinghai* (Mirror Sea).[10] The name *Macau* is thought to be derived from the *A-Ma Temple* (traditional Chinese: 媽閣廟; Jyutping: Maa1 Gok3 Miu6), a temple built in 1448 dedicated to Matsu — the goddess of seafarers and fishermen. It is said that when the Portuguese sailors landed at the coast just outside the temple and asked the name of the place, the natives replied "媽閣" (j="Maa1 Gok3"). [A] The Portuguese then named the peninsula "Macau".[11] The present Chinese name 澳門 (j=Ou3 Mun4) means "Inlet Gates".

**History** [edit]

The history of Macau is traced back to the Qin Dynasty (221–206 BC), when the region now called Macau came under the jurisdiction of Panyu county, in Nanhai prefecture (present day Guangdong). [B] [10] The first recorded inhabitants of the area were people seeking refuge in Macau from invading Mongols during the Southern Song Dynasty.[12] Under the Ming Dynasty (1368–1644 AD), fishermen migrated to Macau from Guangdong and Fujian provinces.

Macau did not develop as a major settlement until the Portuguese arrived in the 16th century.[13] In 1535, Portuguese traders obtained the rights to anchor ships in Macau's harbours and to carry out trading activities, though not the right to stay onshore.[14] Around 1552–1553, they obtained temporary permission to erect storage sheds onshore, in order to dry out goods drenched by sea water;[15] they soon built rudimentary stone houses around the area now called Nam Van. [C] In 1557, the Portuguese established a permanent settlement in Macau, paying an annual rent of 500 taels of silver.[15]

> [A] 在民間，還有一種說法：...路人因不諳外語而聽不明其意，遂以本地粵音俚語「乜溝？」（意即「什麼？」，與「Macau」的譜音相似）...
> Another hypothesis is that "Macau" actually means "What?" in Cantonese, and the natives are simply replying that they did not understand what the Portuguese sailors asked.
>
> [B] 澳門古稱濠鏡澳，與香山縣的歷史關係極其密切
> Macau, also known as 濠鏡澳 in the antiquity, has historically had a connection with Xiangshan County (a place in Guandong).
>
> [C] 惟後期鋪設的碎石道工不佳，大雨過後碎石脫落有礙觀瞻，...
> But the stone pavements were not well constructed, and the stones fell apart after rainfall... (referring to a tourist area)

---

**Fig. 5.** Sentence insertion for the article 'Macau'. We show part of the English edition and three sentences containing new information in Chinese. The alphabets indicate the suggested insertion positions.

# 4   Discussion

## 4.1   An Example of System Output

We apply our framework to the article 'Macau' to identify new information in Chinese and insert sentences into the English edition[6]. The article is a featured article in Chinese but not in English. Figure 5 shows three sentences identified as containing new information using the self-trained SVM approach, and the positions where our label propagation algorithm suggests they should be inserted.

Sentence (A) and (B) are correct insertions. The former provides information about the origin of the name of Macau, while the latter provides information about Macau's geographical relations. However, while being a sentence containing new information for the English edition, Sentence (C) is an example of incorrect insertion. The sentence is irrelevant to the paragraph, but is inserted at that position because of the word 'stone', which is a rare word throughout the documents. Since this sentence refers to a topic that is not present in the English edition, it becomes difficult for the algorithm to find a correct position. Overall, our algorithm works well when sentences contain information that is new but is still related to the topics present in the target document.

---

[6] This is the opposite direction of the experiments, to demonstrate the flexibility of our approach.

## 4.2   Other Issues and Limitations

Several issues deserve further attention. Firstly, we treat sentences as the basic units of information, which has resulted in certain limitations. We plan to study how these can be solved. Nevertheless, the existence of this problem actually motivate our work, because this means that it requires even more effort from human editors to distinguish between new and existing information.

Secondly, our current similarity measure only takes into account lexical similarity and may overlook synonyms. The accuracy of similarity also depends on the result of machine translation. While Google Translate mostly return sufficiently good translations for measuring sentences similarity on the lexical level, to further improve performance we will consider incorporating the translation model into our framework instead of treating it as a black box.

Finally, in this work we do not consider the 'value' of the sentences. As articles are constantly under revision, sentences may be deleted for various reasons. Vandalism is also not uncommon in Wikipedia. Hence, it would be desirable to determine whether a sentence (and the information it contains) is valuable to be inserted into other languages. We can incorporate into our framework methods for vandalism detection [15], or methods for assessing the credibility of the editors who wrote the sentences [7]

## 5   Related Works

While there are no directly comparable works, some authors have studied related problems. For example, Chen et al. [5] propose a method based on sentence features for inserting new information into existing texts in a monolingual setting. Our work differs from theirs in that we consider a cross-lingual setting and can therefore take advantage of the document structures of both the source and target article. Our method also does not require supervised labels.

Adar et al. [2] introduce an automated system called Ziggurat for aligning and complementing infoboxes across different languages in Wikipedia. Tacchini et al. [11] presents some experiments on data fusion across languages using the DBpedia framework. Our proposed framework can be used to handle the texts of the articles and is therefore more general and applicable to other settings.

Sauper and Barzilay [9] propose a method for generating Wikipedia articles by inducing an article template automatically and retrieving relevant texts from the Web. We believe that this method would be complementary to our proposal, because our method relies on the fact that the articles already contain some information. In cases when a topic simply does not exist, an automatically generated article will be a very good starting point for cross-lingual enrichment.

Lapata [8] proposes using a Markov chain to model the structure of a document. Barzilay and Elhadad [4] proposes a method for sentence alignment that involves first matching larger text fragments by clustering and further refine these matches by local similarity. These techniques, however, require a large corpus for training, while our proposed model operates only on the article level and does not require any labels.

Finally, our work is also related to the task of automatic extraction of parallel sentences from comparable corpora [10], as sentences that are not found to have any correspondence in another language should contain new information. We plan to investigate how methods for this task can be incorporated into our framework to improve performance.

# 6    Conclusions

We propose a new task, 'cross-lingual document enrichment', of which the goal is to assist editors in bridging the information gap within multi-lingual document collections. Our contributions include (1) a framework for addressing this task in terms of two sub-tasks: new information identification and cross-lingual sentence insertion; and (2) a proof-of-concept system using a novel combination of NLP and machine learning techniques. While there are other ways for improvement, our system already demonstrates the ability to significantly alleviate the load for human editors. In addition to investigating the issues mentioned in Section 4.2, we will also carry out evaluations of larger scale on various datasets. We believe that this is a promising research direction for NLP to impact the creation and maintenance of vast multi-lingual document collections.

# References

1. Abney, S.: Bootstrapping. In: 40th Annual Meeting of the Association for Computational Linguistics (2002)
2. Adar, E., Skinner, M., Weld, D.S.: Information arbitrage across multi-lingual Wikipedia. In: WSDM 2009, pp. 94–103 (2009)
3. Bai, B., Weston, J., Grangier, D., Collobert, R., Sadamasa, K., Qi, Y., Cortes, C., Mohri, M.: Polynomial Semantic Indexing. In: NIPS (2009)
4. Barzilay, R., Elhadad, N.: Sentence alignment for monolingual comparable corpora. In: EMNLP 2003, pp. 25–32 (2003)
5. Chen, E., Snyder, B., Barzilay, R.: Incremental text structuring with online hierarchical ranking. In: EMNLP-CoNLL 2007, pp. 83–91 (2007)
6. Hecht, B., Gergle, D.: The Tower of Babel meets Web 2.0: user-generated content and its applications in a multilingual context. In: CHI 2010, pp. 291–300 (2010)
7. Javanmardi, S., Lopes, C., Baldi, P.: Modeling user reputation in Wikipedia. Journal of Statistical Analysis and Data Mining 3(2), 126–139 (2010)
8. Lapata, M.: Probabilistic text structuring: experiments with sentence ordering. In: ACL 2003, Morristown, NJ, USA, pp. 545–552 (2003)
9. Sauper, C., Barzilay, R.: Automatically generating Wikipedia articles: A structure-aware approach. In: ACL 2009, pp. 208–216 (2009)
10. Smith, J., Quirk, C., Toutanova, K.: Extracting Parallel Sentences from Comparable Corpora using Document Level Alignment. In: ACL 2010 (2010)
11. Tacchini, E., Schultz, A., Bizer, C.: Experiments with Wikipedia cross-language data fusion. In: Proceedings of the 5th Workshop on Scripting and Development for the Semantic Web, ESWC 2009 (2009)
12. Wikipedia. Translation of the week (2010),
    http://meta.wikimedia.org/wiki/Translation_of_the_week
    (accessed May 10, 2010)

13. Wikipedia. Wikipedia machine translation project (2010),
    http://meta.wikimedia.org/wiki/Wikipedia_Machine_Translation_Project
    (accessed May 10, 2010)
14. Wikipedia. Wikipedia:translate (2010),
    http://en.wikipedia.org/wiki/Wikipedia:Translation
    (accessed May 10, 2010)
15. Wang, W., McKeown, K.: Got You!: Automatic Vandalism Detection in Wikipedia
    with Web-based Shallow Syntactic-Semantic Modeling. In: COLING 2010,
    pp. 1146–1154 (2010)
16. Zhu, X.: Semi-supervised learning literature survey. Technical Report 1530, Com-
    puter Sciences, University of Wisconsin-Madison (2005),
    http://www.cs.wisc.edu/~jerryzhu/pub/ssl_survey.pdf
17. Zhu, X., Ghahramani, Z., Lafferty, J.: Semi-supervised learning using gaussian
    fields and harmonic functions. In: Proceedings of International Conference on Ma-
    chine Learning (2003)

# Reducing Overdetections in a French Symbolic Grammar Checker by Classification

Fabrizio Gotti[1], Philippe Langlais[1], Guy Lapalme[1],
Simon Charest[2], and Éric Brunelle[2]

[1] DIRO/Univ. de Montréal
C.P. 6128, Succ Centre-Ville
H3C 3J7 Montréal (Québec) Canada
http://rali.iro.umontreal.ca
[2] Druide Informatique
1435 rue Saint-Alexandre, bureau 1040
H3A 2G4 Montréal (Québec) Canada
http://www.druide.com

**Abstract.** We describe the development of an "overdetection" identi-
fier, a system for filtering detections erroneously flagged by a grammar
checker. Various families of classifiers have been trained in a supervised
way for 14 types of detections made by a commercial French grammar
checker. Eight of these were integrated in the most recent commercial
version of the system. This is a striking illustration of how a machine
learning component can be successfully embedded in Antidote, a robust,
commercial, as well as popular natural language application.

## 1 Introduction

Even though most modern writers use, often unknowingly, the grammar checker
embedded in Microsoft Word, few NLP researchers have addressed the problem
of improving the quality of the grammatical error detection algorithms [1,2].
Clément et al. [3] suggest that this could be explained by the lack of an annotated
error corpus and by the close link that exists between a grammar checker and
the proprietary word processor that embeds it.

Bustamante and Léon [4] present a typology of errors often encountered in
Spanish and describe how the GramCheck project dealt with them. They distinguish structural errors (e.g. bad prepositional attachments) from non structural
ones (e.g. subject verb agreement). The former is dealt with by crafting rules
encoding typical errors that are added to the language parsing rules or by using
auxiliary grammars on an ad hoc basis. The latter is dealt by loosening the unification process within the parser. These developments are quite complex and
require a fine tuning of linguistic heuristics used within the parsing process.

Two main approaches to grammar checking have been taken by researchers.
The first approach consists in comparing the sentence to proofread against a
model of proper language use (a positive grammar). For instance, [5] propose
using $n$-grams to create a language model of lemmas and part-of-speech tags
(POS) occurring in proper English text. The second strategy seeks to create

A. Gelbukh (Ed.): CICLing 2011, Part II, LNCS 6609, pp. 390–401, 2011.

negative grammars in order to represent erroneous language constructs, like in [6] or [7]. Both approaches will often use a corpus of correct and faulty sentences to learn sequences of words that are then compared with the text to check. [8] propose the use of grammar error rules derived from a normal grammar's rule so that the relationship between the correct rule and its derived error rules reflects a possible error as well as the correction to apply. Finding a representative training corpus is a challenge for these approaches, although one could use the one described in [9] or derived from it [10], but more important is the fact that regular expressions cannot reliably detect errors between distant words.

Sofkova Hashemi [11] also uses (positive) regular grammars on POS tags to detect grammatical errors. Using a notion of automata subtraction, she builds on the idea that if a coarse grammar (e.g. not taking into account number and gender agreement) can parse a sentence but not a more precise one, then there is probably an error in this sequence and a detection is made.

It is natural to think that a good grammar checker should strive to reach two conflicting goals : detect all errors present, offering a good *recall*, while avoiding false flags (henceforth *overdetections*), offering a good *precision*. But these goals are not equal in the eyes of the end-user. Indeed, a low precision is a major source of dissatisfaction for them: The overdetections give an (often false) impression that the grammar checker is incorrect in all of its suggestions. Indeed, when Microsoft researched their customers in order to properly design the grammar checker incorporated into their Office suite, they concluded [12] that "increasing precision and decreasing the false flag per page rate have had a higher priority than recall for these grammar checkers." This is corroborated by [13].

In this paper, we focus on a new NLP task: the identification of *overdetections*, i.e. grammatical error detections erroneously made by the system on flawless excerpts of text. We hope to demonstrate that this task presents interesting scientific challenges while offering some feedback to a large community of end-users.

We propose to tackle this task by training classifiers in a supervised way in order to recognize these overdetections. Our work is very different from the ones alluded to above because we are positioned *downstream from* the grammar checker. Resorting to a post-processing strategy has several potential benefits. First, the approach we propose can in principle be adapted to another grammar checker or to other types of errors than those we studied here (see Section 2.3). Second, we already mentioned that modifying an existing parser to account for ill-formed input is a difficult enterprise, one that we avoid here. In fact, the task we address is relatively simpler: we do not locate the errors or suggest a correction because this has already been done by the grammar checker.

We present our project in section 2. In section 3, we describe our approach to the overdetection problem. Results are presented in section 4. Contributions and new perspectives are presented in section 6.

## 2   ScoRali

The development of a grammar checker or its improvement is a complex endeavor involving many strategic choices as described by [12]. Here, we describe

SCORALI, resulting from the close collaboration between Druide Informatique (Druide) and researchers from RALI. Druide has been, for many years, actively developing a symbolic French parsing technology called Analytix which is based on rich symbolic description dictionaries and on a dependency parser both built and maintained manually by a team of linguists. The parser can deal with complex syntactic phenomena such as coordination (complete and elliptic), extraposition, correlation, punctuation use and some categorial inference. Particular care has been devoted to the parser robustness in the case of lexical and syntactical errors. A correction module uses the syntactic trees to pinpoint errors and suggest appropriate corrections. This technology is embedded in many commercial products, including Antidote, a writing assistant developed for the French language.

## 2.1   Requirements

The goal of the project was to train classifiers in a supervised way to detect overdetections for 14 common error types processed by Analytix. The error types have been selected by Druide according to their frequency and their overdetection risk (see section 2.3). Used after Analytix processing pipeline, these classifiers would judge the quality of the detections in order to filter out those that would most probably not be appropriate in this context. To be considered worthwhile, a classifier should identify at least 66% of overdetections and not remove more than 10% of correct detections. The classifiers should fit Analytix's processing pipeline.

## 2.2   Methodology

Data preparation[1] was crucial in this project. For each type of error, Druide prepared a sample of about 1000 detections, separated on average into an equal number of overdetections and legitimate detections, produced by Analytix on "real" texts, representing different types of use of the application. Each occurrence was then annotated by a linguist as being an overdetection or not and was associated with the syntactical parse produced by Analytix. This parse gives the position of each word, its grammatical category and about fifty morphosyntactic attributes such as gender, number (before and after correction) and, for verbs, their mode, tense and person. Moreover, all syntactic relations between words in the sentence were given, allowing to rebuild the syntactic parse tree of the sentence including, for each node, its grammatical category and a confidence weight. Each word was associated with a number of semantic-syntactic tags chosen from more than a thousand available.

This data was used as a basis for the features that we extracted for training our classifiers (see section 3). To help us determine the best ones, Druide also provided, for each type of detection, a summary characterization of the most frequent overdetection contexts and a linguistically motivated estimate of what they felt were the most suitable identification features.

---

[1] This data is unfortunately not available to the community.

## 2.3   Types of Errors

After many annotation and development cycles, 14 detection types were studied. Some of them are quite specific with respect to the linguistic phenomena they detect, e.g. the confusions between two words, like QUE/DONT — confusion between *que* ("that") and *dont* ("of that", used with a verb requiring a preposition) — or OU/OÙ — confusion between the conjunction *ou* ("or") and the adverb and pronoun *où* ("where"), which share the same pronunciation and differ only by the grave accent, a common source of error in French texts. An example of a good detection and a incorrect one is shown in Figure 1 for QUE/DONT. In these examples, the underlined word is the site of the detection and * indicates an overdetection. An English translation of the original text and its correction is also provided.

Je comprends ce que tu dis mais pas ce que *[dont]* tu parles.
I understand what you say but not **what** *[of what]* you speak.

Mais bon dieu que *[\*dont]* les adultes s'amusent!
But gosh **what** *[\*of what]* fun these adults have!

**Fig. 1.** An example of a good detection (top) and of an overdetection (bottom) for QUE/DONT. This type of detection is concerned with the confusion between 2 French words, "que" (what) and "dont" (of what).

Other detections are more general, in the sense that a given detection, like PP/VC (confusion between the past participle of a verb and its other conjugations) could encompass numerous different linguistic manifestations, given that it applies to many inflected forms of different verbs, sometimes with intervening words within the ill-formed construct. We give an example of such an overdetection for PP/VC in Figure 2 below.

Roman ou récit, la « Collection blanche » de Gallimard éblouit *[\*éblouí]*.
Novel or story, the « Collection blanche » from Gallimard **dazzles** *[\*dazzled]*.

**Fig. 2.** An example of an overdetection for PP/VC, the confusion between the past participle of a verb and its other forms

It should be pointed out that there are many different types of texts in the training corpus: Some sentences were extracted from Wikipedia articles (including some headers), others from Internet discussion boards or scientific texts, etc. Some were even text messages without any diacritical marks, sometimes resorting to *phonetic* spelling. The quality therefore greatly varies.

One observable consequence of the poor quality of some of these sentences is that some overdetections result from other problems in the same sentence. In the example of Figure 3, the correction of *le*[*the (masc.)*] into *la*[*the (fem.)*] is proposed because the word **marché**[*deal (masc.)*] is misspelled **marche**[*step (fem.)*]. As explained by [12] and [5], the (obviously poor) French of the writer could have been taken into account when making corrections, here.

```
Ils veulent un libéralisme VRAI, accepte le [*la] marche mais...
```
They want a TRUE liberalism, accepts **the (masc.)** *[\*the (fem.)]* step but...

**Fig. 3.** An example of an overdetection for ACCORD: an agreement error either in number or gender. In the translation, we purposely introduced errors in "TRUE" (capitalization), "accept" and "step" (the writer misspelled "deal") to illustrate the errors in the French original text, for the corresponding words.

# 3　Classification of Detections

## 3.1　Feature Extraction

The manual inspection of hundreds of instances of correct and incorrect detections (like those presented in Figures 2 and 3) shows that words in the neighborhood of the detections made by `Analytix` can guide the classification of a given detection. This context can simply be words before and after the detection or, since the training data includes the syntactical parse produced by `Analytix`, head words or dependents. For instance, for the confusion OU/OÙ, it is rather obvious for a French speaker that, whenever the word "là" precedes a potential confusion, a "où" is expected, rather than "ou". Similarly, for the confusion QUE/DONT, if the head word for the site of the confusion is a verb calling for a prepositional object, then "dont" should be used. Other features of the word flagged as a detection are important. For instance, when detecting capitalization errors, it is not advisable to correct a capitalized word when it follows another capitalized word: they probably participate in a named entity.

As a consequence of the previous observations, we selected a set of more than 1200 features per detection, among which:

**The features of the word at the site of the detection.** They are : its case, its length in letters, its gender, its number (for a noun), its tense, its number (for a verb), its part of speech, its position in the sentence, as well as features specific to the parser used by `Analytix`, for instance "verbs ending in -yer than can be confused with a noun".

**The features of the words surrounding the site of the detection.** The same features as those of the previous item, but this time for the words preceding and following the detection, as well as for the head word of the detection, in the parse.

**The features of the detection itself.** Precisely, the certainty with which `Analytix` proposes a correction, and the features of the correction proposed as a replacement for the word detected.

**The features of the sentence in which the detection is found.** That is, its length in words, the numbers of dependency relations identified, and the number of unknown words[2].

**The nature of the dependency links** in which the word detected participates. Namely, the links between the detection and its head word or its possible dependents. Here, we identify the usual relations, like "noun adjunct" or others, more specific to the grammar checker, e.g. the French "tel que" ("for instance").

**Some ad hoc features, specific to each type of detection.** For instance, we attempted to reframe the classification problem at hand as a word sense disambiguation (WSD) task, for the confusion QUE/DONT. Indeed, if we examine the example in Figure 1, we can consider the site of the detection as a placeholder, and "que" and "dont" as potential "semantic" labels. The disambiguation process allows us then to determine which one of those labels to insert at the site of the ambiguity. This is similar in spirit to the strategy adopted in [14] for fixing context-sensitive spelling corrections, that is, spelling mistakes resulting in existing words (e.g. *piece* versus *peace*). Among other strategies, we used an adaptation of the technique described by [15]. Unfortunately, our incursion into the WSD territory meant that we had to build and use external texts for modeling the context of *que* and *dont*, which precluded the integration of this classifier in Analytix.

## 3.2   Classifiers Studied

To create and put to the test the required classifiers, we used the free software package Weka [16], written in Java[3]. This package allows the easy experimentation of numerous families of classifiers and possesses valuable features, like the visualization of data and classifiers as well as the preprocessing of training data. Moreover, it is possible to bypass the graphical interface and launch a classifier from the command line, which proved invaluable in our case when batch-processing thousands of classifier commands.

Weka allows the prototyping of roughly 50 classifiers, grouped into 8 families, e.g. Bayesian classifiers, decision trees, perceptrons, SVM and meta-classifiers. The latter combine other classifiers, by making them vote, for instance. Each of these classifiers is typically controlled by 1 to 20 hyperparameters, discrete or continuous. This causes a combinatorial explosion in the number of possible classifiers. Therefore, our first efforts focused on the exploration of classifiers that are rapid during training and classification, partly to satisfy the specifications for the project. Additionally, we preferred classifiers which were conceptually "simple", in order to facilitate their tuning, design, and eventual implementation within Analytix. These reasons led us to concentrate our efforts on symbolic classifiers (but see section 4.4), namely rules.ConjunctiveRule, rules.DecisionTable,

---

[2] It is noteworthy that most of the features which are numerical counts are doubled: one is the count itself, the other is the count normalized by the length of the sentence.

[3] www.cs.waikato.ac.nz/ml/weka/

rules.JRip (a propositional rule learner, like RIPPER [17]), trees.ADTree (alternating decision trees), trees.DecisionStump, trees.J48 (C4.5 decision trees) and trees.J48graft.

Beyond the selection and parametering of the classifiers, Weka allows diverse pre-processing strategies on the training data. We first filtered features, to remove those which did not vary enough or varied too much among the instances of the training set (these features cannot be used to discriminate). Furthermore, for every type of detection, we tried different filtering strategies for their features, reducing in some case the 1200 features to a mere dozen. Naturally, this simplifies the training and test of the classifiers, but also their eventual implementation. We also varied the cost that Weka attributes to false negatives and false positives. Ultimately, for each of the 14 detection types, we tested about 4000 classifier settings in order to find one which would meet the requirements set by Druide. This exploration was made on a 16 dual-core computer cluster, with a computing time of 5 days for each detection.

# 4   Results

It is impossible to fit all the results obtained on all classifiers tested in this article. For this reason, in this section, we will detail the results for the detection PP/VC, for which we provided an example earlier in Figure 2. We chose this example because it lent itself quite successfully to the approach, and because it is representative of the results we obtained on several other detections. Also, it is an illustration of a detection arising in very varying contexts, which proves challenging. We will nonetheless present a global overview of the results for all the detection types later in this article.

## 4.1   The pp/vc Detection

Figure 4 shows the performance of more than 3000 classifiers for the project SCORALI (it is the scatter plot cluster labeled "System". The $x$-axis represents the percentage of good corrections erroneously flagged as bad corrections by the classifiers (false positives), while the $y$-axis represents the percentage of bad corrections correctly identified as such (true positives). For each classifier, this data was obtained through 10-fold cross-validation on the training set.

Remarkably, the scatter plot cluster is relatively compact, forming a band encompassing the possible compromises each classifier offers. The choice of a classifier is made manually, based on this kind of figure, while striving to meet the requirements set by Druide (section 2.1). In our case, the classifier recommended for Analytix is the one whose data point is circled in the figure. It is a C4.5 decision tree, grafted and pruned, allowing the identification of 77 % of overdetections, at the cost of a loss of 8 % of good corrections. The decision tree classifies 88 % of instances correctly, with a substantial agreement of $\kappa = 0.76$ between all 10 folds of the cross-evaluation. Other classifiers with similar performances were discarded, either because they were too complex to implement or because they used too many features.

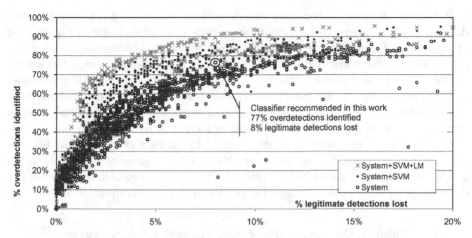

**Fig. 4.** Classifiers tested for the detection PP/VC. Each point represents the performance of a single classifier, i.e. a compromise between the true and false positives. The graph shows 3 scatter plots, one for SCORALI per se (Section 4, labeled "System") and 2 others, resulting from further research (Section 4.4).

## 4.2    An Overview of All Detection Types

We applied the strategy described in the previous section for all 14 types of detections provided by **Druide**. Although the lack of place prevents us from describing each error type, the overall results are presented in Table 1. We first observe that we succeeded in creating classifiers meeting the project's requirements for 9 out of the 14 detection types. They all are decision trees: when a group of classifiers proved equally good at classifying detections, we selected a decision tree among them. A closer inspection of the induced rules used in the decision trees did not allow the identification of features consistently present in all or most of the trees.

It is difficult to explain why certain detections lent themselves well to classification, and others not. Despite our best efforts, QUE/DONT could not find a good classifier, whereas some detections proved easy to process, although it seemed at first that rule induction would be difficult because they occurred in extremely different contexts, for different reasons and often in text of very poor quality.

Naturally, classification rules are induced more easily if **Analytix** overdetects within a certain language construct in a systematic way. This is the case for the detection LA/LÀ, where a manual inspection of the decision tree produced shows that 20 % of the overdetections occur when "la" is an article ending an abruptly truncated sentence, like in the instance "recommandations de **la** *[\*là]*" (recommendations of **the** *[\*there]*). The construct is always the same then: a missing noun adjunct preceded by the article.

It is also obvious that certain overdetections made by `Analytix` are very difficult to classify, even for a human being. The examples shown in Figure 2 for PP/VC and in Figure 3 are striking. In the latter Figure, the instance "accepte **le** *[\*la]* marche", one must really understand the sentence to know that the author meant "accepte le marché" ("accept the deal", where "marché" is masculine) rather than "accepte la marche" ("accept the step", where "marche" is feminine). The overdetection is then due to the missing acute accent on the final letter of "marche". It is highly likely that classifiers have a hard time with these cases, especially when these detections are flagged in varying contexts, for different reasons. It is the case for ACCORD, which can be an agreement error, either in gender or number. It could be interesting to further our research by creating two different detection types: ACCORD for gender and ACCORD in number.

**Table 1.** Complete results of SCORALI, as delivered to `Druide`

| Detection | % fp | % tp | Detection | % fp | % tp | Detection | % fp | % tp |
|---|---|---|---|---|---|---|---|---|
| ÉLISION | 7% | 87% | INV | 8% | 71% | QUE/DONT | 9% | 57% |
| LA/LÀ | 9% | 84% | MAJ | 7% | 71% | OU/OÙ | 9% | 49% |
| PP INV | 6% | 80% | PP/INF | 7% | 68% | ACCORD | 9% | 40% |
| PP/VC | 8% | 77% | CONJUG | 8% | 66% | MODE | 9% | 27% |
| APOS | 6% | 73% | ER/EZ | 9% | 65% | | | |

### 4.3   Tests at `Druide` and Implementation

The eleven best classifiers were therefore converted and integrated to `Analytix`, then tested on a corpus distinct from the one we used for their training, a recommended practice in the industry (see for instance [12]). These tests revealed that 3 classifiers degraded the performance of the grammar checker to a point where they had to be rejected. These classifiers were not necessarily those with the worst performances during training, but had to be removed nonetheless. The 8 remaining classifiers are part of the latest commercial version of `Analytix`. Among these, 3 are used as they are, and five are subject to an ad hoc test determining whether the engine will use them, based on the context of the detection. Finally, the user interface includes an on/off switch for these classifiers: The users are presented with a checkbox labeled "statistical filtering of detections". The user wishing not to miss any good detection can deactivate this setting, but will be presented with more false detections. The classifiers are on by default.

### 4.4   Better Classifiers

As a requirement (see Section 2.1), the classifiers delivered to `Druide` had to be simple to interpret and to embed within `Analytix`. In order to measure the improvements that could be made to SCORALI, we studied the performances offered by SVM classifiers and we added features derived from language models trained on the Canadian Hansard. The gains obtained are shown in Figure 4 by their respective scatter plot cluster, for detection PP/VC. SVMs (labeled "System+SVM") alone allow a gain of about 5% in identification of overdetections, compared to the classifiers delivered to Druide, at the cost of an increase in computational needs and a decrease in expressivity of the model created. The addition of language model features (labeled "System+SVM+LM") increases the number of true positives, for a further 10% gain.

## 5  Related Work

A fair number of studies have been dedicated to spelling correction (e.g. [18,14,19]). Grammar checking, which we believe is a useful component of a writing assistant tool, has received — somehow paradoxically — much less attention. However, we see many advantages to studying grammar checkers on their own. Indeed, we feel this naturally belongs to the field of *grammar engineering*. Behind this expression, we group activities as diverse as making parsing faster and more robust (e.g. [20]), adapting parsers to new domains (e.g. [21]), or simply improving existing parsers (e.g. [22]).

Actually, for certain types of detections, the classifiers we trained proved very useful as error mining tools within Analytix's parsing grammar.[4] For instance, we noticed that Analytix has difficulty recognizing the expression "faire partie" (take part) and makes the overdetection "faire **partie** *[\*partit]*" (take parted), for detection PP/VC. Although the particular classifier for PP/VC did not include such a feature, it would probably be interesting to detect the presence of the verb "faire" in the context leftward of the detection.

Thus, the work we conducted could prove, as a side effect, to be complementary to the studies made on error identification in wide-coverage grammars [22,23,24], but has the advantage of not requiring the modification of the grammar studied, a delicate task which is not always possible in a complex grammar maintained manually, especially in a commercial context.

## 6  Conclusion and Future Work

SCORALI allowed the creation and implementation into Analytix of 8 out of 14 classifiers identifying overdetections, downstream of the grammar checking engine. This successful transfer of technologies from the laboratory to a commercial product entailed the exploration of thousands of different classifiers, as well as a delicate balance between performance and technical constraints.

We think that this paper clearly shows that statistical and symbolic approaches can go hand in hand, and is indeed a very clear illustration of the kind of balancing act that such a combination requires [25].

---

[4] This is one argument in favor of classifiers such as decision trees that can be easily interpreted against other ones such as SVMs.

Despite the fact that the corpus we used in this study can not be released to the scientific community, we hope to have shown that overdetection identification constitutes a "real" task in NLP, one that presents interesting scientific challenges while offering some feedback to a large community of end-users.

This work shows some avenues that we think are worth investigating. For the time being, certain detections seem not to lend themselves to the proposed approach, maybe because they occur in contexts which are too varied, thus defying rule induction. The work we conducted on using more features and more robust classifiers (see Section 4.4) for the PP/VC detection shows that there is room for improvement. This suggests further experiments to see if such gains carry over other detections.

One limitation of our work lies in the simplifying assumption that each detection within a sentence is independent of the other possible detections within the same sentence, although evidence shows that one actual error can trigger an overdetection.

Also, we feel the evolution of SCORALI poses a number of exciting questions. The data used to train the classifiers is not frozen in time: it was generated by Analytix at a given moment in its life cycle. Although our classifiers passed a number of regression tests at Druide, it remains to be seen whether they will withstand the likely changes that will happen over time (within Analytix, in detection statistics, etc.) or whether new training (or adaptation) will be required.

# References

1. Fontenelle, T.: Dictionnaires et outils de correction linguistiques. Rev. franç. de linguistique appliquée X-2, 119–128 (2005)
2. Véronis, J.: Texte: Correcteurs orthographiques en panne? Blog du (July 6, 2005), http://aixtal.blogspot.com/2005/07/texte-correcteurs-orthographiques-en.html
3. Clément, L., Gerdes, K., Marlet, R.: Grammaires d'erreur – correction grammaticale avec analyse profonde et proposition de corrections minimales. In: 16è TALN, Senlis, France (2009)
4. Bustamante, F.R., León, F.S.: Gramcheck: A grammar and style checker. In: 16th COLING, Denmark, pp. 175–181 (1996)
5. Napolitano, D., Stent, A.: TechWriter: An Evolving System for Writing TechWriter: An Evolving System for Writing Assistance for Advanced Learners of English. CALICO 26(3), 611–625 (2009)
6. Rider, Z.: Grammar checking using pos tagging and rules matching. In: Proceedings of the Class of 2005, Senior Conference, Computer Science Department, Swarthmore College, pp. 14–19 (2005)
7. Souque, A.: Vers une nouvelle approche de la correction grammaticale automatique. In: Récital, Avignon, France (2008)
8. Foster, J., Vogel, C.: Parsing ill-formed text using an error grammar. Artif. Intell. Rev. 21(3-4), 269–291 (2004)
9. Foster, J.: Good Reasons for Noting Bad Grammar: Empirical Investigations into the Parsing of Ungrammatical Written English. PhD thesis, Department of Computer Science - University of Dublin (May 2005)

10. Foster, J.: Treebanks gone bad: Parser evaluation and retraining using a treebank of ungrammatical sentences. Int. J. Doc. Anal. Recognit. 10(3), 129–145 (2007)
11. Sofkova Hashemi, S.: Detecting grammar errors in children's writing: A finite state approach. In: 13th Nordic Conference on Computational Linguistics, Uppsala, Sweden (May 2001)
12. Helfrich, A., Music, B.: Design and evaluation of grammar checkers in multiple languages. In: Project notes and demonstration at the 18th COLING, Saarbrücken, Germany, pp. 1036–1040 (2000)
13. Bernth, A.: Easyenglish: a tool for improving document quality. In: Proceedings of the Fifth Conference on Applied Natural Language Processing, Morristown, NJ, USA, pp. 159–165. Association for Computational Linguistics (1997)
14. Golding, A.R., Roth, D.: A winnow-based approach to context-sensitive spelling correction. CoRR cs.LG/9811003 (1998)
15. Yarowsky, D.: Unsupervised Word Sense Disambiguation Rivaling Supervised Methods. In: 33rd Meeting of the ACL, Cambridge, MA, pp. 189–196 (1995)
16. Hall, M., Frank, E., Holmes, G., Pfahringer, B., Reutemann, P., Witten, I.H.: The WEKA Data Mining Software: An Update. SIGKDD Explorations 11(1), 10–18 (2009)
17. Cohen, W.W.: Fast effective rule induction. In: Proceedings of the Twelfth International Conference on Machine Learning, pp. 115–123. Morgan Kaufmann, San Francisco (1995)
18. Damerau, F.: A technique for computer detection and correction of spelling errors. Commun. ACM 7(3), 171–176 (1964)
19. Brill, E., Moore, R.C.: An improved error model for noisy channel spelling correction. In: ACL 2000: Proceedings of the 38th Annual Meeting on Association for Computational Linguistics, Morristown, NJ, USA, pp. 286–293. Association for Computational Linguistics (2000)
20. Kiefer, B., Krieger, H.U., Carroll, J., Malouf, R.: A bag of useful techniques for efficient and robust parsing (1999)
21. Rimell, L., Clark, S.: Adapting a lexicalized-grammar parser to contrasting domains. In: EMNLP 2008: Proceedings of the Conference on Empirical Methods in Natural Language Processing, Morristown, NJ, USA, pp. 475–484. Association for Computational Linguistics (2008)
22. van Noord, G.: Using self-trained bilexical preferences to improve disambiguation accuracy. In: IWPT 2007: Proceedings of the 10th International Conference on Parsing Technologies, Morristown, NJ, USA, pp. 1–10. Association for Computational Linguistics (2007)
23. Sagot, B., de la Clergerie, E.: Fouille d'erreurs sur des sorties d'analyseurs syntaxiques. Traitement Automatique des Langues 49(1), 41–60 (2009)
24. de Kok, D., Ma, J., van Noord, G.: A generalized method for iterative error mining in parsing results. In: Proceedings of the 2009 Workshop on Grammar Engineering Across Frameworks (GEAF 2009), Suntec, Singapore, pp. 71–79. Association for Computational Linguistics (August 2009)
25. Klavans, J.L., Resnik, P. (eds.): The balancing act: combining symbolic and statistical approaches to language. MIT Press, Cambridge (1996)

# Performance Evaluation of a Novel Technique for Word Order Errors Correction Applied to Non Native English Speakers' Corpus

Theologos Athanaselis, Stelios Bakamidis, and Ioannis Dologlou

Institute for Language and Speech Processing (ILSP) / R.C "ATHENA"
Artemidos 6 and Epidavrou, GR-15125,
Athens, Greece
{tathana,bakam,ydol}@ilsp.gr

**Abstract.** This work presents the evaluation results of a novel technique for word order errors correction, using non native English speakers' corpus. This technique, which is language independent, repairs word order errors in sentences using the probabilities of most typical trigrams and bigrams extracted from a large text corpus such as the British National Corpus (BNC). A good indicator of whether a person really knows a language is the ability to use the appropriate words in a sentence in correct word order. The "scrambled" words in a sentence produce a meaningless sentence. Most languages have a fairly fixed word order. For non-native speakers and writers, word order errors are more frequent in English as a Second Language. These errors come from the student if he is translating (thinking in his/her native language and trying to translate it into English). For this reason, the experimentation task involves a test set of 50 sentences translated from Greek to English. The purpose of this experiment is to determine how the system performs on real data, produced by non native English speakers.

**Keywords:** Word order errors; statistical language model; permutations filtering; British National Corpus; non native English speakers' corpus.

## 1 Introduction

Research on detecting erroneous sentences can be mainly classified into three categories. The first category makes use of hand-crafted rules [1],[2],[3]. These methods have been shown to be effective in detecting certain kinds of grammatical errors, but it is expensive to write non-conflicting rules in order to cover the wide range of grammatical errors. The second category focuses on parsing ill-formed sentences [4],[5],[6],[7]. The third category uses statistical techniques to detect erroneous sentences. Instead of asking experts to write hand-crafted rules, statistical approaches [8],[9],[10],[11] build statistical models to indentify sentences containing errors.

There are also other studies on detecting grammar errors at sentence level. More [12] introduced an English grammar checker for non-native English speakers. Heift

A. Gelbukh (Ed.): CICLing 2011, Part II, LNCS 6609, pp. 402–410, 2011.

[13] released the German Tutor, an intelligent language tutoring system where word order errors are diagnosed by string comparison of base lexical forms. Bigert and Knutsson [14] showed how a new text can be compared to known correct text and deviations from the norm flagged as suspected errors. Sjöbergh and Knutsson [15] introduced a method of grammar error recognition by adding errors to several (mostly error free) unannotated texts and applying a machine learning algorithm.

In contrast to existing statistical methods, the proposed method is applicable to any language (language models can be computed in any language) and it is not restricted to a specific set of words. For that reason use of parser and/or tagger is not necessary. Also, it does not require manual collection of written rules since they are outlined by the statistical language model. A comparative advantage of this method is that it avoids the laborious and costly process of collecting word order errors for creating error patterns.

The paper is organized as follows: section 2 describes the method for finding the correct word order in a sentence and a description of the language model. Section 3 shows how permutations are filtered by the proposed method and Section 4 describes the method that is used for searching for valid trigrams in a sentence. An experimental setup and results using non native English speakers' corpus test data is given in section 5. The concluding remarks are discussed in section 6.

## 2   Finding the Correct Word Order in a Sentence

This paper presents a new method for repairing sentences with candidate word order errors that is based on the conjunction of a new association technique with a statistical language model. The best way to reconstruct a sentence with word order errors is to reorder the words' sequence. However, the question is how it can be achieved without knowing the Part of Speech (POS) tag of each word. Many techniques have been developed in the past to cope with this problem using a grammar parser and rules. This paper deals with a new technique for handling word order errors using all the possible words permutations of the sentence. The process of repairing sentences with word order errors incorporates the followings tools:

- a fast algorithm for filtering the sentence's permutations
- and a statistical Language Model (LM) based on N-grams.

The concept is based on the following two axioms: the first axiom concerns the assumption that taking into account all the permutations of a sentence with word order errors, it is absolutely certain that the correct sentence will be included in the set of the permutations. The second axiom relies on the assumption that the number of valid trigrams (sentence's trigrams that are included in the language model) increases as the number of word order errors declines. Therefore, the system provides as output a list of N-best sentences according to the number of valid trigrams (see Fig. 1). The following steps summarise the main algorithmic steps of the proposed method for repairing sentences' word order errors.

1. The basic units for the reconstruction method are the sentence's words $W_1W_2...W_{n-2}W_{n-1}W_n$ (where $W_i$ is the i-th word).
2. Construction of an association matrix for extracting valid bigrams
3. Construction of a network with valid bigrams in order to form possible permuted sentences of length N
4. Decomposition of each permuted sentence into a set of trigrams
5. Evaluation of the sentences according to the number of valid trigrams
6. If more than one sentence has the same number of valid trigrams the sum of trigrams' log probability is taken into account.

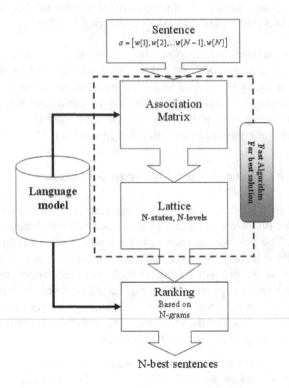

**Fig. 1.** System's architecture

## 2.1 Language Model

The language model (LM) that is used subsequently is the standard statistical N-grams [16]. The N-grams provide an estimate of $P(W)$, the probability of observed word sequence $W$. Assuming that the probability of a given word in an utterance depends on the finite number of preceding words, the probability of N-word string can be written as:

$$P(W) = \prod_{i=1}^{N} P(w_i \mid w_{i-1}, w_{i-2}, ..., w_{i-(N-1)}) \tag{1}$$

One major problem with standard N-gram models is that they must be trained from some corpus, and because any particular training corpus is finite, some perfectly acceptable N-grams are bound to be missing from it. That is, the N-gram matrix for any given training corpus is sparse; it is bound to have a very large number of cases of putative "zero probability N-grams" that should have some non zero probability. Some part of this problem is endemic to N-grams; since they can not use long distance context, they always tend to underestimate the probability of strings that happen no tot have occurred nearby in their training corpus. There are some techniques that can be used in order to assign a non zero probability to these zero probability N-grams. In this work, the language model has been trained using BNC and consists of trigrams with Good-Turing discounting [17] and Katz back off [18] for smoothing. BNC contains about 6.25M sentences and 100 million words. The figure below depicts the number of bigrams of the LM (Language Model) with respect to their logarithmic probabilities. The 80% of the LM's bigrams are between -5,2 and -1,6.

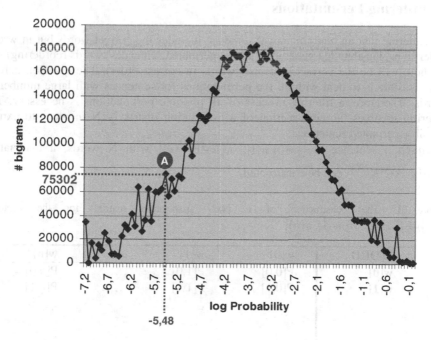

**Fig. 2.** The distribution of bigrams with respect to the log probabilities. The A symbol corresponds to 75302 single bigrams with log probability -5,48.

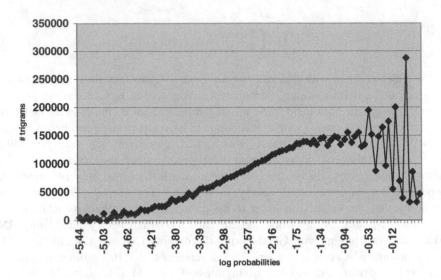

**Fig. 3.** The distribution of trigrams with respect to the log probabilities

## 3 Filtering Permutations

Considering that an ungrammatical sentence includes the correct words but in wrong order, it is plausible that generating all the permuted sentences (words reordering) one of them will be the correct sentence (words in correct order). The question here is how feasible is to deal with all the permutations for sentences with large number of words. Therefore, a filtering process of all possible permutations is necessary. The filtering involves the construction of a association matrix NxN in order to extract possible permuted sentences.

Given a sentence $a = [w[0], w[1], ... w[n-1], w[n]]$ with N words, a association

matrix $A \in R^{NXN}$ can be constructed,

**Table 1.** The construction of a NxN association matrix, for the sentence $a = [w[0], w[1], ... w[n-1], w[n]]$

| WORD | w[0] | w[1] | ....... | w[n] |
|------|------|------|---------|------|
| **w[0]** | P[0,0] | P[1,0] | ....... | P[n,0] |
| **w[1]** | P[0,1] | P[1,1] | ....... | P[n,1] |
| . | . | . | | . |
| . | . | . | | . |
| . | . | | | . |
| **w[n]** | P[0,n] | P[1,n] | ....... | P[n,n] |

The size of the matrix depends on the length of the sentence. The objective of this association matrix is to extract the valid bigrams according to the language model. The element $P[i, j]$ indicates the validness of each pair of words $(w[i]w[j])$ according to the list of language model's bigrams. If a pair of two words $(w[i]w[j])$ cannot be found in the list of language model bigrams then the corresponding $P[i, j]$ is taken equal to 0 otherwise it is equal to one. Hereafter, the pair of words with $P[i, j]$ equals to 1 is called as valid bigram. Note that, the number of valid bigrams is $M$ lower than the size of the association matrix which is $N^2$, since all possible pairs of words are not valid according to the language model. In order to generate permuted sentences using the valid bigrams all the possible words' sequence must be found. This is the search problem and its solution is the domain of this filtering process.

As with all the search problems there are many approaches. In this paper a left to right approach is used. To understand how it works the permutation filtering process, imagine a network of $N$ layers with $N$ states. The factor $N$ concerns the number of sentence's words. Each layer corresponds to a position in the sentence. Each state is a possible word. All the states on layer 1 are then connected to all possible states on the second layer and so on according to the language model. The connection between two states $(i, j)$ of neighboring layers $(N \ 1, N)$ exists when the bigram $(w[i]w[j])$ is valid. This network effectively visualizes the algorithm to obtain the permutations. Starting from any state in layer 1 and moving forward through all the available connections to the $N$-th layer of the network, all the possible permutations can be obtained. No state should be "visited" twice in this movement.

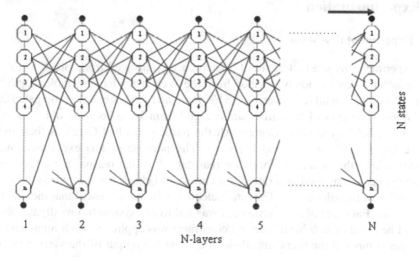

**Fig. 4.** Illustration of the lattice with N-layers and N states

## 4 Searching for Valid Trigrams

The prime function of this approach is to decompose any input sentence into a set of trigrams. To do so, a block of words is selected. In order to extract the trigrams of the input sentence, the size of each block is typically set to 3 words, and blocks are normally overlapped by two words. Therefore, an input sentence of length N, includes N-2 trigrams.

The second step of this method involves the search for valid trigrams for each sentence. A probability is assigned to a valid trigram, which is derived by the frequency of its occurrences in the corpus.

In the third step of this method the number of valid trigrams per each permuted sentence is calculated. Considering that the sentence with no word-order errors has the maximum number of valid trigrams, it is expected that any other permuted sentence will have less valid trigrams. Although some of the sentence's trigrams may be typically correct, it is possible not to be included into the list of LM's trigrams. The plethora of LM's trigrams relies on the quality of corpus. The lack of these valid trigrams does not affect the performance of the method since the corresponding trigrams of the permuted sentence will not be included into LM as well. The criterion for ranking all the permuted sentences is the number of valid trigrams. The system provides as an output, a sentence with the maximum number of valid trigrams. In case where two or more sentences have the same number of valid trigrams a new distance metric should be defined. This distance metric is based on the total log probability of the sentence's trigrams. The total log probability is computed by adding the log probability of each valid trigram, whereas the probability of non valid trigrams is assigned to a negative number. Therefore the sentence with the maximum total log probability is the system's response.

## 5 Experimentation

### 5.1 Experimental Scheme

The experiment involves a test set of 50 sentences translated from Greek to English. The sentences were randomly selected from the Web. They have variable length from 10 to 12 words. A total of 40 persons (51% male and 49% female) participated in the experiment. The ages of the participants ranged from 16 to 36 years old. 60% of the participants held a university degree. All the participants had Greek as their mother tongue and English as a second language. The purpose of this experiment was to determine how the system performs on real data. For that reason, a native English-speaking person was used as a rater. The aim was to check whether our system could fix real life word order errors. The rater selected only the sentences that include word order errors. Each one of these sentences was fed to our system to investigate whether it could be fixed or not. No reordering constraint was applied to each input sentence. The rater compared the users' translations against the output of the system to check their correctness.

A total of 320 answers were gathered. This means that every sentence was answered on average 320 / 50 = 6,4 times. From 320 answers, more than half of them included word order errors (189). These sentences were free from other grammatical errors since they have been corrected by the English rater.

## 5.2 Experimental Results

The findings from the experimentation (Fig. 5) show that, according to the rater, 73% of the users' answers have been repaired using our system. For the rest of the sentences (27%), our system did not manage to correct the users' word order errors.

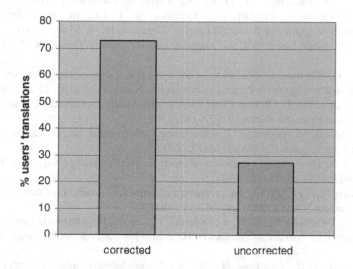

**Fig. 5.** The percentage of corrected and uncorrected sentences

# 6  Conclusions

The findings from the experiments show that in the case of the non native English speakers' corpus, 73% of the users' answers were repaired using our system. This implies that the input sentence provides more valid trigrams compared with the rest of the permuted sentences and can be fixed using the proposed system. False alarms (27%) detecting ill formed sentences are similar to repairing, because both use the same mechanism.

The novelty of the proposed method is the use of a technique for filtering the initial number of permutations and the elimination of grammatical rules to repair sentences with word-order errors. Another aspect of the method is the ability to use it to distinguish different writing styles. The findings show that most of the sentences can be repaired by this method independently from the type of word order errors. Further consideration involves an evaluation of the proposed method using real data with misspellings. Fixing word order errors combined with misspellings certainly invites research. Also, another important issue is to clarify whether the incorporation of grammar information can improve the system's performance. Five simple improvements over basic language models such as variable length N-grams, caching, skipping, clustering and sentence mixture models will be tested and compared.

# References

1. Michaud, L., McCoy, K.F., Pennington, C.A.: An Intelligent Tutoring System for Deaf Learners of Written English. In: Proc. 4th International ACM SIGCAPH Conference on Assistive Technologies (ASSETS), Arlington, pp. 92–100 (2000)
2. Bender, E.M., Flickinger, D., Oepen, S., Walsh, A., Baldwin, T.: Arboretum: Using a Precision Grammar for Grammar Checking in CALL. In: Proc. InSTIL/ICALL Symposium on Computer Assisted Learning, Venice, Italy, pp. 83-86 (2004)
3. Naber, D.: A rule based style and grammar checker. Master Thesis, Bielefeld University (2003)
4. Fouvry, F.: Constraint Relaxation with Weighted Feature Structures. In: Proc. 8th International Workshop on Parsing Technologies, Nancy, France, pp. 23-25 (2003)
5. Vogel, C., Cooper, R.: Robust Chart Parsing with Mildly Inconsistent Feature Structures. Nonclassical Feature Systems 10, 127–136 (1995)
6. Lee, J.: Seneff. S.: Automatic Grammar Error Correction for Second-Language Learners. In: Interspeech, paper 1299-Wed3A3O.1 (2006)
7. Atwell, E.S.: How to detect grammatical errors in a text without parsing it. In: Proceedings of the 3rd EACL, pp. 38–45 (1987)
8. Chodorow, M., Leacock, C.: An unsupervised method for detecting grammatical errors. In: Proceedings of NAACL 2000, pp. 140–147 (2000)
9. Izumi, E., Uchimoto, K., Saiga, T., Supnithi, T., Isahara, H.: Automatic Error Detection in the Japanese Learners' English Spoken Data. In: Proc. ACL, Sapporo, Japan, pp. 145–148 (2003)
10. Costa-Jussà, M.R., Fonollosa, J.A.R.: An Ngram-based reordering model. Computer Speech & Language 23(3), 362–375 (2009)
11. Sun, G., Liu, X., Cong, G., Zhou, M., Xiong, Z., Lee, J., Lin, C.: Detecting Erroneous Sentences using Automatically Mined Sequential Patterns. In: Proceedings of the ACL, Prague, pp. 81–88 (2007)
12. More, J.: A grammar Checker based on Web searching. Digithum [online article]. Iss. 8. UOC (2006)
13. Heift, T.: Intelligent Language Tutoring Systems for Grammar Practice. Zeitschrift für Interkulturellen Fremdsprachenunterricht (Online) 6(2), 15 (2001)
14. Bigert, J., Knutsson, O.: Robust error detection: A hybrid approach combining unsupervised error detection and linguistic knowledge. In: Proceedings of Robust Methods in Analysis of Natural language Data (ROMAND 2002), pp. 10–19 (2002)
15. Sjöbergh, J., Knutsson, O.: Faking errors to avoid making errors: Very weakly supervised learning for error detection in writing. In: the Proceedings of RANLP 2005, Borovets, Bulgaria (2005)
16. Young, S.J.: Large Vocabulary Continuous Speech Recognition. IEEE Signal Processing Magazine 13(5), 45–57 (1996)
17. Good, I.J.: The population frequencies of species and the estimation of population parameters. Biometrika 40(3 and 4), 237–264 (1953)
18. Katz, S.M.: Estimation of probabilities from sparse data for the language model component of a speech recogniser. IEEE Transactions on Acoustics, Speech and Signal Processing 35(3), 400–401 (1987)

# Correcting Verb Selection Errors for ESL with the Perceptron

Xiaohua Liu[1,3], Bo Han[2,*], and Ming Zhou[3]

[1] School of Computer Science and Technology,
Harbin Institute of Technology, Harbin, 150001, China
xiaoliu@microsoft.com
[2] Department of Computer Science and Software Engineering,
The University of Melbourne, Victoria 3010, Australia
b.han@pgrad.unimelb.edu.au
[3] Microsoft Research Asia, Beijing, 100190, China
mingzhou@microsoft.com

**Abstract.** We study the task of correcting verb selection errors for English as a Second Language (ESL) learners, which is meaningful but also challenging. The difficulties of this task lie in two aspects: the lack of annotated data and the diversity of verb usage context. We propose a perceptron based novel approach to this task. More specifically, our method generates correction candidates using predefined confusion sets, to avoid the tedious and prohibitively unaffordable human labeling; moreover, rich linguistic features are integrated to represent verb usage context, using a global linear model learnt by the perceptron algorithm. The features used in our method include a language model, local text, chunks, and semantic collocations. Our method is evaluated on both synthetic and real-world corpora, and consistently achieves encouraging results, outperforming all baselines.

**Keywords:** verb selection, perceptron learning, ESL.

## 1 Introduction

Learners of English as a second language (ESL) are a large and growing section of the world's population. They tend to make various errors in English writing, among which, verb selection errors can be quite confusing and misleading. For example, in the sentence (written by a Chinese), *"I often **play with** my friends at school"*, the intended meaning of *"to play with"* is *"to have fun with one's friends"*. However, *"play with"* in English is often understood as *"to play (a game) with"*, *"to play with (among young children)"*, or *"to treat somebody or something frivolously"*; thus it deviates in a subtle way from the meaning intended by the Chinese speaker.

Therefore, designing such a device that can automatically detect and correct verb selection errors made by ESL learners is meaningful. However, this task is

---

* This work has been done while the author was visiting Microsoft Research Asia.

A. Gelbukh (Ed.): CICLing 2011, Part II, LNCS 6609, pp. 411–423, 2011.

challenging, mainly for two reasons. First of all, the conventional data driven approach requires a large volume of annotated training corpus, where every verb selection error is marked with correction suggestion. Unfortunately, no such data is available and manually annotating them is prohibitively unaffordable. Secondly, verb usage is sensitive to its context, which is hard to be represented by categorized linguistic rules. This, in turn, increases the need of a large annotated data to represent the fine grained knowledge about verb usage. For example, consider this sentence *"It's raining outside. Please _ the raincoat with you."* At the first glance, both *"wear"* and *"take"* seem to fill the blank, since they both form collocations with *"raincoat"*; however, once *"with you"* becomes part of the context, *"wear"* no longer fits, and *"take"* wins. In this case, an incorrect identification of verb usage context or the absence of knowledge about *"take"* can lead to verb selection errors.

Brockett et al. [1] use phrasal Statistical Machine Translation (SMT) techniques to correct ESL writing errors and achieve promising results for correcting countability errors. They use rules to introduce errors to sentences written by native speakers, thus to get engineered data to train the model. However, unlike our work, correcting verb selection errors is not their focus. Yi et al. [10] leverage the web to detect and correct verb-noun collocation errors. Though their method requires no training data, the results are not good because of the noisy nature of the web. In contrast, we combine various linguistic features with a global linear scoring function, rather than solely depend on the web frequency, to decide which verb is the best. Liu et al. [7] propose to utilize outputs from semantic role labeling (SRL) system to correct verb selection errors. Our work is a development of the work in [7], in the sense that more verbs are studied in our work. However difference exists: SRL related features are not used in our work, for the reason that, SRL, which depends on a couple of other components including syntactic parsers, though good at capturing the verb usage context, is too heavy to be easily integrated into a real system.

In this paper, we present a lightweight novel approach based on the perceptron learning. Firstly, our method manipulates well-formed English with predefined confusion sets to generate correct/incorrect pairs, which are used to train a global discriminative linear model. Secondly, our method uses various linguistic features, including a language model, local text, chunks, and semantic collocations, to represent the verb usage context. All the features are integrated by the learnt linear model. Finally, we choose perceptron algorithm to train the model, which is simple and fast. Experiment results show that our method achieves encouraging results on both the synthetic and real-world corpora, outperforming baselines based on SMT and the web, respectively.

Our contributions can be summarized as follows:

1. We propose to use confusing sets to automatically generate correct/incorrect pairs for training.
2. We propose to combine various linguistic features to fully represent the diversified verb usage context.

3. We propose to use perceptron learning to train a global discriminative linear model, which performs well on both synthetic and real-world corpora.

Our paper is organized as follows. In the next section, we review related work. In Section 3, we describe our method. In Section 4, we evaluate our method. Finally, Section 5 concludes and presents the future work.

## 2    Related Work

Our work belongs to the line of research of ESL error detection and correction, which enjoys a long history. Michaud et al. [8] design hand-crafted error production rules for a writing tutor aiming at deaf students. Dale et al. [4] use error templates for a word processor. Unlike these rule-based methods, ours is a data driven approach.

Gamon et al. [5] use contextual spelling techniques and language modeling for correcting preposition errors and achieve promising results. Tetreault and Chodorow [9] use a Maximum Entropy model with a large set of features, to combat preposition choice errors. Lee and Seneff [6] use template matching on the parsing tree to correct verb form errors. Differently, our approach targets the challenging verb selection error correction task, which they do not cover.

There are only a handful of methods that can be directly applied to our task. Brockett et al. [1] propose to use phrasal SMT techniques to identify and correct ESL errors. They design several heuristic rules to derive synthetic data from high quality newswire corpus, which, together with its original counterpart, are used to train a SMT model. Their method achieves good results on correcting countability errors. Moreover, it is reported that their method is generic and can be directly applied to other error types. Yi et al. [5] introduce a method that uses web frequency to identify and correct determiner and verb-noun collocation errors. Their approach first generates queries of different granularities, which are then submitted to the search engine, and finally the related snippets are retrieved as references to identify and correct errors. Unlike these methods, which use either the parallel data or the web to get the correction knowledge, our method use predefined confusing sets to control the quality of correction candidates.

Liu et al. [7] is the most recent related work, which uses SRL to represent the verb usage context. But their work focuses on only fifty verbs. Inspired by [7], we use a similar way to generate the training data. However, we study more verbs, and use various lighter linguistic features to model the verb usage context.

## 3    Our Method

Our method simultaneously detects and corrects verb selection errors in the following way: Firstly, correction candidates are generated with predefined confusion sets; and secondly, a linear scoring function is applied to all candidates and the one with the highest score is selected as the suggested correction. Core components, including the confusion sets generation, features, and perceptron based learning, will be discussed separately.

## 3.1   Task Definition

Formula 1 formally defines our task, where $s$ denotes the input sentence for proofing, $s^*$ denotes the suggested correction, $GEN(s)$ is the set of correction candidates, and $score(s)$ is the linear model trained using the perceptron learning algorithm described in Section 3.4.

$$s^*(s) = argmax_{s' \in GEN(s)} score(s')$$  (1)

We refer to every verb in $s$ as a checkpoint. For example, "*sees*" is a checkpoint in "*Jane sees TV every day.*" Correction candidates are generated by replacing each checkpoint with its confusions. Table 1 presents examples of checkpoints.

**Table 1.** Examples of checkpoints

| Input | *Reading can increase my vocabulary, and expand my eyesight* |
|---|---|
| | *Reading can enlarge my vocabulary, and expand my eyesight* |
| **Correction** | ... |
| **candidates** | *Reading can enlarge my vocabulary, and broaden my eyesight* |
| | ... |

Let $\varnothing(s) \in R^d$ denote the global feature vector for the correction candidate $s$, where $d$ is the total number of features, and $w \in R^d$ the feature weight vector, then $score(s)$ is computed as follows:

$$score(s) = w \cdot \varnothing(s)$$  (2)

Finally, $score(s)$ is applied to every candidate $s'$, and $s^*$, the one with the highest score, is selected as the proofing output, shown in Table 2.

**Table 2.** Examples of correction candidate scoring

|  | Correction candidate | Score |
|---|---|---|
| $s^*$ | Jane watches TV every day. | 10.8 |
| | Jane looks TV every day. | 0.8 |
| | Jane reads TV every day. | 0.2 |
| | ... | ... |

## 3.2   Verb Confusion Set Generation

In our work, we focus on 500 common verbs. For every verb we construct a confusion set, which comes from the following sources:

1. Encarta treasures. We extract all the synonyms of verbs from the Microsoft Encarta Dictionary, and this source forms the major part of the confusion sets.

2. English-Chinese Dictionaries. Many errors made by ESL learners are linked to their first language background. For example, some Chinese people incorrectly say *"see newspaper"*, because the translation of *"see"* co-occurs frequently with *"newspaper"* in Chinese. Therefore English verbs sharing more than two (Chinese) meanings are collected. For example, *"see"* and *"read"* are in a confusion set because they share the meanings of both " *kàn* " ("to see", "to read") and " *lǐng huì* " ("to grasp").

3. SMT translation table. We extract paraphrasing verb expressions from a phrasal SMT translation table learnt from parallel corpora (Chiang [2]). This may help us use the implicit semantics of verbs that SMT can capture but a dictionary cannot, such as the fact that the verb *agree* can have a meaning similar to the verb *support*.

We drop verbs in any confusion set that goes beyond the 500 target verbs. Among the 500 verb confusion sets, *take* has the largest confusion set, namely 180, and the average size of confusion set is 29.8. Every verb itself is in its own confusion set, which allows correction candidates with one or more checkpoints unchanged.

## 3.3  Features

Given a correction candidate, various features are produced with respect to every *checkpoint* and its context.

We intuitively use the information in text windows containing *checkpoints*, such as language model scores of strings in the text windows. For example, in the sentence *"I want [to see/watch TV] tonight."*, *"[to see/watch TV]"* is a text window related to the verb *"see/watch"*. The language model score of *"see TV"* is significantly lower than its counterpart *"watch TV"*, and thus it helps *"watch"* stand out in this case.

**Table 3.** An example of analyzed sentences to illustrate features

| Sentence | I have opened an American investment bank account in Boston. | | | | | | | | | | |
|---|---|---|---|---|---|---|---|---|---|---|---|
| Tokens | I | have | opened | an | American | investment | bank | account | in | Boston | . |
| POS | PRP | VBP | VBN | DT | JJ | NN | NN | NN | IN | NNP | . |
| Chunks | NP | VP | | | NP | | | | | PP | . |
| Miscellany | | Active voice; present perfect; | | | | | | | | | |

However, local contextual features cannot capture long-distance dependences. For example, in the sentence *"Computers have become [played] an extremely important role in business."*, the *"to play...role in"* collocation is beyond the scope of conventional n-gram models.

Therefore, we consider features of four different levels (local features, syntactic features, semantic related features, and others), to find out what kinds of features are important in verb selection. Table 4 shows such features with examples.

**Table 4.** Examples of four different level features

| Local contextual: trigrams |  |
|---|---|
| 1.1 | 1_1_I_have_opened |
|  | 1_1_have_opened_a |
|  | 1_1_opened_an_American |
| **Local contextual: trigrams of POS tags** |  |
| 1.2 | 1_2_PRP_VBP_opened |
|  | 1_2_VBP_opened_DT |
|  | 1_2_opened_DT_JJ |
| **Syntactic: chunks** |  |
| 2.1 | 2_1_I_have_opened |
|  | 2_1_opened_an_American_investment_bank_account |
| 2.2 | 2_2_I_opened |
|  | 2_2_opened_account |
| **Semantic: top N latent semantic features** |  |
| 3 | 3_PRP_opened |
|  | 3_opened_bank |
|  | 3_opened_NN |
|  | ... |
| **Others** |  |
| 4 | 4_1_active_voice |
|  | 4_2_present_perfect |
|  | ... |

The local contextual features are related to tokens and POS tags, as illustrated by type 1.1 and 1.2 features in Table 4. Syntactic features are related to chunks. We use a chunk parser instead of a dependency parser with the following consideration: For ill-formed sentences, the dependency parser tends to generate erroneous outputs or just give no results; while the chunk parser, we believe, is more reliable: Verb substitution can change the sentence's semantics, and lead to incorrect collocations during parsing, but it hardly affects the sentence's chunking structure.

The complexity rises up when it comes to the semantic features, as semantic features are correlated with multiple factors like the agent of the verb, the sentence voice, idiomatic usages and fixed collocations. However, semantic knowledge does matter for verb selection and deserves deep mining. For example, consider the following sentences:

1. *The **book costs** forty **dollars** at Amazon.*
2. ***I spent** forty **dollars** on the book.*

Both "*cost*" and "*spend*" can collocate with "*dollars*", however, the agent determines the right verb.

We combine features acquired from the training data and using the top $N$ (experimentally set to 10,000 in our work) frequent features as the implicit semantic features. This approach, to some extent, helps us mine a certain amount of underlying features such as "*I spend dollars*" or "*book costs dollars*". However, this solution also leads to a feature explosion even on a small-scale training data. And even worse, not all the highly frequent features make sense, like "*I spend forty*" and "*cost Amazon*". Mining such kinds of useless features increases the computational burden and should be avoided. We use several heuristic rules to filter such features:

1. We require that the VP can only collocate with NPs and PPs in the same clause. The reason is that features of the verb collocating with an NP or PP in a different clause will not actually form sound collocations. For example, consider "*My teacher **said** that **Chinese** is not easy to learn*". Obviously, features like "*say Chinese*" introduce noise to the native expression "*speak Chinese*", and these features should be dropped.
2. We use only the last NN for combinations. This is because, in most cases, when there are multiple NNs in the NP, the last NN is the functional word for the verb-noun collocation. For example, for the sentence listed in Table 4, both "*opened investment*" and the "*opened bank*" are eliminated. This restriction is important, because if we don't enforce such a restriction, these cases will introduce noisy features. For example, in "*His only hobby is to see that boring TV program.*", the incorrect collocation "*see TV*" could be extracted.
3. We normalize all the words to lower-case spelling. Moreover, numbers, dates, and NNPs are replaced with special symbols to alleviate data sparseness.
4. All features must include the target verb; otherwise, they might be of no use for verb selection.

## 3.4   Perceptron Learning

We use the generalized perceptron learning algorithm [3] to train our model, because: 1) It can freely incorporate various features; 2) it is more lightweight than other sophisticated algorithms; and 3)it enables us to handle various other types of word choice errors with only a little more additional efforts in defining features associated with the context.

Algorithm 1 shows the training process, where $s^i$ denotes the $i^{th}$ correct sentence in the training data, $T$ and $N$ represent the training iteration and the total number of training examples, respectively. The training process is to update $w$ , when the output differs from the training sample oracle. For example, when the output is "*I want to look TV*" and $s^i$ is "*I want to watch TV*", $w$ will be updated.

We use the averaged perceptron algorithm [3] to alleviate overfitting on the training data. The averaged perceptron parameter vector is defined as:

$$w' = \frac{1}{T \cdot N} \sum_{i=1...N, r=1...T} w^{i,r} \tag{3}$$

Here, $w^{i,r}$ is the parameter vector immediately after the $i^{th}$ sentence in the $r^{th}$ iteration.

**Algorithm 1.** The generalized perceptron training algorithm, adapted from Collins [3] .

**Require:** Training examples $s^i, i = 1...N$.

1: Initialization: $w = 0$
2: **for** $r = 1...T, i = 1...N$ **do**
3:     Calculate $o = argmax_{s \in GEN(s^i)} score(s)$
4:     **if** $s^i \neq o$ **then**
5:         $w = w + \varnothing(s^i) - \varnothing(o)$
6:     **end if**
7: **end for**
8: **return** $w$

## 4   Experiments

In this section, we evaluate our method on a synthetic and a real-world corpus, respectively. Experiment results show that our method achieves encouraging results, outperforming both the SMT-based and the web-based system. The contributions of each set of features are presented and how the training data size affects the performance is studied as well.

### 4.1   Baselines

We consider two baselines: The SMT-based system and the web-based system. Liu et al. [7] is not considered as a baseline, because it requires SRL, which falls out of our interests. For the web-based baseline, we use Bing[1] as the primary search engine to obtain web statistics with the $mfreq$ parameter set to 0.01, as determined by the preliminary test. As for SMT-based system, we build up a typical phrasal SMT system with some features disabled to avoid introducing undesirable errors. For example, word re-ordering, though potentially leading to better lexical selection, is disabled, since it can introduce errors concerning word order, and our task is restricted to verb substitution.

### 4.2   Data Preparation

The training corpus is from LDC (2005T12). We randomly select newswires from New York Time as the training data. We then use OpenNLP[2] to extract sentences from the newswire text, and parse them into corresponding tokens, POS tags and chunks as illustrated in Table  4. We assume that the newswire data is of high quality and free of linguistic errors. We gather 5 million sentences in total.

    To generate correction candidates for training, we manipulate the newswire data by replacing verbs with those in their confusion sets. However, the generated

---

[1] http://www.bing.com/
[2] http://opennlp.sourceforge.net/

candidates may sometimes contain reasonable output. For instance, in *"President Qi Huaiyuan **conveyed** [**expressed**] a warm welcome and best wishes from Chinese Premier Li Peng and his wife Zhu Lin"*, both *"convey"* and *"express"* (which are in the same confusion set) are acceptable. To address this kind of flaw, we train a trigram language model using the SRI Language Modeling Toolkit on the English Gigaword corpus, and calculate the logarithms language model score of the original sentence and its artificial manipulations, and only keep those manipulations with a language model score that is t (experimentally set to 5) times lower than that of the original sentence.

We construct two test datasets. Ten thousand samples are randomly selected from the previous unused LDC dataset for the automatic in-domain test. The target verbs are replaced by confusion counterparts and the language model-based pruning strategy is applied. We also collect 186 samples from web blogs and from the Chinese Learner English Corpus (CLEC) written by Chinese ESL users to investigate the generalization over all methods. These samples are revised to remove other types of errors and corrected by an English native speaker, forming the out-of-domain test dataset.

## 4.3  Evaluation Metric

Following Yi et al. [10], we use the following metrics: Revised precision ($RP$), recall of the correction ($RC$) and false alarm ($FA$).

$$RP = \frac{\text{\# of corrected modified checkpoints}}{\text{\# of modified checkpoints}} \tag{4}$$

$RP$ reflects how many checkpoint modifications are suitable corrections. We ignore paraphrasing scenarios. For example, in *"**send** (**express**) my best wishes to your brother"*, the checkpoint *"send"* is the correct usage; we will not count *"express"* as a correction in the open test, since offering an alternative expression doesn't improve the performance and would even confuse users. In the in-domain test, we regard both the original newswire sentence and corrections with a larger language model score than the original newswire sentence as correct modifications.

$$RC = \frac{\text{\# of corrected modified checkpoints}}{\text{\# of total errors}} \tag{5}$$

$RC$ indicates the coverage of the proposed methods. In our data set, a sentence may have multiple checkpoints, but each contains only one verb selection error.

$$FA = \frac{\text{\# of incorrectly modified checkpoints}}{\text{\# of checkpoints}} \tag{6}$$

$FA$ is related to the cases that the system brings noise to users, when a correct verb is mistakenly replaced by an erroneous one. These false suggestions will be considered low-quality by the users, and thus should be avoided as much as possible.

## 4.4  Results and Analysis

Table  5 shows the results, where each row denotes a system, and cells in each row denotes the revised precision, recall and false alarm, respectively. Note that our method makes no attempts of correcting what is correct in the out-of-domain test.

Table 5. Experimental results

| System | In-domain test(%) | Out-of-domain test(%) |
|--------|-------------------|-----------------------|
| Web    | 51.5;4.8;17.5     | 42.2;17.9;4.2         |
| SMT    | 55.1;3.2;22.6     | 30.0;25.5;23.3        |
| Ours   | 56.3;7.1;18.2     | 71.0; 21.0; 0.0       |

We observe that all systems have high false alarm rates, which suggest that verb selection is a challenging task for all the methods. The SMT-based approach turns out to perform worst. This is partially because the insertions and deletions of words are recorded as false positives in the automatic evaluation, and we guess the quantity and quality of training data also restrict the performance of the SMT-based system. Compared with the web-based proofing method, our approach can handle more errors; however, our method also tends to change other correct verbs, increasing the number of false alarms slightly.

In the out-of-domain test, our approach achieved the best results in terms of all metrics. For fairness, we relax the restrictions of the SMT-based method, counting correct candidates with minor modifications as valid. This greatly reduces the number of false alarms, and the SMT-based approach thus achieves the highest recall among all methods. The deficiency of web proofing links to two factors: Firstly, web statistics are so unreliable that they offer less supportive information when the queries are too specific; secondly, when the queries are too general, web retrieved results have too much noise; and finally, the web-based proofing method may change the semantics. For example, for the sentence "*The policy will **add** [enlarge] the gap between the rich and the poor*", web-based proofing suggests the phrase "*close the gap*", which is opposite to the intended meaning. In contrast, confusion sets in our method ensure that the meaning of the output won't stray too far from the original meaning.

There are some cases out of the reach for all of these methods. For instance, consider "*Everyone **doubts** (suspects) that Tom is a spy.*" Both of the two verbs can be followed by a clause. However, in English "doubt" often co-occurs with ideas, phenomena, etc., but seldom with "*Tom is a spy*". Such deep semantics out of the sentence (or so-called or background knowledge) are not considered by all these methods. In addition, long distance interference also impedes the correction process. For example, consider the following case:

1. Input: "*I have* **received** *[answered] a telephone call from my boss just now*"
2. Output: "*I have* **made** *[answered] a telephone call from my boss just now*"

The collocation of "*make a telephone call*" is sound. However, "*make a telephone call from the boss*" is contradictory, since "*make a call*" is not compatible with "*from*". This reflects the fact that the verb selection is sensitive to its context, which may be too subtle to be represented by conventional features. We leave it to our further study for this issue.

## 4.5 Feature Contribution

Table 6 shows the performance of our method with different feature sets. From Table 6, we can see that the local contextual features achieve an encouraging precision, but a lower recall. We also observe that when syntactic features are introduced, the performance goes down significantly, with many false positives. This suggests that syntactic chunk features are not strong enough to represent the verb usage context. The best performance is achieved when all feature sets are combined, suggesting that semantic features are crucial for verb selection. In spite of this, the recall is not satisfying. We guess the low recall is partially owing to the insufficient mining of the semantic features. We plan to investigate more about this issue in future.

**Table 6.** Results with incremental feature sets

| Feature set(s) | Out-of-domain Test(%) |
|---|---|
| Local Contextual | 53.8; 12.3; 0.0 |
| Local Contextual+Syntactic | 30.0;5.7;1.0 |
| Local Contextual+Syntactic+Semantic | 71.0; 21.0; 0.0 |

## 4.6 Performance vs. Size of Training Data

We further investigate the impact of training data size. Figure 1 shows $RP$, $RC$, and $FA$ with 1, 2, 3, 4, and 5 million sentences of training data, respectively. It can be seen that, when we enlarge the training data, FA drops first and then experiences a slight increase, however, $RC$ increases steadily along with the data size. This observation indicates that more training data are needed to learn the knowledge about verb selection, i.e., to improve $RC$. However, how much data is sufficient and how to further reduce the false positives are still unknown, and we leave those questions to our future work. Interestingly, we observe that the $RP$ remains almost stable, but drops a little when the size of training data is increased from 4M to 5M. This can be explained in part by the noise in the training data.

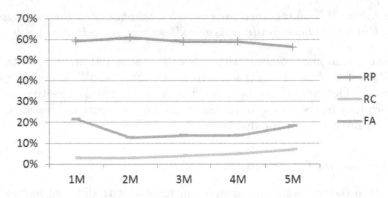

**Fig. 1.** Impact of the training data size. Horizontal and vertical axes represent the size of training data and the performance (in terms of $RP$, $RC$ or $FA$), respectively.

## 5    Conclusions and Future Work

The task of detection and correction verb selection errors for ESL leaners is meaningful, but challenging. The difficulties lie in the lack of annotated data and the diversified verb usage context. We propose a novel approach that uses the perceptron algorithm to learn a global discriminative linear model, which is trained on the engineered data, to integrate various linguistically motivated features to represent the verb usage context. Experiment results show that our method outperforms baselines based on the phrasal SMT and the web respectively.

In future, we plan to explore more features that can better represent complex context, such as reliable verb-noun collocations. We are interested in how to automatically construct confusion sets. We also hope to apply our method to correct other types of ESL errors, such as errors related to nouns.

**Acknowledgments.** We thank the anonymous reviewers for their invaluable comments on an earlier draft of the paper.

## References

1. Brockett, C., Dolan, W.B., Gamon, M.: Correcting esl errors using phrasal smt techniques. In: Proceedings of the 21st International Conference on Computational Linguistics and the 44th Annual Meeting of the Association for Computational Linguistics, ACL-44, pp. 249–256. Association for Computational Linguistics, Morristown (2006), http://dx.doi.org/10.3115/1220175.1220207
2. Chiang, D.: Hierarchical phrase-based translation. Computational Linguistics 33 (2007)
3. Collins, M.: Discriminative training methods for hidden markov models: theory and experiments with perceptron algorithms. In: Proceedings of the ACL 2002 Conference on Empirical Methods in Natural Language Processing, EMNLP 2002, vol. 10, pp. 1–8. Association for Computational Linguistics, Morristown (2002), http://dx.doi.org/10.3115/1118693.1118694

4. Dale, R., Moisl, H., Somers, H. (eds.): A Handbook of Natural Language Processing: Techniques and Applications for the Processing of Language as Text. Marcel Dekker Inc., New York (2000)
5. Gamon, M., Gao, J., Brockett, C., Klementiev, R.: Using contextual speller techniques and language modeling for esl error correction. In: Proceedings of IJCNLP 2008 (2008)
6. Lee, J., Seneff, S.: Correcting misuse of verb forms. In: Proceedings of ACL 2008: HLT, pp. 174–182. Association for Computational Linguistics, Columbus (June 2008), http://www.aclweb.org/anthology/P/P08/P08-1021
7. Liu, X., Han, B., Li, K., Stiller, S.H., Zhou, M.: Srl-based verb selection for esl. In: Proceedings of the 2010 Conference on Empirical Methods in Natural Language Processing, EMNLP 2010, pp. 1068–1076. Association for Computational Linguistics, Morristown (2010),
http://portal.acm.org/citation.cfm?id=1870658.1870762
8. Michaud, L.N., McCoy, K.F., Pennington, C.A.: An intelligent tutoring system for deaf learners of written english. In: Proceedings of the Fourth International ACM Conference on Assistive Technologies, Assets 2000, pp. 92–100. ACM, New York (2000), http://doi.acm.org/10.1145/354324.354348
9. Tetreault, J.R., Chodorow, M.: The ups and downs of preposition error detection in esl writing. In: Proceedings of the 22nd International Conference on Computational Linguistics, COLING 2008, vol. 1, pp. 865–872. Association for Computational Linguistics, Morristown (2008),
http://portal.acm.org/citation.cfm?id=1599081.1599190
10. Yi, X., Gao, J., Dolan, W.B.: A web-based english proofing system for english as a second language users. In: Proceedings of IJCNLP 2008 (2008)

# A Posteriori Agreement as a Quality Measure for Readability Prediction Systems

Philip van Oosten[1,2], Véronique Hoste[1,2], and Dries Tanghe[1,2]

[1] LT³ Language and Translation Technology Team, University College Ghent,
Groot-Brittanniëlaan 45, 9000 Ghent, Belgium
[2] Ghent University, Krijgslaan 281, 9000 Ghent, Belgium

**Abstract.** All readability research is ultimately concerned with the research question whether it is possible for a prediction system to automatically determine the level of readability of an unseen text. A significant problem for such a system is that readability might depend in part on the reader. If different readers assess the readability of texts in fundamentally different ways, there is insufficient a priori agreement to justify the correctness of a readability prediction system based on the texts assessed by those readers. We built a data set of readability assessments by expert readers. We clustered the experts into groups with greater a priori agreement and then measured for each group whether classifiers trained only on data from this group exhibited a classification bias. As this was found to be the case, the classification mechanism cannot be unproblematically generalized to a different user group.

## 1 Introduction

In the most general terms, the goal of authoring a text is to get a message across to an intended audience. The readability of a text, then, can be defined as the relative ease of that audience to understand the author's message. It is intuitively clear that, even when defined in such general terms, the inherent subjectivity of the concept of readability cannot be ignored. The ease with which a given reader can correctly identify the message conveyed in a text is, among other things, inextricably related to the reader's background knowledge of the subject at hand [11].

The domain of readability research has at its primary research goal the design of a method to automatically predict the readability of a text. In recent years, a tendency seems to have arisen to explicitly address the subjective aspect of readability. [14] ultimately base their readability prediction method exclusively on the extent to which readers found a text to be "well-written". [10] take the assessments supplied by a number of experts as their gold standard, and test their readability prediction method as well as assessments by novices against these expert opinions. Similarly, [13] compile a gold standard for readability prediction by collecting assessments by expert and naive readers.

Subjective assessment entails the problem of reliably aggregating data that were obtained from various sources. This is a recurring issue in Natural Language

A. Gelbukh (Ed.): CICLing 2011, Part II, LNCS 6609, pp. 424–435, 2011.

Processing, and is routinely caused by several contributors making different decisions regarding some manual annotation task. [2] give a good overview of the standard practice that has arisen within the NLP domain, viz. to calculate some measure of inter-annotator agreement. If this measure is high enough, the data are deemed acceptable to serve as a gold standard.

In readability research, however, this practice does not seem to have gained much ground. Given that many readability prediction methods (e.g. [6,5,17]) were developed before it became commonplace, it is not surprising that inter-annotator agreement played no great part in the development of those readability formulas. However, even recent publications such as [14] and [10] make no mention of the issue, and uncritically average out results collected from different readers. This should be done with great caution indeed: [1] claimed that if the data on which readability formulas are based were not aggregated on the school grade level but considered at the individual level, their predictive power would drop from around 80% to an estimated 10%.

We aim to determine whether a readability prediction system can be generalized to a broader audience, even when lacking a priori agreement measures. This is done by evaluating the accuracy of different readability systems on different groups of experts with a large a priori agreement. Poor performance would then imply that the annotation behaviour of the expert group deviates from the larger group of annotators, which leads to the conclusion that the readability system is not appropriate for the general public. To compose the groups of experts, we used a simple clustering technique, combining experts with similar annotations together. Classification accuracy is used to measure the deviations between an expert group and the rest, i.e. the concatenation of the other expert groups.

Instead of calculating inter-annotator agreement prior to training a readability prediction system, we verify whether the classification accuracies of systems trained on a single cluster and the concatenation of the other clusters differ for the same test set.

The remaining sections of this article contain details on how we composed our data set (section 2), a discussion of the issue of determining inter-annotator agreement in our data set and a proposed approach to locate generalization problems (section 3), experimental results (section 4) and conclusions and further work (section 5).

## 2 Annotation Process and Data Set

### 2.1 Training Corpus

Readability research is often concerned with the readability prediction of texts for relatively unaccomplished readers. The goal, then, is to identify reading material suited to the reading competence of a given individual [6,17,16,18]. Training data for the readability prediction system can then be drawn from textbooks intended for different competence levels [16,7]. However, since our system must be applicable to generic Dutch text, such educational material is insufficient, and we assembled a new training corpus.

We selected 105 texts from the Lassy corpus [12], which is a corpus annotated with lexical and syntactic features. From the selected texts, fragments of one or more paragraphs were used for readability assessment. The length of the fragments ranged from 81 to 306 tokens, resulting in a total amount of just under 17K assessed tokens. In order to develop a generically applicable system that can predict readability across text domains, we attempted to construct a cross-domain training corpus. Therefore, the texts in the corpus were selected manually from several sources, such as children's literature, Wikipedia, newspaper articles and technical reports. Each of the text fragments received on average 22 individual assessments, with a standard deviation of 9.12. As different annotators applied different scoring strategies, it is impossible to give an overall description of the way in which assessments were distributed in the range of possible values.

## 2.2   The Expert Readers Annotation Tool

The corpus was assessed for readability by a number of experts, who are professionally involved with the Dutch language. The experts used a password protected web application to assess the texts.

In the application, multiple texts can be placed underneath each other in a column, that visually represents an overview of the ratings an expert assigned during the current session and helps the annotators to build up a frame of reference against which to assess newly loaded texts.

An annotator can load texts and assign a score between 0 (easy) and 100 (difficult) to them. Previously assigned scores can be revised.

A batch of texts with accompanying scores can be sent to the database by pressing a button. The texts are then removed, except if the annotators indicated they wanted to keep them available, so as to maintain a frame of reference across batches. When a user submits the current assessments, all scores in the batch are logged.

Texts are provided to the annotators randomly, with equal probability of providing a text from each text type, and independent from which texts were previously provided. However, a text can never appear twice in the same batch.

Apart from the readability scores and the rankings in the batches, the experts can also enter comments on what makes each text more or less readable. That allows for qualitative analysis. We did not ask more detailed questions about certain aspects of readability, because we wanted to avoid influencing the text properties experts pay attention to. Neither did we inform the experts in any way how they should judge readability. Any presumption about which features are important readability indicators was thus avoided. We do not know which experts based their assessments on which text properties, and which relative weights they attributed to them. Yet our main interest is to design a system that is robust enough to model readability as generally as possible.

## 2.3 Data Provided through the Application

The assessments of the experts are stored in a database. For each expert, all the batches, containing texts and corresponding scores are available. A qualitative survey reveals that different experts sometimes employ a different scoring strategy. For example, some people only use scores that are multiples of 10, while others use the full range of possible scores. This is not a trivial observation: such a difference in score assignment compromises the possibility to use the scores directly for regression.

The batches can also be seen as rankings of texts. We further consider the *text pairs* that can be extracted from the batches. From each batch, we extract all pairs of texts that differ in score and for which at least two other texts are ranked between the pair. In this way, we can reasonably assume that the expert evaluated the lower-ranked text as more readable than the text with the higher score.

## 3  Detecting Disagreement between Annotators

In this article, we use one particular type of readability prediction system as a working example: a binary classifier which is able to predict which of two given texts is the more readable one. To construct such a readability prediction system, a possible approach would be to first determine inter-annotator agreement for each text pair. The text pairs for which reasonable agreement [2] is found can then serve as the basis for a gold standard, which can then be used to train a binary classifier. More generally, composing a gold standard prior to performing supervised learning experiments is the standard practice.

However, in the data set provided by our experts, not everyone has assessed all of the same texts, let alone text pairs. It is therefore not possible to determine the agreement for all text pairs with sufficient accuracy, prior to training a binary classifier: there are too many missing values. Not all annotators spent the same amount of time assessing texts and some assessed more texts per batch than others. Therefore, not all annotators contribute the same amount of text pairs and there is not always an overlap between the texts they have seen. We also want to be able to maximally employ minor contributions. Furthermore, we found disagreement concerning some text pairs, and we want to examine whether those disagreements are incidental or whether they betray a more fundamental controversy in readability assessment.

We can identify two possible causes for the disagreements: there is no clear difference in readability between the two texts in the text pair; or different experts have contrasting opinions on what factors constitute readability. Overcoming both issues would require more experiments and a qualitative analysis. Further in this article, we don't attempt to distinguish between these issues, but we perform a quantitative analysis to uncover their effects.

As explained above, a readability prediction system can be developed by merging all the text pairs into a gold standard and training a classifier. In order to merge the training data with an acceptable degree of reliability, there should be sufficient agreement between different experts' assessments of the same text pairs. An estimate of the classification accuracy, for example through cross-validation, then indicates how well the trained system works. However, since inter-annotator agreement could be too low to speak of a gold standard, we also need to investigate in further detail to what extent the resulting system can be generalized. That means that apart from achieving a high classification accuracy, it is also important that the eventual system delivers results that are acceptable for all experts. To facilitate a priori agreement and to be able to check a posteriori whether no expert views were excluded, we created groups of experts who provided the most similar annotations.

### 3.1  Preparation of the Data Sets

Figure 1 gives a schematic overview of how the data sets used for classification are prepared. Each block in the figure represents the execution of a set of commands. If an arrow points from one block to another, the former is executed before the latter and output from the former is passed as input to the latter.

**Expert Readers Data.** In this node, data are extracted from the Expert Readers Application database. Annotators who provided 25 text pairs or fewer are excluded.

**Create Proximity Matrix.** For our experiments, we need groups of experts who have a shared view on readability. To divide the experts in those groups, we need a *proximity* measure: a metric to indicate to what extent the judgements of different experts are similar. The metric should allow us to distinguish experts who agree on how to order texts from those who disagree. Precision of the text pairs of one annotator with regard to the other meets this requirement.

In general, precision and recall are calculated by the following formulas: $P = \frac{TP}{TP+FP}$ and $R = \frac{TP}{TP+FN}$, where $TP$ is the number of *true positives* (i.e. text pairs on which both annotators agree), and $FP$ is the number of *false positives* (i.e. text pairs on which the annotators disagree). *Negatives* with regard to a particular pair of experts would be the text pairs that only one of the two experts has reviewed. Since the annotation procedure does not require all annotators to see the same text pairs, no sensible distinction can be made between *true negatives*

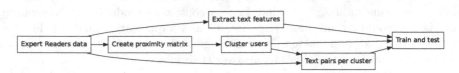

**Fig. 1.** Outline of how the data sets are composed from the expert assessments

and *false negatives*. Therefore, the number of false negatives ($FN$) cannot be determined in this context, and we cannot calculate meaningful recall figures. The proximity between two experts is therefore the precision: the number of ordered text pairs that both annotators agree on, divided by the total number of text pairs that appear in the data sets of both annotators. The result of this block is a square symmetric matrix with proximity measures.

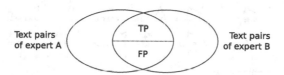

**Fig. 2.** True and false positives for the text pairs of two experts. Since the experts have not annotated all text pairs, there is no sensible notion of true and false negatives.

**Cluster Users.** Using the proximity degrees between all experts, it is possible to divide them into groups, so that the assessments of each expert correspond more to those of every other expert within the same group, than to those of other experts. We thereby make groups of experts with high a priori agreement. To create the groups, we use a simple agglomerative clustering algorithm [8]. Initially, a cluster is created for each individual expert. Subsequently, the two clusters with the highest degree of proximity are merged into a single one, until there is only one cluster left. The proximity between clusters is calculated as the minimal proximity between any of the members of each of the clusters. In this way, the agreement between all experts per cluster is maximized. The dendrogram in figure 3 shows the result of the clustering algorithm. Finally, in order to divide the experts into similar groups, we branch the dendrogram, keeping only the greatest possible clusters of experts among which the precision is higher than a given cut-off value. We experimented with different cut-off values, ranging from 0.5 to 0.9. If, for example, the precision is more than 0.5, there are at least as many text pairs about which each pair of annotators agree, as there are pairs about which they disagree.

**Text Pairs Per Cluster.** Given the set of experts in each of the clusters, their text pairs are merged into a single set. The set of text pairs for the cluster is simply the union of the text pairs of the annotators in the cluster.

**Extract Text Features.** For all the texts in the corpus, a number of features are extracted that can be used as training material for a classification algorithm. These are primarily indicators for lexical complexity, such as mean word length in number of characters [5,17] and number of syllables [6], TF-IDF, log-likelihood and mutual information [9] computed against a large reference corpus [15], as well as character bigram and trigram frequencies. Additionally, some syntactic information is encoded in the form of proportions of different part of speech classes as tagged by the Tadpole parser [3].

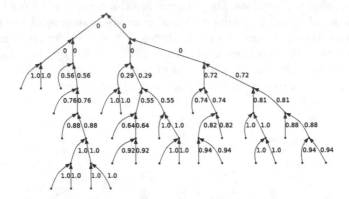

**Fig. 3.** A dendrogram showing the result of clustering the annotators. The edge labels show the precision between the child nodes. Each node represents a set of annotators. The leaf nodes represent the individual annotators.

## 3.2   Train and Test

Given the feature vector $V_a$ for text $T_a$ and the vector $V_b$ for text $T_b$, we construct a single vector $V_{ab}$ for the text pair $T_{ab}$ by calculating the difference in feature values: the values from $V_b$ are subtracted from the corresponding values from $V_a$. The class label 1 is assigned to $V_{ab}$ if $T_a$ is assessed as more readable than $T_b$, and label -1 is assigned if the opposite is true. Additionally, the vector $V_{ba}$ is constructed with the reverse values and the reverse class label. Self-evidently, such corresponding vectors are never distributed over the training and test data, as that would amount to contamination of the test data. A simplified example of a feature vector is shown in table 1. These feature vectors can serve as training data for a binary classifier, which can then be used to predict which of two texts is more readable than the other (see [18] for a similar procedure).

**Table 1.** An example of two feature vectors for a text pair. The vectors are truncated.

| Class | Feature 1 | Feature 2 | Feature 3 | Feature 4 | Feature 5 | Feature 6 |
|---|---|---|---|---|---|---|
| 1 | 0.46 | 0.01 | 2.92 | -0.01 | 0.2 | 0.03 |
| -1 | -0.46 | -0.01 | -2.92 | 0.01 | -0.2 | -0.03 |

For each cluster, two data sets are generated. One set contains the feature vectors of the text pairs as assessed by the annotators in the cluster, and the other set contains those of the concatenation of the other clusters (the *complement*). The two data sets are then split up to perform 10-fold cross validation. An outline of the experiments per fold is shown in figure 4. The folds are created by splitting up the text sets (rather than the sets of text pairs) in 10 parts, since splitting only the text pair sets could result in contamination of the test sets.

Text pairs of which at least one text is assigned to the test fold are added to the test set. The rest of the text pairs are added to the training set. This division of text pairs is done both for the cluster and for the complement.

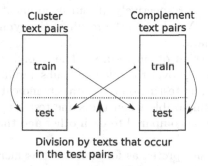

Fig. 4. Division in training data and test data per fold. A classifier is trained for both training sets and both classifiers are then tested on both test sets, so that the classification accuracy can be compared per test set.

To avoid that the amount of available data in either of the training sets might skew the classification results, we downsample the greater training set by randomly selecting an amount of text pairs that is equal to the amount of pairs present in the smaller training set.[1]

For each fold in each cluster, this results in two data sets that serve as training data for a binary classifier, and two test sets. We call the corresponding data sets the cluster training set, cluster test set, complement training set and complement test set. Both training sets are used to train a binary classifier [4]. We call a classifier trained on a cluster training set a *cluster classifier* and a classifier trained on a complement training set a *complement classifier*. Both of these resulting classifiers are tested on both test sets to obtain the classification accuracy.

The goal of this experiment is to measure the influence of diverging annotation strategies on classification accuracy. If different annotation strategies have no influence on classification performance, both cluster and complement classifiers should perform equally well on cluster and complement test sets, or one of the classifiers should outperform the other for both test sets. If, however, each classifier performs better on the test set corresponding with its training set, that indicates a bias between training and test set, revealed by the combination of the feature set and the learning method that is used. If such a bias is found, the generalization ability of the learning method with the given feature set is questionable.

---

[1] In order to prevent that a particular downsampling of the training data might yield anomalous results, 10 random downsamplings have been performed and tested for each of the 10 test folds. We consider the mean classification accuracy over these 10 downsamplings within a fold as the classification accuracy of the test fold.

## 4    Results

Classification accuracy is given by the formula $CA = \frac{TP+TN}{TP+TN+FP+FN}$, where $TP, TN, FP$ and $FN$ are the number of true positives, true negatives, false positives and false negatives, respectively. In our experiments, we always observed $TP = TN$ and $FP = FN$, which is the result of the symmetric construction of the training and test data.

Table 2 gives an overview of our results.[2] The second to fourth column of the subtables show the average over the folds and subsamplings of the classification accuracies for training and testing on the data sets indicated in the header rows. Accuracies are only comparable when the same test set is used, so the second and third columns can be compared to each other and the two last columns are comparable.

The results can be interpreted as follows. If a classifier is generalizable, that implies that the test sets are not biased towards the classifier trained on the corresponding training set. The upshot of this is that the classification accuracies should indicate that either the cluster classifier or the complement classifier performs better on both test sets.

To clarify, we consider cluster 5 at cut-off level 0.8. Here, we see that the complement classifier performs better on both test sets than the classifier trained on cluster 5 itself. When testing on the cluster test set, we observe a higher classification accuracy for the complement classifier: 0.72 versus 0.64 for the cluster classifier. Similarly, when testing on the complement test set, the complement classifier achieves higher accuracy than the cluster classifier: 0.64 versus 0.58. When taking only the results for cluster 5 at cut-off level 0.8 into account, then, it would be plausible that the classification results can be generalized.

However, for 5 out of 11 clusters at cut-off level 0.8, the situation is more problematic. When we consider cluster 2 at cut-off level 0.8, we observe a different situation: each classifier achieves higher accuracy on the test set corresponding with its own training set. The cluster classifier performs better on the cluster test set (0.71 versus 0.65), while the complement classifier performs better on the complement test set (0.68 versus 0.61). This indicates a bias in the classifiers to the test set corresponding with their own training set, which compromises the generalizability. We observe the same situation for a further 4 clusters out of 11 at cut-off level 0.8, and at cut-off level 0.5, the bias even manifests itself for all clusters.

We consider the average of the classification accuracies over all clusters as the criterion to decide whether a posteriori agreement is sufficient to call the results generalizable. If the mean cluster classification accuracy is higher for the cluster test set and the mean complement classification accuracy higher for the complement test set, a posteriori agreement is insufficient. It then seems that in

---

[2] We also computed results at cut-off level 0.9, but since too many individual annotators appear as expert groups, the cluster test sets often became too small to calculate meaningful results (cfr. figure 3). Therefore, only results at cut-off levels 0.5 to 0.8 are shown.

**Table 2.** Classification accuracy for each cluster and complement, at different cut-off levels. The average of the cluster averages is given in the last row. The greater of each pair of comparable accuracies is shown in bold.

| Train Test | Cluster Cluster | Compl. Cluster | Compl. Compl. | Cluster Compl. |
|---|---|---|---|---|
| 1 | **0.75** | 0.63 | **0.68** | 0.52 |
| 2 | **0.78** | 0.75 | **0.67** | 0.63 |
| 3 | **0.68** | 0.63 | **0.70** | 0.62 |
| 4 | **0.71** | 0.67 | **0.68** | 0.61 |
| 5 | **0.68** | 0.65 | **0.72** | 0.61 |
| Mean | **0.72** | 0.66 | **0.69** | 0.60 |

(a) Cut-off 0.5

| Train Test | Cluster Cluster | Compl. Cluster | Compl. Compl. | Cluster Compl. |
|---|---|---|---|---|
| 1 | **0.71** | 0.62 | **0.68** | 0.59 |
| 2 | 0.60 | **0.70** | **0.64** | 0.58 |
| 3 | 0.77 | **0.78** | **0.66** | 0.62 |
| 4 | **0.68** | 0.65 | **0.72** | 0.60 |
| 5 | **0.77** | 0.63 | **0.68** | 0.53 |
| 6 | **0.68** | 0.63 | **0.70** | 0.62 |
| 7 | **0.74** | 0.74 | **0.67** | 0.61 |
| Mean | **0.71** | 0.68 | **0.68** | 0.59 |

(b) Cut-off 0.6

| Train Test | Cluster Cluster | Compl. Cluster | Compl. Compl. | Cluster Compl. |
|---|---|---|---|---|
| 1 | **0.71** | 0.64 | **0.68** | 0.59 |
| 2 | 0.80 | **0.84** | **0.67** | 0.64 |
| 3 | 0.61 | **0.75** | **0.64** | 0.58 |
| 4 | **0.66** | 0.61 | **0.70** | 0.59 |
| 5 | 0.77 | **0.81** | **0.66** | 0.63 |
| 6 | **0.79** | 0.63 | **0.69** | 0.52 |
| 7 | **0.68** | 0.65 | **0.72** | 0.61 |
| 8 | **0.73** | 0.72 | **0.67** | 0.61 |
| Mean | **0.72** | 0.71 | **0.68** | 0.60 |

(c) Cut-off 0.7

| Train Test | Cluster Cluster | Compl. Cluster | Compl. Compl. | Cluster Compl. |
|---|---|---|---|---|
| 1 | 0.69 | **0.70** | **0.68** | 0.61 |
| 2 | **0.71** | 0.65 | **0.68** | 0.61 |
| 3 | **0.79** | 0.72 | **0.67** | 0.59 |
| 4 | 0.81 | **0.83** | **0.67** | 0.64 |
| 5 | 0.64 | **0.72** | **0.64** | 0.58 |
| 6 | **0.71** | 0.68 | **0.67** | 0.56 |
| 7 | **0.68** | 0.61 | **0.71** | 0.59 |
| 8 | 0.76 | **0.80** | **0.67** | 0.62 |
| 9 | 0.71 | **0.72** | **0.68** | 0.61 |
| 10 | 0.74 | **0.79** | **0.66** | 0.61 |
| 11 | **0.78** | 0.63 | **0.69** | 0.52 |
| Mean | **0.73** | 0.71 | **0.67** | 0.59 |

(d) Cut-off 0.8

general, classifiers expose a bias towards the test set corresponding to the training set the classifier was trained on. For our experiments, that observation holds for all cut-off levels, as can be seen in the last row of the subtables of table 2. As a consequence, a posteriori agreement is insufficient to call a classifier as outlined in this article generalizable to a broader audience.

Although the accuracies on different test sets are incomparable, it seems that the complement classifier consistently performs better on the cluster test set than the cluster classifier on the complement test set. That may indicate that the complement classifier generally has a stronger prediction ability, even after subsampling. Further research is required to verify that hypothesis.

It seems that an increased cut-off level results in more clusters for which the complement classifier performs better on the cluster test set than the cluster classifier. With cut-off 0.5, this is nowhere the case, for 0.6 for 2 clusters,

3 clusters for 0.7 and and 6 for cut-off 0.8. Due to a redivision in folds per cut-off level, the classification accuracies are incomparable across levels. However, future work will establish whether this trend generally holds.

## 5    Conclusions and Further Work

NLP-problems customarily require some sort of inter-annotator agreement to be determined prior to performing classification experiments. The degree of agreement can then be seen as a quality measure for a data set. However, in a domain that is as potentially sensitive to annotator bias as readability, standard inter-annotator agreement statistics seem inadequate, as it is not unproblematic to simply average out the available data. Furthermore, in a data set consisting of a large number of sources supplying only a partial assessment of the data, such agreement measures quickly become more or less meaningless due to the relative sparsity of overlapping data points. To overcome these issues, we have developed a method to determine the generalizability of the classification method *after* training and testing. Determining a posteriori agreement is useful for data sets with low a priori agreement or when determining a priori agreement is problematic.

For the learning method and data set used in this article, we found insufficient a posteriori agreement, so further analysis is needed in order to determine whether a way to find consensus among experts is crucial, or whether a different combination of learning methods and feature sets must be used.

Future work includes further development of readability prediction systems and methodologies. We will extend the feature set used to predict readability and perform experiments with a range of classification and regression methods. We will also further extend our data set by collecting more assessments from experts, and by adding new texts to our corpus. We will use the method outlined in this article to assess the quality of the newly collected data, as well as the overall accuracy. Apart from the difference in classification accuracy, we will look into other informative measures to determine the generalizability of readability prediction systems.

## Acknowledgements

This research was funded by the University College Ghent Research Fund.

## References

1. Anderson, R.C., Davison, A.: Conceptual and Empirical Bases of Readability Formulas. Tech. Rep. 392, University of Illinois at Urbana-Champaign (October 1986)
2. Beigman Klebanov, B., Beigman, E.: From Annotator Agreement to Noise Models. Computational Linguistics 35(4), 495–503 (2009)
3. van den Bosch, A., Busser, B., Canisius, S., Daelemans, W.: An efficient memory-based morphosyntactic tagger and parser for dutch. In: van Eynde, F., Dirix, P., Schuurman, I., Vandeghinste, V. (eds.) Proceedings of CLIN17, pp. 99–114 (2007)

4. Chang, C.C., Lin, C.J.: LIBSVM: a library for support vector machines (2001), http://www.csie.ntu.edu.tw/~cjlin/libsvm
5. Coleman, M., Liau, T.L.: A computer readability formula designed for machine scoring. Journal of Applied Psychology 60, 283–284 (1975)
6. Flesch, R.: A new readability yardstick. Journal of Applied Psychology 32(3), 221–233 (1948)
7. Heilman, M.J., Collins-Thompson, K., Callan, J., Eskenazi, M.: Combining lexical and grammatical features to improve readability measures for first and second language texts. In: Proceedings of HLT (2007)
8. Jain, A.K., Murty, M.N., Flynn, P.J.: Data clustering: a review. ACM Comput. Surv. 31(3), 264–323 (1999)
9. Jurafsky, D., Martin, J.H.: Speech and Language Processing. Prentice-Hall, Englewood Cliffs (2008)
10. Kate, R.J., Luo, X., Patwardhan, S., Franz, M., Florian, R., Mooney, R.J., Roukos, S., Welty, C.: Learning to Predict Readability using Diverse Linguistic Features. In: Proceedings of Coling23 (2010)
11. McNamara, D.S., Kintsch, E., Songer, N.B., Kintsch, W.: Are good texts always better? Interactions of text coherence, background knowledge, and levels of understanding in learning from text. Tech. rep., University of Colorado (1993)
12. van Noord, G.J.: Large Scale Syntactic Annotation of written Dutch (LASSY) (January 2009), http://www.let.rug.nl/vannoord/Lassy/
13. van Oosten, P., Tanghe, D., Hoste, V.: Towards an Improved Methodology for Automated Readability Prediction. In: Proceedings of LREC7 (2010)
14. Pitler, E., Nenkova, A.: Revisiting readability: A unified framework for predicting text quality. In: EMNLP, pp. 186–195. ACL (2008)
15. Schuurman, I., Hoste, V., Monachesi, P.: Cultivating Trees: Adding Several Semantic Layers to the Lassy Treebank in SoNaR. In: Proceedings of TLT7. Groningen, The Netherlands (2009)
16. Schwarm, S.E., Ostendorf, M.: Reading level assessment using support vector machines and statistical language models. In: Proceedings of ACL43, pp. 523–530. Association of Computational Linguistics, Ann Arbor (June 2005)
17. Staphorsius, G.: Leesbaarheid en leesvaardigheid. De ontwikkeling van een domeingericht meetinstrument. Cito, Arnhem (1994)
18. Tanaka-Ishii, K., Tezuka, S., Terada, H.: Sorting texts by readability. Computational Linguistics 36(2), 203–227 (2010)

# A Method to Measure the Reading Difficulty of Japanese Words

Keiji Yasuda, Andrew Finch, and Eiichiro Sumita

National Institute of Information and Communications Technology
3-5, Hikaridai, Keihanna Science City, Kyoto, 619-0289, Japan
{keiji.yasuda,andrew.finch,eiichiro.sumita}@nict.go.jp

**Abstract.** In this paper, we propose an automatic method to measure the reading difficulty of Japanese words. The proposed method uses a statistical transliteration framework, which was inspired by statistical machine translation research. A Dirichlet process model is used for the alignment between single kanji characters and one or more hiragana characters. The joint probability of kanji and hiragana is used to measure the difficulty. In our experiment, we carried out a linear discriminate analysis using three kinds of lexicons: a Japanese place name lexicon, a Japanese last name lexicon and a general noun lexicon. We compared the discrimination ratio given by the proposed method and the conventional method, which estimates a word difficulty based on manually defined kanji difficulty. According to the experimental results, the proposed method performs well for scoring Japanese proper noun reading difficulty. The proposed method produces a higher discrimination ratio with the proper noun lexicons (14 points higher on the place name lexicon and 26.5 points higher on the last name lexicon) than the conventional method.

## 1   Introduction

The Japanese writing system uses two sets of phonograms (hiragana and katakana) and one set of logograms ( "kanji," or Chinese characters). A single kanji has one or more possible readings, which are categorized as "onyomi" (Sino-Japanese reading) or "kunyomi" (Japanese reading). In some cases, a single kanji character can have more than 10 different readings.

A Japanese single word is written using one or more kanji characters. Depending on the anteroposterior characters or even on context, the reading of the kanji can change. Consequently, some Japanese words are very difficult to read even for native Japanese speakers. In this paper, we propose a method to measure the reading difficulty of Japanese words, which could have practical application in language learning.

The difficulty of Japanese word readings can be attributed to two factors. The first factor is the difficulty of the kanji. There are 50,000 kanji characters in total. Only 2136 characters have been designated for everyday use, and rarely used kanji characters can be difficult to read. The second factor is the irregularity

A. Gelbukh (Ed.): CICLing 2011, Part II, LNCS 6609, pp. 436–445, 2011.

**Table 1.** Examples of the irregular readings

| Example from lexicons | | | Dictionary information | | |
|---|---|---|---|---|---|
| Category | Word | Reading | Kanji | Onyomi | Kunyomi |
| Place name | 熊耳 | Kumagami | 熊 | YUU | KUMA |
| Person name | 十 | Yokotate | 耳 | JI,NI | MIMI |
| | | | 十 | JYU,JI | TOU,TO |

of kanji reading. Japanese words can sometimes have very irregular readings, which makes reading kanji difficult. Examples of difficult readings are shown in Table 1. In the first example, the second character is read as "$GAMI$." However, readings of the character in dictionaries are "$JI$", "$NI$" and "$MIMI$". The second example in the table is a more extreme case. This is a Japanese last name, written using a single kanji character, that means "ten." The most common reading of the character is "$JYU$," but here, it is read as "$YOKOTATE$". "$YOKO$" means "vertical" and "$TATE$" means "horizontal," so the reason that the character in read "YOKOTATE" is that the character consists of a horizontal bar ("$YOKO$") and vertical bar ("$TATE$").

There has been some research dealing with the difficulty of kanji[1] that manually defined kanji difficulty to correspond to the "Japanese Language Examination". Related work has been done on Japanese word familiarity[2]. This research manually annotated the familiarity of Japanese words to build a language resource. However, neither of them can easily be applied to measuring Japanese word reading difficulty. The former method can not measure Japanese word reading difficulty because it only defines a kanji character's difficulty. The latter method has very low coverage for proper nouns. Additionally, there is a high cost involved in expanding the lexicon because it requires works by human annotation.

Comparing to the above mentioned research results, the proposed method has following merits:

1. The method can automatically score the reading difficulty of Japanese words, no human annotator is required.
2. The scoring scheme takes into account the reading irregularity factor, which has not been dealt with by the aforementioned related research.

Section 2 describes the proposed method of scoring Japanese word reading difficulty. Section 3 details the experiments scoring reading difficulty of Japanese words using several kinds of lexicons, and also describes how we evaluated our method using discrimination analysis. Section 4 concludes the paper.

## 2    Proposed Method

Figure 1 shows the framework of the proposed method. Our method requires a large sized lexicon in both kanji and hiragana. Since a Japanese word can be

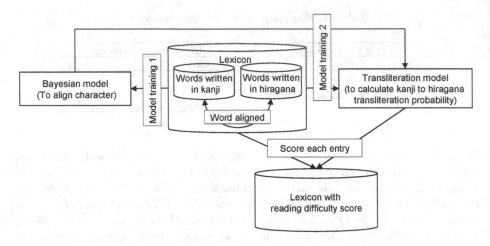

**Fig. 1.** Framework of the proposed method

written with multiple kanji characters and the reading of a single kanji character can be expressed by multiple hiragana characters, the proposed method first
aligns single kanji character to a sequence of hiragana characters. This character alignment is based on a Dirichlet process model trained using Bayesian
inference[3]. Secondly, using the alignment results, we train a statistical transliteration model to score reading difficulty. Details on the character alignment and
the scoring method are explained in 2.1 and 2.2, respectively.

## 2.1  Character Alignment

A character sequence-pair is a tuple $(\overline{\mathbf{k}}, \overline{\mathbf{h}})$ consisting of a sequence of kanji characters together with a sequence of hiragana characters $(\overline{\mathbf{k}}, \overline{\mathbf{h}}) = (<k_1, k_2, \ldots k_i>,$
$<h_1, h_2, \ldots, h_j>)$.

The training lexicon probability is simply the probability of all possible derivations of the lexicon given the set of sequence-pairs and their probabilities.

$$p(\overline{\mathbf{k}}_1^M, \overline{\mathbf{h}}_1^N) = P(k_1, k_2, \ldots, k_M, h_1, h_2, \ldots, h_N)$$
$$= \sum_{\gamma \in \Gamma} P(\gamma)$$

where $\gamma = ((\overline{\mathbf{k}}_1, \overline{\mathbf{h}}_1), \ldots, (\overline{\mathbf{k}}_j, \overline{\mathbf{h}}_j), \ldots, (\overline{\mathbf{k}}_J, \overline{\mathbf{h}}_J))$ is a derivation of the lexicon
characterized by its co-segmentation, and $\Gamma$ is the set of all derivations (co-
segmentations) of the lexicon.

The probability of a single derivation is given by the product of its component
character sequence-pairs.

$$p(\gamma) = \prod_{j=1}^{J} P((\overline{\mathbf{k}}_j, \overline{\mathbf{h}}_j)) \tag{1}$$

The lexicon for our experiments is segmented into kanji and hiragana word-pairs. We therefore constrain our model such that both kanji and hiragana character sequences of each character sequence-pair in the derivation of the lexicon are not allowed to cross a word segmentation boundary. Equation 1 can therefore be arranged as a product of word-pair $w$ derivations of the sequence of all word-pairs $\mathcal{W}$ in the lexicon.

$$p(\gamma) = \prod_{w \in \mathcal{W}} \prod_{(\overline{k}_j, \overline{h}_j) \in \gamma_w} P((\overline{k}_j, \overline{h}_j)) \tag{2}$$

where $\gamma_w$ is a derivation of kanji and hiragana word-pair $w$.

The Dirichlet process model we use in our approach is a simple model that resembles the cache models used in language modeling [4]. Intuitively, the model has two basic components: a model for generating an outcome that has already been generated at least once before, and a second model that assigns a probability to an outcome that has not yet been produced. Ideally, to encourage the re-use of model parameters, the probability of generating a novel sequence-pair should be considerably lower then the probability of generating a previously observed sequence-pair. This is a characteristic of the Dirichlet process model we use and furthermore, the model has a preference to generate new sequence-pairs early on in the process, but is much less likely to do so later on. In this way, as the cache becomes more and more reliable and complete, so the model prefers to use it rather than generate novel sequence-pairs. The probability distribution over these character sequence-pairs (including an infinite number of unseen pairs) can be learned directly from unlabeled data by Bayesian inference of the hidden cosegmentation of the lexicon.

For the *base measure* that controls the generation of novel words, we use a joint spelling model that assigns probability to new words according to the following joint distribution:

$$\begin{aligned} G_0((\overline{k}, \overline{h})) &= p(|\overline{k}|)p(\overline{k}||\overline{k}|) \times p(|\overline{h}|)p(\overline{h}||\overline{h}|) \\ &= \frac{\lambda_k^{|\overline{k}|}}{|\overline{k}|!}e^{-\lambda_k}v_k^{-|\overline{k}|} \times \frac{\lambda_h^{|\overline{h}|}}{|\overline{h}|!}e^{-\lambda_h}v_h^{-|\overline{h}|} \end{aligned} \tag{3}$$

where $|\overline{k}|$ and $|\overline{h}|$ are the length in characters of the kanji and hiragana sides of the character sequence-pair; $v_k$ and $v_h$ are that vocabulary (alphabet) sizes of the kanji and hiragana languages respectively; and $\lambda_k$ and $\lambda_h$ are the expected lengths of kanji and hiragana. In our experiments, we set $|\overline{k}| = 1$, in other words, we only allow to align single kanji character and one or more Hiragana characters.

The generative model is given in Equation 4 below. The equation assigns a probability to the $k^{\text{th}}$ character sequence-pair $(\overline{k}_j, \overline{h}_j)$ in a derivation of the lexicon, given all of the other phrase-pairs in the history so far $(\overline{k}_{-j}, \overline{h}_{-j})$. Here $-j$ is read as: "up to but not including $j$".

$$p((\overline{k}_j, \overline{h}_j))|(\overline{k}_{-j}, \overline{h}_{-j})) = \frac{N((\overline{k}_j, \overline{h}_j)) + \alpha G_0((\overline{k}_j, \overline{h}_j))}{N + \alpha} \tag{4}$$

In this equation, $N$ is the total number of character sequence-pairs generated so far, $N((\overline{\mathbf{k}}_j, \overline{\mathbf{h}}_j))$ is the number of times the phrase-pair $(\overline{\mathbf{k}}_j, \overline{\mathbf{h}}_j)$ has occurred in the history. $G_0$ is the base measure and $\alpha$ is a concentration parameter that determines how close the resulting distribution over sequence-pairs is to $G_0$.

For the model training, a blocked version of a Gibbs sampler are used. Details of the algorithm are explained in [4,5,3].

## 2.2 Scoring the Reading Difficulty of Japanese Word

To score the irregularity of a kanji reading, the proposed method builds a table of probabilities given by Equation 5,

$$p(h, k) = p(h|k)p(k) \tag{5}$$

where $k$ and $h$ are a single kanji character and the hiragana character sequence corresponding to $k$ respectively. In this formula, $p(h|k)$ is the probability of hiragana sequence $h$ given kanji character $k$. $p(k)$ is the occurrence probability of kanji character $k$ in the training lexicon. Intuitively these correspond respectively to the irregularity of a kanji reading, and the kanji difficulty.

The probability table is made by calculating $p(h, k)$ for all of the aligned pairs created by the character alignment. Using the table, the proposed method computes the score of the reading difficulty according to Equation 6,

$$S_{proposed} = \min_{i}^{M} \log p(h_i, k_i) \tag{6}$$

where $M$ is the number of the kanji characters in target Japanese word, and $S_{proposed}$ is the reading difficulty of the Japanese word.

# 3   Experiments

In this section, we describe the evaluation experiments of the proposed method. Discrimination analysis is carried out for the evaluation.

## 3.1   Conventional Method

As we explained in section 1, there has been some research investigating Japanese word familiarity and kanji difficulty. Since the previous research does not widely cover proper nouns, most of the words in our lexicon are words that have not been used in research. Therefore, we decided to use kanji difficulty rank as the baseline method, and extended kanji difficulty rank to word difficulty rank by Equation 7,

$$S_{conventional} = \min_{i}^{M} R_{kanji}(k_i) \tag{7}$$

**Table 2.** Training lexicons

| Lexicon type | Japanese place name | Japanese last name | Japanase general noun |
|---|---|---|---|
| # of lexicon entries | 82 K | 130K | 45 K |
| Total number of kanji characters | 308 K | 282 K | 103 K |
| Size of kanji character set | 2,459 | 3,671 | 4,345 |

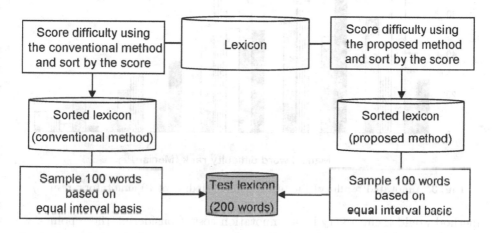

**Fig. 2.** Test lexicon sampling method

where $R_{kanji}(k_i)$ is the difficulty rank of $i$-th kanji character decided by [1]. Ranking ranged from 1 (difficult) to 4 (easy) [1], and "out of rank." To calculate $S_{conventional}$, we treated "out of rank" as 0 because most of "out of rank" kanji characters appeared to be more difficult than those of rank 1.

## 3.2 Experimental Settings

Table 2 shows the details of the lexicons used for the experiments. As shown in the table, we used three kinds of lexicons: Japanese place name, Japanese last name and Japanese general noun. These lexicons were extracted from the "Zip Code Data"[6] released by Japan Post Service Co., Ltd., the "Japanese Last Name Reading Dictionary"[7] and the "ipadic"[8], respectively.

To carry out linear discrimination analysis, we need a test lexicon with manually annotated difficulty ranking. Since word with irregular readings do not occur as often as regular readings, a simple random sampling method may not be able to pick an irregular reading sample. The ideal way to build a test lexicon is to manually annotate the difficulty ranking for the entire lexicon, then randomly sample a certain number of words from each difficulty rank. However, the ideal

---

[1] These ranks are corresponding to the class of Japanese Language Examination (class 1 to 4).

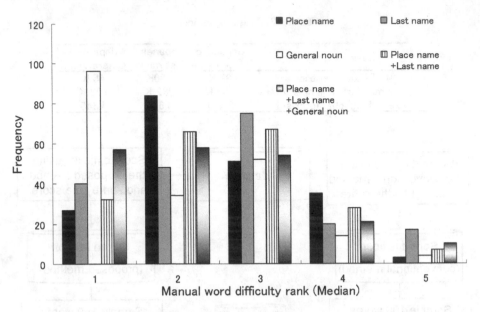

**Fig. 3.** Annotated results of the test lexicons (Median of 10 annotators' results)

method would incur a very large annotation cost. Considering these points, we used the test lexicon sampling method shown in Figure 2. First, the selection method sorts lexicon entries by using $S_{proposed}$ and $S_{conventional}$ to obtain two sorted lexicons. Secondly. 100 indices are collected at equal intervals from the ordered list of indices of the sorted lexicons. Then the words at those indices in both lexicons are drawn as our sample.

All of the words in the test lexicons were manually annotated for Japanese word reading difficulty as: 1 (very easy), 2 (easy), 3 (medium), 4 (difficult) and 5 (very difficult).

Figure 3 shows the histogram of the manual rank in each test. The median of 10 different annotators' results are used. The test lexicon of "Place name + Last name" (striped bars) and "Place name + Last name + General noun" (gradient bars) are not made by concatenating the corresponding test lexicons, they are obtained by concatenating the corresponding training lexicons, then applying the test lexicon sampling processing shown in Figure 2.

As shown in the figure, although, distribution is not an ideal uniform distribution, there are multiple samples in each rank.

## 3.3 Discrimination Analysis

We carried out discrimination analysis in order to compare the proposed method to the conventional method. Using the five manually annotated ranks, we carried out five-class linear discrimination analysis. Equation 8 is the discrimination function we used.

$$\operatorname{argmin}_{i=1}^{5}\{\operatorname{abs}(S - \mu_i)\} \tag{8}$$

**Table 3.** Average difficulty score $(\mu_i)$

| | Coventional method | | | Proposed method | | |
|---|---|---|---|---|---|---|
| | Place name | Last name | General noun | Place name | Last name | General noun |
| $\mu_1$ | 1.33 | 1.28 | 1.85 | −8.09 | −7.47 | −8.94 |
| $\mu_2$ | 1.23 | 1.21 | 1.56 | −8.02 | −7.98 | −9.07 |
| $\mu_3$ | 0.98 | 1.08 | 1.04 | −9.69 | −9.06 | −11.79 |
| $\mu_4$ | 1.09 | 1.05 | 0.36 | −12.42 | −16.10 | −22.63 |
| $\mu_5$ | 1.33 | 1.24 | 0.25 | −18.96 | −23.50 | −29.98 |

**Table 4.** Evaluation results (Discrimination ratio)

| | Place name | Last name | General noun | Place name +Last name | Place name +Last name +General noun |
|---|---|---|---|---|---|
| Conventional method (closed) | 24.00% | 14.50% | 49.50% | 26.00% | 18.00% |
| Prposed method (closed) | 40.00% | 43.50% | 48.50% | 41.50% | 34.00% |
| Human (closed) | 57.03% | 54.40% | 67.30% | 57.85% | 59.85% |
| Conventional method (10 cross validation) | 13.00% | 15.50% | 50.00% | 24.50% | 16.00% |
| Prposed method (10 cross validation) | 27.00% | 42.00% | 37.00% | 40.50% | 33.00% |
| Human (10 cross validation) | 56.70% | 54.40% | 67.30% | 57.20% | 59.10% |

where $S$ is the automatic difficulty score of the target word. For $S$, $S_{proposed}$ and $S_{conventional}$ are used for the proposed method and conventional method, respectively. For the manual difficulty rank, the median of the results from 10 annotators was used, and $\mu_i$ is the average difficulty score of group of manual difficulty rank $i$.

Table 3 shows the $\mu_i$ on the three test lexicons. For the proper discrimination, it is favorable for $\mu_i$ to be arrayed (decrease monotonically with $i$). Comparing the conventional method to the proposed method, the conventional method has more unarrayed values (underlined) than the proposed method.

## 3.4   Experimental Results

Table 4 shows the discrimination ratio $(R_{lex})$ computed by the following equation:

$$R_{lex} = \frac{\sum_{i=1}^{5} c_i}{\sum_{i=1}^{5} n_i} \tag{9}$$

where $n_i$ and $c_i$ are the number of the test words of rank $i$ and the number correctly discriminated to rank $i$ respectively.

**Table 5.** Detailed evaluation results of 10 cross validation

| Correct rank | Conventional method | | | Proposed method | | |
|---|---|---|---|---|---|---|
| | Place name | Last name | General noun | Place name | Last name | General noun |
| 1 | 3.70% | 17.50% | 72.92% | 25.93% | 65.00% | 52.08% |
| 2 | 0.00% | 2.08% | 0.00% | 26.19% | 22.92% | 26.47% |
| 3 | 49.02% | 24.00% | 50.00% | 37.25% | 49.33% | 23.08% |
| 4 | 0.00% | 25.00% | 14.29% | 17.14% | 0.00% | 0.00% |
| 5 | 0.00% | 0.00% | 50.00% | 0.00% | 58.82% | 75.00% |

The bold parts of the table represent the results of 10-fold cross validation tests and the non-bold parts represent the results of a closed experiment. The rows labeled as "Human" in the table are the results of the discrimination using one of ten human annotation results instead of the automatic score. The "Human" discrimination ratio is the averaged value of 10 annotaters. This value indicates the upper bound of the discrimination. As shown in the table, the proposed method gives a higher discrimination ratio than the conventional method for all, except the general noun lexicon.

As shown in Figure 3, the distribution of the manual reading difficulty rank is not a uniform distribution. With this kind of test lexicon, outputting the most frequent rank gives high discrimination results. In order to remove the distribution bias, we evaluate according to the discrimination ratio for each rank $(R_i)$ computed by the following equation:

$$R_i = \frac{c_i}{n_i} \tag{10}$$

Table 5 shows $(R_i)$ for the 10-fold cross validation. In the table, the underlined value indicates that method obtained a higher $R_i$ under the same conditions. As shown in the table, the proposed method has more underlined numbers than the conventional method. Additionally, in most of the cases (11 out of 15), the $R_i$ of the proposed method exceeds the chance level which is 20%.

As shown in Figure 4 and 5, the proposed method works well for a proper noun lexicon (place and last name). However, the conventional method surpasses the proposed method for the general noun lexicon. The reason is as follows.

1. Most of Japanese native speakers can handle irregular readings of general noun because acquiring the knowledge is an important task of Japanese language learning. Therefore, irregularity of the reading has very weak influence to the reading difficulty of words.
2. Meanwhile, the readings of proper nouns are hardly dealt in Japanese language class. Thus, the irregularity of the reading has a strong influence to the reading difficulty of words.

# 4   Conclusions

We propose a method to automatically measure the reading difficulty of Japanese words expressed in kanji characters. The proposed method calculates the irregularity of a kanji reading based on statistical modeling.

For the experiments, we used three kinds of lexicons; a Japanese place name lexicon; a Japanese last name lexicon, and a Japanese general noun lexicon. Discrimination analysis was carried out for the evaluation. According to the experimental results, the proposed method gives a higher discrimination ratio for the proper noun lexicon (14 points higher on the place name lexicon and 26.5 points higher on the last name lexicon) than the conventional method. However, the conventional method surpasses the proposed method on the general noun lexicon (13 points).

We therefore conclude the proposed method performs well for scoring Japanese proper noun reading difficulty.

# References

1. Kawamura, Y.: Two new tools for analyzing japanese textbooks: Vocabulary and kanji level checker. Learning Japanese in the Network Society, 71–88 (2002)
2. Hiroshi, S., Kaname, K., Tomoko, K., Shigeaki, A.: Fundamental vocabulary selection based on word familiarity. Transactions of the Japanese Society for Artificial Intelligence 19, 502–510 (2004-11-01)
3. Finch, A., Sumita, E.: A bayesian model of bilingual segmentation for transliteration. In: Proceedings of IWSLT (2010) (will be appeared)
4. Goldwater, S., Griffiths, T.L., Johnson, M.: Contextual dependencies in unsupervised word segmentation. In: ACL-44: Proceedings of the 21st International Conference on Computational Linguistics and the 44th Annual Meeting of the Association for Computational Linguistics, Morristown, NJ, USA, pp. 673–680. Association for Computational Linguistics (2006)
5. Mochihashi, D., Yamada, T., Ueda, N.: Bayesian unsupervised word segmentation with nested pitman-yor language modeling. In: ACL-IJCNLP 2009: Proceedings of the Joint Conference of the 47th Annual Meeting of the ACL and the 4th International Joint Conference on Natural Language Processing of the AFNLP, Morristown, NJ, USA, vol. 1, pp. 100–108. Association for Computational Linguistics (2009)
6. Japan Post Service Co., Ltd.: Zip code data (2010) http://www.post.japanpost.jp/zipcode/download.html
7. Nichigai Assiciates Inc.: Japanese lastname reading dictionary (1998), http://www.nichigai.co.jp
8. Asahara, M., Matsumoto, Y.: ipadic version 2.7.0 User's Manual (2003)

# Informality Judgment at Sentence Level and Experiments with Formality Score

Shibamouli Lahiri, Prasenjit Mitra, and Xiaofei Lu

The Pennsylvania State University, University Park PA 16802, USA
shibamouli@cse.psu.edu, pmitra@ist.psu.edu, xxl13@psu.edu

**Abstract.** Formality and its converse, informality, are important dimensions of authorial style that serve to determine the social background a particular document is coming from, and the potential audience it is targeted to. In this paper we explored the concept of formality at the sentence level from two different perspectives. One was the Formality Score (F-score) and its distribution across different datasets, how they compared with each other and how F-score could be linked to human-annotated sentences. The other was to measure the inherent agreement between two independent judges on a sentence annotation task. It gave us an idea how subjective the concept of formality was at the sentence level. Finally, we looked into the related issue of document readability and measured its correlation with document formality.

## 1 Introduction

Writing style is an important dimension of human languages. Two documents can provide the same content, but they may have been written using very different styles [9]. Authors from different social, educational and cultural backgrounds tend to use different writing styles [4]. With the evolution of Web 2.0, user-generated content has given rise to a variety of writing styles. Blog posts, for example, are written differently from the way academic papers are written. Twitter chats manifest yet another kind of writing style. Wikipedia articles use their own style guide[1].

One prominent dimension of writing style is the formality of a document. Academic papers are usually considered more formal than online forum posts. The notions of formality and contextuality at the document level have been illustrated by Heylighen and Dewaele [7]. They proposed a frequentist statistic known as the Formality Score (F-score) of a document, based on the number of deictic and non-deictic words (cf. Section 2). F-score is a coarse-grain measure, but it works well when used to classify documents according to their authorial style [15].

Classifying sub-document units such as sentences as formal or informal is more difficult because they are typically much smaller than a document and provide much less information. For example, the sentence "She doesn't like the piano" may be considered informal because it contains the colloquial usage "doesn't". But some native English speakers may think that the usage of "doesn't" is quite appropriate and formal. So we note that the notion of formality at the sentence level is subjective. On the other

---

[1] http://en.wikipedia.org/wiki/Wikipedia:Manual_of_Style

A. Gelbukh (Ed.): CICLing 2011, Part II, LNCS 6609, pp. 446–457, 2011.

hand, the sentence "She does not like the piano" is more formal than the sentence "She doesn't like the piano". So instead of classifying a sentence as formal or informal, we might actually be better off by assigning a formality score to a sentence, which would then reflect its *degree of formality*. A question that immediately arises is whether we can use the F-score of a sentence for this purpose.

As pointed out in [7], a frequentist statistic such as F-score should not be applied directly to measure the formality of a small text sample, e.g., a sentence. In this paper we look into the F-score distribution at the sentence level for four independent corpora and observe that these distributions broadly follow the corpus-level F-score trend. Moreover, the sentence-level F-score distribution on a human-annotated dataset shows a clear distinction between sentences labeled formal and sentences labeled informal. These observations indicate that the sentence-level F-score may be used as a feature in designing a formality score for sentences.

The second experiment reported in this paper is an inter-annotator agreement study for constructing a gold-standard dataset for the binary sentence classification task. Two independent annotators, both native speakers of English, judged sentences as formal or informal according to their own perception and intuition. Annotation judgments on two different datasets show poor agreement. We reason that this negative result is because of the arbitrariness of the notion of informality in two different judges' minds. A take-home message from this study is to either carefully design an annotation guideline or to adopt a Likert-style labeling scheme instead of a binary one, and let the judges discuss their results among themselves to improve agreement.

Apart from the formality of a document, we also consider the related issue of its readability. Traditional readability tests like the Flesch Reading Ease Score measure how difficult it is to read a piece of text. As a document becomes more formal, it starts introducing more context (cf. Section 2). So the document usually becomes longer, with more intricate sentence structure. Intuition suggests that such context insertion would typically mean a corresponding increase in reading time, i.e., reading difficulty. Document-level correlation between F-score and readability tests justifies our intuition. We found moderate correlation in all cases.

This paper is organized as follows. Section 2 discusses the background on F-score. Section 3 describes our experiments. Section 4 gives related work and Section 5 concludes the paper. The complementary code and data are available at http://www.CICLing.org/2011/software/251.

## 2   Background

The seminal study on measuring text formality by Heylighen and Dewaele [7] considers two different variants of formal expressions - *surface formality* and *deep formality*. Surface formality is the case when language is formalized for its own sake, e.g., a marital vow. Deep formality on the other hand represents the case when language is formalized so that the meaning is communicated clearly and as completely as possible. Complete communication of meaning involves putting in more background information so that no question regarding a document may go unanswered. This background information is known as "context". So we observe that as more context is inserted into a document, the language tends to become more (deeply) formal. Conversely, as a document is gradually robbed of its context, the language tends to become more *contextual*. Heylighen and Dewaele also argued that surface formality emerges from deep formality, so the latter is sufficient to characterize both.

As an example of deep formality, consider the sentence "She likes the piano". This sentence can be made more formal by saying "Ms Muffet likes the piano". Here "Ms Muffet" is a part of the context of the first sentence. However, we can make the sentence even more (deeply) formal by saying "Ms Muffet likes the piano beside the door". Note that in the last sentence we added more context than there was in the second sentence. This context-addition and resulting formalization process can be continued ad infinitum, because it is impossible to fully specify the meaning of a text in itself without some unsaid background assumptions. Since context-addition is always possible, we cannot make a hard judgment that one document is strictly formal and another one is strictly informal. We can say that document A is more formal than document B. This is known as the *continuum of formality*.

Informality is introduced by deixis and implicature. Deixis indicates a set of words that anchors to another set of words for contextual information [11]. For example, in the sentence "She likes the piano", the word "she" anchors to "Ms Muffet". Four types of deixis have been recognized - time, place, person and discourse [11]. Time deixis can be seen in the words "today", "now", "then", etc. These words anchor to specific time points. Place deixis is exemplified in the place-anchoring words "here", "there", "around", etc; person (or object) deixis gives us words like "this", "that", "he", "she", etc; and discourse deixis engenders words like "therefore", "hence", "notwithstanding", etc. Detailed word correlation studies indicate some categories of words are *deictic* (pronouns, verbs, adverbs, interjections), some others are *non-deictic* (nouns, adjectives, prepositions, articles) and the rest are *deixis-neutral* (conjunctions) [7].

In deixis, there are some anchor words that explicitly relate to the context information. In implicature, the context information must be inferred from background knowledge. As an example, consider the sentence "Einstein rocks!" In this sentence the context information - why Einstein rocks - is absent. Only when we couple this sentence with the background knowledge that Einstein was a great scientist, do we come to appreciate the full meaning. But quantifying the impact of implicature is more difficult because we need to call upon the background information - something which is not present in the document. Therefore only deictic and non-deictic words were considered in the definition of F-score:

F = (noun frequency + adjective freq. + preposition freq. + article freq. - pronoun freq. - verb freq. - adverb freq. - interjection freq. + 100)/2

where the frequencies are taken as percentages with respect to the total number of words in the document [7]. Note that as the number of deictic words increases and non-deictic words decreases, F-score becomes lower, indicating a more contextual (informal) document. The reverse happens in the case of a more formal document. F-score of a document can range from zero to 100.

Note that the definition of F-score is valid for sentences as well. But sentences are much smaller than documents, so we cannot directly use F-score for measuring sentence-level formality. However, we would like to observe if F-score can be used as a feature for designing a sentence-level formality score. To address this question, we look into the sentence-level F-score distributions on unlabeled as well as labeled corpora. In the next section we describe the results of our exploratory analysis.

## 3   Experiments

### 3.1   Datasets

We compiled four different datasets - blog posts, news articles, academic papers and on-line forum threads. Each dataset has 100 documents. For the blog dataset, we collected most recent posts from the top 100 blogs listed by Technorati[2] on October 31, 2009. For the news article dataset, we collected 100 news articles from 20 news sites (five from each). These articles are mostly from "Breaking News", "Recent News" and "Local News" categories, with no specific preference to any of the categories. The news sites we used were CNN, CBS News, ABC News, Reuters, BBC News Online, New York Times, Los Angeles Times, The Guardian (U.K.), Voice of America, Boston Globe, Chicago Tribune, San Francisco Chronicle, Times Online (U.K.), news.com.au, Xinhua, The Times of India, Seattle Post Intelligencer, Daily Mail and Bloomberg L.P. For the academic paper dataset, we randomly sampled 100 papers from the CiteSeerX[3] digital library. For the online forum dataset, we sampled 50 random documents crawled from Ubuntu Forums[4] and 50 random documents crawled from TripAdvisor New York forum[5]. The blog, news, paper and forum datasets have 2110, 3009, 161406 and 2569 sentences respectively. The overall F-scores of these datasets are 65.24, 66.51, 68.62 and 58.52 respectively[6].

### 3.2   Sentence Level F-score Distributions

We recall from Section 2 that the F-score of a document uses deixis information as a measure of formality. Since a sentence can be thought of as a small document, deixis is present at the sentence level as well. It is therefore of interest to explore how the sentence-level F-score distributions compare with each other, and whether they bear any consistent form across different datasets. Apart from shedding light on the variation of sentence-level deixis and its types in various corpora, such an exploratory analysis would also allow us to observe if the sentence-level F-score distributions follow a specific trend. Figure 1(a) gives us the histogram of sentence-level distributions on four datasets (cf. Section 3.1) and Table 1 outlines some of their key properties. We note from

---

[2] http://www.technorati.com
[3] http://citeseerx.ist.psu.edu
[4] http://ubuntuforums.org/
[5] http://www.tripadvisor.com/ShowForum-g28953-i4-New York.html
[6] F-score computation involves part-of-speech tagging. We used CRFTagger [16] in all our experiments.

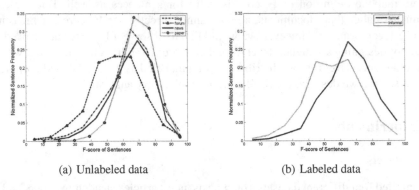

(a) Unlabeled data                    (b) Labeled data

**Fig. 1.** Histogram of sentence-level F-score distributions on different datasets

Figure 1(a) that the sentence-level F-score distribution of a higher-F-scored dataset is shifted towards the high formality zone and the sentence-level F-score distribution of a lower-F-scored dataset is shifted towards the low formality zone. Moreover, as the corpus-level F-score increases more and more, the sentence-level F-score distributions shift more and more to the higher formality zone.

**Table 1.** Properties of Sentence-level F-score Distributions

| Dataset | Mean | SD | Median | QD | Skewness | Kurtosis |
|---------|------|-----|--------|-----|----------|----------|
| Forum | 56.74 | 15.82 | 57.14 | 9.58 | -0.12 | 3.37 |
| Blog | 65.02 | 15.01 | 66.67 | 9.38 | -0.64 | 4.27 |
| News | 65.18 | 13.34 | 66.67 | 8.93 | -0.59 | 3.57 |
| Paper | 69.29 | 10.44 | 70 | 6.70 | -0.53 | 3.96 |

**Table 2.** Multiple Comparison Test between all groups with Tukey-Kramer's HSD correction

| Group 1-Group 2 | $F_{mean}^{Group1} - F_{mean}^{Group2}$ | Confidence Interval | Conclusion |
|-----------------|------------------------------------------|---------------------|------------|
| Blog-Forum | 8.28 | [7.48, 9.09] | $F_{mean}^{blog} > F_{mean}^{forum}$ |
| Blog-News | -0.16 | [-0.93, 0.62] | $NOT(F_{mean}^{news} > F_{mean}^{blog})$ |
| Blog-Paper | -4.27 | [-4.87, -3.67] | $F_{mean}^{paper} > F_{mean}^{blog}$ |
| Forum-News | -8.44 | [-9.18, -7.70] | $F_{mean}^{news} > F_{mean}^{forum}$ |
| Forum-Paper | -12.55 | [-13.10, -12.01] | $F_{mean}^{paper} > F_{mean}^{forum}$ |
| News-Paper | -4.11 | [-4.62, -3.61] | $F_{mean}^{paper} > F_{mean}^{news}$ |

Table 1 gives the Mean, Median, Standard Deviation (SD), Quartile Deviation (QD), Skewness and Kurtosis of the distributions. Note that the standard and quartile deviations for paper sentences are the smallest, so these sentences vary least in terms of F-score, while those from the forum dataset vary the most. One possible reason for such a high variation in forum sentences (along with low kurtosis) is that they come from

**Table 3.** Confidence Intervals obtained using different multiple comparison tests

| Group 1 | Group 2 | Confidence Intervals | | | |
|---------|---------|------|------------|-------------|---------|
| | | LSD | Bonferroni | Dunn-Šidák | Scheffé |
| Blog | Forum | [7.67, 8.90] | [7.46, 9.11] | [7.46, 9.11] | [7.41, 9.16] |
| Blog | News | [-0.75, 0.44] | [-0.96, 0.64] | [-0.95, 0.64] | [-1.00, 0.69] |
| Blog | Paper | [-4.73, -3.81] | [-4.89, -3.65] | [-4.88, -3.65] | [-4.92, -3.62] |
| Forum | News | [-9.00, -7.88] | [-9.20, -7.68] | [-9.20, -7.69] | [-9.24, -7.64] |
| Forum | Paper | [-12.97, -12.14] | [-13.11, -11.99] | [-13.11, -11.996] | [-13.15, -11.96] |
| News | Paper | [-4.50, -3.73] | [-4.63, -3.59] | [-4.63, -3.60] | [-4.66, -3.56] |

different types of users - some are information seekers, typically issuing sentences with less context (lower F-score), while others are information providers, issuing sentences with more context (higher F-score). On the other hand the paper sentences are somewhat "homogenized" and "compressed" into the higher end of formality continuum, because they all tend to follow the strict norms of written English.

To test whether these distributions are significantly different from each other, we performed a two sample Kolmogorov Smirnov test on each pair of distributions. At significance level $\alpha = 0.001$, all pairs (except the blog-news pair) were found to be significantly different from each other. Similar results were obtained in the pairwise comparison between distribution means ($F_{mean}$). We first performed a one-way ANOVA[7] on the null hypothesis:

$$F_{mean}^{paper} = F_{mean}^{news} = F_{mean}^{blog} = F_{mean}^{forum}$$

where $F_{mean}^{i}$ denotes the mean sentence-level F-score of dataset $i$. The ANOVA results reject this null hypothesis at significance level $\alpha = 0.001$, which indicates that at least two of the group means are significantly different from each other. Pairwise comparison between the group means were performed next with multiple testing correction. The results are shown in Tables 2 and 3. Each pairwise test is equivalent to an unpaired two-sample one-tailed t-test for comparing the means of two groups, with the addition of correction and adjustments for multiple comparison problem. Table 2 has six rows. Each row gives the groups of one pair, the difference in $F_{mean}$s between the two groups and the confidence interval of this difference using Tukey-Kramer's HSD correction. Note that if this confidence interval contains zero, then we conclude that the group means are not significantly different from each other. Otherwise, the sign of the group mean difference indicates whether group 1 has larger $F_{mean}$ than group 2, or vice versa.

In Table 3, we report the confidence intervals obtained using other multiple comparison tests, e.g., Fisher's least significant difference (LSD) method, Bonferroni's method, Dunn-Šidák's method and Scheffé's method, respectively. The confidence intervals follow the same trend as in Table 2, and they lead to the same conclusions - all group means are significantly different from each other (except the Blog-News pair) and the group means satisfy

---

[7] We used MATLAB for all our significance tests.

1. $F_{mean}^{paper} > F_{mean}^{news}$
2. $F_{mean}^{news} > F_{mean}^{forum}$
3. $F_{mean}^{paper} > F_{mean}^{blog}$
4. $F_{mean}^{blog} > F_{mean}^{forum}$

The reason why $F_{mean}^{news}$ was not significantly different from $F_{mean}^{blog}$ is that the blog posts were collected from the top 100 list of Technorati. Since blog is a bridging genre [6], many blog posts may actually be modified news articles. It is especially true with a generic blog search engine like Technorati, which indexes all kinds of blogs. This is also the reason why sentence-level F-score distribution for blogs was not found to be significantly different from that for news articles in the Kolmogorov-Smirnov Test. Note that the sentence-level F-score distribution for two very similar corpora may not be significantly different from each other. For example, if we modify a large dataset by introducing a few non-deictic words here and there, then the overall F-score will slightly increase, but the sentence-level F-score distribution will remain virtually the same.

### 3.3 Sentence Level F-score on Annotated Data

The results of Section 3.2 indicate that unless two corpora are very similar in their deixis content, their sentence-level F-score distributions will be different. But this observation in itself is not sufficient for declaring F-score as a sentence-level feature. We would also need to link F-score with the human notion of formality at the sentence level. In this experiment we labeled a 50-document dataset (7488 sentences) from the Splog Blog Collection[8]. A graduate student labeled each sentence as formal or otherwise according to whether or not the sentence contains informal/slang words and expressions, grammatical inconsistencies, visual cues like smileys and character repetition, etc. This student was not given any background on F-score at the time of the annotation, thereby eliminating bias. Among 7488 sentences, 4185 were labeled formal, and 3303 were labeled informal.

**Table 4.** Properties of Sentence-level F-score Distributions - Labeled Data

| Dataset | Mean | SD | Median | QD | Skewness | Kurtosis |
|---------|------|-----|--------|------|----------|----------|
| Formal | 65.82 | 15.85 | 66.67 | 9.92 | -0.31 | 3.55 |
| Informal | 56.65 | 16.58 | 57.14 | 10.99 | -0.19 | 3.18 |

The sentence-level F-score histogram of these sentences is shown in Figure 1(b) and the distribution properties are given in Table 4. Figure 1(b) and Table 4 show that the two distributions are different from each other with formal sentences shifted towards relatively higher F-score zones and informal sentences shifted towards relatively lower F-score zones. This is an important finding, because it indicates that the human-labeled sentences form a clear split in terms of F-score distribution.

---

[8] http://ebiquity.umbc.edu/resource/html/id/212/Splog-Blog-Dataset

A two-sample Kolmogorov-Smirnov test showed that at significance level $\alpha = 0.001$, the two distributions were different from each other. A one-tailed two-sample unpaired t-test for comparing the group means led to the same conclusion, where the confidence interval of the group mean difference was found to be [7.93,10.41]. Note that this interval does not contain zero, so the two group means are significantly different. This observation allows us to reason that F-score can be useful as one of the sentence-level features for capturing formality.

## 3.4   Inter-annotator Agreement Study

Designing a sentence-level formality score is complicated by the fact that different people have different notions regarding what should be considered formal or what should be considered informal. The concept of formality as native speakers perceive, is fairly subjective. It is therefore of importance to measure by how much two independent judges differ on a set of sentences, when no specific instructions are given as to what constitutes a formal or an informal sentence. If this "inherent" agreement is high, then we are able to establish a reliable gold-standard sentence-annotated dataset. If on the other hand this agreement is low, then we get an idea of how subjective the idea of sentence-level formality really is. In that case we can either employ a set of annotation instructions for improving agreement, or we can change the labeling scheme and let the annotators discuss among themselves to minimize disagreement. Note that the issue of constructing a gold-standard sentence-annotated dataset assumes importance because a sentence-level formality score can only be evaluated on such a hand-crafted corpus.

**Table 5.** Confusion Matrix and Inter-annotator Agreement

| Blog Posts | | | | |
|---|---|---|---|---|
| | C | NC | Raw Agreement | 0.692 |
| C | 168 | 172 | Kappa | 0.164 |
| NC | 480 | 1300 | Jaccard | 0.205 |

| News Articles | | | | |
|---|---|---|---|---|
| | C | NC | Raw Agreement | 0.756 |
| C | 71 | 383 | Kappa | 0.019 |
| NC | 352 | 2204 | Jaccard | 0.088 |

In this section we describe the results of an inter-annotator agreement study aimed at measuring the inherent agreement between two native English speakers regarding the concept of sentence-level formality. We enlisted help from four undergraduate students, who independently labeled each sentence of the blog and news datasets (cf. Section 3.1) as formal or informal.[9] Two students worked on the blog dataset and the other two worked on the news dataset. Students were requested to mark each sentence they considered informal as "C" and each sentence they considered formal as "NC" (Table 5).

---

[9] Students were remunerated with extra course credit at the end of the annotation.

They were not allowed to discuss among themselves or see each other's annotations. Since the purpose of this study was to measure "inherent" agreement between two native speakers of English, we did not specify what constitutes a formal sentence or an informal sentence. In other words, we did not have an annotation guideline or a rubric.

After the annotation process was over, we computed Cohen's Kappa and Jaccard Similarity along with raw agreement scores based on the confusion matrices (Table 5). Jaccard Similarity was computed as:

$$Jaccard = \frac{\#CC}{\#CC + \#CNC + \#NCC}$$

where $\#CC$, $\#CNC$ and $\#NCC$ denote the number of sentences in the top left, top right and bottom left cells of the confusion matrix, respectively. The agreement results are shown in Table 5. The raw agreement values are moderately high, but both Cohen's Kappa and Jaccard Coefficient indicate poor agreement. The reason behind this apparent paradox lies in the fact that the number of NCNC sentences - sentences both annotators considered formal, is very high (Table 5, NC row and NC column).

These findings imply a negative result in terms of inherent agreement at the sentence level regarding the notion of informality. The very low Kappa values obtained across two independent datasets show that there is hardly any agreement. This stance is bolstered by equally low values of Jaccard Coefficient obtained in both cases. So, coming up with a reliable gold-standard set of annotated sentences without some annotation guidelines is difficult. One way to improve agreement is to do several rounds of annotation and let the judges discuss after each round to converge into a common labeling scheme [2]. But this procedure as observed in [2], improves agreement only marginally, and that also when the initial agreement is already quite high. Another way to improve agreement is to design a detailed annotation guideline. However, design of such a guideline may entail loss of generalizability across multiple datasets and bias the study somewhat from the experimenter's perspective, so this approach needs to be carefully investigated before being put into effect.

The take-home message from this experiment is clear: formal/informal-type gold-standard sentence set construction will prove to be difficult because of the poor inter-annotator agreement. The poor agreement is not also very unexpected, because as we discussed in Section 2, the formality continuum is present at the sentence level as well. The binary annotation process forces the judges to do an arbitrary thresholding in this continuum and declare sentences "formal" when they are above this threshold and "informal" when they are below. This thresholding can be very different for two different persons and thereby yield poor agreement values. An alternative is to adopt a Likert-style labeling scheme [12], where instead of labeling sentences as formal/informal, judges provide a formality rating. Our future work includes working on this alternative. We also plan to let judges discuss among themselves for minimizing disagreement and coming up with a consistent set of annotation guidelines across multiple datasets.

## 3.5   F-score and Readability

An important observation with F-score is that it captures *deep formality* (cf. Section 2). As we go on adding context to a document, its deep formality increases. However,

**Table 6.** Overall F-score and Readability on different datasets

| Dataset | F-score | FRES | ARI | FKRT | CLI | GFI | SMOG |
|---|---|---|---|---|---|---|---|
| Forum | 58.52 | 77.71 | 7.90 | 6.05 | 9.43 | 9.83 | 9.38 |
| Blog | 65.24 | 61.04 | 11.63 | 9.47 | 11.43 | 13.83 | 12.03 |
| News | 66.51 | 56.21 | 13.13 | 10.78 | 12.47 | 15.50 | 13.46 |
| Academic Paper | 68.62 | 48.41 | 15.86 | 12.62 | 14.20 | 18.00 | 15.15 |

**Table 7.** Correlation of F-score with Readability measures

| Readability Measure | Pearson's $\rho$ | Spearman's $\rho$ | Kendall's $\tau$ | Quadrant Correlation |
|---|---|---|---|---|
| ARI | 0.45 | 0.57 | 0.41 | 0.48 |
| CLI | 0.46 | 0.61 | 0.44 | 0.52 |
| FKRT | 0.49 | 0.60 | 0.43 | 0.48 |
| FRES | -0.50 | -0.64 | -0.46 | -0.54 |
| GFI | 0.53 | 0.61 | 0.44 | 0.53 |
| SMOG | 0.54 | 0.62 | 0.46 | 0.55 |

adding context usually involves introducing new words, which increases the length of the document. Although in certain cases new words replace old words, so the document length remains unchanged, we expect that as more and more context information is added, a document tends to become longer. Longer documents take more time to process than shorter ones, so we expect that the overall *reading difficulty* of a document starts increasing as we go on adding more and more context. In other words, as the deep formality of a document increases, its reading difficulty also increases. Since the reading difficulty of a document is measured by *readability tests* and deep formality by F-score, we expect that there should be some correlation between F-score and readability tests.

To test the presence of such a correlation, we measured corpus-level F-score and readability scores on four datasets (cf. Section 3.1). Six standard readability tests were performed. These are Flesch Reading Ease Score (FRES), Automated Readability Index (ARI), Flesch-Kincaid Readability Test (FKRT), Coleman-Liau Index (CLI), Gunning fog Index (GFI) and SMOG (Simple Measure of Gobbledygook) [14]. Results are shown in Table 6, which indicates a clear trend in F-score and readability tests. All the readability tests (except FRES) show positive correlation with F-score. Pearson and rank correlation tests between document-level F-score and readability scores (Table 7) show moderate correlation values[10] in all cases. Pearson, Spearman and Kendall correlation values were found to be statistically highly significant with p-value $< 0.0001$. The negative correlation with FRES can be explained by the fact that FRES actually measures "reading ease" as opposed to "reading difficulty". This result justifies our intuition that context addition (F-score) and reading difficulty (readability tests) are correlated, but since the correlation is not very high, we believe there are factors other than readability that get into play when more context is inserted into a document, so the reading difficulty does not increase as much. This point merits further investigation.

---

[10] http://pathwayscourses.samhsa.gov/eval201/eval201_4_pg9.htm

## 4 Related Work

In this section we give a very brief sketch of the related studies. The presence of formality as a prominent dimension of language variation was first noted by Biber [1]. Formality of a language is largely determined by four factors - time, place, context and person. The four factors have been arrived at in the study of *registers* in sociolinguistics [17]. Registers denote a form of language variation that occurs both as a result of difference in speaker identity and as a result of difference in situation (context) [5]. Zampolli [19] and Hudson [8] arrived at the dimension of formality based on their own style analyses, but they could not explain it theoretically. Heylighen and Dewaele [7] were the first to summarily assess the causes of formality and design a document-level formality score, called the F-score. F-score uses the idea of *context* [10] and is much in the same spirit as the lexical density [18]. While F-score has not yet been applied to the sub-document level, a recent study by Brooke, et al. [3] looks into the notion of formality at the word level. They used publicly available formal and informal word lists as seed sets and analyzed large corpora to evaluate the effectiveness of several different approaches for measuring word-level formality. While our goal is different in the sense that we want to measure sentence-level formality, we can still use the word-level scores as features. The sentences are somewhat more difficult to deal with, because we cannot have a seed set of sentences without human annotation. Some of the results reported in this paper constitute the first step towards the creation of such a gold standard.

## 5 Conclusion

We have four principal contributions in this paper:

1. Exploratory analysis and comparison of sentence-level F-score distributions of four different datasets
2. Linking F-score with the human perception of sentence-level formality using F-score distribution on an annotated dataset
3. An inter-annotator agreement study to measure the inherent agreement between two independent native speakers of English on the notion of sentence-level formality
4. Correlation between F-score and readability tests

Our future work includes the design of a sentence-level formality score. Such a score would require, among other things, syntactic, semantic and pragmatic considerations [7]. Even more challenging is the problem of formality assessment at sub-sentence level. While there has been work on local emotion detection [13], it remains open whether similar techniques can be exploited in sub-sentence level formality judgment.

## References

1. Biber, D.: Variation Across Speech and Writing. Cambridge University Press, Cambridge (1988)
2. Brants, T., Skut, W., Uszkoreit, H.: Syntactic annotation of a German newspaper corpus. In: Proceedings of the ATALA Treebank Workshop, Paris, France, pp. 69–76 (1999)

3. Brooke, J., Wang, T., Hirst, G.: Automatic acquisition of lexical formality. In: Proceedings of the 23rd International Conference on Computational Linguistics (COLING) (2010)
4. Chambers, J.K., Schilling-Estes, N., Trudgill, P.: The handbook of language variation and change. Blackwell, Malden (2006)
5. Halliday, M.: Comparison and translation. In: Halliday, M., McIntosh, M., Strevens, P. (eds.) The linguistic sciences and language teaching. Longman, Harlow (1964)
6. Herring, S.C., Scheidt, L.A., Wright, E., Bonus, S.: Weblogs as a bridging genre. IT & People 18(2), 142–171 (2005)
7. Heylighen, F., Marc Dewaele, J.: Formality of language: definition, measurement and behavioral determinants. Tech. rep. (1999)
8. Hudson, R.: About 37% of word-tokens are nouns. Language 70(2), 331–339 (1994)
9. Karlgren, J.: Stylistic experiments for information retrieval (2000)
10. Leckie-Tarry, H., Birch, D.: Language and context: a functional linguistic theory of register. In: Birch, D. (ed.) Pinter Publishers, London (1995)
11. Levelt, W.J.M.: Speaking: From Intention to Articulation. MIT Press, Cambridge (1989)
12. Likert, R.: A technique for the measurement of attitudes. Archives of Psychology 22(140), 1–55 (1932)
13. Mao, Y., Lebanon, G.: Isotonic Conditional Random Fields and Local Sentiment Flow. In: Advances in Neural Information Processing Systems (2007)
14. McLaughlin, H.G.: SMOG grading - a new readability formula. Journal of Reading, 639 646 (May 1969)
15. Nowson, S., Oberlander, J., Gill, A.J.: Weblogs, genres and individual differences. In: Proceedings of the 27th Annual Conference of the Cognitive Science Society, pp. 1666–1671 (2005)
16. Phan, X.H.: CRFTagger: CRF English POS Tagger (2006),
    http://crftagger.sourceforge.net/
17. Reid, T.B.: Linguistics, structuralism, philology. Archivum Linguisticum 8
18. Ure, J.N.: Lexical density and register differentiation. In: Perren, G.E., Trim, J.L.M. (eds.) Applications of Linguistics: Selected Papers of the 2nd International Congress of Linguistics, Cambridge 1969. Cambridge University Press, Cambridge (1971)
19. Zampolli, A.: Statistique linguistique et dépouillements automatiques. Lexicologie, 325–358

# Combining Word and Phonetic-Code Representations for Spoken Document Retrieval

Alejandro Reyes-Barragán[1], Manuel Montes-y-Gómez[1,2], and Luis Villaseñor-Pineda[1]

[1] Laboratory of Language Technologies,
National Institute of Astrophysics, Optics and Electronics (INAOE),
Luis Enrique Erro #1, Sta. María Tonantzintla, Puebla, Mexico
{alejandroreyes,mmontesg,villasen}@inaoep.mx
[2] Department of Computer and Information Sciences,
The University of Alabama at Birmingham (UAB),
1300 University Boulevard, Birmingham, Alabama, USA

**Abstract.** The traditional approach for spoken document retrieval (SDR) uses an automatic speech recognizer (ASR) in combination with a word-based information retrieval method. This approach has only showed limited accuracy, partially because ASR systems tend to produce transcriptions of spontaneous speech with significant word error rate. In order to overcome such limitation we propose a method which uses word and phonetic-code representations in collaboration. The idea of this combination is to reduce the impact of transcription errors in the processing of some (presumably complex) queries by representing words with similar pronunciations through the same phonetic code. Experimental results on the CLEF-CLSR-2007 corpus are encouraging; the proposed hybrid method improved the mean average precision and the number of retrieved relevant documents from the traditional word-based approach by 3% and 7% respectively.

## 1 Introduction

The large amount of information existing in spoken form, such as TV and radio broadcasts, recordings of meetings, lectures and telephone conversations, has motivated the development of new technologies for its searching and browsing. Particularly, spoken document retrieval (SDR) refers to the task of finding segments from recorded speech that are relevant to a user's information need [1].

The traditional approach for SDR consists in a simple concatenation of an automatic speech recognition (ASR) system with a standard word-based retrieval method [2]. The main inconvenience of this approach is that it greatly depends on the accuracy of the recognition output. It is well known that recognition errors usually degrade the effectiveness of a SDR system, and that, unfortunately, current ASR methods have word error rates that vary from 20% to 40% in accordance to the kind of discourse.

A. Gelbukh (Ed.): CICLing 2011, Part II, LNCS 6609, pp. 458–466, 2011.

With the aim of reducing the impact of recognition errors on the retrieval performance, we investigated the helpfulness of using phonetic codifications[1] for representing documents' content. The idea of using phonetic codifications on this task was motivated by two facts. On the one hand, transcriptions errors are not randomly generated; words/phases are commonly substituted by others with similar pronunciation. For instance, the speech utterance "Unix Sun Workstation" would be incorrectly transcribed into "unique set some workstation". On the other hand, phonetic codifications allow characterizing phonetically similar words through the same code. For instance, using Soundex codes, the words "unique" and "Unix" are both represented by the code U52000, whereas the words "some" and "sun" are represented by S50000.

In this paper we propose a retrieval approach that uses word and phonetic-code based representations in cooperation. In particular, we focus on two main concerns. First, we evaluate the usefulness of different phonetic codifications algorithms, namely, Soundex [3], NYSIIS [4], Phonix [5], DMetaphone [6] and DM [7], and second, we analyze the synergy between word and phonetic-code representations. Our results on the CLEF-CLSR-2007 corpus suggest that NYSIIS codes are the more appropriate, and that the combination of word and phonetic-code representations is relevant for SDR and particularly useful for handling short queries.

The rest of the paper is organized as follows. Section 2 introduces some related work on SDR. It particularly presents the major approaches for handling with transcription errors. Section 3 describes our proposed approach for SDR, using word and phonetic code representations in conjunction. Section 4 presents the experimental results on CLEF-CLSR-2007 corpus. Finally, section 5 shows our conclusions.

# 2  Related Work

Due to the limited accuracy of current speech recognizers, several works on SDR have focused on proposing different methods for reducing the impact of transcription errors on the retrieval performance. In general, these methods are of two types: dependent and independent from the ASR system.

From the first kind, we can distinguish two main methods. The first one considers the transcription of speech utterances into phoneme or syllable sequences instead of word sequences by using a phoneme/syllable recognizer [8, 9, 10]. On the other hand, the second method proposes making use of more than the top-1 transcription hypothesis. Particularly, it considers using the n-best hypotheses or the complete word-lattice used internally by the recognizer [11]. As expected, these methods have the disadvantage of requiring access to the inside of the ASR system.

From the second group, we can also differentiate two main methods. One of them proposes using multiple recognizers [12, 13]. It is supported on the idea that independently developed recognizers tend to make different kinds of errors, and, therefore, that by combining their outputs it might be possible to recover some of them. The second method from this approach proposes reducing the effect of transcription errors by adding some related extra terms to the queries and/or

---

[1] Phonetic codification methods were initially propose for identifying the variants of personal names and for obtaining a canonical or normalized representation of them [20]. Traditionally, these kinds of methods are considered as a kind of approximate string matching technique.

documents [14, 15]. These extra terms can be found by analyzing the transcribed corpus and locating relevant terms based on co-occurrence. However, it has been shown that it is better to use a parallel written corpus, since transcriptions contain recurrent errors and may cause erroneous words to appear as expansion terms.

Similar to the above method, the one proposed in this paper also aims to tackle recognition errors by expanding documents and queries. However, different to this previous approach, it does not achieve this expansion by including some extra words; instead, it proposes to enrich the representation of documents and queries by adding the phonetic codes from the original terms. The purpose of this alternative representation is to reduce the impact of the transcription errors by characterizing words with similar pronunciations through the same phonetic code.

In addition to the previous difference, the proposed method has the advantage of being more portable; it does not require using any external resource (such as a parallel text collection), and, moreover, some phonetic codifications (e.g., Soundex) may be applied with minimal modifications to languages other than English.

Finally, it is important to mention that in a previous work [16] we proposed using Soundex codes to enrich the representation of transcriptions. However, this paper goes several steps forward. First, it presents the evaluation on the use of five different phonetic codifications, namely, Soundex [3], NYSIIS [4], Phonix [5], DMetaphone [6] and DM [7], and second, it explores different ways to combine word and code representations in order to find a reasonable tradeoff between precision and recall.

# 3   Proposed Method

As we previously mentioned, the proposed approach for SDR relies on the use of an expanded representation of automatic transcriptions which combines words and phonetic codes. The following subsections describe in detail two main issues regarding this approach: one the one hand, how to construct the expanded representation, and, on the other hand, how to use this representation through the retrieval process.

## 3.1   Constructing the Combined Representation

The construction of the expanded representation considers the following steps:

1. Remove unimportant tokens from transcriptions. Mainly, we consider eliminating a set of common stop words.

2. Compute the phonetic codification for each word from each transcription using a given codification algorithm. A general description and comparison of the codification algorithms used in our experiments can be found in [17], for further details we refer to [3, 4, 5, 6, 7].

3. Combine transcriptions and their phonetic codifications in order to form the expanded document representations. By this combination documents are represented by a mixed bag of words and phonetic codes. Correspondingly, queries need to be represented by their words and phonetic codes.

In order to clarify this procedure, Table 1 illustrates the construction of the expanded representation for the transcription segment "...just your early discussions was roll wallenberg's uh any recollection of of uh where he came from and so...", which belongs to the transcription (spoken document) with id=VHF31914-137755.013 from the CLEF CL-SR 2007 corpus.

**Table 1.** Example of an expanded document representation using Soundex codes

| Automatic transcription | ...just your early discussions was roll wallenberg uh any recollection of of uh where he came from... |
|---|---|
| Preprocessed transcription | ... early discussions roll wallenberg recollection came ... |
| Phonetic codification | ... E64000 D22520 R40000 W45162 R24235 C50000... |
| Expanded representation | {early, discussions, roll, wallenberg, recollection, came, E64000, D22520, R40000, W45162, R24235, C50000} |

## 3.2  Using the Combined Representation

Reports on the TREC's SDR track [14, 1] concluded that traditional word-based representations are good enough for SDR; however, they also indicated that this basic representation has difficulties to effectively handle complex queries, such as small queries or queries containing a large number of out-of-vocabulary (OOV) words.

On the other hand, [16] showed that phonetic codes help to improve retrieval recall but, due to the large number of word coalitions they generate, they tend to decrease precision rates.

Based on this previous evidence, we propose not to use the combined representation in all cases, but only to handle complex queries. We consider the following two criteria for determining –presumably– complex queries.

- *By query length*: a complex query has a length shorter than a given specified threshold.
- *By percentage of OOV words*: a complex query has a percentage of OOV words greater than a given specified threshold.

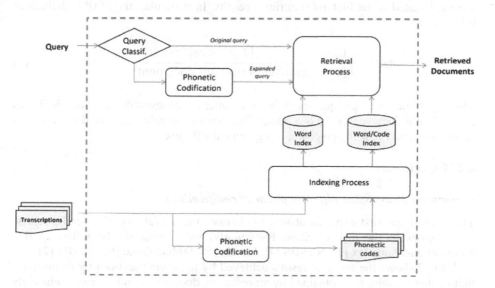

**Fig. 1.** Architecture of the proposed method

Figure 1 shows the general architecture of the proposed method. As noticed, it uses two different indexes, one based on words and other on the combination of words and phonetic codes. Both indexes are built offline using the whole document collection. It also includes a module for the online analysis of queries, which allows selecting presumably complex queries that require to be phonetically codified. Finally, it considers a retrieval module which uses the word index or the combined index depending on the form of the given question.

# 4 Evaluation

## 4.1 Experimental Setup

This section presents some experiments for evaluating the usefulness of the proposed representation. In all experiments, we used the training dataset from the CLEF CL-SR 2007 task [18]. This dataset includes 8,104 transcriptions of English interviews as well as a set of 63 queries.

It is important to mention that for each interview this collection provide three automatic transcriptions (having different word error rates) as well as some sets of automatically and manually extracted keywords. However, in order to get our experiment closer to a real scenario, we decided not to use any set of keywords and to consider only one automatic transcription, namely, the ASR06 with 25% word error rate (WER).

In all the experiments, indexing and retrieval was done by means of the Lemur search engine [19], which was configured to run as a traditional vector space model with tfxidf weights.

On the other hand, the evaluation was carried out using the *MAP* (mean average precision) and the number of relevant retrieved documents (*RelRET*). Both measures were calculated at the first 1000 retrieval results. In particular, the *MAP* is defined as follows:

$$MAP = \frac{1}{|Q|} \sum_{\forall q \in Q} \frac{\sum_{r=1}^{N}(P(r) \times rel(r))}{\text{number of relevant document}} \tag{1}$$

where $Q$ is the set of test queries, $N$ is the number of retrieved documents, $r$ indicates the rank of a document, $rel()$ is a binary function on the relevance the document at a given rank, and $P()$ is the precision at a given cut-off rank.

## 4.2 Experiments

### Experiment 1: assessing different phonetic codifications

The goal of our first experiment was to evaluate the usefulness of several phonetic codifications in the task of SDR. Particularly, we considered the following five codifications: Soundex [3], NYSIIS [4], Phonix [5], DMetaphone [6] and DM [7].

Table 2 shows the retrieval results achieved by these codifications by themselves; that is, these results were obtained by representing documents and queries exclusively by their phonetic codes. This table also shows the results corresponding to the traditional word-based indexing, which is our main baseline.

As expected, due to the generalization caused by the phonetic codifications, their results were worse than those achieved by the word-based indexing. In particular, word-based indexing improved by 5% the *MAP* obtained by the NYSIIS-based representation, which turned out to be the best phonetic-code representation.

An important finding was the number of relevant retrieved documents obtained by the NYSIIS-based representation. It got 1734 relevant documents for the 63 queries, outperforming by almost 10% the result from the word-based indexing. In addition, we noticed that these two representations (using words and NYSIIS-codes) are complementary, since they together may get 1820 relevant documents, and, therefore, they are good candidates for being combined.

**Table 2.** Results achieved by a phonetic-code-based indexing

| | | Indexed by | | | | |
|---|---|---|---|---|---|---|
| | Words | Soundex codes | NYSIIS codes | Phonix codes | DMetaphone codes | DM Codes |
| *MAP* | 0.062 | 0.051 | 0.059 | 0.047 | 0.037 | 0.038 |
| *RelRET* | 1578 | 1529 | 1734 | 1539 | 1567 | 1483 |

### Experiment 2: using the combined representation for all queries

As suggested by the previous results, we evaluated the effectiveness of applying a combined representation of words and NYSIIS-codes for handling all queries. Table 3 shows the results from this experiment.

**Table 3.** Results from the combined representation (used for handling all queries)

| | *MAP* | *RelRET* |
|---|---|---|
| Words | 0.062 | 1578 |
| Words + NYSIIS codes | 0.063 | 1701 |
| % of improvement over word-indexing | 1.6% | 7.8% |

The obtained results demonstrate the potential of the combined representation, which outperformed the *MAP* and *RelRET* of the traditional approach by 1.6% and 7.8% respectively. However, a deeper analysis showed us that the combined representation produced worse results than the word-based representation in more than a third part of the queries.

### Experiment 3: handling complex queries with the combined representation

The goal of this experiment was to validate the proposed method (refer to Section 3.2), which suggests not to use the combined representation in all cases, but only to handle complex queries. In particular, through this experiment we aimed to evaluate our two different criteria for selecting complex queries.

Table 4 shows the results from this experiment. The first two rows correspond to baseline results: word-based and combined representations. Then, there are the results achieved by applying the proposed combined representation to handle queries of length less than a given threshold. We used three different thresholds, 8, 11 and 14, which correspond to the average minus a standard deviation, the average and the average plus a standard deviation of the lengths from all training queries. Finally, the last three rows show the results obtained by using the proposed combined representation to handle queries having a percentage of OOV words greater than a given specified threshold. We used three different thresholds, 7%, 20% and 33%, which correspond to the average minus a standard deviation, the average and the average plus a standard deviation of the percentage of OOV words from all training queries.

The results from Table 4 once again indicate that using a combined representation is a better alternative than using the traditional word-based indexing. In particular, best results were obtained when the combined representation was used to manage short queries with length lesser than the average length. Using this configuration, the baseline *MAP* and *RelRET* were outperformed by 3.2% and 6.9% respectively.

One important conclusion from this experiment is that the selective use of the combined representation did not show a great advantage over its arbitrary usage, which may point to the necessity of better criteria for evaluating queries complexity.

**Table 4.** Results from the combined representation (used for handling only complex queries)

|  | MAP | RelRET |
| --- | --- | --- |
| Words | 0.062 | 1578 |
| Words + NYSIIS codes (used in all queries) | 0.063 | 1701 |
| Query length ≤ 8 | 0.062 | 1670 |
| Query length ≤ 11 | **0.064** | **1687** |
| Query length ≤ 14 | 0.064 | 1686 |
| % of OOV words ≤ 7% | 0.061 | 1660 |
| % of OOV words ≤ 20% | 0.063 | 1676 |
| % of OOV words ≤ 33% | 0.063 | 1674 |

# 5  Conclusions

In this paper we have proposed a retrieval method specially suited for SDR. This method relies on the idea of using word and phonetic-code based representations in collaboration in order to tackle the effects caused by the transcription errors.

One important contribution of this paper is the evaluation of the usefulness of five different phonetic codifications. Regarding this aspect, our results indicate that NYSIIS is the best phonetic codification for the SDR task. However, they also suggest that phonetic-code-based representations must be used in conjunction with traditional word-based indexing in order to be effective.

The second contribution of this paper is the analysis of the synergy between word and phonetic-code representations. Our results in that direction indicate that the combination of word and phonetic-code representations is relevant for SDR, since using this combination it was possible to outperform the baseline *MAP* and *RelRET* results by 3.2% and 6.9% respectively. Although the combined representation appeared to be more useful for handling short queries, the experimental results suggest that the selective use of the combined representation is not clearly superior to its arbitrary usage.

As future work we plan to explore the usage of the proposed combined representation at character n-gram level. In this way, we think it will be possible to carry away the word segmentation imposed by the ASR process, and, therefore, it will be easier to tackle the problems of word insertions and deletions.

**Acknowledgments.** This work was done under partial support of CONACYT (project grant CB-2008-106013-Y, and scholarship 204467). We would also like to thank the CLEF organizing committee for the resources provided.

# References

1. Garofolo, J.S., Auzanne, C.G.P., Voorhees, E.M.: The TREC Spoken Document Retrieval Track: A Success Story. In: NIST, pp. 107–129 (1999), Special publication 500-246
2. Comas, P.R., Turmo, J.: Spoken document retrieval based on approximated sequence alignment. In: Sojka, P., Horák, A., Kopeček, I., Pala, K. (eds.) TSD 2008. LNCS (LNAI), vol. 5246, pp. 285–292. Springer, Heidelberg (2008)
3. Odell, K.M., Russell, R.C.: Soundex phonetic comparison system. [U.S. Patents 1261167 (1918), and 1435663 (1922)]
4. Taft, R.L.: Name Search Techniques, Albany, New York: New York State Identification and Intelligence System. Technical Report, State of New York (1970)
5. Gadd, T.: PHONIX: The algorithm. In: Program: Automated Library and Information Systems, pp. 363–366 (1990)
6. Philips, L.: The double-metaphone search algorithm. C/C++ User's Journal 18(6) (2000)
7. Mokotoff, G., Sack, S.A.: Where once we walked: a guide to the Jewish communities destroyed in the Holocaust. Avotaynu, Teaneck (1991)
8. Whittaker, E.W.D., Van Thong, J.M., Moreno, P.J.: Vocabulary Independent Speech Recognition Using Particles, Trento, Italy (2001)
9. Siegler, M.: Integration of continuous speech recognition and information retrieval for mutually optimal performance. Ph.D. dissertation. Carnegie Mellon University, Carnegie Mellon (1999)
10. Ng, C., Wilkinson, R., Zobel, J.: Experiments in spoken document retrieval using phoneme N-grams, vol. 32(1-2), pp. 61–77. Elsevier Science Publishers B. V, Amsterdam (September 2000)
11. Zhang, L., et al.: Topic indexing of spoken documents based on optimized N-best approach, Shanghai, November 20-22, vol. 4, pp. 302–305 (2009)
12. Siegler, M., et al.: Experiments in Spoken Document Retrieval at CMU. National Institute for Standards and Technology, Gaithersburg (1997) NIST-SP 500-240
13. Nishizaki, H., Nakagawa, S.: Japanese spoken document retrieval considering OOV keywords using LVCSR system with OOV detection processing, pp. 157–164. Morgan Kaufmann Publishers Inc., San Diego (2002)
14. Allan, J.: Robust techniques for organizing and retrieving spoken documents, vol. 2003, pp. 103–114. Hindawi Publishing Corp., New York (January 2003)

15. Wang, J., Oard, D.W.: CLEF-2005 CL-SR at maryland: Document and query expansion using side collections and thesauri. In: Peters, C., Gey, F.C., Gonzalo, J., Müller, H., Jones, G.J.F., Kluck, M., Magnini, B., de Rijke, M., Giampiccolo, D. (eds.) CLEF 2005. LNCS, vol. 4022, pp. 800–809. Springer, Heidelberg (2006)
16. Alejandro Reyes-Barragán, M., Villaseñor-Pineda, L., Montes-y-Gómez, M.: A soundex-based approach for spoken document retrieval. In: Gelbukh, A., Morales, E.F. (eds.) MICAI 2008. LNCS (LNAI), vol. 5317, pp. 204–211. Springer, Heidelberg (2008)
17. Christen, P.: A Comparison of Personal Name Matching Techniques and Practical Issues. In: Proceedings of the Sixth IEEE International Conference on Data Mining (September 2006)
18. Pecina, P., Hoffmannová, P., Jones, G.J.F., Zhang, Y., Oard, D.W.: Overview of the CLEF-2007 cross-language speech retrieval track. In: Peters, C., Jijkoun, V., Mandl, T., Müller, H., Oard, D.W., Peñas, A., Petras, V., Santos, D. (eds.) CLEF 2007. LNCS, vol. 5152, pp. 674–686. Springer, Heidelberg (2008)
19. Ogilvie, P., Callan, J.: Experiments Using the Lemur Toolkit (2002)
20. Gálvez, C.: Identificación de Nombres Personales por Medio de Sistemas de Codificación Fonética. Encontros Bibli, Florianópolis, Santa Catarina, Brasil, vol. 11(22), pp. 105–116 (2006)

# Automatic Rule Extraction
# for Modeling Pronunciation Variation*

Zeeshan Ahmed and Julie Carson-Berndsen

CNGL, School of Computer Science and Informatics
University College Dublin, Ireland
zeeshan.ahmed@ucdconnect.ie, julie.berndsen@ucd.ie

**Abstract.** This paper describes the technique for automatic extraction of pronunciation rules from continuous speech corpus. The purpose of the work is to model pronunciation variation in phoneme based continuous speech recognition at language model level. In modeling pronunciation variations, morphological variations and out-of-vocabulary words problem are also implicitly modeled in the system. It is not possible to model these kind of variations using dictionary based approach in phoneme based automatic speech recognition. The variations are automatically learned from annotated continuous speech corpus. The corpus is first aligned, on the basis of phoneme and letter, using a dynamic string alignment algorithm. The DSA is applied to isolated words to deal with intra-word variations as well as to complete sentences in the corpus to deal with inter-word variations. The pronunciation rules *phonemes → letters* are extracted from these aligned speech units to build pronunciation model. The rules are finally fed to a phoneme-to-word decoder for recognition of the words having different pronunciations or that are OOV.

## 1 Introduction

Pronunciation variations and treatment of out-of-vocabulary (OOV) words in automatic speech recognition (ASR) have always emerged as one of the biggest drawbacks for an ASR system. Pronunciation variation can occur because of dialect, native and non-native speaker, age, gender, emotions, position of words in the utterance etc.

Pronunciation variations can be incorporated at different levels in ASR system as explained in [1]. There are three levels at which pronunciation variations can be modeled: the lexicon, the acoustic model, the language model. To deal with pronunciation variations at the lexicon level, different variants of word pronunciation are added to the lexicon. At the acoustic level, context dependent phone modeling [2,3] has been widely used to capture the phone variations

---

* This research is supported by the Science Foundation Ireland (Grant 07/CE/I1142) as part of the Center for Next Generation Localization (www.cngl.ie) at University College Dublin. The opinions, findings and conclusions or recommendations expressed in this material are those of the authors and do not necessarily reflect the views of Science Foundation Ireland.

A. Gelbukh (Ed.): CICLing 2011, Part II, LNCS 6609, pp. 467–476, 2011.

within particular contexts. There are still many problems with context dependent phone modeling for pronunciation variation as explained in [4]. At the language model level, the intra-word pronunciation variations are tackled with sophisticated phonotactic models [5,6,7,8] and the inter-word pronunciation variations are handled with grammar network or statistical language models [9,10].

The paper is concerned with modeling pronunciation variations at the language model level and targets a phoneme based continuous speech recognition (CSR) system. The motivation behind the work is to make ASR recognize large vocabulary as well as handle pronunciation variants of a word. Currently, the number of words recognized by a typical ASR system is limited to the number of words in a pronunciation dictionary. The approach presented separates the acoustic modeling and the linguistic modeling in an ASR. It restricts an acoustic engine of ASR to recognize only phonemes of a language. While, the responsibility of modeling words and sentences is delegated to the pronunciation modeling component presented in the paper. The ultimate objective of the pronunciation modeling is to convert the sequence of phonemes output by acoustic model of CSR system into proper words of the language, given that the phoneme sequence contains pronunciation variation, morphological variation and OOV words.

The work is the extension of the work presented in [11]. The technique presented in [11] is applicable to isolated words only. In this paper, the technique is further extended to deal with continuous speech. The dynamic string alignment (DSA) algorithm, pronunciation extraction and decoding techniques are thoroughly revised for continuous speech.

In this approach, the continuous speech corpus is used to learn pronunciation variations. The objective is to collect the *phoneme → letter* pronunciation rules with occurrence probabilities from the corpus. The probability specifies the likelihood of translating a sequence of *phonemes* into a sequence of *letters* in the corpus. The approach attempts to extract as much variety of rules as possible from the corpus in order to model different variations. The rules are fed to a phoneme-to-word decoder to identify correct words of the language. The advantage of the technique is its ability to predict new words (OOV) from extracted rules in addition to handling pronunciation variation.

To learn the rules, the continuous speech corpus is first aligned on the basis of phonemes and letters. The alignment is done automatically because these kind of alignments are usually not available in the speech corpus and manual alignment is difficult and time consuming. The alignment technique uses the DSA algorithm with limited linguistic knowledge related to pronunciation of a language. The alignment is performed on sentences as well as words in the corpus. The sentence alignment is done to capture inter-word pronunciation variations.

The aligned speech units are then analyzed to find out the *phoneme → letter* rules that can be used in word decoding. The pronunciation rules are extracted on the basis of manually developed extraction rules as explained in section 3.

The rest of the paper is structured as follows. The next section outlines phoneme/letter alignment. Section 3 explains how pronunciation rules are extracted from alignments. Section 4 represents main features of decoding process. Section 5 presents the results and conclusion is presented in section 6.

## 2   Phoneme/Letter Alignment

Automatic letter-to-phoneme alignment has been targeted from the perspective of speech synthesis for a long time. The techniques developed previously include pronunciation by analogy [12] , constraint satisfaction [13], Hidden Markov Models [14], decision trees [15], and neural networks [16]. The detailed discussion on letter-to-phoneme alignment can be found in [17].

All of these alignment techniques were developed from the perspective of speech synthesis and only for alignment within word. In this paper, the alignment problem is similar to letter-to-phoneme alignment but concerned with the other direction namely phoneme-to-letter. Furthermore, the sentences in continuous speech are also considered during alignment.

### 2.1   Dynamic String Alignment(DSA)

To align phonemes and letters, the DSA algorithm is employed. The DSA algorithm uses the dynamic programming (DP) [18] technique to optimally align two string sequences. The technique for DSA in this paper is similar to [17]. However, instead of using Expectation Maximization (EM) algorithm to derive letter-phoneme association matrix $A$, basic linguistic knowledge about the pronunciation of words is used. The scoring table is created manually based on the classification with respect to vowels and consonants for a language. For the case of English, the scoring table is shown in Table 1. Table 1 replaces the association matrix $A$ used by the Damper DSA algorithm [17].

**Table 1.** Alignment Scoring

| | |
|---|---|
| Exact Match | 4 |
| Vowel to Vowel | 4 |
| Consonants to Consonants (ambiguous mapping like c and k etc.) | 2 |
| No match | -2 |

In this table, *Exact Match* is the case when a phone like /p/ matches "p", /b/ matches "b", /k/ matches "k" etc. For vowel to vowel matching, the score is also 4. The score is 2 when ambiguous consonants like /k/ matches "c", /s/ matches "z", /f/ matches "ph" etc. If there is no match, the score is -2. There is also a penalty of -1 for horizontal and vertical movement in the alignment matrix. All of these scores have been obtained based on various experiments with different settings.

In most of the cases in English words, the number of letters are greater than their corresponding phonemic representation because some phones align to more than one letter. Due to this mismatch in number, dynamic alignment produces

appeared → ax p iy r d

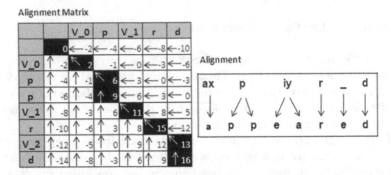

**Fig. 1.** Dynamic String Alignment of word ("appeared",/ax p iy r d/)

ambiguous results for these cases. During the experiment, it was found that the sequences are best aligned when number of phonemes are equal to number of letters. To overcome this problem, the contiguous sequence of vowels is treated as single vowel in both phoneme sequence and letter sequence during the alignment process. Once the sequences are aligned, vowels are restored in their original position.

Figure 1 shows the alignment matrix generated for aligning *abbreviate* with phoneme sequence /ax b r iy v iy ey t/. Before aligning these two strings, the strings are preprocessed by replacing the vowel sequence with a $V$ symbol. After the preprocessing, the two strings become $V_1bbrV_2vV_3tV_4$ and /$V_1$ b r $V_2$ v $V_3$ t/. The $V$ symbol represent the replaced vowel sequence. The subscript associated with $V$ specifies position of vowel sequence in the string. The phoneme sequence and letter sequence subscript are independent of each other. After the alignment has been generated the symbols are replaced with original values as shown in figure 1.

## 2.2    Alignment in Continuous Speech

The previous section presented the approach to align phonemes and letters within the word. In this section, this process is further extended to continuous speech to learn inter-word variation.

To align sentences in continuous speech, the phoneme and word annotations of sentences and time information are used. The sentence is segmented into words using the timing information. The alignment process is then applied on segmented words. After the word alignment, the aligned words are rejoined to form an aligned sentence. The word boundaries are preserved in the aligned sentence. Table 2 shows an annotated sentence in the speech corpus with the corresponding phoneme representation. The first part of the table shows the original sentence with transcription. The second part shows the segmented word with corresponding alignment generated from the process. The words were segmented using the

**Table 2.** Continuous Speech Alignment

| she had your dark suit in greasy wash water all year |
|---|
| sh ix hv eh d jh ih d ah k s ux q en g r ix s ix w ao sh w ao dx axr ao l y ih axr |

| Word | Pronunciation | Generated Alignment |
|---|---|---|
| she | sh ix | sh/sh ix/e |
| had | hv eh d jh | hv/h eh/a d_jh/d |
| your | ih | ih/y_o_u /r |
| dark | d ah k | d/d ah/a /r k/k |
| suit | s ux q | s/s u_i/ux q/t |
| in | en | e/i n/n |
| greasy | g r ix s ix | g/g r/r ix/e_a s/s ix/y |
| wash | w ao sh | w/w ao/a sh/sh |
| water | w ao dx axr | w/w ao/a dx/t ax/e r/r |
| all | ao l | ao/a l/l_l |
| year | y ih axr | y_ih_ax/y_e_a r/r |

| Aligned Sentence |
|---|
| $/$ sh/sh ix/e #/# hv/h eh/a d_jh/d #/# ih/y_o_u /r #/# d/d ah/a /r k/k #/# s/s ux/u_i q/t #/# e/i n/n #/# g/g r/r ix/e_a s/s ix/y #/# w/w ao/a sh/sh #/# w/w ao/a dx/t ax/e r/r #/# ao/a l/l_l #/# y_ih_ax/y_e_a r/r $/$ |

corpus annotation. The last row of the table shows the sentence which is finally aligned. In the aligned sentence '#/#' are used to keep the word boundaries and '$/$' mark the start and end of the sentence.

# 3 Pronunciation Extraction

The pronunciation model is based on the following Bayes' Theorem

$$Pr(L|P) = \frac{Pr(L) * Pr(P|L)}{Pr(P)} \tag{1}$$

where, $L$ stands for a letter or sequence of letters and $P$ stands for a phoneme or sequence of phonemes. The goal is to maximize the posterior probability i.e $Pr(L|P)$ given the prior probability $Pr(P|L)$ and word model probability $Pr(L)$. Here, the prior probability $Pr(P|L)$ is called pronunciation model. The model is calculated using the probability estimation technique on the corpus as follow.

$$Pr(P|L) = \frac{freq(P, L)}{freq(L)} \tag{2}$$

The pronunciation extraction process uses the alignment generated in the last step to estimate pronunciation model. The process analyzes the generated alignment and extracts the *phoneme* → *letter* probabilistic pronunciation rules based on different cases. The pronunciation rules can be either one-to-one, one-to-many, many-to-one or many-to-many. Some of the rules are also comprised of contextual information to model variations at word boundaries. As a result of the

**Table 3.** Extracted Pronunciation Rules

| Phoneme | Letter | Probability |
|---------|--------|-------------|
| ah | a | 1.00 |
| ao | a | 1.00 |
| d | d | 1.00 |
| dx | t | 1.00 |
| e | e | 0.50 |
| e | i | 0.50 |
| eh | a | 1.00 |
| g | g | 1.00 |
| hv | h | 1.00 |
| ix | e | 0.33 |
| ix | y | 0.33 |
| k | k | 1.00 |
| n | n | 1.00 |
| q | t | 1.00 |
| r | r | 1.00 |
| s | s | 1.00 |
| sh | sh | 1.00 |
| w | w | 1.00 |

a: Case 1

| Phoneme | Letter | Probability |
|---------|--------|-------------|
| ix | e a | 0.33 |
| ux | u i | 1.00 |
| y ih e | y e a | 1.00 |

b: Case 2

| Phoneme | Letter | Probability |
|---------|--------|-------------|
| l | l l | 1.00 |

c: Case 3

| Phoneme | Letter | Probability |
|---------|--------|-------------|
| ah k | a r k | 1.00 |
| eh d jh <R> ih | a d | 1.00 |
| jh <L> ih <R> d | y o u r | 1.00 |

d: Case 4 and 5.

pronunciation extraction process, a pronunciation table is generated. Each row of the table is a triple $< P, L, prob >$, which is defined as the phoneme sequence $P$ is translated to letter sequence $L$ with $prob$ probability in training corpus. The probability is used for scoring the most like recognized words during recognition process. Following are the cases used for pronunciation rule extractions.

**Case 1:** If the alignment is one-to-one i.e single phoneme aligns to single letter, and the alignment score is greater than or equal to 2 according to Table 1, the rule is included in the pronunciation table. Table 3(a) shows the rules extracted as result of applying this case to the alignment in Table 2.

**Case 2:** If the alignment is not one-to-one and contains vowel(s) on both phoneme and letter sequences then alignment is included in the pronunciation table. The Table 3(b) shows the extracted rules from the alignment in Table 2.

**Case 3:** If the alignment is double letter alignment like $/p/$ aligning to $pp$ as in $(apple, /ae\ p\ el/)$ then alignment is included in pronunciation table. Table 2(c) shows the result of applying this case to the alignment in Table 2.

**Case 4:** If the alignment does not fall into any of the above cases, it is considered ambiguous. The ambiguity is resolved by joining left and right alignments into the rule. If the left and right alignments are also ambiguous then the rule is further extended to the left and right. The alignment with left and right alignment is included in the pronunciation table as one rule.

**Case 5:** This case is the extension of case 4. If the left or right alignment is the word boundary in the aligned sentence, the left and right alignment are not consider the part of the rule. Instead, the left and right alignment are considered as context in phoneme sequence. The alignment is included in the pronunciation table with left and right phoneme as context. Table 2(d) shows the result of applying case 4 and case 5 to the alignment in Table 2. The <L> and <R> symbols denote the left and right context. The rule

$$jh < L > ih < R > d \rightarrow your$$

is stated as; *ih* is translated to *your*, when the left of *ih* is *jh* and right is *d* in phoneme sequence.

## 4   Word Decoding

For evaluation of the pronunciation model, we use our own custom built decoder for recognition of words. The decoding technique is an extension of the isolated word decoding as described in [11]. The decoder uses the pronunciation model (as outlined in this paper), a word model for modeling spellings of the words and a language model to recognize the words in continuous speech.

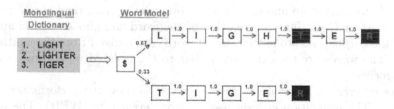

**Fig. 2.** Word model example for three-word dictionary

The word model is in the form of a probabilistic finite state machine (FSM). This model is like the prefix tree representation of the word spelling as shown for the three-word English dictionary in Figure 2. This model is employed so that only valid words of the language are recognized. Here, the letters of the word form states in the FSM and the transitions between these letters represent next possible letters in the word. All the words which have the same prefix have the same previous states as shown in Figure 2. There is a default start state $ for each word. The accepting states are highlighted in dark. The word model is not tightly integrated with pronunciation rules of words which facilitates the freedom of adding new words (without pronunciation) to the vocabulary on the fly.

The decoder has been integrated with SRILM language modeling toolkit [9] and all the manipulation of the language model parameters are delegated to SRILM. The decoding approach applied is similar to stack-based decoding [19]. Two stacks are used for the purpose of decoding: one for recognition of words from the phoneme sequence and another for recognition of the sentence from the words. The detailed discussion of decoding approach is beyond the scope of this paper. However, further details of decoding for the case of isolated words can be found in [11].

# 5   Evaluation

The model was evaluated on TIMIT continuous speech corpus [20]. TIMIT contains a total of 6300 sentences, 10 sentences spoken by each of 630 speakers from 8 major dialect regions of the United States. The corpus consists of three parts; the speech wave forms, the annotation of continuous speech (word level and phoneme level) and the pronunciation dictionary. The paper is concerned only with the annotation of the speech and pronunciation dictionary.

The corpus is divided into a training set and a test set. The training set is used for learning the pronunciation rules and results are reported on the test set. The pronunciation dictionary is used for building an FSM for modeling word spellings. The complete corpus text is used to build the language model. The language model is the standard SRILM tri-gram language model with Good-Turing discounting and Katz backoff for smoothing.

The pronunciation dictionary is used to learn the intra-word pronunciation variations of isolated words. The annotations of the continuous speech are used to learn both intra-word and inter-word pronunciation variations. From the duration (start and end time) information of the word and phoneme level annotation of the corpus, the pronunciations of words are derived from the continuous speech. The words are then treated similar to the isolated words during alignment process.

Three different experiments were carried out to show the performance of the approach. The evaluation criteria used is word error rate (WER). The results were calculated on recognized output using NIST Speech Recognition Scoring Toolkit (SCTK)[1]. In the first experiment, only the pronunciation dictionary is used to convert the phonemes into words. In the second experiment, the pronunciation dictionary is extended with words extracted from continuous speech corpus annotation. In the final experiment, pronunciation extraction process is applied on both pronunciation dictionary as well as extracted words to obtain pronunciation rules. The rules are then used for recognition instead of the pronunciation dictionary and extracted word pronunciations.

Due to the underspecification of pronunciation rules, some pronunciation rules are explicitly added to the model. During the rule learning process, it was found that vowels are very confusing i.e any vowel in set of phonemes of a language can be translated to any vowel in set of letters of a language. Therefore, during the learning phase, if any combination of *phonemevowel* → *lettervowel* rule is not present in the corpus, it is introduced to the pronunciation model.

The results in Table 4 show that the dictionary alone is not able to recognize the words in the test set. This is due to the fact that test set contains many

---

[1] http://www.itl.nist.gov/iad/mig/tools/

**Table 4.** Experimental Results

| S.No | Experiment | Results | | | | | |
|---|---|---|---|---|---|---|---|
| | | Cor. | Sub. | Del. | Ins. | W.E.R | Sent. Err. |
| 1 | Dictionary | 0.1 | 0.0 | 99.9 | 0.0 | 99.9 | 100.0 |
| 2 | Extended Dictionary | 15.6 | 4.1 | 80.3 | 7.8 | 92.2 | 100.0 |
| 3 | Pronunciation Rules | 73.3 | 18.1 | 8.5 | 10.0 | 36.7 | 64.9 |

pronunciation variations. The dictionary extended with pronunciations extracted from training set is also not sufficient because variations of the words found in the training set do not cover test set variations. However, the pronunciation rule approach is very promising because it caters for pronunciation variations in the test set effectively by learning pronunciation rules from training set.

# 6   Conclusion

The paper presented an approach to modeling pronunciation variation in continuous speech. The approach uses a DSA technique to align letter and phoneme in continuous speech. The pronunciation rule extractor uses these alignments to extract pronunciation rules from aligned units. The pronunciation rules are then fed to a phoneme-to-word decoder for recognition. The decoder uses a pronunciation model, a word model and a statistical language model for decoding.

The technique for phoneme-to-letter conversion presented in this paper is quite novel. Most importantly, it is independent of the underlying acoustic model thus making the acoustic engine robust by reducing the recognition set to the phonemes of the language rather than words. The technique is data-driven which attempts to induce rules from training set to model pronunciation variation, morphological variation and OOVs. The technique is scalable where new words (without pronunciation) can be added to the dictionary on the fly. Currently, error introduced by acoustic engine is not explicitly handled in the approach . In future, we plan to address this issue.

# References

1. Strik, H., Cucchiarini, C.: Modeling pronunciation variation for ASR: A survey of the literature, pp. 225–246 (1999)
2. Bahl, L.R., Souza, P.V.D., Gopalakrishnan, P.S., Nahamoo, D., Picheny, M.A.: Context dependent modelling of phones in continuous speech using decision trees. In: Proceedings DARPA Speech and Natural Language Processing Workshop, pp. 264–270 (1991)
3. Odell, J.J.: The Use of Context in Large Vocabulary Speech Recognition. PhD thesis, University of Cambrige,UK (1995)
4. Jurafsky, D., Ward, W., Jianping, Z., Herold, K., Xiuyang, Y., Sen, Z.: What kind of pronunciation variation is hard for triphones to model? In: Proc. ICASSP, pp. 577–580 (2001)

5. Safra, S., Lehtinen, G., Huber, K.: Modeling pronunciation variations and coarticulation in CSR. In: Proc. ESCA Workshop Modeling Pronunciation Variation Automatic Speech Recognition, pp. 125–130 (1998)
6. Cremelie, N., Martens, J.P. In search of pronunciation rules. In: Proc. ESCA Workshop Modeling Pronunciation Variation Automatic Speech Recognition, pp. 23–28 (1998)
7. Creutz, M., Hirsimäki, T., Kurimo, M., Puurula, A., Pylkkönen, J., Siivola, V., Varjokallio, M., Arisoy, E., Saraçlar, M., Stolcke, A.: Morph-based speech recognition and modeling of out-of-vocabulary words across languages. ACM Trans. Speech Lang. Process. 5, 1–29 (2007)
8. Kelly, R.: Learning Multitape Finite-State Machines from Mulilevel Annotations of Speech. PhD thesis, University College Dublin, Ireland (2005)
9. Stolcke, A.: Srilm - an extensible language modeling toolkit. In: International Conference on Spoken Language Processing, Denver, Colorado (2002)
10. Clarkson, P., Rosenfeld, R.: Statistical language modeling using the CMU-cambridge toolkit. In: Proceedings of the 5th European Conference on Speech Communication and Technology (Eurospeech), pp. 799–802 (1997)
11. Ahmed, Z., Carson-Berndsen, J.: Modeling pronunciation of OOV words for speech recognition. In: Thirteenth Australasian International Conference on Speech Science and Technology, Melbourne, Australia (2010)
12. Marchand, Y., Damper, R.I.: A multistrategy approach to improving pronunciation by analogy. Computational Linguistics 26, 195–219 (2000)
13. van den Bosch, A., Canisius, S.: Improved morpho-phonological sequence processing with constraint satisfaction inference. In: SIGPHON 2006: Proceedings of the 8th Meeting of the ACL Special Interest Group on Computational Phonology and Morphology, Morristown, NJ, USA, pp. 41–49 (2006)
14. Taylor, P.: Hidden markov models for grapheme to phoneme conversion. In: Proceedings of the 9th European Conference on Speech Communication and Technology (2005)
15. Black, A.W., Lenzo, K., Pagel, V.: Issues in building general letter to sound rules. In: The Third ESCA Workshop in Speech Synthesis, pp. 77–80 (1998)
16. Sejnowski, T.J., Rosenberg, C.R.: Parallel networks that learn to pronounce english text. Complex System, 145–168 (1987)
17. Damper, R.I., Marchand, Y., Marsters, J.D., Bazin, A.: Aligning letters and phonemes for speech synthesis. In: 5th ISCA Speech Synthesis Workshop (2004)
18. Bellman, R.E.: Dynamic Programming. Princeton University, Princeton (1957)
19. Jelinek, F.: A fast sequential decoding algorithm using a stack. IBM Journal of Research and Development 13, 675–685 (1969)
20. Garofolo, J.S., Lamel, L.F., Fisher, W.M., Fiscus, J.G., Pallett, D.S., Dahlgren, N.L.: DARPA TIMIT Acoustic Phonetic Continuous Speech Corpus CDROM (1993)

# Predicting Word Pronunciation in Japanese

Jun Hatori[1,*] and Hisami Suzuki[2]

[1] Department of Computer Science, University of Tokyo
7-3-1 Hongo, Bunkyo, Tokyo, 113-0033 Japan
`hatori@is.s.u-tokyo.ac.jp`
[2] Microsoft Research
One Microsoft Way, Redmond, WA 98052, USA
`hisamis@microsoft.com`

**Abstract.** This paper addresses the problem of predicting the pronunciation of Japanese words, especially those that are newly created and therefore not in the dictionary. This is an important task for many applications including text-to-speech and text input method, and is also challenging, because Japanese kanji (ideographic) characters typically have multiple possible pronunciations. We approach this problem by considering it as a simplified machine translation/transliteration task, and propose a solution that takes advantage of the recent technologies developed for machine translation and transliteration research. More specifically, we divide the problem into two subtasks: (1) Discovering the pronunciation of new words or those words that are difficult to pronounce by mining unannotated text, much like the creation of a bilingual dictionary using the web; (2) Building a decoder for the task of pronunciation prediction, for which we apply the state-of-the-art discriminative substring-based approach. Our experimental results show that our classifier for validating the word-pronunciation pairs harvested from unannotated text achieves over 98% precision and recall. On the pronunciation prediction task of unseen words, our decoder achieves over 70% accuracy, which significantly improves over the previously proposed models.

**Keywords:** Japanese language, pronunciation prediction, substring-based transliteration, letter-to-phone.

## 1  Introduction

This paper explores the problem of assigning pronunciation to words, especially when they are new and therefore not in the dictionary. The task is naturally important for the text-to-speech application [27], and has been researched in that context as letter-to-phoneme conversion, which converts an orthographic character sequence into phonemes. In addition to speech applications, the task is also crucial for those languages that require pronunciation-to-character conversion to input text, such as Chinese and Japanese, where users generally type in the pronunciations of words,

* This work was conducted during the first author's internship at Microsoft Research.

A. Gelbukh (Ed.): CICLing 2011, Part II, LNCS 6609, pp. 477–492, 2011.
© Springer-Verlag Berlin Heidelberg 2011

which are then converted into the desired character string via the software application called pinyin-to-character or kana-kanji conversion (e.g. [8] [9]).

Predicting the pronunciation of words is particularly challenging for Japanese. Japanese orthography employs four sets of characters: *hiragana* and *katakana*, which are syllabary systems thus phonemic; *kanji*, which is ideographic and represents morphemes, and Roman alphabet. Kanji characters typically have multiple possible pronunciations, making the prediction of their pronunciation difficult. In many cases, you need to know the word to know its pronunciation: after all, the pronunciation is an idiosyncratic property of the word. Therefore, one goal of this paper is to propose an effective method for exploring textual resources to learn the pronunciation of words. At the same time, we are also motivated to find out how predictable the pronunciations of kanji words are. Native speakers of the language can take an educated guess at predicting a pronunciation of an unseen word; can a machine replicate such sophisticated performance?

Our approach to the problem of pronunciation prediction therefore consists of two parts: we first try to model the intuition that a fluent speaker has on how to pronounce words by a statistical model via the task of *pronunciation modeling*; we then use the model to harvest word-pronunciation pairs from the web in the task of *pronunciation acquisition*. In this paper, the pronunciation modeling task is considered as a simplified machine translation (MT) task, i.e., a substring-based monotone translation, inspired by recent work on string transduction research. Our model, trained discriminatively using the features that proved useful in related tasks, outperforms a strong baseline as well as an average human performance, while making the types of errors that are considered acceptable by human. For the pronunciation acquisition task, we use a classifier to validate word-pronunciation pairs extracted automatically from text, exploiting the convention of Japanese text that the pronunciation is often inserted in parentheses immediately following the word with a difficult or unusual pronunciation. Our classifier achieves over 98% precision and recall when Wikipedia was used as the source corpus.

There are several contributions of this paper. We believe that this is the first work to address the problem of word pronunciation prediction for Japanese in a comprehensive manner. We apply the state-of-the-art technology developed for related problems to solve this problem, with modifications that are motivated by the specific problem at hand. The use of unannotated corpus for the extraction of pronunciation in Japanese is also novel and proved effective.

The rest of the paper is organized as follows. Section 2 gives some background, including the task description and related work. Section 3 introduces our approach to the pronunciation modeling task, along with experimental results. Section 4 deals with the task of pronunciation acquisition from corpora, which takes advantage of the prediction model described in Section 3. We conclude with comments on future work in Section 5.

# 2 Background

## 2.1 Pronunciation Prediction: Task Description

We define the task of pronunciation prediction as converting a string of orthographic characters representing a word (or a phrase corresponding to an entity) into a sequence of hiragana, which straightforwardly maps to pronunciation.[1] The problem is trivial if the word is spelled entirely in non-kanji characters, so we only target the cases where at least one character in the word is spelled in kanji. Let us take an example of the name of the recently appointed prime minister of Japan, Naoto Kan (菅直人). Our goal is to convert this string intoかんなおと, which is pronounced as [ka-N-na-o-to].[2] How ambiguous is this name to pronounce? According to the kanji pronunciation dictionary we have, the first character has three pronunciations, the second fourteen and the third twelve:[3] therefore, there are $3 \times 14 \times 12 = 504$ possible ways to pronounce this word. Naturally, some pronunciations are more common than others, especially given some contextual information. For example, 直人 is a common first name, pronounced as [nao-to] or [nao-hito] or maybe [tada-hito]; other pronunciations are highly unusual. Given that直人 is probably a first name, 菅 may be a last name, pronounced as [kan] or [suga], though it is fairly uncommon as a last name. Kanji characters typically have two types of pronunciations called *on-yomi* (literally 'sound pronuncation') and *kun-yomi* (literally 'meaning pronunciation'), corresponding to their origin (Chinese and Japanese, respectively), and they tend not to mix within a word, exemplified in 運転手 ([uN-teN-shu] 'driver', all on-yomi) vs. 手紙 ([te-gami] 'letter', all kun-yomi). Using these types of knowledge, one might guess that the name is reasonably pronounced as [kaN-nao-hito], [kaN-nao-to], [suga-tada-hito] and so forth. Eventually, the correct pronunciation can only be obtained by knowing the word, i.e., by identifying this string as a dictionary entry. The problems we try to solve in this paper is therefore twofold: one is to increase the dictionary coverage by learning word-pronunciation pairs automatically from text through *pronunciation acquisition*, secondly, for those words for which a dictionary entry is still missing, we would like to build a model to predict pronunciation that is not only highly accurate, but also makes reasonable mistakes when it fails – using the直人 example above, we hope to generate one of the three reasonable pronunciations. We focus on the task of predicting *word* pronunciation in this paper – selecting the right pronunciation for the words in a sentence is a related but independent task of *pronunciation disambiguation*, for which the pronunciation prediction task discussed in this paper will serve as an essential component.

## 2.2 Related Work

The task of pronunciation prediction is inspired by previous research on string transduction. The most directly relevant one is the work on letter-to-phoneme

---

[1] To be precise, additional operations are required to adjust the hiragana string for speech or text input applications, but we do not deal with this problem here.

[2] A hyphen is used to indicate a character boundary of the preceding string; [N] is used to indicate the pronunciation of the moraic nasal ん.

[3] This kanji pronunciation dictionary was available to us prior to the current research. It lists the pronunciations for about 6,000 kanji characters, with 2.5 pronunciations on average per character. The possible pronunciations for the three letters here are: 菅（すが,すげ,かん）, 直（ちょっ,すなお,す,ただし,ちょく,ね,ひた,ただ,のう,じき,なお,すぐ,じか,なおし）, 人（ひと,じん,と,り,たり,ど,にん,びと,うど,ぴと,うと,とな）.

conversion, where many approaches have been proposed for a variety of languages. The methods include joint n-gram models (e.g. [1] [2] [4]), discriminatively trained substring-based models (e.g. [11] [12]) which are themselves influenced by the phrasal statistical MT (SMT) models [15], and minimum description length-based methods [24]. The joint n-gram estimation method has also been applied to predicting pronunciation in Japanese (e.g., [21] [22]).

Similar techniques to the letter-to-phoneme task have also been applied to transliteration, which converts the words in one language into another that uses a different script, maintaining phonetic similarity. Early works on this task used the source-channel model based on one-to-one (or more) character alignment (e.g. [14]). Later they were extended to use many-to-many alignments using substring operations in the style of phrasal SMT (e.g. [28]), demonstrating improved accuracy over the character-based models. The components of the model proposed by [28] are themselves generative models, which can also be used in a SMT-style discriminative framework, where the weights on the component generative models are discriminatively trained. [5] proposed such a hybrid model, further improving the accuracy of transliteration. Joint n-gram models have also been applied to the task of transliteration (e.g. [17]).

In contrast to the wealth of literature in string transduction research, the task of pronunciation acquisition has attracted much less attention in the past. [10] describes a method in which they learn English pronunciations from the web using IPA (e.g., 'beet /bit/') and ad-hoc (e.g., 'bruschetta (pronounced broo-SKET-uh)') transcriptions by first extracting candidate pairs using a letter-to-phoneme model, which are then validated using SVM classifiers. Our approach is similar to theirs, with modifications in the method of generating candidates, to be explained in Section 4. [29] proposed a method to use the web for assigning word pronunciation in Japanese, but their focus is on disambiguating known word pronunciations rather than learning new word-pronunciation pairs. [16] and [26] discuss the methods of disambiguating new word pronunciation using speech data.

# 3  Substring-Based Pronunciation Prediction

This section describes our substring-based approach to pronunciation modeling. As mentioned above, the pronunciation of a kanji is dependent on those of the surrounding characters, which motivates a substring-based alignment and decoding over a character-based approach. We also assume that the task is basically monotone and without insertion/deletion, with kanji–hiragana alignments of 1–n (source–target, $n \geq 1$) characters.[4] We adopt a discriminative learning framework that uses

---

[4] This is a slight oversimplification – we are aware of the cases where these assumptions do not hold. The monotonicity assumption breaks in the pronunciation of a kanji sequence that reflects the Chinese SVO word order, as in 不弓引 [yumi-hika-zu] (a place name, which originally means 'not draw a bow', in which the correct alignment is assumed to be 不-zu (not), 弓-yumi (bow) and 引-hika (draw). Also, hiragana insertions occurs quite commonly as in 一関 [ichi-no-seki] (a place name, meaning 'first checkpoint'), where 'no' is a genitive marker inserted between the kanji characters.

component generative models as real-valued features, which is the standard method for statistical MT [23], and is reported to work comparably or better on a transliteration task than a discriminative model that uses sparse indicator features [5].

## 3.1 Model and Features

We adopt a linear model of pronunciation prediction: given the target character (hiragana) sequence $t$ and the source (kanji) sequence $s$, we define features over $s$ and $t$, $f_i(s, t)$ for $i = 1, \cdots, n$ The features are arbitrary functions that map $< s, t >$ to real values, and the model parameters are a vector of $n$ feature weights, $\lambda = (\lambda_1, \cdots, \lambda_n)$. The score of $t$ with respect to $s$ is given by

$$\text{Score}(s, t, \lambda) = \lambda \cdot \mathbf{f}(s, t) = \sum_{i=1}^{n} \lambda_i f_i(s, t).$$

For the features, we use those that are motivated by MT and transliteration research: the translation probabilities in both directions, $P(t|s)$ and $P(s|t)$, the target character language model probability $P(t)$, the operation count, which corresponds to the number of phrases in phrasal SMT, and the ratio of the source and target character length. Crucially, the estimation of the first three of these probabilities requires a set of training corpus with source and target alignment at the substring level. We take an unsupervised approach in generating such training data: we used an automatic word aligner developed for MT for obtaining these alignments, as detailed in Section 3.3 below.

## 3.2 Training and Decoding

For the training of the parameters of the linear model, we used averaged perceptron training. Let $d$ stand for a *derivation* that describes a substring operation sequence converting $s$ into $t$. Given a training corpus of such derivations $D = \{d_i, \cdots, d_n\}$ obtained from the substring-aligned text, the perceptron iterates the following two steps for each training sample $d_i \in D$:

$$\text{Decode: } d^* = \underset{d \in D(src(d_i))}{\text{argmax}} \ \lambda \cdot \mathbf{f}(d)$$
$$\text{Update: } \lambda \leftarrow \lambda + \mathbf{f}(d_i) - f(d^*),$$

where $D(src(d))$ are all possible derivations with the same source side as d. For decoding, we used a monotone phrasal decoder similar to the one employed in phrasal SMT [31], a stack decoder with the beam size of 20, which was set using a development data.

## 3.3 Experiments

**Data and settings.** As mentioned above, we need a parallel data of kanji words with their pronunciation in our approach. An obvious source of such data is a dictionary: we used UniDic [6], a resource available for research purposes, which is updated on the regular basis and includes 625K word forms as of the version 1.3.12 release

(July 2009). Since we focus on the prediction of new words which are mostly nouns, we used the noun (including proper noun) portion of the dictionary, containing 195K words in total.

Though UniDic is a lexical resource that is constantly refreshed, we also investigated into a dictionary-free approach, where we exploit a large body of unannotated text to collect words' pronunciation. Specifically, our approach takes advantage of the convention of Japanese text that the pronunciation of those words that are difficult or unusual to pronounce [5] are often indicated in parentheses immediately following the word in question, as shown in Figure 1.

---

新潟県（にいがたけん）は、本州日本海側に位置する
旧国名から越佐（えっさ）と表現することもある。
ふたご座流星群（ふたござりゅうせいぐん、学名 Geminids）は…
名取市立館腰（たてこし）小学校
一力亭（うどん）

---

**Fig. 1.** Examples of parenthetical pronunciation expression from Wikipedia. Strings in boldface indicate the words corresponding to the pronunciation in parentheses; the regular expression (described below) extracts the underlined substrings.

We used a simple regular expression-based pattern matching to extract word-pronunciation candidate pairs from Japanese Wikipedia. It extracts a substring of hiragana characters in a pair of parentheses, preceded by any character string bounded by a punctuation character or a beginning of a sentence. Additional heuristics consist of the constraints based on kana characters (i.e. no kana character is allowed in the word string unless it also appears in the pronunciation string.) and length ratio (e.g. the pronunciation string cannot be shorter than the word string. [6]). Note that the extraction method runs the risk of extracting too much pre-parenthetical material: as seen in the second to last example of Figure 1, たてこし indicates the pronunciation of only the last two characters (館腰). Another more substantial source of noise comes from the cases where the hiragana characters in parentheses do not indicate the pronunciation at all, as in the last example of Figure 1: 一力亭 [ichi-riki-tei] is a name of a restaurant, followed by the kind of food they serve (うどん [u-do-N] 'noodle') which happens to be written in hiragana. Though the extracted word-pronunciation data is therefore quite noisy, we will demonstrate that the use of this data greatly enhances the accuracy of the prediction. Note that in spite of the use of simple heuristics, the annotator found that more than 90% of extracted instances are valid word-pronunciation pairs (as mentioned in the last paragraph of Section 3.4), while the heuristics were weak enough to cover most pronunciation candidates in Wikipedia.

---

[5] This contrasts with the dictionary, where the pronunciations of all words are found. As is explained below, we used Wikipedia, which is a cross between a dictionary and free text: pronunciations are always given in parentheses for each title word, in addition to the words that occur in the free text portion of the articles.

[6] This is because a kanji character normally corresponds to one or more hiragana characters. While we are aware of some exceptional cases in non-compositional pronunciation, as in 啄木鳥 [kera] 'woodpecker', they are negligibly rare (<10 cases in 195K nouns in UniDic.).

The parallel data extracted from Wikipedia in this manner as well as from the UniDic entries is then aligned at the substring level. Our method for this follows [5]: we use a phrase-based word aligner originally developed for MT, similar to the word aligner described in [32], by considering each character as a word. We also used hard substring length limits for the same purpose: 1 for the input and 4 for the output strings, reflecting the fact that word pronunciation is typically composed of the pronunciation of individual kanji characters.[7] The aligner generates only monotonic alignments, and does not allow alignments to a null symbol in either source or target side. The same restriction is applied during decoding as well.

We extracted a total of 463,507 word-pronunciation pairs from Japanese Wikipedia articles as of January 24, 2010. After removing duplicates, we reserved 5,000 pairs for development and testing (of which we used 200 for development and 2,000 for final evaluation), and used the rest for training, i.e., for generating training derivations upon which the features of the linear model were computed. The translation probabilities, $P(s|t)$ and $P(t|s)$, are estimated by maximum likelihood on the operations observed in the training corpus with one important modification: recall that these operations, estimated using the character aligner in an unsupervised manner, are minimal non-decomposable operations, and therefore does not capture any contextual information. In order to remedy this, we re-align the training data by using composed operations which are constructed from operation sequences attested in the training data to maximize $P(s|t)$ and $P(t|s)$, respectively, thereby removing the substring length limit employed in the character alignment phase.[8] Figure 2 shows an example of an alignment before and after the composition. This process offers an additional benefit of noise reduction of the training data, as we removed the operations that occurred less than $C$ times ($C$ is set using the development data, $C=2$ in our case), removing the training examples that are not reachable from the remaining operations. This reduced our data size for perceptron training to 427,644 pairs, a reduction of 6.7%. More detail on the relation between the data size and the accuracy of the prediction task is discussed in the next subsection. For the target character language

**Fig. 2.** Alignment before (=character level indicated by the lines) and after (=substring level by the boxes) composition for 益子祇園祭 ([mashi-ko-gi-on-matsuri], 'Mashiko Gion Festival')

---

[7] There are exceptions to this: occasionally, a pronunciation is assigned to a kanji string in a non-compositional manner (e.g., 今日 [kyou] 'today') . This is handled by the use of composed operations, to be explained below.

[8] We do however impose a limit on the length of composition (3 in our case). This two-path alignment approach also follows [5]. Since the phrase length limit for original operations are 1 for the source and 4 for the target, resulting composed operations can capture up to 3–12 (source–target) character alignments.

model, we used a 4-gram language model with Kneser-Ney smoothing and the BOS (beginning-of-string) and EOS (end-of-string) symbols, and trained it with the same training data as described above.

**Baseline.** We describe two baseline models that we used for comparison in the experiment. The first is KyTea, a publicly-available Japanese word segmentation and pronunciation prediction tool,[9] which achieves the state-of-the-art performance on the task of Japanese pronunciation prediction. According to [20] and the KyTea manual, the program first performs word segmentation, after which the pronunciation of each word is independently selected using a linear SVM classifier, choosing among the pronunciations that have appeared in the training data. When they encounter an unknown word, the output is the combination of the most frequent pronunciations of each kanji character. We ran KyTea (Version 0.11) with the default settings and with "the high-performance SVM model" available from the website, which is mainly trained on the Balanced Corpus of Contemporary Written Japanese (BCCWJ; [19]) and UniDic.

Our second baseline, the joint n-gram model, was proposed by [1], which has also been used for Japanese [21]. In this model, n-gram statistics are learned over the sequences of *pairs* of letters and phonemes, instead of the sequences of phonemes. While [21] used KyTea to extract word-pronunciation pairs from the annotated BCCWJ and newswire corpus to learn bigram statistics, we learned our n-gram statistics from the alignments obtained from the Wikipedia training set as described above. Note that even though we describe this approach as a baseline, it crucially relies on the paired substrings extracted in an unsupervised manner using the proposed approach. In that sense, the effectiveness of this baseline also incorporates a novel contribution of this work. We implemented the joint trigram model with Kneser-Ney smoothing, after adding the BOS (beginning-of-string) and EOS (end-of-string) symbols. The decoder performs the exact inference with Viterbi search over the probability space.

### 3.4 Results and Discussion

Table 1 shows the comparison of the proposed method against the baseline models in terms of the whole word accuracy on the pronunciation prediction task, evaluated on the 2,000 Wikipedia test pairs. All models other than KyTea were trained on the combination of Wikipedia-derived and UniDic pairs. As is observed from the table,

**Table 1.** Word-level accuracy (in %) of pronunciation prediction on Wikipedia test data

| Model | Accuracy |
|---|---|
| KyTea | 57.8 |
| Joint Bigram | 66.4 |
| Joint Trigram | 70.0 |
| Proposed | 71.7 |

---

[9] http://www.phontron.com/kytea/

the proposed method outperforms all the baseline models, with a 1.7% improvement over the joint trigram model, which is statistically significant with the significance level of $p < 0.01$ by the McNemar's test. Though falling short of the best model, the joint n-gram models are quite competitive, suggesting that the proposed method of unsupervised training data generation using word alignment techniques is beneficial for the task. The advantages of using an MT-inspired framework are therefore twofold: word alignment techniques for training data generation, and the linear combination of relevant feature functions for the best accuracy on the prediction task. KyTea performs poorly on this data set, suggesting that using a unigram model trained on manually created resources does not work well on the words that appear in Wikipedia.

It is noteworthy that the joint trigram and proposed models both outperformed the average human performance [10] (~65%) on the same data set. Since the current experimental setting allows only one pronunciation to be correct disregarding the context in which the word is used, it is possible that many errors are actually not errors but are acceptable in other contexts. An error analysis on the output of the proposed model confirms this speculation: about half of the errors were judged acceptable or correct upon human verification. Considering this fact, the performance of the proposed model is considered to be approaching the upper bound of this task, hence, the improvement of 1.7% is quite meaningful.

**Table 2.** Accuracy (in %) of pronunciation prediction models evaluated on the Wikipedia test set with respect to various training data sets

|  | Joint Trigram | Proposed |
|---|---|---|
| UniDic | 46.9 | 47.5 |
| Wikipedia | 68.5 | 70.8 |
| Wikipedia+UniDic | 70.0 | 71.7 |

Table 2 shows the accuracy of the joint trigram and proposed models as a function of different training data source. The models trained with UniDic performed poorly, with the accuracy lower than 50%. This suggests that the alignments learned solely from a static dictionary resource are insufficient to predict the pronunciation of new words in Wikipedia. The use of Wikipedia-derived instances as training data improved the accuracy dramatically, achieving more than 20% improvement over the models trained on UniDic. Combining the two resources further improved the accuracy. The proposed model consistently outperforms the joint trigram baseline in all training data settings.

We also examined how incorrect alignments in the training data affect the pronunciation prediction performance. We did this by setting a cutoff threshold of the alignment scores, and removing those alignments with the scores lower than the

---

[10] The human performance is measured as follows: the source strings are presented to two native speakers with no context, and they are asked to assign guessed pronunciation. They both had graduate-level education.

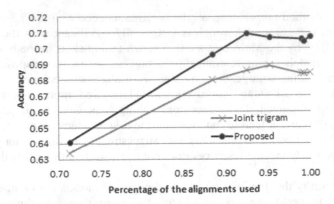

**Fig. 3.** Accuracy of pronunciation prediction models on the test set w.r.t. training data noise filtering

threshold from the training set. The alignment scores are output by the character aligner discussed above as the log probability of the alignment, and are normalized by the length of the source (i.e. kanji) sequence to avoid the preference for shorter sequences. Figure 3 shows the accuracy of the models with respect to the cutoff threshold. The x-axis corresponds to the percentage of the instances used in model training, while the y-axis indicates the word-level accuracy. The nodes in the graph correspond to the score threshold of −1, −2, −3, −4, −6, −8, −10, and −Infinity. The best performance is achieved when 90–95% of the training data is used, which is consistent with the observation that out of 100 words for which we manually verified the alignments, 11 instances contained alignments that appeared improper, and 7 instances were not word-pronunciation pairs though the aligner forced an alignment, which means that these instances are noise. Comparing the proposed vs. joint trigram-based methods, the former appears slightly more robust to noise: though it is rather difficult to see in the graph, the trigram model gained the maximum of 0.45% improvement by the noise filtering, while the proposed model gained the maximum of only 0.2% improvement. This may be attributed to the fact that the proposed model incorporates a noise filtering mechanism by way of minimum operation counts threshold C, as mentioned above.

## 4   Word Pronunciation Acquisition Task

This section describes our approach to harvesting word-pronunciation pairs from a large corpus with minimal supervision. We formulate this problem as a binary classification task: each candidate pair is determined to be a word-pronunciation pair or not, using a discriminatively trained classifier. Using the word-pronunciation candidate pairs in Figure 1 as an example, our goal is to classify the first four examples as positive and the last example as negative. A special treatment is required for the second to last example: recall that in this example, 名取市立館腰（たてこし），

only the last two characters (館腰) of the extracted word string corresponds to the pronunciation in hiragana in parentheses. For the modeling task in Section 3, we ignored this problem and treated these cases simply as noise. For the pronunciation acquisition task, we generate additional candidate pairs from these cases, so that from the string above, we generate the following pairs:

〈腰, たてこし〉, 〈館腰, たてこし〉, 〈立館腰, たてこし〉, 〈市立館腰, たてこし〉

These expanded candidate pairs share the word strings to the right, their length bounded maximally by the number of hiragana characters in the pronunciation (assuming one-to-one or -many mapping between kanji and hiragana), and minimally by the length of hiragana pronunciation string divided by 3 (assuming that *on average*, a kanji character maps to up to 3 hiragana characters at the word level). Each of these candidate pairs are then submitted to the classifier to be validated as a desired word-pronunciation pair or not.

The task of pronunciation acquisition formulated in this manner, along with the sub-problem of the boundary detection, can be viewed as a very similar task to bilingual dictionary creation for Chinese MT (e.g. [3] [18]). The goal of this line of work is to exploit parenthetical expressions in the web text to extract Chinese-English phrase translation candidates, which are then validated using a classifier. Because Chinese text is similar to Japanese in that the word boundaries are not marked explicitly using white spaces, the same boundary detection problem exists as in '我的磁石(magnet)', in which only the underlined part corresponds to the translation of 'magnet'. Despite these similarities, the process of aligning the input and output sequences and using the alignment for the validation task is much more complex in the bilingual dictionary creation task, as it cannot be reformulated as a monotone substring mapping problem, and requires additional steps to generate word translation pairs from extracted string pairs.

Coming back to our pronunciation acquisition task, given that the orthography-pronunciation mapping is basically monotone without insertion or deletion, one can think of a very simple method for validating word-pronunciation candidate pairs which does not use a classifier. A finite-state acceptor can be used to search for a path through the hiragana pronunciation string, from right to left, emitting the corresponding kanji sequence using a fixed kanji pronunciation dictionary. When a valid path is found at the end of the pronunciation string, it returns success, with the emitted kanji character sequence, performing the left boundary detection as the same time. This method is expected to have near 100% precision, as the pronunciations for each kanji character is already validated in the dictionary. The recall, however, suffers from an incomplete coverage of the pronunciation dictionary. The pronunciation of kanji characters in Japanese reflects the effect of various morpho-phonological processes [30] as exemplified in Figure 4. These sound changes are reflected in the pronunciation (hiragana) orthography, but are often missing from the dictionary. We compare the performance of this baseline against the proposed method below.

[Rendaku (sequential voicing)]

神(かみ) kami 'god' + 棚(たな) tana 'shelf' → 神棚(かみだな) kami-dana, 'altar'

[Renjo (liaison)]

反(はん) han 'counter'+ 応(おう) ou 'response'→ 反応(はんのう) han-nou, 'reaction'

[Vowel alteration]

雨(あめ) ame, 'rain' + たれ tare, 'drop' → 雨だれ(あまだれ) ama-dare, 'raindrop'

[Onbin (historical alterations around vowels)]

月(げつ) getsu 'moon'+ 光(こう) kou 'light' → 月光(げっこう) geQkou 'moonlight' (Q indicates germination)

**Fig. 4.** Examples of morpho-phonological alterations in Japanese. Underlined characters indicate the changes in hiragana pronunciation strings.

## 4.1 Model and Features

As we saw in Section 3.3, not all pairs extracted from Wikipedia are valid word-pronunciation pairs. Also, there exists the boundary detection problem from the extracted instances. Thus, the problem we need to solve is twofold: to determine if a set of instances generated from a single original string contains a true word-pronunciation pair, and if it does, to determine which generated instance is the correct pair (i.e. boundary detection). One approach to this dual problem is to construct a ranker, which ranks each instance according to the likelihood of being a correct word-pronunciation pair. This is the approach employed by [3]. However, in our case, our preliminary experiments showed that the ranker approach is not necessary and a simple binary classifier which treats all instances separately will do, as the average number of distinct candidate pairs generated from a single extracted string is not very large (2.3 per extracted instance on average in our training data).

With this observation and the computational cost taken into consideration, we propose to use a binary classifier with MART (Multiple Additive Regression Trees; [7]), a widely used classification framework based on additive trees. The MART classifier was shown to outperform an averaged perceptron classifier with a substantial margin in our preliminary experiment on the development set.

Table 3 shows the list of features we used in the MART classifier. *LR* is the length ratio of pronunciation sequence to the kanji sequence. Since a kanji sequence is in

**Table 3.** Features used in the binary classifier

| Feature | Description |
| --- | --- |
| LR | Length ratio (hiragana / kanji) |
| Align | Log of alignment score |
| Dist | Minimum value of phonologically-motivated edit distance to 20-best outputs of the transducer in Section 3 |
| MT | Features from the transducer described in Section 3.1: $p(s|t)$, $p(t|s)$, $p(t)$, operation count, phrase count, word count |

principle never longer than the corresponding pronunciation sequence, [11] and one kanji character usually corresponds to one to three kana characters, this feature is expected to remove the pairs with obviously non-standard length ratio. *Align* is the log of the alignment score. *Dist* is the smallest edit distance to the 20-best pronunciation prediction results of the transducer described in Section 3. The process of generating this feature is as follows: first, the pronunciation candidate and the transducer outputs are transformed into Roman alphabets (e.g. 'かいぎ' ⇒ 'kaigi'), and the Levenshtein distances between the candidate and transducer outputs are calculated. The model uses the minimum of the edit distance values as the feature value. Crucially, we define some special zero-cost character replacement rules to address the Japanese morpho-phonological transformations as described in Figure 4, trying to capture as many pronunciation variants as possible. The followings are the rules we used; most of these alterations are described in [30].

- Rendaku: [k,t,s,sh] ⇔ [g, d, z, j]
- Vowel change: [e] ⇔ [a], [i] ⇔ [o], [o] ⇔ [a]
- On-bin: [mu] ⇒ [m, n], [tta] ⇒ [ta]
- Chinese dialect variation:[12] [k,s,t,n,h,m,r] ei ⇔ [k,sh,ch,n,h,m,r] ou/you

Although these rules are specific to Japanese pronunciations, the framework based on the extended edit distance with phonologically-motivated rules is generally applicable to any language with phonological transformations. Finally, *MT* indicates the features from the transduction model described in Section 3. They consist of the translation probabilities $P(s|t)$ and $P(t|s)$, target language model probability $P(t)$, operation count, word count, and phrase count, as described in Section 3.

## 4.2 Experiments

**Data and settings.** We use the same Wikipedia data sets used in Section 3. From the training portion, we extracted 3,000 instances, and expanded them to generate the instances differing in the left boundary of the word, based on the heuristics described above. This resulted in 6,872 instances in total. Roughly 90% of the original instances before expansion were positive instances; after expansion, the number of negative instances increases, making the training instances more balanced. We then manually labeled each instance as positive or negative, indicating whether the pair is the correct word-pronunciation pair or not.

## 4.3 Results and Discussion

Table 4 shows the comparison of the models with various feature sets. The baseline is the finite-state acceptor method with a fixed dictionary, applied to the paired data

---

[11] Again, there are some exceptions to this such as <似而非, えせ> [ese]'pseudo'. However, they are very rare: we only found 14 cases in the 195K entries of UniDic nouns.

[12] This reflects the variations in the original Chinese pronunciations. For example, 清 has two common on-yomi pronunciations: [sei], which was imported around 7–8c from Tang dynasty, and [shou], imported earlier around 5–6c.

**Table 4.** Performance (in %) of yomi classification evaluated on labeled Wikipedia pairs, with five-fold cross validation

|                          | Prec. | Recall | F1   |
|--------------------------|-------|--------|------|
| Baseline                 | 99.8  | 80.8   | 89.3 |
| MART(LR)                 | 56.8  | 62.9   | 59.7 |
| MART(LR+Align)           | 94.3  | 90.9   | 92.5 |
| MART(LR+Align+Dist)      | 97.8  | 96.2   | 97.0 |
| MART(LR+Align+Dist+MT)   | 98.5  | 98.0   | 98.2 |

after candidate expansion. We used the dictionary mentioned in Footnote 4 for this purpose, which includes around 15K distinct kanji-pronunciation pairs. As expected, this baseline achieves a very high precision of 99.8%, but the recall is only around 80%, showing that the kanji pronunciation dictionary we used, though reasonably large, does not have satisfactory coverage for this task. The model with the LR feature performed poorly, with approximately 60% precision and recall, which is to be expected, as this feature by itself is a weak feature. The LR+Align model performs quite reasonably, with both the precision and recall over 90%. Adding the linguistically motivated edit distance features (LR+Align+Dist+MT) achieves a nice performance gain, with the F1 score achieving 97%. Finally, with all the features, we achieve the best performance of both precision and recall exceeding 98%.

Using this classifier, we were able to obtain ~420K word-pronunciation pairs from Wikipedia with 98.5% precision. To our knowledge, this is the largest Japanese word-pronunciation lexicon that is automatically generated.

# 5  Conclusion and Future Work

We have presented our approach to the task of Japanese pronunciation prediction. We have shown that the proposed pronunciation prediction model achieves the performance that is coming close to the level of human performance, and also that the model can be used effectively to harvest word-pronunciation pairs from unannotated text. We believe that the accuracy we achieve is sufficiently high to be used in realistic applications, such as text-to-speech and text input method. Measuring the contribution of this research in such application scenarios is one direction of future research.

We also plan to apply the proposed pronunciation acquisition technique to a much larger web-scale corpus. This step will be important for acquiring the pronunciation of the words other than nouns, as the Wikipedia data we used was dominated by noun-pronunciation pairs. The pronunciation extraction of the parts-of-speech that inflect (verbs and adjectives) are expected to be more challenging, as the parenthetical pronunciation aids are inserted within a word, rather than at the word boundary, as in 醒 (さ) めた [sa-me-ta] 'awake'. Although newly created words tend to be nouns, predicting the pronunciation of non-nouns will be important when we use the methods proposed in this paper for the task of predicting pronunciations at the sentence level.

# References

1. Bisani, M., Ney, H.: Investigations on joint-multigram models for grapheme-to-phoneme conversion. In: The Proceedings of the International Conference on Spoken Language Processing (2002)
2. Bisani, M., Ney, H.: Joint-sequence models for grapheme-to-phoneme conversion. Speech Communication 50(5), 434–451 (2008)
3. Cao, G., Gao, J., Nie, J.-Y.: A system to mine large-scale bilingual dictionaries from monolingual web pages. In: MT Summit XI (2007)
4. Chen, S.F.: Conditional and joint models for grapheme-to-phoneme conversion. In: The Proceedings of the European Conference on Speech Communication and Technology (2003)
5. Cherry, C., Suzuki, H.: Discriminative Substring Decoding for Transliteration. In: EMNLP (2009)
6. Den, Y., Ogiso, T., Ogura, H., Yamada, A., Minematsu, N., Uchimoto, K., Koiso, H.: The development of an electronic dictionary for morphological analysis and its application to Japanese corpus linguistics. Japanese linguistics 22, 101–122 (2007) (in Japanese)
7. Friedman, J.H.: Greedy function approximation: a gradient boosting machine. Ann. Statist. 29(5), 1189–1232 (2001)
8. Gao, J., Goodman, J., Li, M., Lee, K.-F.: Toward a unified approach to statistical language modeling for Chinese. ACM Transactions on Asian Language Information Processing 1(1), 3–33 (2002a)
9. Gao, J., Suzuki, H., Wen, Y.: Exploiting headword dependency and predictive clustering for language modeling. In: EMNLP 2002 (2002b)
10. Ghoshal, A., Jansche, M., Khudanpur, S., Riley, M., Ulinski, M.: Web-derived Pronunciations. In: ICASSP (2009)
11. Jiampojamarn, S., Kondrak, G., Sherif, T.: Applying many-to-many alignments and hidden markov models to letter-to-phoneme conversion. In: HLT-NAACL (2007)
12. Jiampojamarn, S., Cherry, C., Kondrak, G.: Joint Processing and Discriminative Training for Letter-to-Phoneme Conversion. In: ACL (2008)
13. Jiampojamarn, S., Cherry, C., Kondrak, G.: Integrating Joint n-gram Features into a Discriminative Training Framework. In: The Proceedings of NAACL (2010)
14. Knight, K., Graehl, J.: Machine Transliteration. Computational Linguistics 24(4) (1998)
15. Koehn, P., Och, F., Marcu, D.: Statistical phrase-based translation. In: NAACL (2003)
16. Kurata, G., Mori, S., Suitoh, N., Nishimura., M.: Unsupervised lexicon acquisition from speech and text. In: The Proceedings of ICASSP (2007)
17. Li, H., Zhang, M., Su, J.: A joint source-channel model for machine transliteration. In: ACL (2004)
18. Lin, D., Zhao, S., Durme, B.V., Pasca, M.: Mining parenthetical translations from the web by word alignment. In: ACL 2008 (2008)
19. Maekawa, K.: Compilation of the KOTONOHA-BCCWJ Corpus. Nihongo no kenkyu (Studies in Japanese) 4(1), 82–95 (2008) (in Japanese)
20. Mori, S., Neubig, G.: Automatically improving language processing accuracy by using kana-kanji conversion logs. In: The Proc. of the 16th Annual Meeting of the Association for NLP (2010) (in Japanese)
21. Mori, S., Sasada, T., Neubig, G.: Language Model Estimation from a Stochastically Tagged Corpus. Technical Report, SIG, Information Processing Society of Japan (2010) (in Japanese)

22. Nagano, T., Mori, S., Nishimura, M.: An n-gram-based approach to phoneme and accent estimation for TTS. Transactions of Information Processing Society of Japan 47(6), 1793–1801 (2006) (in Japanese)

23. Och, F.J.: Minimum Error Rate Training for Statistical Machine Translation. In: ACL (2003)

24. Reddy, S., Goldsmith, J.: An MDL-based approach to extracting subword units for grapheme-to-phoneme conversion. In: NAACL (2010)

25. Sasada, T., Mori, S., Kawahara, T.: Extracting word-pronunciation pairs from comparable set of text and speech. In: The Proceedings of the 9th Annual Conference of the International Speech Communication Association (2008)

26. Sasada, T., Mori, S., Kawahara, T.: Domain adaptation of statistical kana-kanji conversion system by automatic acquisition of contextual information with unknown words. In: The Proceedings of the 15$^{th}$ Annual Meeting of the Association for NLP (2009) (in Japanese)

27. Schroeter, J., Conkie, A., Syrdal, A., Beutnagel, M., Jilka, M., Strom, V., Kim, Y.-J., Kang, H.-G., Kapilow, D.: A perspective on the next challenges for TTS research. In: The Proceedings of the IEEE 2002 Workshop on Speech Synthesis (2002)

28. Sherif, T., Kondrak, G.: Substring-based transliteration. In: ACL (2007)

29. Sumita, E., Sugaya, F.: Word Pronunciation Disambiguation using the Web. In: NAACL (2006)

30. Vance, T.J.: An introduction to Japanese phonology. State University of New York Press (1987)

31. Zens, R., Ney, H.: Improvements in Phrase-Based Statistical Machine Translation. In: HLT-NAACL (2004)

32. Zhang, H., Quirk, C., Moore, R.C., Gildea, D.: Bayesian learning of non-compositional phrases with synchronous parsing. In: ACL (2008)

# A Minimum Cluster-Based Trigram Statistical Model for Thai Syllabification

Chonlasith Jucksriporn and Ohm Sornil

Department of Computer Science
National Institute of Development Administration
Bangkok 10240, Thailand
chonlasith@gmail.com, osornil@as.nida.ac.th

**Abstract.** Syllabification is a process of extracting syllables from a word. Problems of syllabification are majorly caused from unknown and ambiguous words. This research aims to resolve these problems in Thai language by exploiting relationships among characters in the word. A character clustering scheme is proposed to generate units smaller than a syllable, called Thai Minimum Clusters (TMCs), from a word. TMCs are then merged into syllables using a trigram statistical model. Experimental evaluations are performed to assess the effectiveness of the proposed technique on a standard data set of 77,303 words. The results show that the technique yields 97.61% accuracy.

**Keywords:** Thai Syllabification, Thai Minimum Cluster, Trigram Model.

## 1 Introduction

Syllabification is a process to extract syllables from a word. This process is essential to many natural language processing tasks, especially for a text-to-speech system. Thai language with its unique characteristics both syntactically and semantically complicates the problem.

Many approaches were proposed to handle this task for Thai language, such as dictionary-based and rule-based methods. The idea is to group characters and produce syllables from the results. The main problem of Thai syllabification is how to handle unknown words, ambiguous words that can be differently pronounced, and proper names. For example, "กร" (korn), meaning a hand, can be pronounced as "กะ-ระ" (ka-ra) in a word "กรณี" (ka-ra-nee), meaning a case or a situation; or "รัตน" (rat-ta-na), meaning gems, can be pronounced as "รัด" (rat) in "สมเด็จพระเทพรัตนราชสุดาสยามบรมราชกุมารี" (som-det-phra-tep-pha-rat-rat-su-da-sa-yam-bo-rom-ma-rat-cha-ku-ma-ree), the name of a Thai princess.

This paper proposes a novel technique to resolve these problems by using minimum character clustering and a trigram statistical model. The minimum character clustering technique is used to reduce time spent in cluster segmentation by grouping consecutive characters into clusters whose characters cannot be split further. These clusters will then be used in segmentation process instead of iterating though each

A. Gelbukh (Ed.): CICLing 2011, Part II, LNCS 6609, pp. 493–505, 2010.

character one by one. Each cluster is very close to a Thai syllable; however, some may not correspond to a syntactic syllable. Finally, the clusters are merged using a trigram model with maximum likelihood estimation (MLE). The best segmentation is chosen from the segmentation which gives the best outcome. The training data is created from BEST 2010 (a Thai word segmentation contest); it contains 82,309 words with 2,910 distinct clusters.

The proposed merging method is compared with a longest matching algorithm working in both forward and backward directions with some help of trigram statistics. Our approach is trained using the Royal Institute word dictionary which consists of 36,689 words and idioms. A set of 89,640 clusters (8,321 different clusters) is extracted from the dictionary.

## 2   Previous Researches

This section discusses previous researches related to Thai word syllabification. Pooworawan et al. (1986) proposed a dictionary-based approach using forward longest matching search. This method reads an input text from left to right one character at a time and concatenates the character to a read buffer. In each loop, the method tries to match the string in the read buffer with a dictionary. If the syllable is found, the current read buffer is marked as a backtrack position and a segmentable position. Otherwise, the loop keeps executing or backtracks if no syllable can be extracted. The method gave 98.649% of accuracy from a data set of 296 words. The errors were caused from unknown words. Kongsupanich (1997) proposed a dictionary-based approach by removing each character from a Thai sentence backward and searching a dictionary for the remaining characters. This technique is not suited for Thai sentence syllabification since a sentence can be very long and has no explicit delimiter.

Theeramunkong et al. (2000) proposed a clustering technique called Thai Character Cluster (TCC) in their information retrieval research. The purpose of this technique is to create inseparable clusters from Thai character string using a rule-based strategy. A TCC is smaller than a word and larger than a character. For example, a word "อัตรา" (at-tra), meaning a rate, will be clustered as 2 clusters, "อัต" (at) and "รา" (ra). Inrut et al. (2001) proposed an enhanced version of TCC called Enhanced Thai Character Cluster (ETCC) by providing a set of additional predefined rules. These rules are applied to TCC clusters to create syllable-like clusters which may become larger than a TCC. The purpose of ETCC is to speed up the longest matching algorithm.

Paludkong (2006) employed a dictionary-based longest matching approach, but working in both forward and backward directions. This research utilizes ETCC to divide a sentence into small chunks of characters. This method extracts a string which is not longer than the longest word in the dictionary by reading each cluster from the input and concatenates the cluster to a read buffer. The extracted string is compared to the dictionary to find the syllabification. If the string is not found, a cluster is removed from the string and the iteration is started over. The method gave 98.88% of accuracy on a data set of 5,000 words.

A dictionary-based approach in general is fast and simple but limited to only known syllables and unambiguous words. This makes the results rely heavily on the coverage of the dictionary and ambiguities of the context.

Lorchirachoonkul et al. (1989) proposed an algorithm to handle unknown words. The algorithm segments a string of characters into tokens and builds syllables from the tokens using a precedence matrix. A precedence matrix is a table storing sequence information between 2 tokens. Syllables are then created using the table. This algorithm cannot extract some frequently used word, such as "เป็น" (pen), meaning being. The experiments and results for this part were not clearly stated, but about 80% effectiveness were mentioned in the research.

Some approaches use a combination of the dictionary-based and other strategies to overcome the weaknesses of using the dictionary-based method alone. Khruahong (2003) proposed a technique that uses a dictionary-based and a pattern-based strategy, called suited-syllable-structure mapping (3S-mapping). This technique creates clusters of characters using ETCC and merges them together using a predefined set of rules. The consecutive clusters are grouped whose length is not greater than the length of the longest word and compared with the rules. If an applicable rule is found, the syllable is extracted. Otherwise, the last cluster is removed from the group. Other hybrid approaches were also proposed. The results showed 99.12% accuracy on a data set of 4,000 words. Aroonmanakun (2002) proposed a technique that uses a rule-based strategy with trigram statistics for extracting syllables in his word segmentation research. A set of patterns are constructed manually and used to create possible segmentations. The best syllable segmentation will be selected by a trigram statistical model. The accuracy varies between 81 and 98%, depending on the dictionary used in segmentation, on data sets with about 18,000 to 22,500 words.

## 3  Proposed Method

### 3.1  Overview

The method proposed in this research was motivated by how humans pronounce a word. When an arbitrary Thai word is read, we use our experience to determine how it should be pronounced. The experience is composed of knowledge which came from learning and similarity of word compositions. A new word is usually derived from existing words which act as root words. This helps us understand or guess its meaning and pronunciation. Thus, the idea of our method is to group characters into minimum character clusters and merge them into larger clusters, if possible.

The characters are grouped according to their positions and functions, as shown in Table 1. For example, a leading vowel is vowel that if it is in the cluster, it must be the first character, thus some character groups may identify cluster delimiter(s). Some clusters may not correspond to Thai syntactic rules, but they can facilitate the applications of the clusters. For example, "อัตรา" (at-tra) will be grouped into "อั" (o ang and maihan-akat, which will not be pronounced) and "ครา" (tra), instead of "อัต" (at) and "รา" (ra) or "อัตร" (at) and "า" (sara a, this cannot be pronounced). The first segmentation is easier to read if we simply define a behavior for a cluster that ends with "ั" (maihan-akat); in this case, we append the first character of the second cluster which is "ต" (to tao) to the first cluster which gives the correct pronunciation "อัต-ตรา" (at-tra).

Candidate segmentations will be created from the clusters, and a trigram statistical model will be used to select the best one. This will be described in section 3.5.

## 3.2 Thai Character Classification

This section classifies Thai characters into 7 groups, based on its position and function in a word, as listed in Table 1. Some characters identify cluster boundaries. For example, a leading vowel must be the first character in the cluster, if exists. The special mark "ิ", if exists, must be the last character in the cluster.

**Table 1.** Thai characters based on their positions

| Group | Characters |
|-------|------------|
| Consonants | ก,ข,ฃ,ค,ฅ,ฆ,ง,จ,ฉ,ช,ซ,ฌ,ญ,ฎ,ฏ,ฐ,ฑ,ฒ,ณ,ด,ต,ถ,ท,ธ,น,บ,ป,ผ,ฝ,พ,ฟ,ภ,ม,ย,ร,ฤ, ล,ฦ,ว,ศ,ษ,ส,ห,ฬ,อ,ฮ |
| Leading vowels | เ,แ,ไ,ใ,โ |
| Trailing vowels | ะ,า,ๅ,ำ |
| Upper vowels | ◌ั, ◌ิ, ◌ี, ◌ึ, ◌ื, ◌็ |
| Lower vowels | ◌ุ,◌ู,◌ฺ |
| Tonal marks | ◌่,◌้,◌๊,◌๋ |
| Other special marks | ◌์,◌๎,ๆ,๚ |

## 3.3 The Silence Mark

When the silence mark (◌์) character is attached to any consonant in a word, that consonant will not create a sound, such as "รักษ์" (ruck) will be pronounced as "รัก" (ruck). Sometimes the silence mark mutes 2 previous consonants, for example, "ลักษณ์" (luck) will be pronounced as "ลัก" (luck). Accordingly, a cluster that contains the silence mark should contain at least 3 characters to make it pronounceable.

## 3.4 Thai Minimum Cluster

Thai Minimum Cluster (TMC) is an inseparable cluster of Thai character sequence in a word which is based on character groups in Table 1. The goal is to create a minimum cluster that can be pronounced as one syllable (or two syllables, if it is not possible to create one-syllable cluster). The rules are defined based on Thai writing system with an exception: the vowel "◌ั" (maihan-akat) is a variant of a vowel "ะ" (sara a). In Thai writing system, the vowel "◌ั" (maihan-akat) always requires at least one following consonant, but this is not adopted in this research because providing a consonant to follow "◌ั" (maihan-akat) may lead to a wrong pronunciation.

For example, as mentioned in section 3.1, grouping "อัตรา" into "อ" (o ang and maihan-akat, which cannot be pronounced) and "ตรา" (tra) explicitly shows how this word is really pronounced, while grouping as "อัต" (at) and "รา" (ra) and "อัตร" (at) and "า" (sara a, this cannot be pronounced) causes wrong pronunciations.

The rules for generating TMCs are adapted from the Thai writing system and the authors' observations from carefully analyzing word patterns and their pronunciations from the Royal Institute dictionary. These rules are listed as follows:

1. A single Thai character is allowed since it can be pronounced like it is appended by "ะ" (sara a) or "อ" (o ang) implicitly.
2. A leading vowel always precedes at least one consonant.
3. An upper or a lower vowel always follows at least one consonant.
4. A tonal mark may follow a consonant or an upper or a lower vowel.
5. A trailing vowel always follows at least one consonant, except "ะ" (sara a, a short sound) which is able to follow "า" (sara a, a long sound) to produce a compound vowel "เ-าะ" (sara o).
6. The special mark "็" (maitaikhu) and "์" (karan or silence mark) always follow at least one consonant.
7. A cluster containing "์" (karan) must have at least 4 characters.
8. The special mark "็" (maitaikhu) needs at least one following consonant. The cluster "ก็" (ko) is only exception for this rule.
9. A combination of rules is allowed, for example, a combination of rules 2 and 5 creates a cluster like "เกา" (kao).
10. A cluster consisting of a leading vowel and an upper vowel need at least one consonant after the upper vowel.
11. A cluster consisting of an upper vowel "ั" (maihan-akat) with a tonal mark needs at least one consonant after the tonal mark.
12. A cluster consisting of a tonal mark with no preceding vowel needs at least one consonant or one trailing vowel after the tonal mark unless the cluster is "บ่" (bo).
13. A cluster consisting of a special mark "็" (maitaikhu), no leading vowel, and followed by a consonant "อ" (o ang) needs one consonant after the consonant "อ" (o ang).
14. There are 3 clusters containing the special mark "ฯ" (pai-yannoi) which are "ฯ" (pai-yannoi), "ฯลฯ" (pai-yanyai), and "ฯพณฯ" (pana than). The word "ฯพณฯ" (pana than) has been deprecated since 1944 but can be found in some documents.
15. The special mark "ๆ" (maiyamok) always be a singleton.

Figure 1 shows the non-deterministic finite automata (NFA) diagram of the clustering rules. In the figure, C is a consonant, U is an upper vowel, L is a lower vowel, R is a trailing vowel, F is a leading vowel and T is a tonal mark. Since a cluster containing "์" (karan) produces no sound, it will be merged with the previous

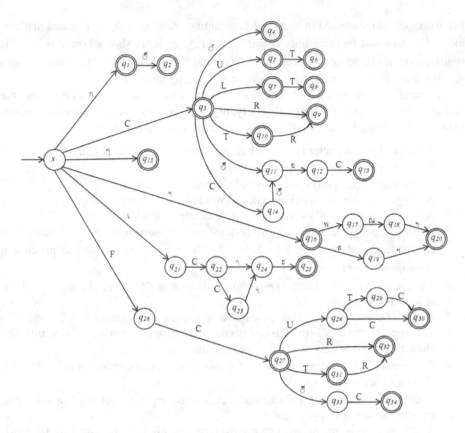

**Fig. 1.** Minimum Clustering NFA

clusters to create a cluster that can be pronounced. A resulting TMC can be as short as a single character or as long as the input word, and a TMC can be smaller or larger than a TCC.

### 3.5   Trigram Statistical Model for Cluster

The probability of a cluster sequence is used to estimate the best cluster sequence by calculating a production of every conditional probability for each cluster given previous clusters and multiply them together to find the probability of the sequence. A sequence of $n$ clusters is denoted as $s_1, s_2, ..., s_n$ or $s_1^n$. The joint probability of each cluster in a word, $P(s_1, s_2, ..., s_n)$ or $P(s_1^n)$, can be formulated as follows:

$$P(s_1^n) = P(s_1)P(s_2 \mid s_1)P(s_3 \mid s_1 s_2)...P(s_n \mid s_1 s_{n-1}) \tag{1}$$

With the Markov assumption, this equation can be simplified to N-grams models. N-gram models are probabilistic models to predict a probability of next cluster, given previous $N$-$1$ clusters instead of all previous clusters. Then a computation of $P(s_n \mid s_{n-N+1}^{n-1})$ can be used instead of $P(s_n \mid s_1^{n-1})$ or written as follows:

$$P(s_n \mid s_1^{n-1}) \approx P(s_n \mid s_{n-N+1}^{n-1}) \qquad (2)$$

An estimation of the above equation is called *maximum likelihood estimation* (MLE). The parameters of an N-gram model can be obtained by learning from a corpus. If $C(s)$ is the frequency of $s$ in the corpus, the computation of $P(s_n \mid s_{n-N+1}^{n-1})$ can be written as follows:

$$P(s_n \mid s_{n-N+1}^{n-1}) = \frac{C(s_{n-N+1}^{n-1} s_n)}{C(s_{n-N+1}^{n-1})} \qquad (3)$$

This paper utilizes a trigram model to find the best segmentation of a word. The most probably segmented sequence $s_1, s_2, ..., s_n$ of a word $W = c_1 c_2 ... c_m$ can be generally formulated as:

$$seg(W) = \underset{S - s_1 s_2 ... s_k}{argmax} \prod_i^k P(s_n \mid s_{n-N+1}^{n-1}) \qquad (4)$$

where $P(s_n \mid s_{n-N+1}^{n-1})$ denotes the language model probability of $s_i$ in word $W$, a cluster sequence $s_{i-N+1} s_{i-N+2} ... s_{i-1}$ is the N-clustered history of $s_i$, and $N$ is the order of the N-gram model in use, which is 3. Thus, $P(s_i \mid s_{i-N+1}^{i-1})$ becomes $P(s_i \mid s_{i-2} s_{i-1})$.

To segment a word consisting of $m$ characters, $W = c_1 c_2 ... c_m$, into $k$ clusters, $C_{n-1}^{k-1}$ different segmentations are possibly produced. Considering $k$ varies from 1 to $m$, a word $W$ has a total of $2^{n-1}$ different segmentations. This may causes a high computation cost, even only some clusters in $W$ are considerable.

To reduce the number of possible clusters, a unigram table is extracted from the training corpus, and only the clusters existing in the unigram table are selected from $W$.

To create a trigram, consider the word $W$ consisting of $n$ clusters delimited by the predefined separators. $W$ is processed as follows:

1. $W$ is split into clusters at the delimiters, and these clusters are used to produce trigrams. The strings which create $k+3$ clusters produce $k+1$ overlapping trigrams, while the group containing 3 clusters or less is taken as the only trigram produced. All produced trigrams are attributes of $W$.
2. The first trigram produced from each cluster from $W$ is marked at the left end by additional symbol "!" while the last is marked at the right end by the same symbol, and the result is included as an additional attribute.

For example, a word "กรณียกิจ" (ka-ra-ne-ya-kit means duty) creates a group of clusters "ก-ร-ณี-ย-กิจ" (ka-ra-ne-ya-kit), separated by delimiter "-". This word produces 3 overlapping trigrams which are "ก-ร-ณี" (ka-ra-ne), "ร-ณี-ย" (ra-ne-ya), and "ณี-ย-กิจ"

(ne-ya-kit). Two additional trigrams are added: "!ก-ร-ณี" (ka-ra-ne) and "ณี-ย-กิจ!" (ne-ya-kit), to identify boundaries of the trigrams. The probabilities of the trigrams are calculated by equation (3).

To select the best segmentation from all possible segmentations, the probability of each candidate is evaluated by calculating the joint probability of each trigram in the clusters. For example, "กรณี" (ka-ra-ne means case) creates clusters "ก-ร-ณี" (ka-ra-ne). The word produces only one overlapping trigram "ก-ร-ณี" (ka-ra-ne) and two additional trigrams: "!ก-ร-ณี" (ka-ra-ne) and "ก-ร-ณี!" (ka-ra-ne). The calculation will evaluate the probability from the trigram with the boundary mark, if possible. From the above example, "!ก-ร-ณี" (ka-ra-ne) is used to find the probability from the trigram statistics records. The segmentation which gives the highest result will be chosen to be the best segmentation.

## 4   Experimental Evaluations

Since this research will compare the proposed cluster merging method with a dictionary-based technique, we now describe the dictionary-based method used in our study. The method merges TMCs using a cluster dictionary which contains the clusters extracted from a word dictionary. The merge process performs both forward and backward longest matching instead of only one direction. Using both directions helps us identify the range of wrong segmentations. If the results from both directions are identical, the result from the forward longest matching is chosen as the result of the segmentation. Otherwise, statistics information is used to find the best segmentation. Trigram statistics of clusters are employed for this purpose. We used trigram statistics to determine which segmentation is the best segmentation by comparing each trigram generated from each segmentation output to the statistics from the dictionary. Since a trigram is trained from a dictionary, and its clusters are not distributed normally, we use its existence instead of its probability. The trigram statistics for clusters are also extracted. The count of a trigram existence can be formulated as follows:

$$seg(W) = \mathop{argmax}_{S = s_1 s_2 \ldots s_k} \sum_i^k E(s_{i-2} s_{i-1} s_i) \tag{5}$$

where $E(s_{i-2} s_{i-1} s_i)$ is the evaluation function based on existence. If the clusters are in the trigram statistics for clusters, the evaluation gives 1 as the result. Otherwise, the result is 0. The segmentation giving the better result is chosen as the best segmentation. In case of a tie, the results from the backward longest matching are selected since, from the experiments, the backward results gave a higher accuracy.

## 5   Results

Experiments are performed to compare the dictionary-based and the proposed trigram statistics-based merges. Each experiment performs 2 tests: self-test and open test. The self-test is testing on training data while the open test is testing on unknown data. All data are clustered into TMCs and merged to create the best segmentations.

The dictionary approach uses a cluster dictionary, created from the Royal Institute word dictionary (1999). A total of 36,689 words from the dictionary are arranged into 89,640 clusters (with 8,321 distinct clusters) by a manual segmentation. A sample data from BEST 2010 (a Thai word segmentation competition) are used in the open test. From the self-test results, TMCs are created correctly 95.27%. The major problem is word ambiguities. Some of these ambiguous words are rarely found in a common document. Once two of rare clusters, "ขุป" (up) and "บริ" (bri), are removed from the cluster dictionary, the accuracy increases to 95.58%. In the open test, the clusters are created correctly 97.60%. The main problem is also the word ambiguities. The results show that ambiguity is the main problem of the dictionary-based segmentation.

In the trigram statistics-based approach, trigram statistics of clusters are created from a corpus of 82,309 words. This corpus is composed of small corpuses used in BEST 2010 in the encyclopedia category. The corpus is manually segmented into 115,099 clusters with 2,910 distinct clusters. The testing data for the self-test is the training corpus. The same testing data used in the open test for the dictionary-based approach are used in the open test for the trigram statistics-based approach. In self-test, TMCs are created correctly 100.0% with no error. In the open test, the result is 97.61% correct. The results of both approaches are shown in Table 2.

**Table 2.** Test results of merging methods

| Method | Self-test | Open test |
|---|---|---|
| Dictionary-based | 95.27% (34,954/36,689) | 97.03% (75,008/77,303) |
| Trigram-based | 100.0% (82,309/82,309) | 97.61% (75,454/77,303) |

The main problem of the trigram-based approach is cluster coverage. Most errors found in the open test are proper names, such as herb names, person names, location names. These words, especially person names, sometimes contain ambiguous pronunciations. The training corpus does not contain these proper names. This makes the syllabification not able to recognize these words. Other noticeable error related to the cluster coverage problem is the last cluster that contains only one consonant. Words which end with a single consonant are not frequently found in a common document. These words are usually used as affixes. An additional experiment for the trigram-based approach is to merge the single consonant cluster in the last position to the preceding one, to check whether merging the last single consonant can reduce the error from the coverage problem. The results are shown in Table 3. They show that the correctness slightly increases to 97.94%.

**Table 3.** Results of the trigram-based method with the last single character merged and not merged into the preceding cluster

| Method | Without Merge | Merge |
|---|---|---|
| Trigram-based | 97.61% (75,454/77,303) | 97.94% (75,714/77,303) |

The experiment also tried including the testing data into the corpus and retest with the testing data on both of merging last single character and no merging to check if the previous results really came from the problem of cluster coverage. Total 159,620 words from corpus were clustered into 231,251 clusters which given 3,692 distinct clusters. The result is shown in Table 4.

**Table 4.** Results of the trigram-based method with testing data included in the training corpus

| Method | Without Merge | Merge |
|---|---|---|
| Trigram-based | 100.0% (77,303/77,303) | 99.99% (77,295/77,303) |

From Table 4, both results are reasonably improved. Expanding trigram coverage helps increase the segmentation accuracy. The remaining errors in the case of merging are from abbreviations and rare words.

Both abbreviations with delimiters and abbreviations without delimiter exist in Thai language. The first one can be easily detected while the latter is not. The errors from segmentation with merge in Table 4 came from the latter abbreviations. The segmentation is unable to detect the abbreviations and treats them as words.

The errors with rare words are from words that are usually used as affixes. When they are used as suffixes, they are usually appended by "ิ์" (karan or silence mark) or "ะ" (sara a), for example, "ศิลป" (sin-la-pa), meaning art, and "นว" (na-wa), meaning new or nine). Using them as a non-affix is hardly found in a common document.

# 6   Conclusion

Syllabification is a process to extract syllables from a word. Coping with unknown words and ambiguous words are the main issues in syllabification. This paper proposes an approach to Thai syllabification using a minimum cluster-based trigram statistical model with maximum likelihood estimation. A string is divided into clusters called Thai Minimum Clusters (TMCs). These TMCs are then merged to create the final result.

The proposed method is evaluated on standard corpuses. The results show that it improves the performance in the presences of above problems, especially with an appropriate training corpus. It helps determine the best segmentation when ambiguity is found. The conditional probability of sequence of cluster represents the segmentation which has many possibilities of being correct. For example, using dictionary-based approach, a word "ภูมิภาค" (poo-mi-pak), meaning a region, was segmented as "ภูมิ-ภาค" (poom-pak) because a cluster "ภูมิ" (poom) exists in the dictionary as in the word "ภูมิใจ" (pook-jai), meaning proud; or a word "สระบุรี" (sa-ra-bu-ri), a province in Thailand, was segmented as "สระ-บุ-รี" (sa-bu-ri) since a cluster "สระ" (sa), meaning a pond, is also a word. Trigram statistics model overcomes these ambiguities. The model contains the relationships of the clusters "ง" (poo), "มิ" (mi), and "กด" (pak), and also "ส" (sa), "ระ" (ra), "บุ" (bu), and "ริ" (ri) which results in highest probabilities.

We conclude that the trigram statistics for Thai minimum clusters with maximum likelihood estimation works well for the syllabification purpose. It can reduce the problem of ambiguity and unknown clusters if trained with a proper corpus with good coverage.

# References

1. Christopher, D.M., Prabhakar, R., Hinrich, S.: Introduction to Information Retrieval. Cambridge University Press, England (2008)
2. Trigram Algorithm, http://ii.nlm.nih.gov/MTI/trigram.shtml (accessed September 28, 2010)
3. Mao, J., Cheng, G., He, Y., Xing, Z.: A Trigram Statistical Language Model Algorithm for Chinese Word Segmentation. In: Preparata, F.P., Fang, Q. (eds.) FAW 2007. LNCS, vol. 4613, pp. 271–280. Springer, Heidelberg (2007)
4. Kanchanacheewa, N.: Principles of Thai Language หลักภาษาไทย. Thai Wattana Panich Co., Ltd., Thailand (1996)
5. Khruahong, S., Nitsuwat, S., Limmaneepraserth, P.: Thai Syllable Segmentation for Text-to-Speech Synthesis by Using Suited-Syllable-Structure Mapping. In: International Conference on Computer Science and Information Technology (2003)
6. Lorchirachoonkul, V., Khuwinphunt, C.: Thai Soundex Algorithm and Thai Syllable Separation Algorithm. Research Report. School of Applied Statistics, National Institute of Development Administration, Bangkok (1982)
7. Thai Script, http://en.wikipedia.org/wiki/Thai_script (accessed October 2, 2010)
8. Aroonmanakun, W.: Collocation and Thai Word Segmentation. In: Proceedings of the Fifth Symposium on Natural Language Processing & the Fifth Oriental COCOSDA Workshop, Pathumthani, pp. 68–75 (2002)
9. Aroonmanakun, W., Rivepiboon, W.: A Unified Model of Thai Romanization and Word Segmentation. In: Proceedings of the 18th Pacific Asia Conference on Language, Information and Computation, Tokyo, pp. 205–214 (2004)
10. Poowarawan, Y.: Dictionary-based Thai Syllable Separation. In: Proceeding of Ninth Electronics Engineering Conference, Khon Kaen (1986)
11. Theeramunkong, T., Sornlertlamvanich, V.: Character Cluster Based Thai Information Retrieval. In: Proceedings of the Fifth International Workshop on Information Retrieval with Asian Languages, Hong Kong, pp. 75–80 (2000)
12. Inrut, J., Yuanghirun, P., Paludkong, S., Nitsuwat, S., Limmaneepraserth, P.: Thai Word Segmentation Using Combination of Forward and Backward Longest Matching Techniques. In: International Symposium on Communications and Information Technology, Chiang Mai, pp. 37–40 (2001)
13. Kongsupanich, S.: The Transformation of Thai Morphemes to Phonetic Symbols for Thai Speech Synthesis System. Master Thesis. Faculty of Engineering, King Mongkut's Institute of Technology Ladkrabang, Bangkok (1997)
14. Paludkong, S.: Developing Thai-Vernacular-to-Romanization Transcriptor Using Ratchabandittayasatan Method. Master Thesis, King Mongkut's Institute of Technology North Bangkok, Bangkok (2006)

## Appendix: Thai Characters

Thai characters, their transcriptions and translations (if possible) are listed in Table 5.

**Table 5.** Thai characters and their transcriptions

| Character | Transcription |
| --- | --- |
| ก | ko kai (chicken) |
| ข | kho khai (egg) |
| ฃ | kho khuat (bottle) [obsolete] |
| ค | kho khwai (water buffalo) |
| ฅ | kho khon (person) [obsolete] |
| ฆ | kho ra-khang (bell) |
| ง | ngo ngu (snake) |
| จ | cho chan (plate) |
| ฉ | cho ching (cymbals) |
| ช | cho chang (elephant) |
| ซ | so so (chain) |
| ฌ | cho choe (tree) |
| ญ | yo ying (woman) |
| ฎ | do cha-da (headdress) |
| ฏ | to pa-tak (goad, javelin) |
| ฐ | tho than (base) |
| ฑ | tho montho (Mandodari, character from Ramayana) |
| ฒ | tho phu-thao (elder) |
| ณ | no nen (samanera) |
| ด | do dek (child) |
| ต | to tao (turtle) |
| ถ | tho thung (sack) |
| ท | tho thahan (soldier) |
| ธ | tho thong (flag) |
| น | no nu (mouse) |
| บ | bo baimai (leaf) |
| ป | po pla (fish) |
| ผ | pho phueng (bee) |
| ฝ | fo fa (lid) |
| พ | pho phan (tray) |
| ฟ | fo fan (teeth) |
| ภ | pho sam-phao (sailboat) |
| ม | mo ma (horse) |
| ย | yo yak (giant) |

**Table 5.** (*continued*)

| Character | Transcription |
|---|---|
| ร | ro ruea (boat) |
| ล | lo ling (monkey) |
| ว | wo waen (ring) |
| ศ | so sala (pavilion) |
| ษ | so rue-si (hermit) |
| ส | so suea (tiger) |
| ห | ho hip (chest) |
| ฬ | lo chu-la (kite) |
| อ | o ang (basin) |
| ฮ | ho nok-huk (owl) |
| เ | sara e |
| แ | sara ae |
| ไ | sara ai mai malai |
| ใ | sara ai mai muan |
| โ | sara o |
| ะ | sara a (short sound) |
| า | sara a |
| ํา | sara am |
| ๅ | lak-khang-yao |
| ิ | sara i (short sound) |
| ี | sara i |
| ึ | sara ue (short sound) |
| ื | sara ue |
| ุ | sara u (short sound) |
| ู | sara u |
| ํ | nikhahit |
| ฺ | pinthu (virama) |
| ่ | mai ek |
| ้ | mai tho |
| ๊ | mai tri |
| ๋ | mai chattawa |
| ็ | mai taikhu |
| ์ | karan (thanthakat) |
| ๆ | mai ya-mok |
| ฯ | pai-yan noi |

# Automatic Generation of a Pronunciation Dictionary with Rich Variation Coverage Using SMT Methods

Panagiota Karanasou and Lori Lamel

Spoken Language Processing Group, LIMSI-CNRS
91403 Orsay, France
{pkaran,lamel}@limsi.fr

**Abstract.** Constructing a pronunciation lexicon with variants in a fully automatic and language-independent way is a challenge, with many uses in human language technologies. Moreover, with the growing use of web data, there is a recurrent need to add words to existing pronunciation lexicons, and an automatic method can greatly simplify the effort required to generate pronunciations for these out-of-vocabulary words. In this paper, a machine translation approach is used to perform grapheme-to-phoneme (g2p) conversion, the task of finding the pronunciation of a word from its written form. Two alternative methods are proposed to derive pronunciation variants. In the first case, an n-best pronunciation list is extracted directly from the g2p converter. The second is a novel method based on a pivot approach, traditionally used for the paraphrase extraction task, and applied as a post-processing step to the g2p converter. The performance of these two methods is compared under different training conditions. The range of applications which require pronunciation lexicons is discussed and the generated pronunciations are further tested in some preliminary automatic speech recognition experiments.

**Keywords:** pronunciation lexicon, G2P conversion, SMT, pivot paraphrasing.

## 1 Introduction

Grapheme-to-phoneme conversion (g2p) is the task of finding the pronunciation of a word given its written form. Despite several decades of research, it remains a challenging task with many applications in human language technologies. Predicting pronunciations and variants, that is, alternative pronunciations observed for a linguistically identical word, is a complicated problem that depends on a number of diverse factors such as the linguistic origin of the speaker and of the word, the education and the socio-economic level of the speaker and the conversational context. Several approaches have been proposed in the literature to generate pronunciations. The simplest technique is manual creation, often relying on dictionary look-up in multiple resources, but making a pronunciation

A. Gelbukh (Ed.): CICLing 2011, Part II, LNCS 6609, pp. 506–517, 2011.
© Springer-Verlag Berlin Heidelberg 2011

dictionary by hand requires specific linguistic skills and necessarily has limited coverage. Rule-based conversion systems, which have a predominantly one-to-one correspondences between letters and predicted phonemes, still require specific linguistic knowledge and do not always capture the irregularities of a natural language even if exception rules or lists are included.

In contrast to knowledge-based approaches, data-driven approaches are based on the idea that given enough examples it should be possible to predict the pronunciation of an unseen word simply by analogy. A variety of machine learning techniques have been applied to this problem in the past including neural networks [16] and decision trees [4] that predict a phoneme for each input letter using the letter and its context as features, but do not consider -or consider very limited- context in the output. Other techniques allow previously predicted phonemes to inform future decisions such as HMM in [18] but they do not take into account the input's letter context. Joint-sequence models have been proposed [2], [3], that achieve better performance by pairing letter substrings with phoneme substrings, allowing context to be captured implicitly by these groupings. Other methods using many-to-many correspondences, as the one proposed in [7] report high accuracy.

Another machine learning approach that has been tried recently is to view g2p conversion as a statistical machine translation (SMT) problem. Moses, a publicly available phrase-based statistical machine translation toolkit [9], has been used for g2p conversion of French [11] and Italian [6] and other languages [14]. The aim of this work is to generate pronunciations with variants for the English language. It should be noted that English is a difficult language for g2p conversion, since there is a loose relationship between letters and sounds. In a first step, Moses is used as a g2p converter. Then, two options are explored to generate pronunciation variants. In the first one, variants are derived from the generated n-best list. The second method is based on the idea that paraphrases in one language can be identified using a phrase in another language as a pivot. In the case of multiple pronunciation generation, sequences of modified phonemes in the variants are identified using a sequence of graphemes in the corresponding word as a pivot. This method can also be used independently to generate alternative pronunciations from a canonical pronunciation of a word, thereby enriching the dictionary. Here it is used as an alternative solution to generate variants in a post-processing step to the g2p converter. It focuses on local variations, which are the most common variations found in multiple pronunciations of a word and permits more generalization in variant generation as will be explained later. To the best of our knowledge, this is the first application of a pivot approach to the generation of pronunciation variants.

The remainder of the paper is organized as follows. Section 2 presents the two methods used in this study. Section 3 describes the experimental framework, the corpora used and the training conditions. Section 4 presents an evaluation of the automatic generation of multiple pronunciations, while in Section 5 some applications requiring such pronunciation dictionaries are discussed and preliminary speech recognition experiments are reported as an illustration of an applicative task. Finally, conclusions and discussions for future work are reported in Section 6.

## 2  Methodology

This section first describes Moses as a g2p converter, and then presents two methods to generate variants. When Moses is used for g2p conversion, a pronunciation dictionary is used in the place of an aligned bilingual text corpora. The orthographic transcription is considered as the source language and the pronunciation as the target language. This method has the desired properties of a g2p system: To predict a phoneme from a grapheme, it takes into account the local context of the input word and of the output pronunciation from a phrase-based model and allows sub-strings of graphemes to generate phonemes. The phoneme sequence information is additionally modeled by a phoneme n-gram language model (LM) that corresponds to the target language model in machine translation. In this study, a phoneme-based 5-gram LM was built on the pronunciations in the training set using the SRI toolkit [17].

Moses also calculates distortion models, but this is not necessary as g2p conversion is a monotonic task. Finally, the combination of all components is fully optimized with a minimum error training step (tuning) on a development set. The tuning strategy used was that of the standard Moses training framework, based on maximizing the BLEU score.

### 2.1  Generation of n-best Lists by Moses

In addition to the best hypotheses, Moses can also output an n-best translation list. This is a ranked list of translations of a source string with the distortion, the translation and the language model weights, and an overall score for each translation. The 2-, 5- or 10-best translations (i.e. pronunciation variants) per word are kept. Some words have fewer possible variants, in which case all variants are taken.

### 2.2  Pivot Paraphrasing Approach

This is an alternative method based on [1] for the generation of pronunciation variants, added as a post-processing step to the Moses-g2p converter. Paraphrases are alternative ways of conveying the same information. The analogy with multiple pronunciations of the same word is easily seen, the different pronunciations being alternate phonemic expressions of the same orthographic information. In [1], a paraphrase probability is defined that allows paraphrases extracted from a bilingual parallel corpus to be ranked using translation probabilities. These are then reranked taking contextual information into account. For the problem of automatic pronunciation variant generation, this bilingual corpus corresponds to the corpus of word-pronunciation pairs already used by

Moses for g2p conversion and the paraphrases are phonemic phrases extracted from the translation table in that task. For each phonemic phrase in the translation table, we find all corresponding graphemic phrases and then look back to find what other phonemic phrases are associated with the set of graphemic ones. These phonemic phrases are plausible paraphrases.

In the following, $f$ is a graphemic phrase and $e_1$ and $e_2$ phonemic phrases. The paraphrase probability $p(e_2 \mid e_1)$ is assigned in terms of the translation phrase table probabilities $\phi(f \mid e_1)$ and $\phi(e_2 \mid f)$ estimated on the counts of the aligned graphemic-phonemic phrases. Since $e1$ can be translated as multiple graphemic phrases, we sum over $f$ for all the graphemic entries of the phrase translation table:

$$\hat{e}_2 = \arg \max_{e_2 \neq e_1} p(e_2 \mid e_1) \tag{1}$$

$$= \arg \max_{e_2 \neq e_1} \sum_{f} \phi(f \mid e_1)\phi(e_2 \mid f) \tag{2}$$

This returns the single best paraphrase, $\hat{e}_2$, irrespective of the context in which $e_1$ appears. The paraphrased pairs with their probabilities are extracted for the input pronunciations, which are the pronunciations generated by Moses for the words of the test set during the g2p conversion task. The 10-best paraphrases for each input phonemic phrase found in the translation table are extracted with a maximum extent of 4 phonemes. An example of a paraphrase pattern in the dictionary is:

discounted        dIskWntxd       dIskWnxd
discountenance dIskWntNxns dIskWnNxns

The alternative pronunciations differ only in the part that can be realized as either nt or n, while the rest remains the same. The nt and n form a paraphrased pair. The pivot method focuses on local modifications observed between variants of a word. The generation of variants with pivot is a lot faster than the n-best list generation by Moses-g2p. All occurrences of these paraphrased patterns are substituted in the input pronunciations for all the possible combinations (only in the first occurrence, only in the second, in the first and the second, etc.), limiting to 3 the maximum number of occurrences of the same paraphrase in a pronunciation.

At this point, different types of pruning are applied on the generated variants. First, the candidate variants are reranked based on additional phonemic contextual information expressed by a simple language model trained on the pronunciations in the training data. This is the same phoneme-based 5-gram LM used by Moses for the g2p conversion. The SRI toolkit was used for reranking. Then pruning based on the length of extracted paraphrases substituted in the pronunciations is realized. Many errors were observed due to substitutions of unigram paraphrases that the reranking did not manage to handle successfully. It was experimentally found that the quality of the generated variants improves when only 3- and 4-grams paraphrases are substituted. This is normal as more context is taken into account throughout the procedure and some confusions are avoided.

The Levenshtein Distance between each pronunciation and its generated variants was then calculated. This measure should not exceed a threshold since the different pronunciations of a word are usually phonemically very close. Pruning with thresholds of 3 (LD3) and 2 (LD2), meaning that all the variants with edit distances greater than 3 and 2 respectively are pruned, were tried. Finally, the 1-, 4- and 9-best pronunciation variants per input pronunciation were kept and merged with the input pronunciations (1-best pronunciations of Moses-g2p) in order to have 2-, 5- and 10-best pronunciations so as to be able to compare these results with the n-best lists of Moses-g2p.

## 3   Experimental Setup

The LIMSI American English pronunciation dictionary serves as basis of this work. It was decided to use this dictionary as it is reputed to be a high quality dictionary for speech recognition, which will be the domain of application of the proposed methods in Section 5[1]. The dictionary has been created with extensive manual supervision and has 187975 word entries. Each dictionary entry contains the orthographic form of a word and its pronunciations (one or more). The pronunciations are represented using a set of 45 phones [10]. 18% of the words are associated with multiple pronunciations. These mainly correspond to well-known phonemic alternatives (for example the pronunciation of the ending "ization"), and to different parts of speech (noun or verb). Case distinction is eliminated since in general it does not influence the word's pronunciation, the main exceptions being acronyms which may have both a spoken and spelled form, but these are quite rare. Some symbols in the graphemic form are not pronounced, such as the hyphen in compound words. The dictionary contains a mix of common words, acronyms and proper names, the last two categories being difficult cases for g2p converters.

The corpus was randomly split based on the graphemic form of the word into a training, a development (dev) and a test set. The dev set is necessary for the tuning of the Moses model. In order to have a format that resembles the aligned parallel texts used for training machine translation models, the dictionary is expanded so that each entry corresponds to a word-one pronunciation pair. The resulting dev and test sets have 11k and 19k distinct entries.

The g2p converter is trained for two conditions, on the entire training subset using all pronunciations for words with multiple ones or on the same word list but using only one (canonical) pronunciation per word. Since canonical pronunciations are not explicitly indicated in the lexicon, the longest one is taken as the canonical form. In the first training condition, there are 200k entries (distinct

---

[1] Although not publicly available, this dictionary is available by request. It has been used by numerous laboratories. SRI, Philips Aachen, ICSI and Cambridge University have reported improving the performance of their systems using this dictionary.

word-pronunciation pairs) in the training set with on average 1.2 pronuncia-
tions/word. In the second training condition, the training set has 160k entries
with a single pronunciation per word.

## 4   Evaluation

In this study, precision and recall, first introduced in information retrieval [15],
as well as phone error rate (PER) are used to evaluate the predictions of one
or multiple pronunciations. Word $x_i$ of the test set (i=1..w) has j distinct pronunciations $y_{ij}$ ($y_i$ is a set with elements $y_{ij}, j = 1..d_i$). Moreover, our systems
can generate one or more pronunciations $f(x_i)$ ($f(x_i)$ is also a set). Recall (R)
is conventionally defined:

$$R = \frac{1}{w} \sum_{i=1}^{w} \frac{|f(x_i) \cap y_i|}{|y_i|} \tag{3}$$

Precision (Pr) is defined analogously as the number of correct generated pronunciations divided by the total number of generated pronunciations. They are
calculated on all references (canonical pronunciations and variants) to evaluate
the g2p conversion, but also only on the variants in order to specifically evaluate
their correctness. The PER is measured using the Levenshtein Distance (LD)
between the generated pronunciations and the reference pronunciations:

$$PER_{n-best} = \frac{\sum_{i=1}^{w} \sum_{j=1}^{d_i} \min LD(y_{ij}, f(x_i))}{\sum_{i=1}^{w} \sum_{j=1}^{d_i} |y_{ij}|} \tag{4}$$

$$PER_{1-best} = \frac{\sum_{i=1}^{w} \sum_{j=1}^{d_i} \min LD(y_{ij}, f(x_i))}{\sum_{i=1}^{w} |y_{im}|} \tag{5}$$

where $y_{im}$ the pronunciation of the word $x_i$ where the LD is minimum.

The Moses-g2p converter (M-g2p) and the pivot paraphrasing method (P)
were tested for the multiple pronunciation and single pronunciation training
conditions. Table 1 gives recall results compared to all references (top) and only
variants (middle), as well as PER (bottom) with both methods for multiple
pronunciation training. Precision was also calculated, but only recall is presented
because we consider it more important to cover possible pronunciations than to
have too many, since other methods can be applied to reduce the overgeneration
(alignment with audio, manual selection, use of pronunciation probabilities, etc).
The best value that both precision and recall can obtain is 1.

It can be seen in Table 1 that Moses-g2p outperforms the pivot-based method
in terms of recall measured on all references (R-all ref) and on variants only
(R-variants). The best result is a recall on all references of 0.94 when using the
10-best pronunciations generated by Moses-g2p. The PER (bottom) is about 6%

**Table 1.** Recall and PER on all references (canonical prons+variants) and only on variants for Moses-g2p (M-g2p) and Pivot (P) for multiple pronunciation training

| Method | Measure | 1-best | 2-best | 5-best | 10-best |
|--------|---------|--------|--------|--------|---------|
| M-g2p | R-all ref | **0.68** | 0.82 | 0.91 | **0.94** |
| P LD2 | R-all ref | - | 0.74 | 0.80 | 0.84 |
| M-g2p | R-variants | 0.27 | 0.63 | 0.82 | 0.89 |
| P LD2 | R-variants | - | 0.50 | 0.66 | 0.73 |
| M-g2p | PER (%) | **6.13** | 4.00 | 1.97 | **1.17** |
| P LD2 | PER (%) | - | 6.00 | 4.47 | 3.52 |

**Table 2.** Recall on all references (canonical prons+variants) and only on variants for Moses-g2p (M-g2p) and Pivot (P) for canonical pronunciation training

| Method | Measure | 1-best | 2-best | 5-best | 10-best |
|--------|---------|--------|--------|--------|---------|
| M-g2p | R-all ref | 0.68 | 0.79 | 0.88 | 0.91 |
| P LD2 | R-all ref | - | 0.72 | 0.78 | 0.83 |
| M-g2p | R-variants | 0.10 | 0.25 | 0.44 | **0.55** |
| P | R-variants | - | 0.19 | 0.32 | 0.44 |
| P LD3 | R-variants | - | 0.35 | 0.49 | 0.60 |
| P LD2 | R-variants | - | 0.36 | 0.50 | **0.61** |

for the 1-best Moses-g2p pronunciation, and 1.17% if the 10-best pronunciations are considered. The string error rate (SER) is 25%. Since the 1-best pronunciations generated by Moses-g2p are used as input to the pivot post-processing, the corresponding entries in the table are empty for Pivot.

In Table 2 the recall on all references (top) and only on variants (bottom) for single-pronunciation training are shown. For the recall on variants, the results of pivot without LD pruning are presented (P) as well as with LD threshold 3 (P LD3) and LD threshold 2 (P LD2) to show the improvement obtained by the intermediate pruning steps. The PER is not reported in this table since it does not change significantly from that of Table 1.

Comparing the recall on all references (R-all ref) in the two tables, a 3% absolute degradation can be seen in Table 2 for both methods. However, the variant-only recall degrades more severely. For the latter case pivot with LD2 or LD3 pruning outperforms Moses-g2p, managing to generate more correct variants even when no variants are given in the training set. Pivot takes directly the variation patterns from the phrase table of Moses avoiding the overfitting effects of the EM algorithm used by Moses for the construction of a generative model. Moreover, to reduce the overall complexity of decoding, the search space of Moses is typically pruned using simple heuristics and, as a consequence, the best hypothesis returned by the decoder is not always the one with the highest score. We plan to experimentally verify this theoretical error analysis in future work.

**Table 3.** Recall on variants only for generation of 1-, 4- and 9-best variants by Moses-g2p (M-g2p) and Pivot (P) for multiple-pronunciation training

| Method | Measure | 1-best | 4-best | 9-best |
|---|---|---|---|---|
| M-g2p | R | 0.35 | 0.55 | 0.62 |
| P LD2 | R | 0.23 | 0.39 | 0.46 |
| P correct entry LD2 | R | 0.39 | 0.65 | **0.75** |

It should be pointed out that the measures (recall and PER) on all references favors the Moses-based approach since the pivot-based approach aims to generate variants. This is why recall only on variants was also evaluated. However, while the pivot method gives better results than Moses-g2p to variants generation for the single-pronunciation training condition, this is not the case when multiple pronunciations are used for training. Some additional analyses were carried out to investigate this further. When the pivot is used as a post-processing step to the Moses-g2p converter, its input is the output of Moses which has PER of 6%, low enough to be reliable, but the SER is 25% which can plausibly degrade the performance of pivot. To verify this hypothesis, the pivot method was applied to the correct canonical pronunciation of the test set and these results were compared to the previous results of 1-, 4- and 9-best variants generated by pivot as well as to the variants generated by Moses-g2p. In order to more clearly see the influence of variants generated by pivot, the 1-best pronunciation generated by Moses was not retained as had been done previously. This pronunciation was also removed from the n-best list generated by Moses-g2p in order to compare the two methods. Table 3 gives recall results computed on variants in the reference set. It can be seen that pivot, when applied to a correct input, not only outperforms itself applied to a 'noisy' input, but also the Moses-g2p method. This is an important observation, as there are cases where the enrichment of a single-pronunciation dictionary is desired, for example in a conversational speech transcription task.

All results presented in this section are calculated with the complete 45-phone set used in the LIMSI dictionary. However, some exchanges are less important than others. If some errors, such as the confusion between syllabic nasals and a schwa-nasal sequence, are not taken into account (a subset of those proposed in [12]), the overall recall improves by 1-2% absolute for both methods, and the PER is reduced by 0.1-0.2% for Moses-g2p and 0.3-0.4% for pivot.

Last but not least, the reference dictionary is mostly manually constructed and certainly incomplete with respect to coverage of pronunciation variants particularly for uncommon words. The pronunciations of words of foreign origin (mostly proper names) may also be incomplete since their pronunciation depends highly on the speaker's knowledge of the language of origin. This means that some of the generated variants are likely to be correct (or plausible) even if they are not in the references used in the upper evaluation.

# 5    Applications

A pronunciation dictionary with variants can be useful for a number of applications. First, it is an essential element of speech recognition and speech synthesis systems. In fact, the construction of a good pronunciation dictionary is important to ensure acceptable automatic speech recognition performance [10]. Moreover with the wide use of real data there are words not yet included in a recognition dictionary (out-of-vocabulary words), for which a pronunciation rapidly and automatically generated is often required. Another domain of application of the phonetization task in natural language processing is the detection and correction of orthographic errors [19], while the strong relation between phonology and morphology is well known and studied with morphological phenomena of purely phonological origins or guided by phonological constraints, among other interactions [8]. Other applications include computer-aided language learning, pronunciation training and in general e-learning systems.

To further test the pronunciations generated by the Moses-g2p method in an application framework, some preliminary speech recognition experiments were conducted. Similar experiments have been reported for the Italian [6] and French [11] languages, but to our knowledge they have never been tested in a state-of-the-art ASR system for English broadcast data.

The speech transcription system uses the same basic modeling and decoding strategy as in the LIMSI English broadcast news system [5]. The speech recognizer makes use of continuous density HMMs with Gaussian mixture for acoustic modeling and 4-gram statistics estimated on large text corpora. The acoustic models are gender-dependent, speaker-adapted, and Maximum Likelihood trained on about 500 hours of audio data. They cover about 30k phone contexts with 11600 tied states. N-gram LMs were trained on a corpus of 1.2 billion words of texts from various LDC corpora (English Gigaword, BN transcriptions, commercial transcripts), news articles downloaded from the web, and assorted audio transcriptions. The recognition word list contains 78k words, selected by interpolation of unigram LMs trained on different text subsets as to minimize the out-of-vocabulary (OOV) rate on set of development texts. Word recognition was performed in a single real-time decoding pass, generating a word lattice followed by consensus decoding [13] with a 4-gram LM. Unsupervised acoustic model adaptation is performed for each segment cluster using the CMLLR and MLLR techniques prior to decoding.

The Quaero (www.quaero.org) 2010 development data were used in the recognition experiments. This 3.5 hour data set contains 9 audio files recorded in May 2010, covering a range styles, from broadcast news (BN) to talk shows. Roughly 50% of the data can be classed as BN and 50% broadcast conversation (BC). These data are considerably more difficult than pure BN data. The overall word error rate (WER) is 30%, but the individual shows vary from 20% to over 40%. These are competitive WERs on these data.

In Table 4, the n-best pronunciations (1-, 2- and 5-best) generated by the Moses-based system under the two training conditions, are added to the canonical pronunciation of the original recognition dictionary (Baseline longest). The

**Table 4.** WER(%) adding Moses nbest-lists (M1, M2,M5) to the longest pronunciation baseline

| Training condition | M1 | M2 | M5 |
|---|---|---|---|
| Single pronunciation | 38.2 | 38.4 | 40.8 |
| Multiple pronunciations | 37.9 | 38.2 | 39.1 |
| Baseline longest | 41.6 | | |

**Table 5.** WER(%) adding Moses nbest-lists (M1, M2,M5) to the most frequent pronunciation baseline

| Training condition | M1 | M2 | M5 |
|---|---|---|---|
| Single pronunciation | 32.0 | 33.4 | 37.3 |
| Multiple pronunciations | 32.0 | 34.5 | 38.9 |
| Baseline most frequent | 32.9 | | |

results show that using only the longest pronunciation results in a large increase in WER. Adding pronunciations improves over the baseline longest dictionary, up until the 5 best pronunciations. The pronunciations trained under the multiple pronunciation training condition improve more the WER compared with the pronunciations trained with the single pronunciation dictionary. This is because the formers are trained to better model the variants which correspond to the reduced forms, closer to the spoken language most of the times.

In Table 5, the same pronunciations (M1, M2, M5) are added to the most frequent pronunciation of the recognition dictionary (Baseline most frequent). The most frequent pronunciation baseline dictionary has a WER closer to the baseline of the original multiple pronunciation dictionary. In this case adding one pronunciation (trained on a single or multiple pronunciation dictionary) improves the performance of the ASR system, but adding more pronunciations degrades it.

Although the quality of the pronunciations trained on a multiple pronunciation dictionary is higher, measured with recall on all references and on variants, they are submitted to the same confusability effects. What is more, when adding two or five pronunciations to the most frequent baseline, the system with pronunciations trained on a single pronunciation presents lower WERs. An explanation could be that the pronunciations trained under multiple pronunciations can better represent reduced forms and, thus, are closer to the most frequent baseline and easier to be confused. An example of the introduced confusability is that of the multiple pronunciation training outputs for the word *you*. These are the pronunciations /yu/ and /yc/ when the 2-best list is kept. The latter pronunciation (/yc/) is not generated under the single pronunciation training. /yc/ in the phrase *you are* is easily confused with /ycr/, the pronunciation of *your*. Such frequent cases can be responsible for the degradation of the ASR system with pronunciations trained on the multiple pronunciation dictionary, when many alternatives are added.

Nevertheless, in neither case was the performance of the original multiple pronunciation dictionary achieved. This dictionary is a difficult baseline because it is mostly manually constructed and well-suited to the needs of an ASR system. However, we expect that it is possible to obtain additional gains if probabilities are added to the generated pronunciation variants to moderate confusability.

## 6   Conclusion and Discussion

This paper has reported on a fully automatic and language independent generation of pronunciations using Moses, an open-source SMT tool, as a g2p converter and generating pronunciation variants taking directly the n-best lists of Moses or applying a novel pivot-based method. The n-best lists of Moses yield better recall results than the pivot-based method on all references. However, it was shown that the pivot-based method can generate more correct variants. This is an advantage of the pivot method that could be useful in certain cases, especially in the case of limited variation in the training set, for example to generate variants from the output of a rule-based g2p system which, if originally developed for speech synthesis, may not model pronunciation variants or to enrich a dictionary with limited pronunciation variants.

The generated pronunciations were also evaluated in an applicative task. They were used to carry out tests in a state-of-the-art ASR system. These experiments show that Moses provides variants of good quality that even without any further pruning can improve the one pronunciation baselines. Our point in this paper is not, however, to present an ASR system and focus on the improvement of its performance, but to propose data-based approaches for the generation of a pronunciation dictionary with variants. In the future, we plan to further evaluate the pronunciations generated by pivot by measuring their influence in ASR systems for different data sets (broadcast news, conversational speech). Another problem that interests us is generating pronunciations specifically for named entities (proper names, geographical names,etc.), which are very often cases of out-of-vocabulary words and their pronunciations rarely follow regular phonological rules.

**Acknowledgments.** This work is partly realized as part of the Quaero Programme, funded by OSEO, French State agency for innovation and by the ANR EdyLex project.

## References

1. Bannard, C., Callison-Burch, C.: Paraphrasing with bilingual parallel corpora. In: Proc. of ACL (2005)
2. Bisani, M., Ney, H.: Investigations on Joint-Multigram Models for Grapheme-to-Phoneme Conversion. In: ICSLP, pp. 105–108 (2002)
3. Deligne, S., Yvon, F., Bimbot, F.: Variable-length sequence matching for phonetic transcription using joint multigrams. In: Proc. European Conf. on Speech Communication and Technology, pp. 2243–2246 (1995)

4. Dietterich, T.G., Bakiri, G.: Solving Multiclass Learning Problems via Error-Correcting Output Codes. Journal of Artificial Intelligence 2, 263–286 (1995)
5. Gauvain, J.L., Lamel, L., Adda, G.: The LIMSI Broadcast News Transcription System. Speech Comm. 37, 89–108 (2002)
6. Gerosa, M., Federico, M.: Coping with out-of-vocabulary words:open versus huge vocabulary ASR. In: ICASSP (2009)
7. Jiampojamarn, S., Cherry, C., Kondrak, G.: Joint processing and discriminative training for letter-to-phoneme conversion. In: Proc. of ACL-HLT, pp. 905–913 (2008)
8. Kaisse, E.M.: Word-Formation and Phonology. In: Handbook of Word-Formation, Studies in Natural Language and Linguistic Theory, vol. 64, pp. 25–47. Springer, Netherlands (2005)
9. Koehn, P., et al.: Moses: Open source toolkit for statistical machine translation. In: ICSLP (2002)
10. Lamel, L., Adda, G.: On designing pronunciation lexicons for large vocabulary, continuous speech recognition. In: Proc. ICSLP, pp. 6–9 (1996)
11. Laurent, A., Deleglise, P., Meignier, S.: Grapheme to phoneme conversion using an SMT system. In: Interspeech (2009)
12. Lee, K.F., Hon, H.W.: Speaker-Independent Phone Recognition Using Hidden Markov Models. IEEE Trans. ASSP 37(11), 1641–1648 (1989)
13. Mangu, L., Brill, E., Stolcke, A.: Finding Consensus Among Words: Lattice-Based Word Error Minimization. In: Eurospeech, pp. 495–498 (1999)
14. Rama, T., Singh, A.K., Kolachina, S.: Modeling Letter-to-Phoneme Conversion as a Phrase Based Statistical Machine Translation Problem with Minimum Error Rate Training. In: Proc. NAACL-HLT: Student Research Workshop & Doctoral Consortium, pp. 90–95 (2009)
15. Van Rijsbergen, C.J.: Information Retrieval, Butterworths, London, UK (1979)
16. Sejnowski, T., Rosenberg, C.: NETtalk: a parallel network that learns to read aloud. In: Report JHU/EECS-86/01 (1986)
17. Stolcke, A.: SRILM-An extensible language modeling toolkit. Proc. ICSLP 2002 (2002)
18. Taylor, P.: Hidden Markov models for grapheme to phoneme conversion. In: Interspeech, pp. 1973–1976 (2005)
19. van Berkel, B., De Smedt, K.: Triphone analysis:a combined method for the correction of orthographical and typographical errors. In: Proc. of the Second Conf. on Applied Natural Language Processing, pp. 77–83 (1988)

# Author Index